Advances in Industrial Control

Advances in Industrial Control is a series of monographs and contributed titles focusing on the applications of advanced and novel control methods within applied settings. This series has worldwide distribution to engineers, researchers and libraries.

The series promotes the exchange of information between academia and industry, to which end the books all demonstrate some theoretical aspect of an advanced or new control method and show how it can be applied either in a pilot plant or in some real industrial situation. The books are distinguished by the combination of the type of theory used and the type of application exemplified. Note that "industrial" here has a very broad interpretation; it applies not merely to the processes employed in industrial plants but to systems such as avionics and automotive brakes and drive-train. This series complements the theoretical and more mathematical approach of Communications and Control Engineering.

Indexed by SCOPUS and Engineering Index.

Proposals for this series, composed of a proposal form (please ask the in-house editor below), a draft Contents, at least two sample chapters and an author cv (with a synopsis of the whole project, if possible) can be submitted to either of the:

Series Editor

Professor **Michael J. Grimble** Department of Electronic and Electrical Engineering, Royal College Building, 204 George Street, Glasgow G1 1XW, United Kingdom **e-mail:** m.j.grimble@strath.ac.uk

or the

In-house Editor

Mr. **Oliver Jackson** Springer London, 4 Crinan Street, London, N1 9XW, United Kingdom **e-mail:** oliver.jackson@springer.com
Proposals are peer-reviewed.

Publishing Ethics

Researchers should conduct their research from research proposal to publication in line with best practices and codes of conduct of relevant professional bodies and/or national and international regulatory bodies. For more details on individual ethics matters please see:
https://www.springer.com/gp/authors-editors/journal-author/journal-author-helpdesk/publishing-ethics/14214

More information about this series at https://link.springer.com/bookseries/1412

He Cai · Youfeng Su · Jie Huang

Cooperative Control of Multi-agent Systems

Distributed Observer and Distributed Internal Model Approaches

Springer

He Cai
School of Automation Science
and Engineering
South China University of Technology
Guangzhou, Guangdong, China

Youfeng Su
College of Computer and Data Science
Fuzhou University
Fuzhou, Fujian, China

Jie Huang
Department of Mechanical and Automation
Engineering
The Chinese University of Hong Kong
Shatin, Hong Kong, China

ISSN 1430-9491 ISSN 2193-1577 (electronic)
Advances in Industrial Control
ISBN 978-3-030-98376-5 ISBN 978-3-030-98377-2 (eBook)
https://doi.org/10.1007/978-3-030-98377-2

This Springer imprint is published by the registered company Springer Nature Switzerland AG
The registered company address is: Gewerbestrasse 11, 6330 Cham, Switzerland

To my parents

H. Cai

To Ting and Keqing

Y. Su

To my parents, Qingwei, Anne, and Jane

J. Huang

Series Editor's Foreword

The rapid development of new control theory and technology has an impact on all areas of engineering and applications. The *Advances in Industrial Control* monograph series has a focus on applications, desirable because new technological developments constantly provide new challenges. The series provides an opportunity for researchers to present new work on industrial control and applications problems and solutions. It raises awareness of the substantial benefits that advanced control can provide whilst not ignoring the difficulties that can sometimes arise. This monograph is concerned with multi-agent systems that by their very nature require a mathematical treatment, but also have an important role in solving very complex systems engineering problems. The field is interesting because it can involve concepts from both the computer science and control communities, although the latter is the focus of this volume.

Viewed as a control plant, a multi-agent system consists of a group of subsystems of which each can contain a conventional control loop. A typical challenge with the control of multi-agent systems is that the control of some subsystems may not have access to the information from all other subsystems. There is therefore the need for a distributed control approach, which may only use local output measurements. This is common when using classical industrial control design but in so-called "modern" control approaches access to all measured outputs is assumed for most of the advanced control design methods. That is, most problems use a centralized control design philosophy, at least for the fundamental problem description.

In multi-agent systems, the information available to particular subsystems can be described by communication networks among the inputs and outputs of the subsystems. The communication topology can of course change with time for some systems, such as autonomous vehicle controls. The so-called "consensus problem" is important when the initial values of agents are given and the conditions are established under which agents asymptotically agree upon a common value. That is, the final value of the state results from a consensus among the agents.

Formation control problems also arise where, for example, the coordinated control of a set of robots involves following a given trajectory while maintaining a desired spatial distribution. The text covers the consensus problem and the cooperative output regulation problem of multi-agent systems. The former involves the two cases of the leaderless and the leader-following consensus problems.

The authors are respected researchers who formulated the cooperative output regulation problem about a decade ago. This is valuable for applications and is one of the features of the text, which also contains chapters on the leader-following consensus problem of multiple Euler–Lagrange systems and on the leader-following consensus problem of multiple rigid-body systems. It covers the design of distributed control laws for complex multi-agent systems, using the distributed observer and distributed internal model approaches.

The first two chapters include the introduction and the mathematical preliminaries. Graph theory is introduced to describe a distributed control law mathematically including both static and time-varying switching graphs. Of course, the difference between centralized and decentralized control systems is well known but the relation between centralized, decentralized and distributed control systems is not so well understood in the general control community. The role of communication networks in new complex application areas does encourage a more formal study of such architectures.

The third chapter, dealing with consensus problems, describes the leaderless consensus problem and the leader-following consensus problem. Chapter 4 deals with a distributed observer approach for synthesizing a distributed control law for complex multi-agent systems. Euler–Lagrange system models can describe the motion of many industrial systems such as robot manipulators and coupled rigid-body systems. The leader-following consensus of multiple Euler–Lagrange systems is considered in Chapter 5 with applications to both multiple three-link cylindrical robot arms and multiple two-link revolute robot arms. Chapter 6 is on the leader-following consensus of multiple rigid-body systems with potential application to the formation flying of multiple uncertain spacecraft systems.

Chapter 7 describes the output regulation problem for the design of a feedback control law so the output of the plant can asymptotically track a class of reference inputs, in the presence of a class of disturbances whilst ensuring stability of the closed-loop system. The problems considered can handle variations in the parameters that is essential for some applications. Chapter 8 considers the cooperative output regulation problem of linear multi-agent systems using the distributed observer approach. It describes an application to the power tracking problem of a grid-connected, spatially concentrated microgrid. This is very relevant for systems with solar and wind power, and with distributed generators involving say batteries and fly wheels.

Chapter 9 discusses the cooperative robust output regulation of linear multi-agent systems. A distributed internal model approach is used. Chapter 10 considers the more difficult cooperative robust output regulation of nonlinear multi-agent systems over static networks. For nonlinear systems it is difficult to generalize, and specific problems must be analysed. Chapter 11 describes the cooperative robust output regulation of nonlinear multi-agent systems over switching networks. The approach, integrating both the distributed observer and the distributed internal model, can handle the nonlinearity and uncertainty of a system over jointly connected switching networks.

This text considers one of the most active and researched areas of advanced control, namely the use of multi-agent systems. The authors are well established researchers that have provided a self-contained treatment of the topic that has well-proven value in the control of autonomous vehicles and great potential in other industrial application sectors.

Glasgow, UK
January 2022

Michael J. Grimble

Preface

Collective behaviors in living beings are ubiquitous such as flocking of birds, schooling of fish, swarming of locusts, or crowd of human beings. Each individual of these living beings is called an agent. These collective behaviors demand effective communications among different agents and coordinated decision-making. The collective behaviors of living beings have motivated many real-world applications such as coordinated motion of mobile robots, coverage of mobile sensor networks, formation flying of unmanned flight vehicles, control of electric power grids, and biological cell sorting. The mathematical formulation of these applications leads to various cooperative control problems of multi-agent systems such as consensus/synchronization, formation, cooperative output regulation, connectivity preservation, distributed estimation, and so on. From the perspective of the systems and control, a multi-agent system consists of a group of subsystems and each subsystem is a conventional control system. Various cooperative control problems aim to achieve a spectrum of desired collective behaviors or global objectives of the multi-agent system by the coordinated control of each subsystem of the multi-agent system. Typically, the control of some subsystems may not access the information of other subsystems for all time. For example, given a group of autonomous vehicles spreading over a large area, the sensor of each vehicle may only obtain the location and the velocity of its nearby vehicles. Thus, instead of the conventional centralized control scheme, which means the control of every subsystem can access the measured output of all other subsystems, one needs to resort to the so-called distributed control scheme, that is, a control scheme where the control of each subsystem only makes use of the measured output of its neighbors at every time instant. The communication constraints among the inputs and outputs of all subsystems can be described by a time-varying or static communication graph, which is also called the communication network. Thus, the control law for a multi-agent system needs to account for the joint effect of the dynamics of each subsystem and the communication constraints imposed by the communication network. As a result, to design a distributed control law for a multi-agent system, one needs to address some specific issues not encountered in the design of a centralized control law for a conventional control system.

The study of multi-agent systems can be traced back to as early as the 1970s when the consensus problem of multiple vehicles was considered. However, it is not until the beginning of this century that the cooperative control of multi-agent systems became one of the central problems in the field of systems and control. The cooperative control of multi-agent systems was first studied for simple linear multi-agent systems where each subsystem is a single integrator, double integrator, or a harmonic system and then progressed to general linear homogenous multi-agent systems and some weak nonlinear multi-agent systems. Since around 2010, the research on the cooperative control has advanced to the stage of handling complex systems characterized by strong nonlinearity, large uncertainty, heterogeneousness, external disturbances, jointly connected switching communication topologies which can be disconnected at every time instant, and so on. Simpler control objectives such as consensus or formation yield to more sophisticated tasks of simultaneous tracking of a class of reference signals and rejection of a class of external disturbances, thus leading to the so-called cooperative output regulation problem of multi-agent systems.

This book mainly treats two basic cooperative control problems, namely the consensus problem and the cooperative output regulation problem. We first present the solutions of both leaderless consensus problem and leader-following consensus problem for general linear homogeneous multi-agent systems over both static and jointly connected switching communication topologies via distributed static control laws. The solution of the consensus problem is not only interesting on its own, but also lays a foundation for establishing the distributed observer, which leads to a systematic paradigm for designing distributed control laws for other cooperative control problems of multi-agent systems such as the asymptotic tracking and disturbance rejection problem of multiple Euler–Lagrange systems, attitude synchronization of multiple rigid-body systems, and formation flying of multiple rigid flight vehicles. Next, we further present the cooperative output regulation problem of multi-agent systems for both linear and nonlinear systems, which can be viewed as the generalization of the classical output regulation problem from a single system to multi-agent systems. The book can be used as a textbook for graduate students in engineering, sciences, and mathematics and can also serve as a reference book for practitioners and theorists in both industry and academia.

A few features of this book are worth mentioning.

First, the bulk of the book is based on the research outcome of the authors of the book over the last decade. In particular, since the cooperative output regulation problem was formally formulated by the authors about a decade ago, it has gained constant attention over the world. The scope of the research keeps expanding in several directions. Thus, the readers of the book may find it quite useful for them to catch up with the current research of the cooperative control of multi-agent systems.

Second, the readability of the book is emphasized by making the presentation of the book as self-contained as possible. In fact, the detailed proof of almost all theoretical results is provided together with illustrative examples. Readers with some knowledge of the fundamentals of linear algebra, calculus, and linear systems should be able to master the materials of the book.

Third, in addition to providing rigorous mathematical treatment to all results, the book also contains many applications to practical multi-agent systems. In particular, the book devotes a whole chapter to the leader-following consensus problem of multiple Euler–Lagrange systems via three different approaches and another whole chapter to the leader-following consensus problem of multiple rigid-body systems for three scenarios.

Finally, the book offers two systematic approaches for designing distributed control laws for complex multi-agent systems, namely the distributed observer approach and the distributed internal model approach. These two approaches are originally developed for dealing with the cooperative output regulation problem and they can be viewed as the generalization of the feedforward control approach and the internal model control approach of the classical output regulation problem to the cooperative output regulation problem. The distributed observer approach is effective in handling exactly known linear multi-agent systems over jointly connected switching topologies and can also handle some weak nonlinear systems with linearly parameterized uncertainties. But, like the feedforward control approach, it is less capable of dealing with systems with norm-bounded uncertainties or dynamic uncertainties. On the other hand, the distributed internal model approach, like the internal model approach, offers a mechanism for accounting for norm-bounded uncertainties or dynamic uncertainties. Moreover, these two approaches also apply to some other problems such as formation problem and connectivity preservation problem when combined with some other techniques. We expect that these two approaches will enhance the capability of the readers of the book in dealing with complex real-world problems.

Guangzhou, China
Fuzhou, China
Hong Kong, China
January 2022

He Cai
Youfeng Su
Jie Huang

Acknowledgements

The development of this book would not have been possible without the support and help from many people. First of all, the authors are grateful to the Series Editor Michael J. Grimble for his constant encouragement and recommendation. The Springer Editor Oliver Jackson and his assistant Sabine Schmitt, Project Coordinator Subodh Kumar and Sriram Srinivas are extremely helpful and enthusiastic in their service and assistance. Zhiyong Chen, Changran He, Tao Liu, Wei Liu, Zhaocong Liu, and Tianqi Wang have proofread the manuscript.

The bulk of this book is based on the research supported by the National Science Foundation of China under grants 61720106011, 61773122, 61803160, 61973260, 62173092, and 62173149 and the Hong Kong Research Grants Council under grants 14201418, 14201420, and 14201621. The authors would also like to thank Sriram Srinivas, Vidyalakshmi Velmurugan, and Oliver Jackson for their help in typesetting this book.

Contents

Notation

$\mathbb{R}$	The set of all real numbers
$\mathbb{C}$	The set of all complex numbers
$\bar{\mathbb{C}}_+$	The set of all complex numbers with nonnegative real parts
$\mathbb{Q}$	The set of all quaternion: $\{q : q = \text{col}(\hat{q}, \bar{q}), \hat{q} \in \mathbb{R}^3, \bar{q} \in \mathbb{R}\}$
$\mathbb{Q}_u$	The set of all unit quaternion: $\{q : q \in \mathbb{Q}, \|q\| = 1\}$
$\imath$	The imaginary unit $\sqrt{-1}$
$\Re(s)$	The real part of a complex number s
$\Im(s)$	The imaginary part of a complex number s
$\text{col}(x_1, x_2, \ldots, x_N)$	$[x_1^\top, x_2^\top, \ldots, x_N^\top]^\top$
$\mathscr{D}(A_1, A_2, \ldots, A_N)$	block diag $\{A_1, A_2, \ldots, A_N\}$
1_n	col $(1, \ldots, 1) \in \mathbb{R}^n$
$\lambda(A)$	The set of all the eigenvalues of matrix A
$\lambda_i(A)$	The ith eigenvalue of matrix A
$\bar{\lambda}_A$	$\max_i\{\Re(\lambda_i(A))\}$
$\underline{\lambda}_A$	$\min_i\{\Re(\lambda_i(A))\}$
$\text{span}\{A\}$	The linear space spanned by the columns of matrix A
$A > 0 (A \geq 0)$	Matrix A is symmetric and positive definite (semidefinite)
$A < 0 (A \leq 0)$	Matrix A is symmetric and negative definite (semidefinite)
$A^{\frac{1}{2}}$	The square root of a positive semidefinite symmetric matrix A
$\det(A)$	The determinant of matrix A
A^T	The transpose of matrix A
A^H	The conjugate transpose of matrix A
$\|A\|$	The two-norm of matrix A
$\text{vec}(A)$	col $(A_1, \ldots, A_n)$ with A_i being the ith column of matrix A with n columns
$\lfloor x \rfloor$	The largest integer less than or equal to the real number x
$x^\times$	For $x = \text{col}(x_1, x_2, x_3) \in \mathbb{R}^3$, $x^\times = \begin{bmatrix} 0 & -x_3 & x_2 \\ x_3 & 0 & -x_1 \\ -x_2 & x_1 & 0 \end{bmatrix}$
$\mathbf{C}(x)$	$(\bar{x}^2 - \hat{x}^\top \hat{x}) I_3 + 2\hat{x}\hat{x}^\top - 2\bar{x}\hat{x}^\times, x \in \mathbb{Q}$

$\mathbf{Q}(x)$	$\mathrm{col}(x, 0), x \in \mathbb{R}^3$

List of Figures

Chapter 1
Introduction

This chapter is devoted to an introduction to continuous-time multi-agent systems. We first give a general description of multi-agent systems and some of their special forms in Sect. 1.1. Then, in Sect. 1.2, after reviewing the centralized and purely decentralized control schemes, we further formulate the distributed control scheme for multi-agent systems. In Sect. 1.3, we present a few basic cooperative control problems of multi-agent systems with the emphasis on two fundamental cooperative control problems: consensus and cooperative output regulation. Finally, in Sect. 1.4, we close this chapter with the organization of the book.

1.1 Multi-agent Control Systems

A multi-agent control system consists of a group of individual subsystems called agents. For a continuous-time multi-agent control system having N subsystems, a general representation for each of these subsystems is given as follows:

$$\dot{x}_i = f_i(x_i, u_i, v_{0i}, w_i) \tag{1.1a}$$

$$y_{mi} = h_{mi}(x_i, u_i, v_{0i}, w_i), \ i = 1, \ldots, N \tag{1.1b}$$

where, for $i = 1, \ldots, N$, $x_i \in \mathbb{R}^{n_i}$, $u_i \in \mathbb{R}^{m_i}$, $y_{mi} \in \mathbb{R}^{p_{mi}}$, $v_{0i} \in \mathbb{R}^{q_i}$, and $w_i \in \mathbb{R}^{n_{wi}}$ are the state, control input, measurement output, exogenous signal, and unknown constant vector of the ith subsystem of the system (1.1), respectively.

Throughout this book, for convenience, we assume the functions f_i and h_{mi} are globally defined smooth functions satisfying, for all $w_i \in \mathbb{R}^{n_{wi}}$, $f_i(0, 0, 0, w_i) = 0$ and $h_{mi}(0, 0, 0, w_i) = 0$. System (1.1) is said to be homogeneous if $f_1 = f_2 = \cdots = f_N$ and $h_{m1} = h_{m2} = \cdots = h_{mN}$. Otherwise, it is heterogeneous.

H. Cai et al., *Cooperative Control of Multi-agent Systems*, Advances in Industrial Control,
https://doi.org/10.1007/978-3-030-98377-2_1

Typically, the exogenous signal v_{0i} itself is generated by a dynamic system, say, an autonomous system of the following form:

$$\dot{v}_{0i} = f_{0i}(v_{0i}, w_{0i}) \tag{1.2a}$$
$$y_{m0i} = h_{m0i}(v_{0i}, w_{0i}),\ i = 1, \ldots, N \tag{1.2b}$$

where, for $i = 1, \ldots, N$, $v_{0i} \in \mathbb{R}^{q_i}$, $y_{m0i} \in \mathbb{R}^{p_{m0i}}$, and $w_{0i} \in \mathbb{R}^{n_{w0i}}$ are the state, measurement output, and unknown constant vector of the ith subsystem of the system (1.2), respectively, and f_{0i} and h_{m0i} are globally defined smooth functions satisfying, for all $w_{0i} \in \mathbb{R}^{n_{w0i}}$, $f_{0i}(0, w_{0i}) = 0$ and $h_{m0i}(0, w_{0i}) = 0$.

Let

$$\begin{aligned} v_0 &= \text{col}(v_{01}, \ldots, v_{0N}) \\ y_{m0} &= \text{col}(y_{m01}, \ldots, y_{m0N}) \\ w_0 &= \text{col}(w_{01}, \ldots, w_{0N}) \\ f_0(v_0, w_0) &= \text{col}(f_{01}(v_{01}, w_{01}), \ldots, f_{0N}(v_{0N}, w_{0N})) \\ h_{m0}(v_0, w_0) &= \text{col}(h_{m01}(v_{01}, w_{01}), \ldots, h_{m0N}(v_{0N}, w_{0N})). \end{aligned}$$

Then we can put (1.2) in the following compact form:

$$\dot{v}_0 = f_0(v_0, w_0) \tag{1.3a}$$
$$y_{m0} = h_{m0}(v_0, w_0) \tag{1.3b}$$

where $v_0 \in \mathbb{R}^q$ with $q = q_1 + \cdots + q_N$, $y_{m0} \in \mathbb{R}^{p_{m0}}$ with $p_{m0} = p_{m01} + \cdots + p_{m0N}$, and $w_0 \in \mathbb{R}^{n_{w0}}$ with $n_{w0} = n_{w01} + \cdots + n_{w0N}$.

Remark 1.1 Even though (1.3) is derived from (1.2), for the sake of generality, in what follows, we will not assume (1.3) is the compact representation of (1.2). In other words, we consider f_0 and h_{m0} as any globally defined smooth functions of v_0 and w_0 satisfying, for all $w_0 \in \mathbb{R}^{n_{w0}}$, $f_0(0, w_0) = 0$ and $h_{m0}(0, w_0) = 0$. Likewise, let $w = \text{col}(w_1, \ldots, w_N) \in \mathbb{R}^{n_w}$ with $n_w = n_{w1} + \cdots + n_{wN}$. Then we can obtain a more general form for (1.1) as follows:

$$\dot{x}_i = f_i(x_i, u_i, v_0, w) \tag{1.4a}$$
$$y_{mi} = h_{mi}(x_i, u_i, v_0, w),\ i = 1, \ldots, N. \tag{1.4b}$$

Again, for $i = 1, \ldots, N$, we consider f_i and h_{mi} as any globally defined smooth functions of x_i, u_i, v_0 and w satisfying, for all $w \in \mathbb{R}^{n_w}$, $f_i(0, 0, 0, w) = 0$ and $h_{mi}(0, 0, 0, w) = 0$.

For the case where the exogenous signal v_0 is generated by (1.3), we can view (1.3) and (1.4) together as a multi-agent control system of $N + 1$ agents. It will be seen, in Sect. 1.3, that system (1.3) will generate reference signals for the performance outputs of (1.4) to track. For this reason, we call (1.3) and (1.4) together the leader–follower multi-agent control system with (1.3) as the leader system and (1.4) as the follower system.

If v_0 is not generated by a dynamic system, in particular, if (1.4) does not contain the exogenous signal v_0, then we may more precisely call (1.4) a leaderless multi-agent control system of N agents.

Three special forms of system (1.4) are as follows:

1. Euler–Lagrange systems

$$M_i(q_i)\ddot{q}_i + C_i(q_i, \dot{q}_i)\dot{q}_i + G_i(q_i) = u_i, \ i = 1, \ldots, N \tag{1.5}$$

   where $q_i, u_i \in \mathbb{R}^n$ are the generalized coordinates and the generalized force vector of the ith agent, respectively.
2. Rigid-body systems

$$\dot{q}_i = \frac{1}{2} q_i \odot \mathbf{Q}(\omega_i) \tag{1.6a}$$

$$J_i\dot{\omega}_i = -\omega_i^{\times} J_i\omega_i + u_i, \ i = 1, \ldots, N \tag{1.6b}$$

   where $q_i \in \mathbb{Q}_u$, $\omega_i, u_i \in \mathbb{R}^3$, and $J_i \in \mathbb{R}^{3\times 3}$ are the attitude, angular velocity, control input, and inertial matrix of the ith rigid-body system, respectively, all expressed in the body frame. Here, $\mathbb{Q}_u$ denotes the set of all unit quaternion, and the notation $\odot$ is defined later in Sect. 6.1.
3. Nonlinear systems with unity relative degree

$$\dot{z}_i = f_{zi}(z_i, y_i, v_0, w) \tag{1.7a}$$

$$\dot{y}_i = f_{yi}(z_i, y_i, v_0, w) + b_i(w)u_i \tag{1.7b}$$

$$y_{mi} = y_i, \ i = 1, \ldots, N \tag{1.7c}$$

   where $x_i = \text{col}(z_i, y_i) \in \mathbb{R}^{n_{z_i}} \times \mathbb{R}$, $u_i, y_{mi} \in \mathbb{R}$, and $v_0 \in \mathbb{R}^q$ are the state, control input, measurement output, and exogenous signal of the ith agent, respectively, and $b_i(w) \in \mathbb{R}$ is a continuous function of w.

More detailed descriptions of system (1.5) and system (1.6) will be given when we study the leader-following consensus problem of these two classes of systems in Chaps. 5 and 6, respectively. Also, a more detailed description of system (1.7) will be given in Chaps. 10 and 11 when we study the cooperative output regulation problem of this class of nonlinear multi-agent systems.

The class of linear multi-agent systems is also a special case of system (1.4) described as follows:

$$\dot{x}_i = A_i(w)x_i + B_i(w)u_i + E_i(w)v_0 \tag{1.8a}$$

$$y_{mi} = C_{mi}(w)x_i + D_{mi}(w)u_i + F_{mi}(w)v_0, \ i = 1, \ldots, N \tag{1.8b}$$

where, for $i = 1, \ldots, N$, $A_i(w) \in \mathbb{R}^{n_i \times n_i}$, $B_i(w) \in \mathbb{R}^{n_i \times m_i}$, $E_i(w) \in \mathbb{R}^{n_i \times q}$, $C_{mi}(w) \in \mathbb{R}^{p_{mi} \times n_i}$, $D_{mi}(w) \in \mathbb{R}^{p_{mi} \times m_i}$, and $F_{mi}(w) \in \mathbb{R}^{p_{mi} \times q}$, and they are globally defined smooth functions of w.

If (1.8) contains no uncertain parameters, then it simplifies to the following form:

$$\dot{x}_i = A_i x_i + B_i u_i + E_i v_0 \tag{1.9a}$$
$$y_{mi} = C_{mi} x_i + D_{mi} u_i + F_{mi} v_0, \; i = 1, \ldots, N \tag{1.9b}$$

where A_i, B_i, E_i, C_{mi}, D_{mi}, and F_{mi} are all known constant matrices. The homogeneous form of (1.9) is as follows:

$$\dot{x}_i = A x_i + B u_i + E v_0 \tag{1.10a}$$
$$y_{mi} = C_m x_i + D_m u_i + F_m v_0, \; i = 1, \ldots, N \tag{1.10b}$$

where $A \in \mathbb{R}^{n \times n}$, $B \in \mathbb{R}^{n \times m}$, $E \in \mathbb{R}^{n \times q}$, $C_m \in \mathbb{R}^{p_m \times n}$, $D_m \in \mathbb{R}^{p_m \times m}$, and $F_m \in \mathbb{R}^{p_m \times q}$. If (1.10) is not subject to external disturbances, then we obtain the following homogeneous linear systems:

$$\dot{x}_i = A x_i + B u_i \tag{1.11a}$$
$$y_{mi} = C_m x_i + D_m u_i, \; i = 1, \ldots, N. \tag{1.11b}$$

A few special cases of the state equation of (1.11a) are as follows:

1. Single integrators:
$$\dot{x}_i = u_i, \; i = 1, \ldots, N$$
where $x_i, u_i \in \mathbb{R}^n$.
2. Double integrators:
$$\dot{x}_{1i} = x_{2i}$$
$$\dot{x}_{2i} = u_i, \; i = 1, \ldots, N$$
where $x_{1i}, x_{2i}, u_i \in \mathbb{R}^n$.
3. Harmonic oscillators:
$$\dot{x}_{1i} = \sigma_i x_{2i}$$
$$\dot{x}_{2i} = -\sigma_i x_{1i} + u_i, \; i = 1, \ldots, N$$
where $x_{1i}, x_{2i}, u_i \in \mathbb{R}^n$, and $\sigma_i > 0$.

Likewise, (1.3) contains the following linear system as a special case:

$$\dot{v}_0 = \mathbf{S}(w_0) v_0 \tag{1.12a}$$
$$y_{m0} = W_0(w_0) v_0 \tag{1.12b}$$

where $\mathbf{S}(w_0) \in \mathbb{R}^{q\times q}$ and $W_0(w_0) \in \mathbb{R}^{p_{m0}\times q}$ are smooth in w_0. Moreover, if (1.12) is known exactly, then system (1.12) is rewritten as follows:

$$\dot{v}_0 = S_0 v_0 \tag{1.13a}$$
$$y_{m0} = W_0 v_0 \tag{1.13b}$$

where $S_0 \in \mathbb{R}^{q\times q}$ and $W_0 \in \mathbb{R}^{p_{m0}\times q}$ are two known constant matrices.

1.2 Graph and Distributed Control

To introduce the concept of distributed control scheme mathematically, let us first recall the so-called centralized control scheme. Note that system (1.4) is after all a control system with a special form and can be put in the following standard form:

$$\dot{x} = f(x, u, v_0, w) \tag{1.14a}$$
$$y_m = h_m(x, u, v_0, w) \tag{1.14b}$$

where $x = \text{col}(x_1, \ldots, x_N)$, $u = \text{col}(u_1, \ldots, u_N)$, $y_m = \text{col}(y_{m1}, \ldots, y_{mN})$, $f = \text{col}(f_1, \ldots, f_N)$, and $h_m = \text{col}(h_{m1}, \ldots, h_{mN})$. For such a standard control system, a standard dynamic measurement output feedback control law takes the following form:

$$u = k(z, y_m, y_{m0}) \tag{1.15a}$$
$$\dot{z} = g(z, y_m, y_{m0}) \tag{1.15b}$$

where k and g are some globally defined smooth functions. If the dimension of z is equal to zero, then (1.15) is called a static measurement output feedback control law. The control law (1.15) includes the state feedback control as a special case if $y_m = x$.

It can be seen that, for a control law of the form (1.15) to be feasible, one needs to assume that the control of each subsystem of (1.4) is able to access the measurement output of all subsystems as well as the measurement output y_{m0} of the leader system. The control law (1.15) leads to a centralized control scheme as illustrated in Fig. 1.1.

On the other hand, suppose, for $i = 1, \ldots, N$, the control u_i of the ith subsystem of (1.4) is able to access the measurement output of itself only as well as the measurement output y_{m0} of the leader system, then, for each subsystem, one can design

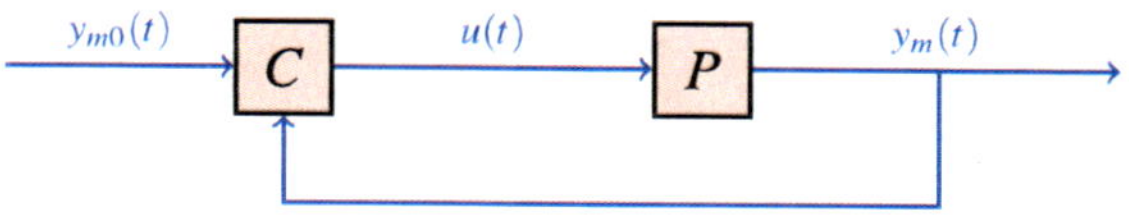

Fig. 1.1 Centralized control scheme

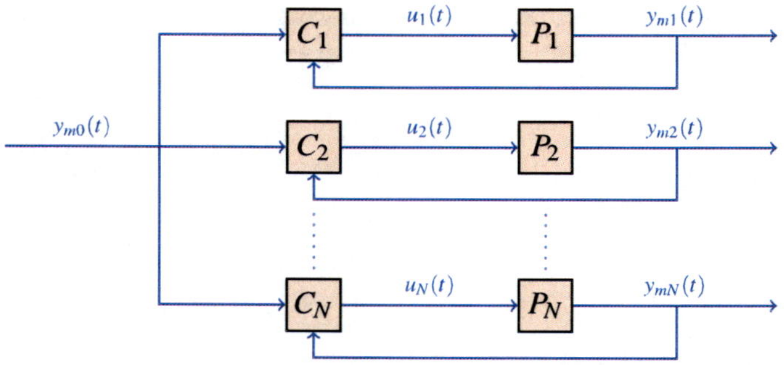

Fig. 1.2 Purely decentralized control scheme

a control law of the following form:

$$u_i = k_i(z_i, y_{mi}, y_{m0}) \tag{1.16a}$$

$$\dot{z}_i = g_i(z_i, y_{mi}, y_{m0}), \ i = 1, \ldots, N. \tag{1.16b}$$

Since, for $i = 1, \ldots, N$, u_i does not rely on the information of other subsystems of (1.4), the control law (1.16) leads to a purely decentralized control scheme as illustrated in Fig. 1.2.

Nevertheless, as mentioned in the preface, a multi-agent system, for example, a group of vehicles, may contain many agents. Some of these agents may stay far away from each other, and therefore, they may not be able to access the information of all the other subsystems. In particular, for a leader–follower multi-agent system, the control of some followers may not be able to access the measurement output of the leader system. Thus, a purely decentralized control law is in general not feasible. To circumvent such difficulty, one has to take advantage of information sharing among the different subsystems, thus leading to the so-called distributed control law. To describe a distributed control law mathematically, we need to introduce some concepts of graph theory.

1.2.1 Static Graphs

A graph $\mathcal{G} = (\mathcal{V}, \mathcal{E})$ consists of a node set $\mathcal{V} = \{1, \ldots, N\}$ and an edge set $\mathcal{E} \subseteq \mathcal{V} \times \mathcal{V}$. For $i, j = 1, 2, \ldots, N, i \neq j$, an edge of $\mathcal{E}$ from node i to node j is denoted by (i, j), where the nodes i and j are called the parent node and the child node of each other, and the node i is also called a neighbor of the node j. By our definition of the edge set, self-edges (i, i) are excluded. Let $\mathcal{N}_i$ denote the subset of $\mathcal{V}$ which consists of all the neighbors of the node i. If the graph $\mathcal{G}$ contains a sequence of edges of the form $(i_1, i_2), (i_2, i_3), \ldots, (i_k, i_{k+1})$, then the set $\{(i_1, i_2), (i_2, i_3), \ldots, (i_k, i_{k+1})\}$ is called a directed path of $\mathcal{G}$ from node i_1 to node i_{k+1}. In this case, node i_{k+1} is said to be reachable from node i_1. If $i_1 = i_{k+1}$, the path is called a cycle. The edge (i, j) is called an undirected edge if $(i, j) \in \mathscr{E} \Leftrightarrow (j, i) \in \mathscr{E}$. The graph is called an undirected graph if every edge in $\mathscr{E}$ is undirected. A graph is called complete if $(i, j) \in \mathscr{E}$ for any $i \neq j, i, j = 1, \ldots, N$. A graph $\mathscr{G}_s = (\mathscr{V}_s, \mathscr{E}_s)$ is a subgraph of $\mathscr{G} = (\mathscr{V}, \mathscr{E})$ if $\mathscr{V}_s \subseteq \mathscr{V}$ and $\mathscr{E}_s \subseteq \mathscr{E} \cap (\mathscr{V}_s \times \mathscr{V}_s)$. Given a set of r graphs $\mathscr{G}_k = (\mathscr{V}, \mathscr{E}_k),\ k = 1, \ldots, r$, the graph $\mathscr{G} = (\mathscr{V}, \mathscr{E})$ where $\mathscr{E} = \bigcup_{k=1}^{r} \mathscr{E}_k$ is called the union of $\mathscr{G}_k,\ k = 1, \ldots, r$, and is denoted by $\bigcup_{k=1}^{r} \mathscr{G}_k$. A tree is a graph in which every node has exactly one parent except for one node, called the root, which has no parent and from which every other node is reachable. A subgraph $\mathscr{G}_s = (\mathscr{V}_s, \mathscr{E}_s)$ of the graph $\mathscr{G} = (\mathscr{V}, \mathscr{E})$ is called a spanning tree of $\mathscr{G}$ if $\mathscr{G}_s$ is a tree and $\mathscr{V}_s = \mathscr{V}$. The graph $\mathscr{G} = (\mathscr{V}, \mathscr{E})$ contains a spanning tree if a subgraph of $\mathscr{G}$ is a spanning tree of $\mathscr{G}$. A graph is called connected if it contains a spanning tree. A matrix $\mathscr{A} = [a_{ij}]_{i,j=1}^{N} \in \mathbb{R}^{N\times N}$ is said to be a weighted adjacency matrix of a graph $\mathscr{G}$ if, for $i, j = 1, \ldots, N$, $a_{ii} = 0, a_{ij} > 0 \Leftrightarrow (j, i) \in \mathscr{E}, a_{ij} = 0 \Leftrightarrow (j, i) \notin \mathscr{E}$, and $a_{ij} = a_{ji}$ if the edge (i, j) is undirected. Let $\mathscr{L} = [l_{ij}]_{i,j=1}^{N} \in \mathbb{R}^{N\times N}$ be such that $l_{ii} = \sum_{j=1}^{N} a_{ij}$ and $l_{ij} = -a_{ij}$ if $i \neq j$. Then $\mathscr{L}$ is called the Laplacian of the graph $\mathscr{G}$. In all numerical examples of this book, we always let the nonzero a_{ij} to be equal to 1. A square real matrix $M = [m_{ij}]_{i,j=1}^{N} \in \mathbb{R}^{N\times N}$ whose off-diagonal entries are nonnegative is called a Metzler matrix. Clearly, $-\mathscr{L}$ is a Metzler matrix with zero row sum. For $\delta > 0$, a δ-graph associated with an $N \times N$ Metzler matrix M is the graph with node set $\{1, \ldots, N\}$ and with an edge from node j to i, $i \neq j$, $i, j = 1, \ldots, N$ if and only $m_{ij} > \delta$. If $\delta = 0$, then the δ-graph associated with a Metzler matrix M is simply called the graph associated with the Metzler matrix M.

Figure 1.3 illustrates four graphs. $\mathscr{G}_1 = (\mathscr{V}_1, \mathscr{E}_1)$ with $\mathscr{V}_1 = \{1, 2, 3, 4, 5\}$, $\mathscr{E}_1 = \{(1, 2), (2, 3), (2, 4), (3, 4), (4, 2), (4, 5), (5, 3)\}$, $\mathscr{N}_1 = \emptyset$, $\mathscr{N}_2=\{1, 4\}$, $\mathscr{N}_3=\{2, 5\}$, $\mathscr{N}_4 = \{2, 3\}$, and $\mathscr{N}_5 = \{4\}$. $\{(1, 2), (2, 3), (3, 4), (4, 5)\}$ is a directed path of $\mathscr{G}_1$ and all nodes $2, \ldots, 5$ are reachable from node 1. In $\mathscr{G}_2$, all edges are undirected and thus $\mathscr{G}_2$ is an undirected graph. By definition, it can be easily verified that $\mathscr{G}_3$ is complete and $\mathscr{G}_4$ is a tree. Note that $\mathscr{G}_4$ is also a subgraph of $\mathscr{G}_1$ and $\mathscr{V}_1 = \mathscr{V}_4$, and thus, $\mathscr{G}_1$ contains a spanning tree and is connected. If we further assume $a_{ij} = 1$ if $a_{ij} > 0$, then the weighted adjacency matrix and its corresponding Laplacian of $\mathscr{G}_1$ are given, respectively, by

$$\mathscr{A}_1 = \begin{bmatrix} 0\,0\,0\,0\,0 \\ 1\,0\,0\,1\,0 \\ 0\,1\,0\,0\,1 \\ 0\,1\,1\,0\,0 \\ 0\,0\,0\,1\,0 \end{bmatrix}, \quad \mathscr{L}_1 = \begin{bmatrix} 0 & 0 & 0 & 0 & 0 \\ -1 & 2 & 0 & -1 & 0 \\ 0 & -1 & 2 & 0 & -1 \\ 0 & -1 & -1 & 2 & 0 \\ 0 & 0 & 0 & -1 & 1 \end{bmatrix}.$$

1.2.2 Switching Graphs

Since the relative positions of agents in a multi-agent system are in general time-varying, the communication topology of a multi-agent system is also in general

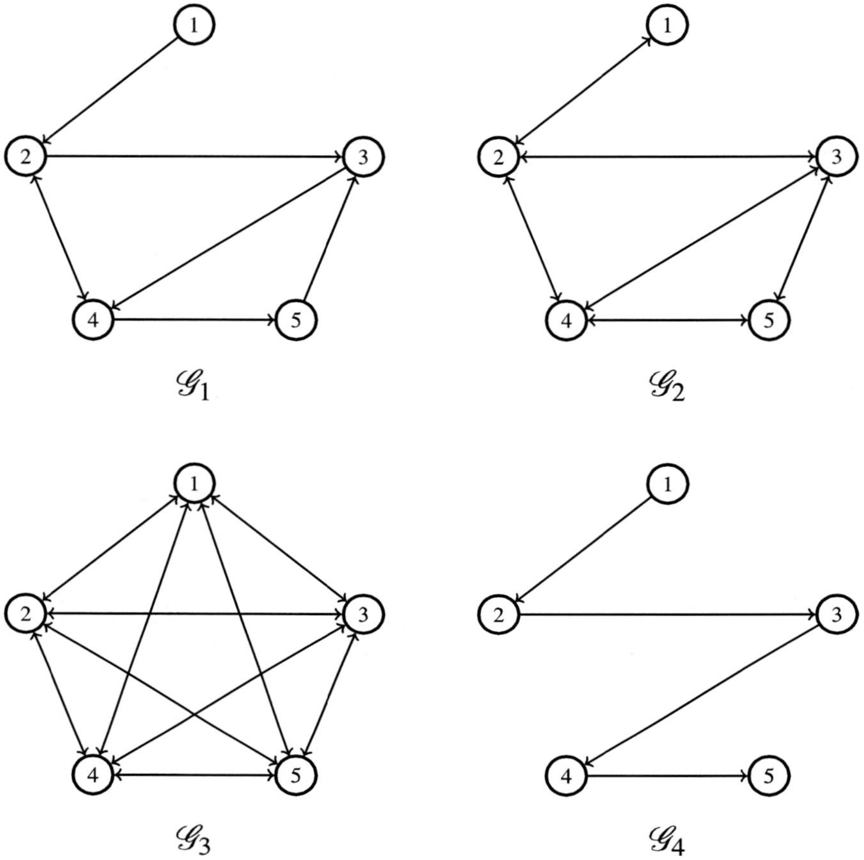

Fig. 1.3 Connected graph $\mathscr{G}_1$, undirected graph $\mathscr{G}_2$, complete graph $\mathscr{G}_3$, and tree/subgraph $\mathscr{G}_4$

time-varying. A time-varying graph is dictated by a time signal described as follows: A time signal $\sigma : [0, \infty) \mapsto \mathscr{P}$, where $\mathscr{P} = \{1, \ldots, \rho\}$ for some positive integer ρ, is said to be a piecewise constant switching signal if there exists a sequence $\{t_i : i = 0, 1, 2, \ldots\}$ satisfying $\lim_{i\to\infty} t_i = \infty$, such that, over each interval $[t_i, t_{i+1})$, $\sigma(t) = p$ for some integer $1 \leq p \leq \rho$. $t_0, t_1, t_2, \ldots$ are called switching instants. The time signal $\sigma(t)$ is said to have a dwell time τ if for $i \geq 1$, $t_i - t_{i-1} \geq \tau$ for some positive constant τ. Given a set of graphs $\mathscr{G}_p = (\mathscr{V}, \mathscr{E}_p)$, $p = 1, \ldots, \rho$, where $\mathscr{V} = \{1, \ldots, N\}$ and $\mathscr{E}_p \subseteq \mathscr{V} \times \mathscr{V}$, and a piecewise constant switching signal $\sigma(t)$ with dwell time τ , we can define a time-varying graph $\mathscr{G}_{\sigma(t)} = (\mathscr{V}, \mathscr{E}_{\sigma(t)})$. As $\sigma(t)$ is a switching signal, $\mathscr{G}_{\sigma(t)}$ is called a switching graph. Let $\mathscr{A}_{\sigma(t)} = [a_{ij}(t)]_{i,j=1}^{N} \in \mathbb{R}^{N\times N}$ denote the weighted adjacency matrix of $\mathscr{G}_{\sigma(t)}$ and $\mathscr{L}_{\sigma(t)}$ denote the Laplacian of $\mathscr{G}_{\sigma(t)}$ corresponding to $\mathscr{A}_{\sigma(t)}$. A switching graph $\mathscr{G}_{\sigma(t)}$ is said to be frequently connected with time period T if, for some $T > 0$, there exists a $t^* \in [t, t+T)$ for any $t > 0$

such that the graph $\mathcal{G}(t^*)$ is connected. For any $t \geq 0, s > 0$, let $\mathcal{G}([t, t+s)) = \bigcup_{t_i \in [t,t+s)} \mathcal{G}_{\sigma(t_i)}$. We call $\mathcal{G}([t, t+s))$ the union graph of $\mathcal{G}_{\sigma(t)}$ over the time interval $[t, t+s)$. A switching graph $\mathcal{G}_{\sigma(t)}$ is said to be jointly connected if there exist a node j and a subsequence $\{i_k : k = 0, 1, 2, \dots\}$ of $\{i : i = 0, 1, 2, \dots\}$ with $t_{i_{k+1}} - t_{i_k} < \nu$ for some $\nu > 0$ and all $k = 0, 1, 2, \dots$ such that the union graph $\mathcal{G}([t_{i_k}, t_{i_{k+1}}))$ contains a spanning tree with the node j as the root.

1.2.3 Distributed Control Scheme

Given a leaderless multi-agent control system of the form (1.4) and a piecewise constant switching signal $\sigma(t)$, one can define a switching graph $\mathcal{G}_{\sigma(t)} = (\mathcal{V}, \mathcal{E}_{\sigma(t)})$[1] with the node set $\mathcal{V} = \{1, \dots, N\}$ where the node i is associated with the ith agent. For each time instant $t \geq 0$, for $i, j = 1, 2, \dots, N$, $i \neq j$, the edge $(j, i) \in \mathcal{E}_{\sigma(t)}$ if and only if, at the time instant t, the control u_i can access the measurement output y_{mj}. Thus, the graph $\mathcal{G}_{\sigma(t)}$ characterizes the communication topology of system (1.4). A control law that satisfies the communication topology is called a distributed control law and, mathematically, this control law is described as follows:

$$u_i = k_i(z_i, z_j, y_{mi}, y_{mj}, j \in \mathcal{N}_i(t)) \tag{1.17a}$$

$$\dot{z}_i = g_i(z_i, z_j, y_{mi}, y_{mj}, j \in \mathcal{N}_i(t)) \tag{1.17b}$$

where, for $i = 1, \dots, N$, $\mathcal{N}_i(t)$ is the neighbor set of the node i of the graph $\mathcal{G}_{\sigma(t)}$ at time t, and k_i and g_i are globally defined smooth functions to be designed. It can be seen that, for $i = 1, \dots, N$, at any time instant $t \geq 0$, u_i depends on $z_j, y_{mj}, j \neq i$, if and only if the node j is a neighbor of the node i at time t. Thus, the control law (1.17) satisfies the communication constraints imposed by the neighbor sets $\mathcal{N}_i(t)$, $i = 1, \dots, N$, which makes it a distributed control law. The distributed control law leads to a distributed control scheme illustrated in Fig. 1.4 with y_{m0} set to zero.

On the other hand, given a leader–follower multi-agent control system described by (1.3) and (1.4), we can also define a switching graph $\bar{\mathcal{G}}_{\sigma(t)} = (\bar{\mathcal{V}}, \bar{\mathcal{E}}_{\sigma(t)})$ with the node set $\bar{\mathcal{V}} = \{0, 1, \dots, N\}$ where the node 0 represents the leader system (1.3), and, for $i = 1, \dots, N$, the node i represents the ith subsystem of (1.4). For each time instant $t \geq 0$, for $i = 1, \dots, N$, $j = 0, 1, 2, \dots, N$, $i \neq j$, the edge $(j, i) \in \bar{\mathcal{E}}_{\sigma(t)}$ if and only if, at the time instant t, the control u_i can access the measurement output y_{mj}. Since the leader system does not have a control input, there are no such edges as $(j, 0)$, $j = 1, \dots, N$.

A distributed control law for the leader–follower multi-agent system (1.3) and (1.4) is described as follows:

[1] A static graph is a special case of a switching graph when $\rho = 1$.

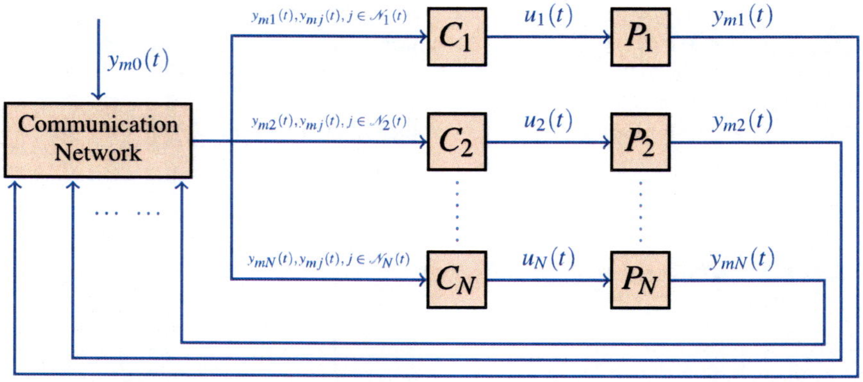

Fig. 1.4 Distributed control scheme

$$u_i = k_i(z_i, z_j, y_{mi}, y_{mj}, j \in \bar{\mathcal{N}}_i(t)) \tag{1.18a}$$

$$\dot{z}_i = g_i(z_i, z_j, y_{mi}, y_{mj}, j \in \bar{\mathcal{N}}_i(t)) \tag{1.18b}$$

where, for $i = 1, \ldots, N$, $\bar{\mathcal{N}}_i(t)$ is the neighbor set of the node i of the graph $\bar{\mathcal{G}}_{\sigma(t)}$ at time t, k_i and g_i are globally defined smooth functions to be designed, and z_0 contains some internal information of the leader system such as the parameters of the leader system. It can be seen that, for $i = 1, \ldots, N$, at any time instant $t \geq 0$, u_i depends on $z_j, y_{mj}, j \neq i$, if and only if the node j is a neighbor of the node i at time t. Thus, the control law (1.18) satisfies the communication constraints imposed by the neighbor sets $\bar{\mathcal{N}}_i(t)$, $i = 1, \ldots, N$, which makes it a distributed control law. This control scheme is also illustrated in Fig. 1.4.

Remark 1.2 The leaderless control law (1.17) can be viewed as a special case of the leader-following control law (1.18) by setting $z_0 = 0$ and $y_{m0} = 0$.

Remark 1.3 If, for $i = 1, \ldots, N$, the dimension of z_i is equal to zero, then the control law (1.17) or the control law (1.18) becomes a static one. If, for $j = 1, \ldots, N$, $y_{mj} = x_j$, then the control law (1.17) or the control law (1.18) is called a state feedback control law.

The centralized control law is a special case of the distributed control law when the graph is complete for all $t \geq 0$. The purely decentralized control is also a special case of the distributed control law when, for all $t \geq 0$, $\bar{\mathcal{N}}_i(t) \equiv \{0\}$, $i = 1, \ldots, N$. Figure 1.5 illustrates the communication topology of three different control schemes for a leader–follower multi-agent system with $N = 3$.

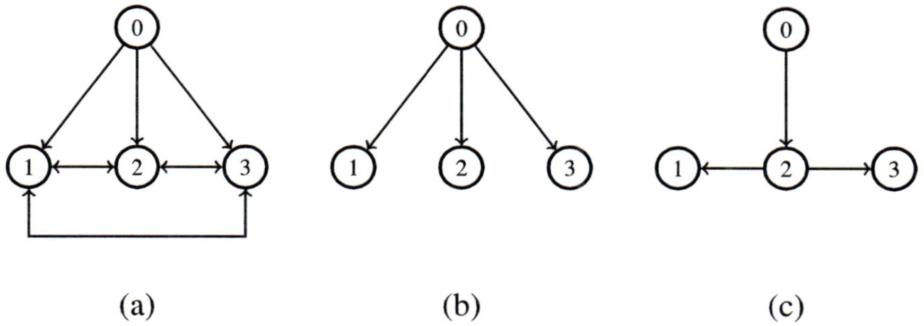

Fig. 1.5 **a** Centralized control; **b** Purely decentralized control; **c** Distributed control

1.3 Cooperative Control

The control of a multi-agent system aims to achieve some global behaviors of all agents by distributed sensing, communication, and computing. There are many global behaviors such as consensus, formation, output regulation, flocking, and rendezvous. Fundamental to all these global behaviors is the consensus problem. In this section, we will briefly describe a few such global behaviors that will be studied in the subsequent chapters. For this purpose, let us further introduce, for each subsystem of (1.4), the performance output y_i as follows:

$$y_i = h_{pi}(x_i, u_i, v_0, w), \ i = 1, \ldots, N \tag{1.19}$$

where, for $i = 1, \ldots, N$, $y_i \in \mathbb{R}^p$ for some integer $p > 0$, and h_{pi} is some globally defined smooth function satisfying, for all $w \in \mathbb{R}^{n_w}$, $h_{pi}(0, 0, 0, w) = 0$. Typically, the performance output y_i is some known function of the measurement output y_{mi} and it represents those quantities of the ith subsystem that need to have desired behaviors through control.

Also, let us introduce the reference output y_0 of the leader system as follows:

$$y_0 = h_0(v_0) \tag{1.20}$$

where $y_0 \in \mathbb{R}^p$, and h_0 is some globally defined smooth function vanishing at the origin. Typically, the reference output y_0 provides the reference signal for the performance output of (1.4) to follow.

Finally, for $i = 1, \ldots, N$, we define the error output (or regulated output) $e_i \in \mathbb{R}^p$ of the ith subsystem of the leader–follower multi-agent system (1.3) and (1.4) as follows:

$$e_i = h_i(x_i, u_i, v_0, w), \ i = 1, \ldots, N \tag{1.21}$$

where, for $i = 1, \ldots, N$, h_i is a globally defined smooth function satisfying, for all $w \in \mathbb{R}^{n_w}$, $h_i(0, 0, 0, w) = 0$. Typically, e_i takes the following form:

$$e_i = y_i - y_0, \ i = 1, \ldots, N. \tag{1.22}$$

In this case, the regulated output e_i represents the tracking error between the performance output y_i of the ith subsystem and the reference output y_0 of the leader system. Nevertheless, in Chap. 6, we will encounter the regulated output e_i not in the form of (1.22). From (1.19) and (1.20), we can easily obtain the expression of h_i in (1.21). In the special case where h_0 is linear, (1.20) reduces to the following linear form:

$$y_0 = F_0 v_0 \tag{1.23}$$

for some constant matrix F_0. If, for $i = 1, \ldots, N$, h_{pi} are also linear, then (1.21) reduces to the following linear form:

$$e_i = C_i(w)x_i + D_i(w)u_i + F_i(w)v_0, \ i = 1, \ldots, N \tag{1.24}$$

for some matrices $C_i(w)$, $D_i(w)$, and $F_i(w)$.

Some typical control objectives of multi-agent systems are as follows.

- **Leaderless Output Consensus Problem:** A distributed control law of the form (1.17) is said to achieve leaderless consensus for the leaderless multi-agent system (1.4) if, for any initial state, the performance output y_i of the closed-loop system satisfies

$$\lim_{t\to\infty} (y_i(t) - y_j(t)) = 0, \ \forall\, i, \ j = 1, \ldots, N. \tag{1.25}$$

  Leaderless output consensus is also called leaderless output synchronization. In the special case where $y_i = x_i$, $i = 1, \ldots, N$, the output consensus problem is also called leaderless state consensus problem or simply consensus problem.
- **Leader-Following Output Consensus Problem:** The leaderless output consensus problem only requires the performance output y_i of all subsystems of (1.4) to synchronize to a common trajectory and does not care what this common trajectory is. In contrast, the leader-following output consensus problem further requires the performance output y_i of all subsystems of (1.4) asymptotically tend to the reference signal $y_0(t)$. In this case, the global objective (1.25) is modified to the following:

$$\lim_{t\to\infty} (y_i(t) - y_0(t)) = 0, \ i = 1, \ldots, N. \tag{1.26}$$

  Obviously, (1.26) implies (1.25). In particular, if $y_i = x_i$, $i = 1, \ldots, N$, then the leader-following output consensus problem is also called the leader-following state consensus problem or simply the leader-following consensus problem.

- **Cooperative Output Regulation Problem:** The consensus problem is first studied for the class of homogeneous linear multi-agent systems of the form (1.11). As will be seen in Chap. 3, under some quite standard assumptions, both the leaderless consensus problem and the leader-following consensus problem of the homogeneous linear systems of the form (1.11) can be dealt with by a distributed static state feedback control law. In a more general setting where the plant is described by (1.4) which can be heterogeneous and may contain external disturbances and uncertain parameters, and the leader system is described by (1.3) which may contain uncertain parameters, the global objectives can be more sophisticated. The cooperative output regulation problem aims to design a control law of the form (1.18) such that, for any initial conditions of the closed-loop system and the leader system, the trajectories of the closed-loop system and the leader system satisfy

$$\lim_{t\to\infty} e_i(t) = 0, \ i = 1, \ldots, N \tag{1.27}$$

where e_i is given by the general form (1.21). In the special case where e_i is given by (1.22), the solvability of the cooperative output regulation problem implies that of the leader-following output consensus problem.
- **Leader-Following Formation Problem:** Formation control aims to design a distributed control law so that a group of agents form a geometric configuration such as a circle or a polygon while moving along a desired trajectory in a two- or three-dimensional space. For example, in a two-dimensional space, given $N > 0$, a desired formation is specified by a reference trajectory $h_0(t) \in \mathbb{R}^2$ and N other vectors $h_{di} \in \mathbb{R}^2$, $i = 1, \ldots, N$, representing the desired relative displacements between the ith agent and the reference trajectory. Thus, let

$$\bar{h}_{di} \triangleq h_{di} + h_0, \ i = 1, \ldots, N.$$

Then, $\bar{h}_{di}$ is the desired trajectory of the ith agent. Let $h_i(t)$ denote the position vector of the ith agent. Then, a control law is said to achieve the leader-following formation if the trajectory of each agent is such that

$$\lim_{t\to\infty} (h_i(t) - \bar{h}_{di}(t)) = 0$$
$$\lim_{t\to\infty} (\dot{h}_i(t) - \dot{\bar{h}}_{di}(t)) = 0, \ i = 1, \ldots, N.$$

Figure 1.6 shows a formation composed of four follower subsystems and one leader system. It will be seen in the example below that, under some conditions, the leader-following formation problem can be converted to a cooperative output regulation problem.

Example 1.1 Consider the leader-following formation problem of four mobile robots, shown in Fig. 1.6. The global objective is to design a distributed controller to make the four mobile robots together with a leader to form a configuration as

Fig. 1.6 An example of the leader-following formation

illustrated in Fig. 1.7, while the leader is moving along a straight line with a constant speed. For simplicity, we assume the motion equations of the mobile robots are the following multiple double integrator systems:

$$\ddot{h}_i = \bar{u}_i, \ i = 1, \ldots, 4 \tag{1.28}$$

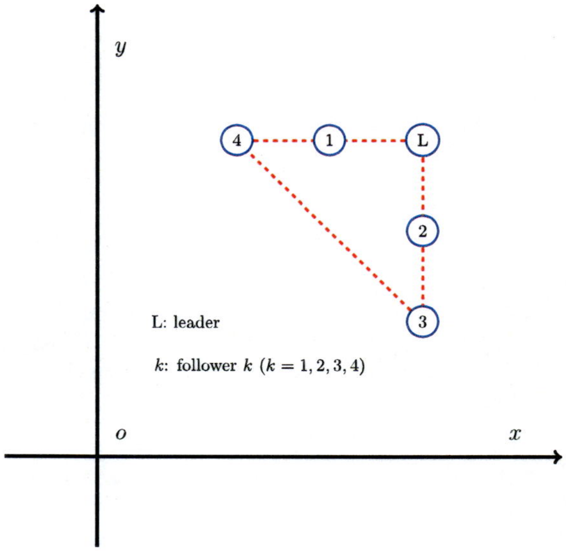

Fig. 1.7 The desired formation pattern

where $h_i(t), \bar{u}_i(t) \in \mathbb{R}^2$ represent the robot hand position off the wheel axis and the control input of the ith mobile robot, respectively. Let $h_i = \mathrm{col}(x_{hi}, y_{hi})$, $\dot{h}_i = \mathrm{col}(\dot{x}_{hi}, \dot{y}_{hi})$, and $\bar{u}_i = \mathrm{col}(\bar{u}_{xi}, \bar{u}_{yi})$. Then, the state-space equations of (1.28) are as follows:

$$\dot{x}_{hi} = p_{xi} \tag{1.29a}$$
$$\dot{y}_{hi} = p_{yi} \tag{1.29b}$$
$$\dot{p}_{xi} = \bar{u}_{xi} \tag{1.29c}$$
$$\dot{p}_{yi} = \bar{u}_{yi}, \ i = 1, \dots, 4. \tag{1.29d}$$

This problem can be formulated as a cooperative output regulation problem as follows. Let the time trajectory of the leader be $h_0(t) = p_0 t + h_{d0}$, where $h_{d0} = \mathrm{col}(x_{d0}, y_{d0})$ and $p_0 = \mathrm{col}(p_x^d, p_y^d)$ for some constants p_x^d, p_y^d, x_{d0}, and y_{d0}. Define the following leader system:

$$\dot{v}_0 = S_0 v_0 = \left(\begin{bmatrix} 0 & 1 \\ 0 & 0 \end{bmatrix} \otimes I_2 \right) v_0 \tag{1.30a}$$
$$y_0 = \left(\begin{bmatrix} 1 & 0 \end{bmatrix} \otimes I_2 \right) v_0. \tag{1.30b}$$

Then it can be seen that $y_0(t) = h_0(t)$ if the initial condition $v_0(0) = \mathrm{col}(x_{d0}, y_{d0}, p_x^d, p_y^d)$.

Let $h_{di} = \mathrm{col}(x_{di}, y_{di})$, $x_i = \mathrm{col}(x_{hi} - x_{di}, y_{hi} - y_{di}, p_{xi}, p_{yi})$, and

$$e_i(t) = h_i(t) - (h_{di} + h_0(t)), \quad i = 1, \dots, 4. \tag{1.31}$$

Then the system composed of (1.29) and (1.31) has the form (1.9) and (1.24) with the state x_i, the input $\bar{u}_i$, the measurement output $y_{mi} = x_i$, the performance output $y_i = h_i - h_{di}$, and the error output e_i. Various matrices in (1.9) and (1.24) are as follows:

$$\begin{aligned} A_i &= \begin{bmatrix} 0 & 1 \\ 0 & 0 \end{bmatrix} \otimes I_2, \ B_i = \begin{bmatrix} 0 \\ 1 \end{bmatrix} \otimes I_2, \ E_i = 0_{4\times 4}, \\ C_{mi} &= I_4, \ D_{mi} = 0_{4\times 2}, \ F_{mi} = 0_{4\times 4}, \\ C_i &= \begin{bmatrix} 1 & 0 \end{bmatrix} \otimes I_2, \ D_i = 0_{2\times 2}, \ F_i = \begin{bmatrix} -1 & 0 \end{bmatrix} \otimes I_2. \end{aligned}$$

It can be verified that, when $\lim_{t\to\infty} e_i(t) = 0$, $\lim_{t\to\infty}(h_i(t) - h_0(t)) = h_{di}$ and $\lim_{t\to\infty} \dot{h}_i(t) = p_0$. Thus, if the cooperative output regulation problem of the system composed of (1.29), (1.30), and (1.31) is solvable with $y_{mi} = x_i$, then the leader and the four mobile robots will asymptotically achieve the formation shown in Fig. 1.7. The detailed solution of this problem will be provided in Chap. 8 via the distributed observer approach.

1.4 Organization of the Book

This book mainly treats two fundamental cooperative control problems of complex multi-agent systems characterized by heterogeneities, external disturbances, nonlinearities, and uncertainties. Two systematic control approaches, namely the distributed observer approach and the distributed internal model approach, will be presented. Several other global objectives, such as formation, rendezvous, and flocking, can also be handled by these two approaches.

The remaining chapters of the book are organized as follows:

- **Chapter 2:** This chapter introduces some fundamental technical results including properties of some typical graphs, stability analysis for perturbed systems, generalized Barbalat's Lemma, and some other results. These results will be used in the subsequent chapters.
- **Chapter 3:** The leaderless and leader-following consensus problems are first formulated. Then, after establishing some stability results of linear switched systems, the two consensus problems for homogeneous linear systems of the form (1.11) are thoroughly studied for various scenarios via the static distributed control laws.
- **Chapter 4:** This chapter first outlines a general framework for synthesizing distributed control laws for multi-agent systems via the distributed observer approach. Three types of distributed observers for a linear leader system, namely the distributed observer for a known linear leader system, the adaptive distributed observer for a known linear leader system, and the adaptive distributed observer for an uncertain linear leader system, are established. The effectiveness of these types of distributed observers is validated by numerical examples.
- **Chapter 5:** This chapter investigates the leader-following consensus problem of multiple uncertain Euler–Lagrange systems of the form (1.5). After a treatment of the asymptotic tracking problem for a single uncertain Euler–Lagrange system, the solution of the leader-following consensus problem of multiple uncertain Euler–Lagrange systems is given by three types of distributed control laws composed of a purely decentralized control law and the three types of distributed observers, respectively.
- **Chapter 6:** This chapter studies the leader-following consensus problem of multiple rigid-body systems of the form (1.6). The attitude parametrization, kinematics, and dynamics of a rigid-body system are first given. Then, following a study of the standard attitude control problem for a single rigid-body system, the solvability of the leader-following consensus problem of multiple rigid-body systems is presented for a few scenarios by various distributed observer-based control laws.
- **Chapter 7:** This chapter treats three types of the output regulation problem of linear systems. The first one studies the output regulation problem for exactly known systems via the so-called feedforward control design method which makes use of the steady-state information for compensation. The second one handles the structurally stable output regulation problem of linear systems with small parameter variations via the so-called p-copy internal model approach. The last one is called the robust output regulation problem which is capable of dealing

with arbitrarily large parametric uncertainties for minimum phase linear systems via the so-called canonical internal model approach.

- **Chapter 8:** This chapter studies the cooperative output regulation problem of linear heterogeneous multi-agent systems of the form (1.9) via the distributed observer approach, which can be viewed as the generalization of the first type of the output regulation problem of linear time-invariant systems studied in Chap. 7. Two types of distributed control laws are considered, which are the compositions of a purely decentralized control law and the distributed observer for a known leader system and the adaptive distributed observer for a known leader system, respectively.
- **Chapter 9:** This chapter studies the cooperative structurally stable output regulation problem of general linear uncertain multi-agent systems of the form (1.8) via the distributed p-copy internal model approach and the cooperative robust output regulation problem for minimum phase linear uncertain multi-agent systems via the distributed canonical internal model approach, respectively. These two problems are the extensions of the second type and the third type of the output regulation problem of linear time-invariant systems studied in Chap. 7, respectively.
- **Chapter 10:** This chapter further studies the cooperative robust output regulation problem for the class of nonlinear multi-agent systems of the form (1.7) over static communication graphs. A nonlinear version of the distributed canonical internal model approach is adopted. The solutions for three different cases regarding the uncertainties of the follower systems and the leader system are given.
- **Chapter 11:** This chapter integrates both the distributed observer approach and the distributed canonical internal model approach, thus leading to an approach capable of handling the nonlinearity and uncertainty of the same system as in Chap. 10 over jointly connected switching communication graphs.
- **Appendix A:** To make the book self-contained, we provide one appendix to summarize some useful results in the Kronecker product, stability analysis of switched systems and nonlinear systems, adaptive control, and a design framework for the nonlinear output regulation problem.

1.5 Notes and References

The consensus problem was studied at least as early as the 1970s in, say, [1, 2]. Since the beginning of this century, the cooperative control of multi-agent systems became one of the central problems in the field of systems and control following the publications of a few celebrated papers, say, [3–6]. The cooperative control of multi-agent systems was first studied for simple linear multi-agent systems where each subsystem is a single integrator [4, 6–9], double integrator [10–13], or harmonic oscillator [14, 15] and then progressed to general linear homogenous multi-agent systems [16–21] and some weak nonlinear multi-agent systems, such as Lipschitz nonlinear systems [22], Euler–Lagrange systems [23–25], and rigid-body systems [26–29], to name just a few. Since around 2010, the research on the cooperative control has advanced to the stage of handling complex systems characterized by strong nonlinearities, large uncertainties, heterogeneities, and jointly connected switching communication topologies which can be disconnected at every time instant. Simpler control objectives such as consensus or formation give way to more sophisticated tasks of simultaneous tracking of a class of reference signals and rejection of a class of external disturbances, thus leading to the so-called cooperative output regulation problem of multi-agent systems. The robust cooperative output regulation problem for uncertain linear multi-agent systems was first studied by the distributed internal model approach over acyclic static communication graphs in [30] and then was extended to the general static communication graph in [31, 32]. Using the distributed observer approach, the cooperative output regulation problem for known heterogeneous linear multi-agent systems was solved in [33] over static communication graphs and in [34] over jointly connected switching communication graphs. The study on the cooperative robust output regulation problem for various nonlinear multi-agent systems can be found in, say, [35–42].

References

1. Chatterjee S, Seneta E (1977) Towards consensus: some convergence theorems on repeated averaging. J Appl Probab 14(1):89–97
2. DeGroot MH (1974) Reaching a consensus. J Am Stat Assoc 69(345):118–121
3. Fax JA, Murray RM (2004) Information flow and cooperative control of vehicle formations. IEEE Trans Autom Control 49(9):1465–1476
4. Jadbabaie A, Lin J, Morse AS (2003) Coordination of groups of mobile autonomous agents using nearest neighbor rules. IEEE Trans Autom Control 48(6):988–1001
5. Ogren P, Egerstedt M, Hu X (2002) A control Lyapunov function approach to multiagent coordination. IEEE Trans Robot Autom 18(5):847–851
6. Olfati-Saber R, Murray RM (2004) Consensus problems in networks of agents with switching topology and time-delays. IEEE Trans Autom Control 49(9):1520–1533
7. Lin Z (2005) Coupled dynamic systems: from structure towards stability and stabilizability, PhD dissertation, University of Toronto, Toronto, Canada
8. Moreau L (2004) Stability of continuous-time distributed consensus algorithms. In: Proceedings of the 41st IEEE conference on decision and control, pp 3998–4003
9. Ren W, Beard RW (2005) Consensus seeking in multiagent systems under dynamically changing interaction topologies. IEEE Trans Autom Control 50(5):655–661
10. Hong Y, Gao L, Cheng D, Hu J (2007) Lyapunov based approach to multiagent systems with switching jointly connected interconnection. IEEE Trans Autom Control 52(5):943–948
11. Hu J, Hong Y (2007) Leader-following coordination of multi-agent systems with coupling time delays. Phys A Stat Mech Appl 374(2):853–863
12. Qin J, Gao H, Zheng W (2011) Second-order consensus for multi-agent systems with switching topology and communication delay. Syst Control Lett 60(6):390–397
13. Ren W (2008) On consensus algorithms for double-integrator dynamics. IEEE Trans Autom Control 53(6):1503–1509
14. Ren W (2008) Synchronization of coupled harmonic oscillators with local interaction. Automatica 44(2):3195–3200
15. Su H, Wang X, Lin Z (2009) Synchronization of coupled harmonic oscillators in a dynamic proximity network. Automatica 45(10):2286–2291
16. Ni W, Cheng D (2010) Leader-following consensus of multi-agent systems under fixed and switching topologies. Syst Control Lett 59(3):209–217

17. Su S, Lin Z (2016) Distributed consensus control of multi-agent systems with higher order agent dynamics and dynamically changing directed interaction topologies. IEEE Trans Autom Control 61(2):515–519
18. Su Y, Huang J (2012) Stability of a class of linear switching systems with applications to two consensus problems. IEEE Trans Autom Control 57(6):1420–1430
19. Su Y, Huang J (2012) Two consensus problems for discrete-time multi-agent systems with switching network topology. Automatica 48(9):1988–1997
20. Tuna SE (2008) LQR-based coupling gain for synchronization of linear systems. arxiv.org/abs/0801.3390
21. You K, Xie L (2011) Network topology and communication data rate for consensusability of discrete-time multi-agent systems. IEEE Trans Autom Control 56(10):2262–2275
22. Song Q, Cao J, Yu W (2010) Second-order leader-following consensus of nonlinear multi-agents via pinning control. Syst Control Lett 59(9):553–562
23. Mehrabain AR, Tafazoli S, Khorasani K (2010) Cooperative tracking control of Euler-Lagrange systems with switching communication network topologies. In: Proceedings of IEEE/ASME international conference on advanced intelligent mechatronics, pp 756–761
24. Mei J, Ren W, Ma G (2011) Distributed tracking with a dynamic leader for multiple Euler-Lagrange systems. IEEE Trans Autom Control 56(6):1415–1421
25. Nuño E, Ortega R, Basañez L, Hill D (2011) Synchronization of networks of nonidentical Euler-Lagrange systems with uncertain parameters and communication delays. IEEE Trans Autom Control 56(4):935–941
26. Bai H, Arcak M, Wen JT (2008) Rigid body attitude coordination without inertial frame information. Automatica 44(12):3170–3175
27. Dimarogonas DV, Tsiotras P, Kyriakopoulos KJ (2009) Leader-follower cooperative attitude control of multiple rigid bodies. Syst Control Lett 58(6):429–435
28. Ren W (2010) Distributed cooperative attitude synchronization and tracking for multiple rigid bodies. IEEE Trans Control Syst Technol 18(2):383–392
29. Sarlette A, Sepulchre R, Leonard NE (2009) Autonomous rigid body attitude synchronization. Automatica 45(2):572–577
30. Wang X, Hong Y, Huang J, Jiang ZP (2010) A distributed control approach to a robust output regulation problem for multi-agent linear systems. IEEE Trans Autom Control 55(12):2891–2895
31. Su Y, Hong Y, Huang J (2013) A general result on the robust cooperative output regulation for linear uncertain multi-agent systems. IEEE Trans Autom Control 58(5):1275–1279
32. Su Y, Huang J (2014) Cooperative robust output regulation of a class of heterogeneous linear uncertain multi-agent systems. Int J Robust Nonlinear Control 24(17):2819–2839
33. Su Y, Huang J (2012) Cooperative output regulation of linear multi-agent systems. IEEE Trans Autom Control 57(4):1062–1066
34. Su Y, Huang J (2012) Cooperative output regulation with application to multi-agent consensus under switching network. IEEE Trans Syst Man Cybern Part B Cybern 42(3):864–875
35. Dong Y, Chen J, Huang J (2018) Cooperative robust output regulation problem for second-order nonlinear multi-agent systems with an unknown exosystem. IEEE Trans Autom Control 63(10):3418–3425
36. Dong Y, Huang J (2014) Cooperative global robust output regulation for nonlinear multi-agent systems in output feedback form. J Dyn Syst Meas Control-Trans ASME 136(3):031001
37. Dong Y, Huang J (2014) Cooperative global output regulation for a class of nonlinear multi-agent systems. IEEE Trans Autom Control 59(5):1348–1354
38. Liu W, Huang J (2017) Cooperative global robust output regulation for nonlinear output feedback multi-agent systems under directed switching networks. IEEE Trans Autom Control 62(12):6339–6352
39. Su Y, Huang J (2013) Cooperative global output regulation of heterogeneous second-order nonlinear uncertain multi-agent systems. Automatica 49(11):3345–3350
40. Su Y, Huang J (2013) Cooperative global output regulation for a class of nonlinear uncertain multi-agent systems with unknown leader. Syst Control Lett 62(6):461–467

41. Su Y, Huang J (2014) Cooperative semi-global robust output regulation for a class of nonlinear uncertain multi-agent systems. Automatica 50(4):1053–1065
42. Su Y, Huang J (2015) Cooperative global output regulation for nonlinear uncertain multi-agent systems in lower triangular form. IEEE Trans Autom Control 60(9):2378–2389

Chapter 2
Preliminaries

This chapter introduces some fundamental technical results including the properties of some typical graphs, stability analysis for perturbed systems, generalized Barbalat's Lemma, and some other results. These results will be used in the subsequent chapters.

2.1 Properties of Some Typical Graphs

In this section, we will establish the properties of two types of graphs associated with the leaderless multi-agent system and the leader–follower multi-agent system, respectively.

2.1.1 Static Graphs

Given a static graph $\mathscr{G} = (\mathscr{V}, \mathscr{E})$ with $\mathscr{V} = \{1, \ldots, N\}$ and $\mathscr{E} \subseteq \mathscr{V} \times \mathscr{V}$, let $\mathscr{A} = [a_{ij}]_{i,j=1}^{N}$ be a weighted adjacency matrix of $\mathscr{G}$ and $\mathscr{L}$ be the corresponding Laplacian. Two frequently used assumptions on the graph $\mathscr{G}$ are listed as follows:

Assumption 2.1 $\mathscr{G}$ is connected.

Assumption 2.2 $\mathscr{G}$ is undirected.

Remark 2.1 Since $-\mathscr{L}$ is a Metzler matrix with zero row sum, by Theorem A.1 in Sect. A.2, the matrix $\mathscr{L}$ has at least one zero eigenvalue with 1_N as an eigenvector, and all the nonzero eigenvalues have positive real parts. Furthermore, the matrix $\mathscr{L}$ has exactly one zero eigenvalue and all the other eigenvalues have positive real parts if and only if Assumption 2.1 holds. Thus, the null space $\mathscr{L}$ is span$\{1_N\}$ if and

H. Cai et al., *Cooperative Control of Multi-agent Systems*, Advances in Industrial Control,
https://doi.org/10.1007/978-3-030-98377-2_2

only if Assumption 2.1 holds. In addition, under Assumption 2.2, the matrix $\mathscr{L}$ is symmetric.

To see the role of Assumptions 2.1 and 2.2, consider the following example:

Example 2.1 Consider the following linear system:

$$\dot{x} = -(\mathscr{L} \otimes I_n)x \tag{2.1}$$

where $n \geq 1$ and $x = \text{col}(x_1, \ldots, x_N)$ with $x_i \in \mathbb{R}^n$, $i = 1, \ldots, N$. Since, under Assumption 2.1, $\mathscr{L}$ has exactly one zero eigenvalue and all the other eigenvalues have positive real parts, the real Jordan canonical form of $\mathscr{L}$ is as follows:

$$J = \left[\begin{array}{c|c} 0 & 0 \cdots 0 \\ \hline 0 & \\ \vdots & \Lambda \\ 0 & \end{array}\right]$$

where all the eigenvalues of $\Lambda \in \mathbb{R}^{(N-1)\times(N-1)}$ coincide with the $(N-1)$ nonzero eigenvalues of $\mathscr{L}$. Let $r \in \mathbb{R}^N$ be the left eigenvector of $\mathscr{L}$ corresponding to the zero eigenvalue such that $r^\mathsf{T} 1_N = 1$. Then there exist $Y \in \mathbb{R}^{N\times(N-1)}$ and $W \in \mathbb{R}^{(N-1)\times N}$ such that

$$V = \begin{bmatrix} 1_N & Y \end{bmatrix}, \quad V^{-1} = \begin{bmatrix} r^\mathsf{T} \\ W \end{bmatrix}$$

and $V^{-1}\mathscr{L}V = J$. Thus, for any $t \geq 0$,

$$e^{-\mathscr{L}t} = Ve^{-Jt}V^{-1} = \begin{bmatrix} 1_N & Y \end{bmatrix} \left[\begin{array}{c|c} 1 & 0 \cdots 0 \\ \hline 0 & \\ \vdots & e^{-\Lambda t} \\ 0 & \end{array}\right] \begin{bmatrix} r^\mathsf{T} \\ W \end{bmatrix} = 1_N r^\mathsf{T} + Ye^{-\Lambda t}W.$$

Thus, for any initial condition $x(0)$, the solution of system (2.1) is

$$x(t) = (e^{-\mathscr{L}t} \otimes I_n)x(0) = ((1_N r^\mathsf{T} + Ye^{-\Lambda t}W) \otimes I_n)x(0).$$

Since all the eigenvalues of Λ have positive real parts, $\lim_{t\to\infty} e^{-\Lambda t} = 0$. As a result,

$$\begin{aligned} \lim_{t\to\infty} x(t) &= \lim_{t\to\infty} (e^{-\mathscr{L}t} \otimes I_n)x(0) = \left((1_N r^\mathsf{T} + Y \lim_{t\to\infty} e^{-\Lambda t}W) \otimes I_n\right)x(0) \\ &= \left((1_N r^\mathsf{T}) \otimes I_n\right)x(0) = 1_N \otimes \left((r^\mathsf{T} \otimes I_n)x(0)\right). \end{aligned} \tag{2.2}$$

That is, for $i = 1, \ldots, N$, all the components $x_i(t)$ of the solution $x(t)$ of (2.1) converge to the same constant vector $(r^\mathsf{T} \otimes I_n)x(0)$ exponentially. In this case, we say all the components of $x(t)$ asymptotically reach the leaderless consensus. If, in addition, Assumption 2.2 is also satisfied, then $\mathcal{L}$ is symmetric. In this case, we can assume Λ is a diagonal matrix whose diagonal elements are such that $0 < \lambda_2 \leq \lambda_3 \leq \cdots \leq \lambda_N$. Thus, λ_2 determines the rate of convergence of the solution of (2.1) to its consensus value.

Remark 2.2 Consider the following multiple single integrator systems:

$$\dot{x}_i = u_i, \ i = 1, \ldots, N \tag{2.3}$$

where $x_i, u_i \in \mathbb{R}^n$. Substituting the following control law:

$$u_i = \sum_{j=1}^{N} a_{ij}(x_j - x_i), \ i = 1, \ldots, N \tag{2.4}$$

into (2.3) gives (2.1). Thus, (2.1) arises from solving the so-called leaderless consensus problem for multiple single integrator systems (2.3) by the distributed control law (2.4). The leaderless consensus problem for general linear multi-agent systems will be studied in detail in next chapter.

To study leader–follower multi-agent systems, we further introduce a graph $\bar{\mathcal{G}} = (\bar{\mathcal{V}}, \bar{\mathcal{E}})$ with $\bar{\mathcal{V}} = \{0, 1, \ldots, N\}$ and $\bar{\mathcal{E}} \subseteq \bar{\mathcal{V}} \times \bar{\mathcal{V}}$. Let $\bar{\mathcal{A}} = [a_{ij}]_{i,j=0}^{N}$ be a weighted adjacency matrix of $\bar{\mathcal{G}}$ and $\bar{\mathcal{L}}$ be the corresponding Laplacian. Let $\Delta = \mathcal{D}(a_{10}, \ldots, a_{N0})$. Partition $\bar{\mathcal{L}}$ as follows:

$$\bar{\mathcal{L}} = \left[\begin{array}{c|c} \sum_{j=1}^{N} a_{0j} & -a_{01} \ \cdots \ -a_{0N} \\ \hline -\Delta 1_N & H \end{array}\right].$$

Let $\mathcal{G} = (\mathcal{V}, \mathcal{E})$ with $\mathcal{V} = \{1, \ldots, N\}$ and $\mathcal{E} = (\mathcal{V} \times \mathcal{V}) \cap \bar{\mathcal{E}}$ be the subgraph of $\bar{\mathcal{G}}$, and $\mathcal{L}$ be the Laplacian of $\mathcal{G}$. Then it can be verified that

$$H = \Delta + \mathcal{L}. \tag{2.5}$$

Assumption 2.3 $\bar{\mathcal{G}}$ contains a spanning tree with the node 0 as the root.

Lemma 2.1 *All the nonzero eigenvalues of H have positive real parts; H is nonsingular if and only if Assumption 2.3 holds; and H is symmetric if and only if Assumption 2.2 holds.*

Proof Note that adding to or removing from $\bar{\mathcal{E}}$ such edges as $(j, 0)$, $j = 1, \ldots, N$, only changes the first row of $\bar{\mathcal{L}}$. Thus, by removing from $\bar{\mathcal{E}}$ all such edges as $(j, 0)$, $j = 1, \ldots, N$, we obtain a subgraph $\hat{\mathcal{G}}$ of $\bar{\mathcal{G}}$ whose Laplacian $\hat{\mathcal{L}}$ is the same as $\bar{\mathcal{L}}$ except that the entries of its first row are all zeros (See Fig. 2.1 for an example illustrating the relationship between $\bar{\mathcal{G}}$, $\hat{\mathcal{G}}$, and $\mathcal{G}$). So the nonzero eigenvalues of H

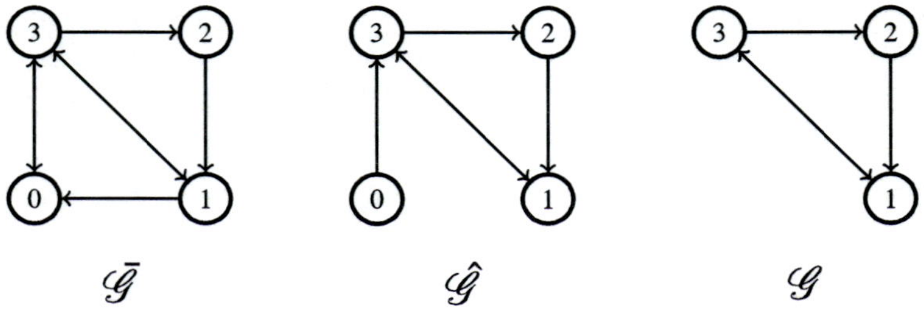

Fig. 2.1 The relationship between $\bar{\mathscr{G}}$, $\hat{\mathscr{G}}$, and $\mathscr{G}$

coincide with those of $\hat{\mathscr{L}}$. Thus, by Remark 2.1, all the nonzero eigenvalues of H have positive real parts. Furthermore, from the lower triangular structure of $\hat{\mathscr{L}}$, we conclude that H is nonsingular if and only if $\hat{\mathscr{G}}$ and hence $\bar{\mathscr{G}}$ contains a spanning tree with the node 0 as the root. Finally, from (2.5), H is symmetric if and only if Assumption 2.2 holds. □

Remark 2.3 For convenience, we call a graph satisfying Assumption 2.3 connected with the node 0 as the root.

2.1.2 *Switching Graphs*

Given a switching graph $\mathscr{G}_{\sigma(t)} = (\mathscr{V}, \mathscr{E}_{\sigma(t)})$ with $\mathscr{V} = \{1, \ldots, N\}$ and $\mathscr{E}_{\sigma(t)} \subseteq \mathscr{V} \times \mathscr{V}$, let $\mathscr{A}_{\sigma(t)} = [a_{ij}(t)]_{i,j=0}^{N}$ be a weighted adjacency matrix of $\mathscr{G}_{\sigma(t)}$ and $\mathscr{L}_{\sigma(t)}$ be the corresponding Laplacian at the time instant t. The two assumptions corresponding to Assumptions 2.1 and 2.2 are as follows:

Assumption 2.4 $\mathscr{G}_{\sigma(t)}$ is jointly connected.

Assumption 2.5 $\mathscr{G}_{\sigma(t)}$ is undirected for any $t \geq 0$.

Remark 2.4 Since $-\sum_{j=i_k}^{i_{k+1}-1} \mathscr{L}_{\sigma(t_j)}$ is a Metzler matrix with zero row sum and the graph associated with $-\sum_{j=i_k}^{i_{k+1}-1} \mathscr{L}_{\sigma(t_j)}$ is the union graph $\bigcup_{j=i_k}^{i_{k+1}-1} \mathscr{G}_{\sigma(t_j)} = \mathscr{G}([t_{i_k}, t_{i_{k+1}}))$, by Theorem A.1, under Assumption 2.4, the matrix $\sum_{j=i_k}^{i_{k+1}-1} \mathscr{L}_{\sigma(t_j)}$ has exactly one zero eigenvalue and all of its other eigenvalues have positive real parts. Also, under Assumption 2.5, the matrix $\sum_{j=i_k}^{i_{k+1}-1} \mathscr{L}_{\sigma(t_j)}$ is symmetric and positive semidefinite.

For clarity, we introduce the following definition:

Definition 2.1 A mapping $F : [t_0, \infty) \mapsto \mathbb{R}^{m \times n}$ is said to be piecewise continuous over $[t_0, \infty)$ if there exists a sequence $\{t_j : j = 0, 1, 2, \ldots\} \subset [t_0, \infty)$ with a dwell time $\tau > 0$ such that $F(t)$ is continuous on each time interval $[t_j, t_{j+1})$, $j = 0, 1, 2, \ldots$, and is said to be uniformly bounded over $[t_0, \infty)$ if there exists a finite positive constant K such that $\sup_{t_0 \le t < \infty} ||F(t)|| \le K$.

Lemma 2.2 *Consider the following linear time-varying system:*

$$\dot{\xi} = -\mu \mathscr{L}_{\sigma(t)} \xi \tag{2.6}$$

where $\xi \in \mathbb{R}^N$, μ is an arbitrary positive constant, and $\mathscr{L}_{\sigma(t)}$ is the Laplacian of the switching graph $\mathscr{G}_{\sigma(t)}$. Under Assumption 2.4, all components of any solution $\xi(t)$ of (2.6) converge to a common value exponentially as $t \to \infty$.

Proof By the definition of Laplacian, for any $\mu > 0$, $-\mu \mathscr{L}_{\sigma(t)}$ is a Metzler matrix with zero row sums for all $t \ge 0$, and $-\mu \mathscr{L}_{\sigma(t)}$ is uniformly bounded and piecewise constant over $[0, \infty)$ since $\sigma(t)$ belongs to a finite set.

Let $T = 2\nu > 0$. For any $t \ge 0$, there exists a positive k such that $[t_{i_k}, t_{i_{k+1}}) \subset [t, t+T)$. Then

$$\int_t^{t+T} -\mu l_{ij}(s) ds \ge \sum_{j=i_k}^{i_{k+1}-1} -\mu l_{ij}(t_j)(t_{j+1} - t_j), \ \forall\, i \neq j$$

where $l_{ij}(t)$ is the (i, j) component of $\mathscr{L}_{\sigma(t)}$ at time t. Since $t_{j+1} - t_j \ge \tau > 0$, the (i, j) component of $\sum_{j=i_k}^{i_{k+1}-1} -\mu \mathscr{L}_{\sigma(t_j)}(t_{j+1} - t_j)$ has the same sign as the (i, j) component of $\sum_{j=i_k}^{i_{k+1}-1} -\mu \mathscr{L}_{\sigma(t_j)}$. Let

$$l_{ij} = \begin{cases} 0, & \text{if } l_{ij}(t) = 0 \text{ for all } t \ge 0; \\ \min_{t \ge 0 \text{ and } l_{ij}(t) > 0} l_{ij}(t), & \text{otherwise} \end{cases}$$

and

$$\delta = \min_{l_{ij} \neq 0,\ i \neq j} \{-\mu l_{ij} \cdot \min\{1, \tau\}\}.$$

Then, for any $t \ge 0$, $k \ge 0$, the three matrices

$$\int_t^{t+T} -\mu \mathscr{L}_{\sigma(s)} ds, \ \sum_{j=i_k}^{i_{k+1}-1} -\mu \mathscr{L}_{\sigma(t_j)}(t_{j+1} - t_j), \ \sum_{j=i_k}^{i_{k+1}-1} -\mu \mathscr{L}_{\sigma(t_j)}$$

have the same δ-graph which is $\bigcup_{j=i_k}^{i_{k+1}-1} \mathscr{G}_{\sigma(t_j)}$. Thus, under Assumption 2.4, the δ-graph of the matrix $\int_t^{t+T} -\mu \mathscr{L}_{\sigma(s)} ds$ has a node that can reach all other nodes. By Theorem A.2, all components of any solution $\xi(t)$ of (2.6) converge exponentially to a common value as $t \to \infty$. □

Remark 2.5 Consider the following time-varying system:

$$\dot{x} = (F(t) \otimes I_l)x \tag{2.7}$$

where $x = \mathrm{col}(x_1, x_2, \dots, x_n)$, $x_i \in \mathbb{R}^l$, $i = 1, \dots, n$, and $F(t)$ is any square matrix of dimension n. Then it is easy to verify that (2.7) is equivalent to

$$\dot{\bar{x}}_k = F(t)\bar{x}_k,\ k = 1, \dots, l$$

where $\bar{x}_k = \mathrm{col}(x_{1k}, \dots, x_{nk})$ with x_{ik} being the kth component of x_i, $i = 1, \dots, n$, respectively. That is, let $\bar{x} = \mathrm{col}(\bar{x}_1, \dots, \bar{x}_l)$. Then, there exists a nonsingular matrix T such that $\bar{x} = Tx$ and

$$\dot{\bar{x}} = (I_l \otimes F(t))\bar{x}. \tag{2.8}$$

Let $\xi = \mathrm{col}(\xi_1, \xi_2, \dots, \xi_N)$, $\xi_i \in \mathbb{R}^l$, $i = 1, \dots, N$. Then, by Lemma 2.2 and Remark 2.5, we have the following corollary:

Corollary 2.1 *Consider the following linear time-varying system:*

$$\dot{\xi} = -\mu(\mathcal{L}_{\sigma(t)} \otimes I_l)\xi. \tag{2.9}$$

Under Assumption 2.4, for any $\mu > 0$, for $i = 1, 2, \dots, N$, all $\xi_i(t)$ converge to a common vector exponentially as $t \to \infty$.

To study leader–follower multi-agent systems over switching graphs, we consider the switching graph $\bar{\mathcal{G}}_{\sigma(t)} = (\bar{\mathcal{V}}, \bar{\mathcal{E}}_{\sigma(t)})$ with $\bar{\mathcal{V}} = \{0, 1, \dots, N\}$ and $\bar{\mathcal{E}}_{\sigma(t)} \subseteq \bar{\mathcal{V}} \times \bar{\mathcal{V}}$. Let $\bar{\mathcal{A}}_{\sigma(t)} = [a_{ij}(t)]_{i,j=0}^N$ be a weighted adjacency matrix of $\bar{\mathcal{G}}_{\sigma(t)}$ and $\bar{\mathcal{L}}_{\sigma(t)}$ be the corresponding Laplacian. Let $\Delta_{\sigma(t)} = \mathcal{D}(a_{10}(t), \dots, a_{N0}(t))$. Partition $\bar{\mathcal{L}}_{\sigma(t)}$ as follows:

$$\bar{\mathcal{L}}_{\sigma(t)} = \left[\begin{array}{c|ccc} \sum_{j=1}^N a_{0j}(t) & -a_{01}(t) & \cdots & -a_{0N}(t) \\ \hline -\Delta_{\sigma(t)}1_N & & H_{\sigma(t)} & \end{array}\right].$$

Similar to the static case, let $\mathcal{G}_{\sigma(t)} = (\mathcal{V}, \mathcal{E}_{\sigma(t)})$ with $\mathcal{V} = \{1, \dots, N\}$ and $\mathcal{E}_{\sigma(t)} = (\mathcal{V} \times \mathcal{V}) \cap \bar{\mathcal{E}}_{\sigma(t)}$ be a subgraph of $\bar{\mathcal{G}}_{\sigma(t)}$, and $\mathcal{L}_{\sigma(t)}$ be the Laplacian of $\mathcal{G}_{\sigma(t)}$ at time instant t. It can be verified that

$$H_{\sigma(t)} = \mathcal{L}_{\sigma(t)} + \Delta_{\sigma(t)}. \tag{2.10}$$

The assumption corresponding to Assumption 2.3 is as follows:

Assumption 2.6 There exists a subsequence $\{i_k : k = 0, 1, 2, \dots\}$ of $\{i : i = 0, 1, 2, \dots\}$ with $t_{i_{k+1}} - t_{i_k} < \nu$ for some $\nu > 0$ and all $k = 0, 1, 2, \dots$ such that the union graph $\bar{\mathcal{G}}([t_{i_k}, t_{i_{k+1}}))$ contains a spanning tree with the node 0 as the root.

Assumption 2.6 is an extension of Assumption 2.3 to switching graphs. An extension of Lemma 2.1 to switching graphs satisfying Assumption 2.6 is as follows:

Lemma 2.3 *The matrix* $-\sum_{j=i_k}^{i_{k+1}-1} H_{\sigma(t_j)}$ *is Hurwitz*[1] *if Assumption 2.6 holds and is symmetric if Assumption 2.5 holds.*

Proof By removing from $\bar{\mathscr{E}}_{\sigma(t)}$ all such edges as $(j, 0)$, $j = 1, \ldots, N$, we obtain a subgraph $\hat{\mathscr{G}}_{\sigma(t)}$ of $\bar{\mathscr{G}}_{\sigma(t)}$ whose Laplacian $\hat{\mathscr{L}}_{\sigma(t)}$ is the same as $\bar{\mathscr{L}}_{\sigma(t)}$ except that the entries of its first row are all zeros. Thus,

$$-\sum_{j=i_k}^{i_{k+1}-1} \hat{\mathscr{L}}_{\sigma(t_j)} = \left[\begin{array}{c|c} 0 & 0_{1\times N} \\ \hline \left(\sum_{j=i_k}^{i_{k+1}-1} \Delta_{\sigma(t_q)}\right) 1_N & -\sum_{j=i_k}^{i_{k+1}-1} H_{\sigma(t_j)} \end{array}\right].$$

It is clear that $\hat{\mathscr{G}}_{\sigma(t)}$ satisfies Assumption 2.6 if and only if $\bar{\mathscr{G}}_{\sigma(t)}$ satisfies Assumption 2.6. Moreover, the graph associated with $-\sum_{j=i_k}^{i_{k+1}-1} \hat{\mathscr{L}}_{\sigma(t_j)}$ is the union graph $\bigcup_{j=i_k}^{i_{k+1}-1} \hat{\mathscr{G}}_{\sigma(t_j)} = \hat{\mathscr{G}}([t_{i_k}, t_{i_{k+1}}))$. By Theorem A.1, under Assumption 2.6, the matrix $-\sum_{j=i_k}^{i_{k+1}-1} \hat{\mathscr{L}}_{\sigma(t_j)}$ has exactly one zero eigenvalue, and all the nonzero eigenvalues have negative real parts. Also, $\sum_{j=i_k}^{i_{k+1}-1} H_{\sigma(t_j)}$ is symmetric if Assumption 2.5 holds.□

Lemma 2.4 *Consider the following linear time-varying system*

$$\dot{\hat{\xi}} = -\mu \hat{\mathscr{L}}_{\sigma(t)} \hat{\xi} \tag{2.11}$$

where $\hat{\xi} = col(\xi_0, \xi_1, \ldots, \xi_N) \in \mathbb{R}^{N+1}$, *$\mu$ is an arbitrary positive constant, and* $\hat{\mathscr{L}}_{\sigma(t)}$ *is the Laplacian of the subgraph* $\hat{\mathscr{G}}_{\sigma(t)}$ *of* $\bar{\mathscr{G}}_{\sigma(t)}$ *by removing all such edges* $(j, 0)$ *for* $j = 1, \ldots, N$. *Under Assumption 2.6, for* $i = 1, \ldots, N$, $\lim_{t\to\infty}(\xi_i(t) - \xi_0(0)) = 0$ *exponentially.*

Proof From the proof of Lemma 2.2, for some $\delta > 0$, $\bigcup_{j=i_k}^{i_{k+1}-1} \hat{\mathscr{G}}_{\sigma(t_j)}$ and $\bigcup_{j=i_k}^{i_{k+1}-1} \bar{\mathscr{G}}_{\sigma(t_j)}$ are the δ-graphs associated with $\int_t^{t+T} -\mu\hat{\mathscr{L}}_{\sigma(s)}ds$ and $\int_t^{t+T} -\mu\bar{\mathscr{L}}_{\sigma(s)}ds$, respectively. Since $\hat{\mathscr{G}}_{\sigma(t)}$ is obtained from $\bar{\mathscr{G}}_{\sigma(t)}$ by removing all such edges $(j, 0)$ for $j = 1, \ldots, N$, under Assumption 2.6, the δ-graph associated with $\int_t^{t+T} -\mu\hat{\mathscr{L}}_{\sigma(s)}ds$, that is, the graph $\bigcup_{j=i_k}^{i_{k+1}-1} \hat{\mathscr{G}}_{\sigma(t_j)}$, has the property that every node $i = 1, \ldots, N$ is reachable from the node 0. By Theorem A.2, all components of any solution $\hat{\xi}(t)$ of (2.11) converge exponentially to a common value as $t \to \infty$.

Since $\hat{\mathscr{G}}_{\sigma(t)}$ contains no such edges as $(i, 0)$, $i = 1, \ldots, N$, at any time t, its Laplacian takes the following form:

$$\hat{\mathscr{L}}_{\sigma(t)} = \left[\begin{array}{c|c} 0 & 0 \ \cdots \ 0 \\ \hline -\Delta_{\sigma(t)} 1_N & H_{\sigma(t)} \end{array}\right].$$

[1] A complex square matrix is called Hurwitz if all of its eigenvalues have negative real parts.

Therefore, $\xi_0(t) \equiv \xi_0(0)$ for any $t \geq 0$. Thus, for $i = 1, \ldots, N$, $\lim_{t\to\infty}(\xi_i(t) - \xi_0(0)) = 0$ exponentially. □

Remark 2.6 For convenience, we call a graph satisfying Assumption 2.6 jointly connected with the node 0 as the common root.

Letting $\xi_0(0) = 0$ shows that all components of any solution $\hat{\xi}(t)$ of (2.11) converge exponentially to the origin as $t \to \infty$ regardless of the initial value of $\xi_i(t)$, $i = 1, \ldots, N$. Thus, we have the following corollary:

Corollary 2.2 *Under Assumption 2.6, for any $\mu > 0$, the origin of the following linear time-varying system*

$$\dot{\xi} = -\mu H_{\sigma(t)}\xi \tag{2.12}$$

is exponentially stable.

By Remark 2.5, (2.7) is exponentially stable if and only if (2.8) is exponentially stable. Thus, Corollary 2.2 can be extended to the following:

Corollary 2.3 *Under Assumption 2.6, for any $\mu > 0$, the origin of the following linear time-varying system*

$$\dot{\xi} = -\mu(H_{\sigma(t)} \otimes I_l)\xi \tag{2.13}$$

is exponentially stable.

2.2 Perturbed Systems

In this section, we establish the stability properties for some perturbed systems. They arise in the stability analysis of the adaptive distributed observers to be studied in Chap. 4.

Lemma 2.5 *Consider the following linear system:*

$$\dot{x} = Ax + F(t), \ t \geq t_0 \geq 0 \tag{2.14}$$

where $x \in \mathbb{R}^n$, $A \in \mathbb{R}^{n\times n}$ is Hurwitz, and $F(t) \in \mathbb{R}^{n\times n}$ is piecewise continuous and uniformly bounded over $[t_0, \infty)$. Then, for any $x(t_0)$, the solution of (2.14) tends to zero asymptotically (exponentially) if $F(t) \to 0$ as $t \to \infty$ (exponentially).

Proof For any $x(t_0)$, the solution of (2.14) is

$$x(t) = e^{A(t-t_0)}x(t_0) + \int_{t_0}^{t} e^{A(t-\tau)}F(\tau)d\tau, \ t \geq t_0.$$

Since A is Hurwitz, there exist $k_1 > 0$ and $\lambda_1 > 0$ such that

$$\|e^{A(t-t_0)}\| \leq k_1 e^{-\lambda_1(t-\tau)}, \ \forall\, t \geq \tau \geq t_0. \tag{2.15}$$

Thus, it suffices to show $\lim_{t\to\infty}\int_{t_0}^{t} e^{A(t-\tau)}F(\tau)d\tau = 0$ (exponentially).

First, consider the case where $\lim_{t\to\infty} F(t) = 0$ exponentially. Then, there exist $k_2 > 0$ and $\lambda_2 > 0$ such that $\|F(t)\| \leq k_2 e^{-\lambda_2(t-t_0)}\|F(t_0)\|$. Without loss of generality, assume $\lambda_2 < \lambda_1$. Then,

$$\begin{aligned}\left\|\int_{t_0}^{t} e^{A(t-\tau)}F(\tau)d\tau\right\| &\leq \int_{t_0}^{t} k_1 e^{-\lambda_1(t-\tau)}k_2 e^{-\lambda_2(\tau-t_0)}\|F(t_0)\|d\tau \\ &\leq k_1k_2\|F(t_0)\|e^{-\lambda_1 t}e^{\lambda_2 t_0}\int_{t_0}^{t} e^{(\lambda_1-\lambda_2)\tau}d\tau \\ &= \frac{k_1k_2}{\lambda_1-\lambda_2}\|F(t_0)\|e^{-\lambda_1 t}e^{\lambda_2 t_0}\left(e^{(\lambda_1-\lambda_2)t} - e^{(\lambda_1-\lambda_2)t_0}\right) \\ &= \frac{k_1k_2}{\lambda_1-\lambda_2}\|F(t_0)\|\left(e^{-\lambda_2(t-t_0)} - e^{-\lambda_1(t-t_0)}\right)\end{aligned}$$

which tends to zero exponentially as $t \to \infty$.

Next, consider the case where $\lim_{t\to\infty} F(t) = 0$ asymptotically. Let $P \in \mathbb{R}^{n\times n}$ be the positive definite matrix such that $PA + A^{\mathsf{T}}P = -I_n$ and let $V(x) = x^{\mathsf{T}}Px$. Then the derivative of $V(x)$ along the solution of (2.14) satisfies

$$\begin{aligned}\dot{V}(x(t)) &= -\|x\|^2 + 2x^{\mathsf{T}}F(t) \\ &\leq -(1-\frac{1}{2})\|x\|^2 + 2\|F(t)\|^2 \\ &= -\alpha(\|x\|) + \sigma(\|F(t)\|)\end{aligned}$$

where $\alpha(s) = \frac{1}{2}s^2$ and $\sigma(s) = 2\,s^2$ are both class $\mathcal{K}_\infty$ functions. By Theorem A.6 of Sect. A.7, (2.14) is input-to-state stable with $F(t)$ as the input. Since $\lim_{t\to\infty} F(t) = 0$, by Remark A.7 of Sect. A.7, $\lim_{t\to\infty} x(t) = 0$. □

To establish Lemma 2.7, we need to introduce the following simple result:

Lemma 2.6 *Suppose* $f_1(t) = (a_0t^n + a_1t^{n-1} + \cdots + a_{n-1}t + a_n)e^{-\delta_1 t}$ *with* $\delta_1 > 0$. *Then for any* $0 < \delta_2 < \delta_1$, *there exists* $b > 0$ *such that* $|f_1(t)| < f_2(t) = be^{-\delta_2 t}$ *for all* $t \geq 0$.

Proof Let

$$\begin{aligned}t^* &> \max\left\{1, \frac{n}{\delta_1-\delta_2}\right\} \\ m_1 &= \sup_{0\leq t\leq t^*}|a_0t^n + a_1t^{n-1} + \cdots + a_{n-1}t + a_n| \\ m_2 &= (n+1)\cdot\max\{|a_0|, |a_1|, \ldots, |a_n|\}.\end{aligned}$$

Since $t^* > 1$, it follows

$$|f_1(t)| \le \bar{f}_1(t) = \begin{cases} m_1 e^{-\delta_1 t}, & t \in [0, t^*) \\ m_2 t^n e^{-\delta_1 t}, & t \in [t^*, \infty). \end{cases}$$

For $t \ge t^*$, let

$$g(t) = \frac{f_2(t)}{\bar{f}_1(t)} = \frac{be^{-\delta_2 t}}{m_2 t^n e^{-\delta_1 t}} = \frac{be^{(\delta_1-\delta_2)t}}{m_2 t^n}$$

and we have

$$\begin{aligned} \dot{g}(t) &= \frac{(\delta_1 - \delta_2)be^{(\delta_1-\delta_2)t} m_2 t^n - be^{(\delta_1-\delta_2)t} m_2 n t^{n-1}}{m_2^2 t^{2n}} \\ &= \frac{be^{(\delta_1-\delta_2)t}[(\delta_1 - \delta_2)t - n]}{m_2 t^{n+1}} > 0. \end{aligned}$$

Therefore, by letting

$$b > \max\left\{ \frac{m_2 (t^*)^n}{e^{(\delta_1-\delta_2)t^*}}, m_1 \right\},$$

we have, for $t \ge t^*$, $f_2(t)/\bar{f}_1(t) = g(t) \ge g(t^*) > 1$, which together with the facts that $b > m_1$ and $0 < \delta_2 < \delta_1$ gives $f_2(t) > \bar{f}_1(t) \ge |f_1(t)|$ for all $t \ge 0$. □

Lemma 2.7 *Consider the following system:*

$$\dot{x} = (\varepsilon A + M(t))\,x + F(t) \tag{2.16}$$

where $x \in \mathbb{R}^n$, $A \in \mathbb{R}^{n\times n}$ *is Hurwitz,* $\varepsilon > 0$, $M(t) \in \mathbb{R}^{n\times n}$, *and* $F(t) \in \mathbb{R}^n$ *are piecewise continuous and uniformly bounded over* $[t_0, \infty)$. *Then*

(i) if $M(t), F(t) \to 0$ *as* $t \to \infty$ *(exponentially), then, for any* $x(t_0)$ *and any* $\varepsilon > 0$, $x(t) \to 0$ *as* $t \to \infty$ *(exponentially);*

(ii) if $M(t) = 0$, $F(t)$ *decays to zero exponentially at the rate of at least* α,[2] *and* $-\varepsilon\bar{\lambda}_A > \alpha$, *then, for any* $x(t_0)$, $x(t) \to 0$ *as* $t \to \infty$ *exponentially at the rate of at least* α.

Proof Part (i). This part can be viewed as a special case of Lemma 2.8 below. The proof is, thus, skipped.

Part (ii). The solution of (2.16) with $M(t) = 0$ is given by $x(t) = x_t(t) + x_s(t)$ where

$$x_t(t) = e^{\varepsilon A t} x(t_0), \; x_s(t) = \int_{t_0}^{t} e^{\varepsilon A(t-\tau)} F(\tau) d\tau, \; t \ge t_0.$$

Since $-\varepsilon\bar{\lambda}_A > \alpha$, $x_t(t)$ decays to zero exponentially at the rate of at least α. Also, there exist $\beta > 0$ such that $\|F(t)\| \le \beta e^{-\alpha(t-t_0)}\|F(t_0)\|$ and a polynomial $p(t)$ of

[2] Given a time function $f(t) : [t_0, \infty) \mapsto \mathbb{R}^{n\times m}$, if there exist $\alpha, \beta > 0$ such that $\|f(t)\| \le \beta e^{-\alpha(t-t_0)}\|f(t_0)\|$ for all $t \ge t_0$, then we say $f(t)$ decays to zero exponentially at the rate of at least α.

degree smaller than n such that $\|e^{\varepsilon At}\| \le p(t)e^{\varepsilon\bar{\lambda}_A t}$. By Lemma 2.6, for some $\beta_s > 0$ and $\alpha < \delta < -\varepsilon\bar{\lambda}_A$, $\|e^{\varepsilon At}\| \le \beta_s e^{-\delta t}$. Therefore,

$$\|x_s(t)\| \le \int_{t_0}^{t} \beta_s e^{-\delta(t-\tau)} \beta e^{-\alpha(\tau-t_0)} \|F(t_0)\| d\tau.$$

Simple calculation gives

$$\|x_s(t)\| \le \frac{\beta_s \beta \|F(t_0)\|}{\delta - \alpha} \left(e^{-\alpha(t-t_0)} - e^{-\delta(t-t_0)}\right),$$

and hence, the proof is completed. □

Lemma 2.8 *Consider the following time-varying system:*

$$\dot{x} = (A(t) + M(t))\, x + F(t) \tag{2.17}$$

where $x \in \mathbb{R}^n$, $A(t)$, $M(t)$, $F(t) \in \mathbb{R}^{n\times n}$ *are piecewise continuous and uniformly bounded over* $[0, \infty)$, *and the following system:*

$$\dot{x} = A(t)x \tag{2.18}$$

is exponentially stable. If $M(t) \to 0$ *as* $t \to \infty$, *and* $F(t) \to 0$ *as* $t \to \infty$ *(exponentially), then, for any* $x(0)$, $x(t) \to 0$ *as* $t \to \infty$ *(exponentially).*

Proof Let $\Phi(t, \tau)$ be the state transition matrix of system (2.18), that is, it is the unique solution of the following matrix equation:

$$\frac{\partial}{\partial t}\Phi(t, \tau) = A(t)\Phi(t, \tau), \;\; \Phi(\tau, \tau) = I_n.$$

Then, differentiating both sides of $\Phi(t, \tau)\Phi(\tau, t) = I_n$ gives

$$\frac{\partial}{\partial t}\Phi(\tau, t) = -\Phi(\tau, t)A(t).$$

Let

$$P(t) = \int_t^{\infty} \Phi(\tau, t)^{\mathrm{T}} Q \Phi(\tau, t) d\tau$$

where $Q > 0$ is some constant matrix. Clearly, $P(t)$ is continuous for all $t \ge 0$. Since system (2.18) is exponentially stable, we have

$$\|\Phi(\tau, t)\| \le \alpha_1 e^{-\lambda_1(\tau - t)}, \;\; \tau \ge t \ge 0$$

for some $\alpha_1 > 0$ and $\lambda_1 > 0$. Then, it can be verified that

$$c_1\|x\|^2 \le x^{\mathsf{T}} P(t)x \le c_2\|x\|^2$$

for some positive constants c_1, c_2. Hence, $P(t)$ is positive definite and uniformly bounded over $[0, \infty)$, and there exists some positive constant c_3 such that $\|P(t)\| \le c_3$ for all $t \ge 0$. Then, for $t \in [t_j, t_{j+1})$, $j = 0, 1, 2, \ldots$,

$$\begin{aligned}
\dot{P}(t) &= \int_t^\infty \Phi(\tau,t)^{\mathsf{T}} Q \left(\frac{\partial}{\partial t}\Phi(\tau,t)\right) d\tau + \int_t^\infty \left(\frac{\partial}{\partial t}\Phi(\tau,t)\right)^{\mathsf{T}} Q\Phi(\tau,t)d\tau - Q \\
&= -\int_t^\infty \Phi(\tau,t)^{\mathsf{T}} Q\Phi(\tau,t)d\tau A(t) - A(t)^{\mathsf{T}} \int_t^\infty \Phi(\tau,t)^{\mathsf{T}} Q\Phi(\tau,t)d\tau - Q \\
&= -P(t)A(t) - A(t)^{\mathsf{T}} P(t) - Q.
\end{aligned}$$

Next, let $U(t) = x^{\mathsf{T}} P(t)x$. Then, along the trajectory of system (2.17), for $t \in [t_j, t_{j+1})$, $j = 0, 1, 2, \ldots$,

$$\begin{aligned}
\dot{U}(t)|_{(2.17)} &= x^{\mathsf{T}} \left(\dot{P}(t) + A(t)^{\mathsf{T}} P(t) + P(t)A(t)\right) x + 2x^{\mathsf{T}} P(t)M(t)x + 2x^{\mathsf{T}} P(t)F(t) \\
&= -x^{\mathsf{T}} Qx + 2x^{\mathsf{T}} P(t)M(t)x + 2x^{\mathsf{T}} P(t)F(t) \\
&\le -\underline{\lambda}_Q\|x\|^2 + 2c_3\|M(t)\|\|x\|^2 + \frac{\|P(t)\|^2}{\varepsilon}\|x\|^2 + \varepsilon\|F(t)\|^2 \\
&\le -\left(\underline{\lambda}_Q - 2c_3\|M(t)\| - \frac{c_3^2}{\varepsilon}\right)\|x\|^2 + \varepsilon\|F(t)\|^2.
\end{aligned}$$

Let $\varepsilon > 0$ be such that $c_4 \triangleq \underline{\lambda}_Q - c_3^2/\varepsilon > 0$. Then

$$\dot{U}(t)|_{(2.17)} \le -\left(c_4 - 2c_3\|M(t)\|\right)\|x\|^2 + \varepsilon\|F(t)\|^2.$$

Since $M(t)$ converges to zero as t tends to infinity, there exist some positive integer l and a constant c_5, such that

$$(c_4 - 2c_3\|M(t)\|) \ge c_5 > 0, \ t \ge t_l.$$

Hence, we have

$$\dot{U}(t)|_{(2.17)} \le -c_5\|x\|^2 + \varepsilon\|F(t)\|^2 \le -\lambda_2 U(t) + \varepsilon\|F(t)\|^2, \ t \ge t_l$$

where $\lambda_2 = c_5/c_2$.

By Lemma 2.5, and the comparison lemma (Theorem A.3), $\lim_{t\to\infty} U(t) = 0$ (exponentially) if $\lim_{t\to\infty} F(t) = 0$ (exponentially). Since $U(t) \ge c_1\|x(t)\|^2$, $\lim_{t\to\infty} x(t) = 0$ (exponentially) if $\lim_{t\to\infty} F(t) = 0$ (exponentially). □

For convenience, we state a special case of Lemma 2.8 as follows.

Corollary 2.4 *Consider the following linear time-varying system:*

$$\dot{x} = A(t)x + F(t), \ t \geq t_0 \geq 0 \tag{2.19}$$

where $x \in \mathbb{R}^n$, $A(t)$, $F(t) \in \mathbb{R}^{n \times n}$ are piecewise continuous and uniformly bounded over $[t_0, \infty)$. Suppose $\dot{x} = A(t)x$ is exponentially stable. Then, for any $x(t_0)$, the solution of (2.19) tends to zero asymptotically (exponentially) if $F(t) \to 0$ as $t \to \infty$ (exponentially).

2.3 Generalized Barbalat's Lemma

Barbalat's Lemma is a well-known technique for the convergent analysis of continuous-time signals and can be stated as follows:

Lemma 2.9 *Suppose $V : [0, \infty) \mapsto \mathbb{R}$ is differentiable and $\dot{V}(t)$ is uniformly continuous on $[0, \infty)$. Then, $\dot{V}(t) \to 0$ as $t \to \infty$ if $V(t)$ has a finite limit as $t \to \infty$.*

Nevertheless, Barbalat's Lemma only applies to signals which are continuous over $[0, \infty)$. To deal with the convergence analysis of switching signals, we need to generalize Barbalat's Lemma to the following form:

Lemma 2.10 *Let $\{t_i : i = 0, 1, 2, \dots\}$ be a sequence satisfying $t_0 = 0$, $t_{i+1} - t_i \geq \tau > 0$ for all $i = 0, 1, 2 \dots$. Suppose $V : [0, \infty) \mapsto \mathbb{R}$ satisfies the following:*

1. *$\lim_{t \to \infty} V(t)$ exists and is finite;*
2. *$V(t)$ is differentiable on each interval $[t_i, t_{i+1})$;*
3. *For any $\varepsilon > 0$, there exists some $\eta > 0$, such that, whenever $|t' - t''| < \eta$ and $t_i \leq t', t'' < t_{i+1}$ for some nonnegative integer i, $|\dot{V}(t') - \dot{V}(t'')| < \varepsilon$.*

Then $\dot{V}(t) \to 0$ as $t \to \infty$.

Proof Suppose that $\dot{V}(t)$ does not approach zero as $t \to \infty$. Then there exists an infinite sequence $\{\tau_{s_k} : k = 1, 2, \dots\}$, $\tau_{s_k} \in [t_{s_k}, t_{s_k+1})$ such that $\lim_{k \to \infty} \tau_{s_k} = \infty$ and $|\dot{V}(\tau_{s_k})| > \varepsilon_0$ for some positive number ε_0.

Since $\lim_{t \to \infty} V(t)$ exists, by Cauchy's convergence criterion, for any $\varepsilon > 0$, there exists a positive number T, such that for any $T_2 > T_1 > T$, $|V(T_2) - V(T_1)| < \varepsilon$, i.e.,

$$\left| \int_{T_1}^{T_2} \dot{V}(t) dt \right| < \varepsilon. \tag{2.20}$$

Let δ be such that $0 < \delta \leq \tau \leq t_{s_k+1} - t_{s_k}$. Then we have either

$$(\tau_{s_k} - \frac{\delta}{2}, \tau_{s_k}] \subset [t_{s_k}, t_{s_k+1}) \tag{2.21}$$

or

$$[\tau_{s_k}, \tau_{s_k} + \frac{\delta}{2}) \subset [t_{s_k}, t_{s_k+1}). \tag{2.22}$$

When (2.21) holds, by condition 3, there exists $0 < \eta \le \delta$ such that, for any $t \in (\tau_{s_k} - \frac{\eta}{2}, \tau_{s_k}]$,

$$|\dot{V}(\tau_{s_k})| - |\dot{V}(t)| \le |\dot{V}(t) - \dot{V}(\tau_{s_k})| \le \frac{\varepsilon_0}{2}.$$

This implies, for any $t \in (\tau_{s_k} - \frac{\eta}{2}, \tau_{s_k}]$,

$$|\dot{V}(t)| \ge |\dot{V}(\tau_{s_k})| - \frac{\varepsilon_0}{2} > \frac{\varepsilon_0}{2}. \tag{2.23}$$

By the continuity of $\dot{V}(t)$ over $(\tau_{s_k} - \frac{\delta}{2}, \tau_{s_k}]$ and (2.23), $\dot{V}(t)$ does not change sign for any $t \in (\tau_{s_k} - \frac{\eta}{2}, \tau_{s_k}]$. Then

$$\left| \int_{\tau_{s_k} - \frac{\eta}{2}}^{\tau_{s_k}} \dot{V}(s) ds \right| = \int_{\tau_{s_k} - \frac{\eta}{2}}^{\tau_{s_k}} |\dot{V}(s)| ds > \int_{\tau_{s_k} - \frac{\eta}{2}}^{\tau_{s_k}} \frac{\varepsilon_0}{2} ds > \frac{\eta \varepsilon_0}{4} > 0. \tag{2.24}$$

Similarly, when (2.22) holds, there exists $0 < \eta \le \delta$ such that, for any $t \in (\tau_{s_k}, \tau_{s_k} + \frac{\eta}{2}]$,

$$\left| \int_{\tau_{s_k}}^{\tau_{s_k} + \frac{\eta}{2}} \dot{V}(s) ds \right| > \frac{\eta \varepsilon_0}{4} > 0. \tag{2.25}$$

Both (2.24) and (2.25) contradict (2.20) for sufficiently large τ_{s_k}. Thus, $\lim_{t \to \infty} \dot{V}(t) = 0$. □

If $V(t)$ is twice differentiable on each interval $[t_i, t_{i+1})$, and $\ddot{V}(t)$ is uniformly bounded over $[0, \infty)$ in the sense of Definition 2.1, then the second and third conditions of Lemma 2.10 are satisfied. Thus, we have the following corollary:

Corollary 2.5 *Let* $\{t_i : i = 0, 1, 2, \dots\}$ *be a sequence satisfying* $t_0 = 0,\ t_{i+1} - t_i \ge \tau > 0$ *for all* $i = 0, 1, 2 \dots$ *Suppose* $V : [0, \infty) \mapsto \mathbb{R}$ *satisfies the following:*

1. $\lim_{t \to \infty} V(t)$ *exists and is finite;*
2. $V(t)$ *is twice differentiable on each interval* $[t_i, t_{i+1})$*;*
3. $\ddot{V}(t)$ *is uniformly bounded over* $[0, \infty)$.

Then $\dot{V}(t) \to 0$ *as* $t \to \infty$.

Applying Corollary 2.5 to a non-increasing and lower bounded continuous function $V(t)$ gives the following result:

Corollary 2.6 *Let* $\{t_i : i = 0, 1, 2, \dots\}$ *be a sequence satisfying* $t_0 = 0,\ t_{i+1} - t_i \ge \tau > 0$ *for all* $i = 0, 1, 2 \dots$ *Suppose* $V : [0, \infty) \mapsto \mathbb{R}$ *is continuous and satisfies the following:*

1. $V(t)$ *is lower bounded;*
2. $\dot{V}(t)$ *is non-positive and differentiable on each interval* $[t_i, t_{i+1})$*;*
3. $\ddot{V}(t)$ *is uniformly bounded over* $[0, \infty)$.

Then $\dot{V}(t) \to 0$ *as* $t \to \infty$.

Proof Since $V(t)$ is lower bounded and continuous and $\dot{V}(t)$ is non-positive, $\lim_{t\to\infty} V(t)$ exists and is finite. Applying Corollary 2.5 completes the proof. □

Remark 2.7 If $\dot{V}(t)$ is uniformly continuous over $[0, \infty)$, then the second and third conditions of Lemma 2.10 are satisfied on one single interval $[t_0, t_1)$ with $t_0 = 0$ and $t_1 = \infty$. Thus, for this special case, Lemma 2.10 reduces to Barbalat's Lemma. For this reason, we call Lemma 2.10 generalized Barbalat's Lemma.

2.4 Some Other Results

This section establishes four lemmas that will be mainly used in Chap. 3. Let us first recall a well-known result in the linear system.

Lemma 2.11 *Given a stabilizable pair* (A, B) *where* $A \in \mathbb{R}^{n\times n}$ *and* $B \in \mathbb{R}^{n\times m}$*, the following algebraic Riccati equation:*

$$A^\top P + PA - PBB^\top P + I_n = 0$$

admits a unique solution $P > 0$.[3]

Lemma 2.12 *Given a stabilizable pair* (A, B) *where* $A \in \mathbb{R}^{n\times n}$ *and* $B \in \mathbb{R}^{n\times m}$*, let* $P > 0$ *be the solution of* $A^\top P + PA - PBB^\top P + I_n = 0$*. Then, for all* $\sigma \geq 1/2$ *and any* $\omega \in \mathbb{R}$*, the matrix* $A - (\sigma + \imath\omega)BB^\top P$ *is Hurwitz.*

Proof By Lemma 2.11, there exists a unique $P > 0$ satisfying the algebraic Riccati equation $A^\top P + PA - PBB^\top P + I_n = 0$. Let $\varepsilon = \sigma - 1/2$. Then we have

$$\begin{aligned} &(A - (\sigma + \imath\omega)BB^\top P)^{\mathsf{H}} P + P(A - (\sigma + \imath\omega)BB^\top P) \\ =&(A - (\sigma - \imath\omega)BB^\top P)^\top P + P(A - (\sigma + \imath\omega)BB^\top P) \\ =&(A - (1/2 + \varepsilon)BB^\top P)^\top P + P(A - (1/2 + \varepsilon)BB^\top P) \\ =&(A - (1/2)BB^\top P)^\top P + P(A - (1/2)BB^\top P) - 2\varepsilon PBB^\top P \\ =&- I_n - 2\varepsilon PBB^\top P. \end{aligned} \tag{2.26}$$

Noting $\varepsilon \geq 0$ when $\sigma \geq 1/2$ completes the proof. □

[3] A proof of this lemma can be found in [16].

Lemma 2.13 *Given a strictly increasing sequence* $\{t_j : j = 0, 1, 2, \ldots\}$ *with* $t_0 = 0$ *and* $\lim_{j\to\infty} t_j = \infty$, *consider the following system:*

$$\dot{\psi} = (M - Q(t))\psi \tag{2.27}$$

where $\psi \in \mathbb{R}^n$, *M is a constant matrix, and* $Q(t)$ *is uniformly bounded over* $[0, \infty)$ *and is constant over each interval* $[t_j, t_{j+1})$, $j = 0, 1, 2, \ldots$. *Suppose the solution* $\psi(t)$ *is uniformly bounded over* $[0, \infty)$ *satisfying* $\lim_{t\to\infty} Q(t)\psi(t) = 0$ *and* $\lim_{t\to\infty} G\psi(t) = 0$ *for some constant matrix G. Then*

$$\lim_{t\to\infty} GM^k\psi(t) = 0, \; k = 1, 2, \ldots.$$

Proof Since $Q(t)$ is constant over each interval $[t_j, t_{j+1})$, $j = 0, 1, 2, \ldots$, by (2.27), over each interval $[t_j, t_{j+1})$, for $k = 1, 2, \ldots$,

$$\psi^{(k)} = (M - Q(t))^k\psi.$$

Since $\psi(t)$ is uniformly bounded over $[0, \infty)$, so is $\psi^{(k)}(t)$ for $k = 1, 2, \ldots$. Thus, each component of $G\psi(t)$ satisfies the three conditions in Corollary 2.5. Thus, $\lim_{t\to\infty} G\dot{\psi}(t) = 0$. Therefore,

$$\lim_{t\to\infty} GM\psi(t) = \lim_{t\to\infty} G(\dot{\psi}(t) + Q(t)\psi(t)) = 0.$$

Now suppose $\lim_{t\to\infty} GM^{k_0}\psi(t) = 0$ for some positive integer k_0. Then, again by Corollary 2.5, we have $\lim_{t\to\infty} GM^{k_0}\dot{\psi}(t) = 0$. Therefore,

$$\lim_{t\to\infty} GM^{k_0+1}\psi(t) = \lim_{t\to\infty} GM^{k_0}(\dot{\psi}(t) + Q(t)\psi(t)) = 0.$$

The proof is completed by induction. □

Lemma 2.14 *Suppose the function* $f(t)$*:* $[0, \infty) \mapsto \mathbb{R}$ *is piecewise continuous and satisfies* $\lim_{t\to\infty} f(t) = 0$. *Then, for any positive number* T_0,

$$\lim_{t\to\infty} \int_t^{t+T_0} f(s)ds = 0.$$

Proof Since $\lim_{t\to\infty} f(t) = 0$, for any $\varepsilon > 0$, there exists $T > 0$ such that whenever $t > T$, $|f(t)| < \varepsilon/T_0$. Thus,

$$\left|\int_t^{t+T_0} f(s)ds\right| \leq \int_t^{t+T_0} |f(s)|ds < \int_t^{t+T_0} \frac{\varepsilon}{T_0}ds = \varepsilon.$$

That is, $\lim_{t\to\infty} \int_t^{t+T_0} f(s)ds = 0$. □

2.5 Notes and References

A good reference for graph theory is [1]. The early results on the leaderless and leader-following consensus problems of linear multi-agent systems were given in [2, 3], respectively. The properties of Remark 2.1 on the leaderless network can be seen in [2, 4, 5]. The concepts of the connected/jointly connected graph vary in different literatures. They were introduced in [4] for undirected graphs and [5] for directed graphs. The definition of uniformly connected graph was given in [6] which provides an equivalent form to jointly connected conditions. In this book, we allow the connected graph to be directed. The δ-graph was introduced in [7]. Lemmas 2.2 and 2.4 and Corollaries 2.2 and 2.3 are based on the results in [8].

Lemmas 2.7 and 2.8 are based on the results in [9, 10], respectively.

Barbalat's Lemma (Lemma 2.9) is well known and can be found from the standard textbooks such as [11, 12]. Generalized Barbalat's Lemma (Lemma 2.10) is a generalization of Corollaries 2.5 and 2.6, which are based on the results in [13].

Lemma 2.12 is adapted from [14], which solves the leaderless consensus problem by simultaneous eigenvalue assignment as will be seen in the next chapter. Lemmas 2.13 and 2.14 are based on the results in [13] for the convergence analysis of piecewise continuous-time signals. The discrete-time version of Lemma 2.13 can be found in [15].

References

1. Godsil C, Royle G (2001) Algebraic graph theory. Springer, New York
2. Olfati-Saber R, Murray RM (2004) Consensus problems in networks of agents with switching topology and time-delays. IEEE Trans Autom Control 49(9):1520–1533
3. Hu J, Hong Y (2007) Leader-following coordination of multi-agent systems with coupling time delays. Phys A Stat Mech Appl 374(2):853–863
4. Jadbabaie A, Lin J, Morse AS (2003) Coordination of groups of mobile autonomous agents using nearest neighbor rules. IEEE Trans Autom Control 48(6):988–1001
5. Ren W, Beard RW (2005) Consensus seeking in multi-agent systems under dynamically changing interaction topologies. IEEE Trans Autom Control 50(5):655–661
6. Lin Z (2005) Coupled dynamic systems: from structure towards stability and stabilizability, PhD dissertation, University of Toronto, Toronto, Canada
7. Moreau L (2004) Stability of continuous-time distributed consensus algorithms. In: Proceedings of the 41st IEEE conference on decision and control, pp 3998–4003
8. Su Y, Huang J (2012) Cooperative output regulation with application to multi-agent consensus under switching network. IEEE Trans Syst Man Cybern Part B Cybern 42(3):864–875
9. Cai H, Lewis FL, Hu G, Huang J (2017) The adaptive distributed observer approach to the cooperative output regulation of linear multi-agent systems. Automatica 75:299–305
10. Liu T, Huang J (2018) Leader-following attitude consensus of multiple rigid body systems subject to jointly connected switching networks. Automatica 92:63–71
11. Khalil H (2002) Nonlinear systems. Prentice Hall, New Jersey
12. Slotine JE, Li W (1991) Applied nonlinear control. Prentice Hall, Englewood Cliffs
13. Su Y, Huang J (2012) Stability of a class of linear switching systems with applications to two consensus problems. IEEE Trans Autom Control 57(6):1420–1430

14. Tuna SE (2008) LQR-based coupling gain for synchronization of linear systems. arxiv.org/abs/0801.3390
15. Su Y, Huang J (2012) Two consensus problems for discrete-time multi-agent systems with switching network topology. Automatica 48(9):1988–1997
16. Kucera V (1972) A contribution to matrix quadratic equations. IEEE Trans Autom Control 17(3):344–347

Chapter 3
Two Consensus Problems

As we pointed out in Chap. 1, depending on the nature of various global objectives of a multi-agent system, there are a variety of cooperative control problems of multi-agent systems such as consensus/synchronization, formation, cooperative output regulation, connectivity preservation, distributed estimation, distributed optimization, and so on. Fundamental to all cooperative control problems of multi-agent systems is the consensus problem. There are two types of consensus problems, namely the leaderless consensus problem and the leader-following consensus problem. The leaderless consensus problem is to design a distributed control law such that the state vector of each subsystem asymptotically synchronizes to the same trajectory regardless of the time profile of the trajectory. In contrast, the leader-following consensus problem is to design a distributed control law such that the state vector of each subsystem is able to asymptotically track a class of prescribed trajectories. Typically, the class of prescribed trajectories are produced by another dynamic system called leader system. In this chapter, we will study both the leaderless consensus problem and the leader-following consensus problem for the class of linear homogeneous multi-agent systems.

In the rest of this chapter, we start from a motivating example in Sect. 3.1, and then formulate the two consensus problems in Sect. 3.2. Sections 3.3 and 3.4 deal with the two consensus problems over the static graphs and switching graphs, respectively. In Sect. 3.5, we present a result dual to Theorem 3.4, which will play a key role in the construction of the distributed observer to be studied in Chap. 4.

H. Cai et al., *Cooperative Control of Multi-agent Systems*, Advances in Industrial Control,
https://doi.org/10.1007/978-3-030-98377-2_3

3.1 A Motivating Example

The two consensus problems arise from many aspects of life such as coordinated manufacturing, attitude synchronization of flight vehicles, swarm motions, flocking of birds, schooling of fish, and so on. In this section, we illustrate the leaderless consensus problem by the cyclic pursuit problem.

The cyclic pursuit problem is also known as the "bugs" problem, which involves N bugs with the bug i moving toward the bug $i+1$ modulo N. Imagining the N bugs as N autonomous agents in the plane, their positions at time instant $t \geq 0$ are denoted by $z_i(t) = \text{col}(x_i(t), y_i(t)) \in \mathbb{R}^2$, $i = 1, \ldots, N$. The kinematics of these agents are described by N single integrators as follows:

$$\dot{z}_i = u_i, \ i = 1, \ldots, N \tag{3.1}$$

where $u_i \in \mathbb{R}^2$ is the input of the ith agent. The communication among these agents is described by a so-called pursuit graph $\mathscr{G} = (\mathscr{V}, \mathscr{E})$, in which $\mathscr{V} = \{1, \ldots, N\}$ with the node i representing the agent i, and $\mathscr{E} = \{(1, N), (2, 1), \ldots, (N, N-1)\}$. Figure 3.1 shows the pursuit graph with $N = 6$.

Consider the following control law:

$$u_i = -\sum_{j=1}^{N} a_{ij}(z_i - z_j), \ i = 1, \ldots, N \tag{3.2}$$

where $[a_{ij}]_{i,j=1}^{N}$ are the entries of the weighted adjacency matrix of $\mathscr{G}$. Clearly, the pursuit graph $\mathscr{G}$ is connected, i.e., it satisfies Assumption 2.1. It can be verified that the closed-loop system composed of the kinematic dynamics (3.1) and the control law (3.2) can be put into the following compact form:

$$\dot{z} = -(\mathscr{L} \otimes I_2)z \tag{3.3}$$

where $z = \text{col}(z_1, \ldots, z_N)$, which is in the form of (2.1) with $n = 2$.

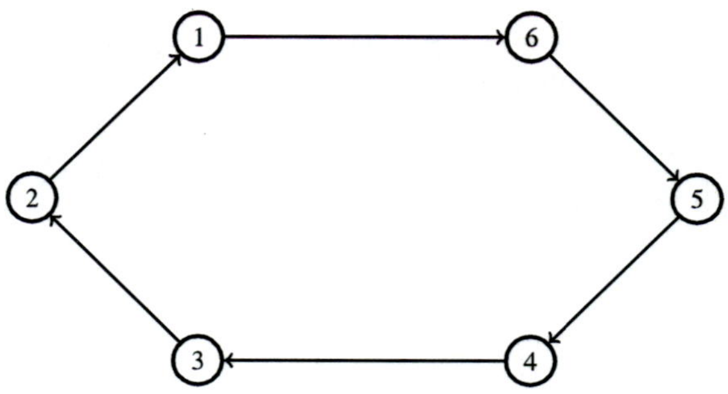

Fig. 3.1 A pursuit graph $\mathscr{G}$ with six bugs

Let all nonzero a_{ij} be equal to 1. Then, the Laplacian matrix of $\mathscr{G}$ is as follows:

$$\mathscr{L} = \begin{bmatrix} 1 & -1 & 0 & \cdots & 0 \\ 0 & 1 & -1 & \cdots & 0 \\ \vdots & \vdots & \vdots & & \vdots \\ -1 & 0 & 0 & \cdots & 1 \end{bmatrix}.$$

As a result, the control law (3.2) reduces to the following simple form:

$$u_i = -(z_i - z_{i+1}), \; i = 1, \ldots, N \tag{3.4}$$

where $z_{N+1} \triangleq z_1$. That is, the velocity of the ith agent is simply proportional to the vector from the ith agent to the $(i+1)$th agent.

It can be seen that $\mathscr{L}$ has a left eigenvector $r = (1/N)1_N$ corresponding to the zero eigenvalue such that $r^{\mathsf{T}} 1_N = 1$. Thus, from Example 2.1 with $r = (1/N)1_N$,

$$\lim_{t\to\infty} \left\{ z(t) - 1_N \otimes \left(\left(\left(\frac{1_N}{N} \right)^{\mathsf{T}} \otimes I_2 \right) z(0) \right) \right\} = 0, \text{ exponentially}$$

i.e.,

$$\lim_{t\to\infty} \left\{ z_i(t) - \frac{1}{N} \sum_{i=1}^{N} z_i(0) \right\} = 0, \text{ exponentially}, \; i = 1, \ldots, N.$$

Define the center of these N agents by the quantity $(1/N) \sum_{i=1}^{N} z_i(t)$. Since

$$\frac{1}{N} \sum_{i=1}^{N} \dot{z}_i = \frac{1}{N} 1_N^{\mathsf{T}} \dot{z} = \left(- \left(\frac{1_N}{N} \right)^{\mathsf{T}} \mathscr{L} \otimes I_2 \right) z = 0$$

for all $t \geq 0$,

$$\frac{1}{N} \sum_{i=1}^{N} z_i(t) = \frac{1}{N} \sum_{i=1}^{N} z_i(0).$$

That is, the center of these N agents remains stationary. Therefore, for any initial condition, every agent exponentially converges to the center of the N agents.

To conduct a simulation, let $N = 6$. Then the pursuit graph is given by Fig. 3.1. Let the initial condition be $z_1(0) = \text{col}(-50, 86)$, $z_2(0) = \text{col}(-100, 0)$, $z_3(0) = \text{col}(-50, -86)$, $z_4(0) = \text{col}(50, -86)$, $z_5(0) = \text{col}(100, 0)$, $z_6(0) = \text{col}(50, 86)$. The motions of these bugs under the control input (3.4) are shown in Fig. 3.2. It is observed that all the six bugs converge to their center at the origin.

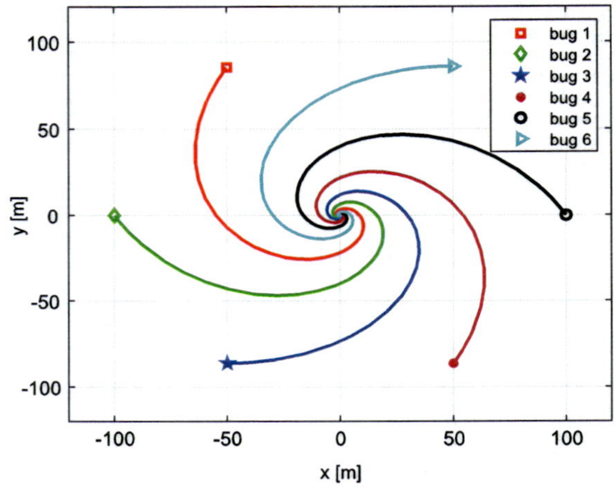

Fig. 3.2 Global motions of six bugs

3.2 Problem Formulation

Consider the following multi-agent system:

$$\dot{x}_i = Ax_i + Bu_i \tag{3.5}$$

where $x_i \in \mathbb{R}^n$ and $u_i \in \mathbb{R}^m$ are the state and control input of the ith agent, and $A \in \mathbb{R}^{n\times n}$ and $B \in \mathbb{R}^{n\times m}$ are constant matrices.

Associated with system (3.5), we can define a switching communication graph $\mathscr{G}_{\sigma(t)} = (\mathscr{V}, \mathscr{E}_{\sigma(t)})$ where $\mathscr{V} = \{1, \ldots, N\}$ and, for $i, j = 1, 2, \ldots, N, i \neq j, (j, i) \in \mathscr{E}_{\sigma(t)}$ if and only if the control u_i can make use of $(x_j - x_i)$ for feedback control at time t. Denote the weighted adjacency matrix of the switching communication graph $\mathscr{G}_{\sigma(t)}$ by $\mathscr{A}_{\sigma(t)} = \left[a_{ij}(t)\right]_{i,j=1}^N$. Then we can define the distributed state feedback control law as follows:

$$u_i = K\sum_{j=1}^{N} a_{ij}(t)(x_j - x_i),\ i = 1, \ldots, N \tag{3.6}$$

where $K \in \mathbb{R}^{m\times n}$ is a gain matrix to be designed.

Problem 3.1 (*Leaderless Consensus Problem*) Given the system (3.5) and a switching communication graph $\mathscr{G}_{\sigma(t)}$, find a feedback gain matrix K for the distributed state feedback control law (3.6) such that, for any initial condition of (3.5), and any $i, j = 1, \ldots, N$, $(x_i(t) - x_j(t)) \to 0$ as $t \to \infty$.

In the problem described above, the steady-state behavior of the solution of each subsystem is immaterial. There is another consensus problem called the leader-following consensus problem where the solution of each subsystem is required to be able to asymptotically track a class of prescribed time signals $x_0(t)$. We assume that the signal $x_0(t)$ is generated by a linear system of the following form:

$$\dot{x}_0 = Ax_0 \tag{3.7}$$

where $x_0 \in \mathbb{R}^n$.

In what follows, we call system (3.5) and system (3.7) the follower system and the leader system, respectively. Associated with system (3.5) and system (3.7), we can define another switching communication graph $\bar{\mathscr{G}}_{\sigma(t)} = \{\bar{\mathscr{V}}, \bar{\mathscr{E}}_{\sigma(t)}\}$ where $\bar{\mathscr{V}} = \{0, 1, \ldots, N\}$ and, for $i = 1, \ldots, N$, $j = 0, 1, \ldots, N$, $i \neq j$, $(j, i) \in \bar{\mathscr{E}}_{\sigma(t)}$ if and only if the control u_i can make use of $(x_j - x_i)$ for feedback control at time t. Clearly, $\mathscr{G}_{\sigma(t)}$ is a subgraph of $\bar{\mathscr{G}}_{\sigma(t)}$ and can be obtained from $\bar{\mathscr{G}}_{\sigma(t)}$ by removing the node 0 from $\bar{\mathscr{V}}$ and all the edges incident on the node 0 at time t from $\bar{\mathscr{E}}_{\sigma(t)}$.

Denote the weighted adjacency matrix of the switching communication graph $\bar{\mathscr{G}}_{\sigma(t)}$ by $\bar{\mathscr{A}}_{\sigma(t)} = \left[a_{ij}(t)\right]_{i,j=0}^{N}$. Then, we consider the distributed state feedback control law for the leader-following consensus problem as follows:

$$u_i = K \sum_{j=0}^{N} a_{ij}(t)(x_j - x_i), \ i = 1, \ldots, N \tag{3.8}$$

where $K \in \mathbb{R}^{m \times n}$ is a gain matrix to be designed.

Problem 3.2 (*Leader-Following Consensus Problem*) Given the leader system (3.7), the follower system (3.5), and a switching communication graph $\bar{\mathscr{G}}_{\sigma(t)}$, find a feedback gain matrix K for the distributed state feedback control law (3.8) such that, for any initial condition of (3.7) and (3.5), and any $i = 1, \ldots, N$, $(x_i(t) - x_0(t)) \to 0$ as $t \to \infty$.

Remark 3.1 In both (3.6) and (3.8), for $i = 1, \ldots, N$, the control u_i only depends on the difference between the state of the subsystem i and the state of its neighboring subsystem j with $j = 1, \ldots, N$ for the leaderless case and $j = 0, 1, \ldots, N$ for the leader-following case. Thus, the control laws (3.6) and (3.8) are said to be in diffusive form, which is a special type of the distributed control law.

3.3 The Two Consensus Problems over Static Networks

In this section, we consider the special case where the communication network is static. In this case, the switching communication graph $\bar{\mathscr{G}}_{\sigma(t)}$ ($\mathscr{G}_{\sigma(t)}$) reduces to the static communication graph $\bar{\mathscr{G}}$ ($\mathscr{G}$). Let $\bar{\mathscr{L}}$ ($\mathscr{L}$) be the Laplacian matrix of $\bar{\mathscr{G}}$ ($\mathscr{G}$).

We need the following standard assumption.

Assumption 3.1 (A, B) is stabilizable.

Let us first establish two lemmas.

Lemma 3.1 *Given $F \in \mathbb{R}^{l\times l}$ for any positive integer l, suppose $\underline{\lambda}_F > 0$. Under Assumption 3.1, let $P > 0$ satisfy $PA + A^{\mathsf{T}}P - PBB^{\mathsf{T}}P + I_n = 0$. Then, the matrix $I_l \otimes A - \mu F \otimes (BB^{\mathsf{T}}P)$ is Hurwitz if $\mu \geq \frac{1}{2}\underline{\lambda}_F^{-1}$.*

Proof Under Assumption 3.1, by Lemma 2.11, there exists a $P > 0$ satisfying $PA + A^{\mathsf{T}}P - PBB^{\mathsf{T}}P + I_n = 0$. Let $T \in \mathbb{C}^{l\times l}$ be such that $TFT^{-1} = J$ is in the Jordan form of F. Then we have

$$I_l \otimes A - \mu F \otimes (BB^{\mathsf{T}}P) = (T^{-1} \otimes I_n)(I_l \otimes A - \mu J \otimes (BB^{\mathsf{T}}P))(T \otimes I_n).$$

Due to the block triangular structure of J, the eigenvalues of $I_l \otimes A - \mu J \otimes (BB^{\mathsf{T}}P)$ coincide with those of $A - \mu\lambda_i(J)BB^{\mathsf{T}}P$, $i = 1, \ldots, l$. Since $\lambda(F) = \lambda(J)$, $\mu\Re(\lambda_i(J)) \geq \mu\underline{\lambda}_J = \mu\underline{\lambda}_F \geq 1/2$. Thus, by Lemma 2.12, all the matrices $A - \mu\lambda_i(J)BB^{\mathsf{T}}P, i = 1, \ldots, l$, are Hurwitz. Therefore, the matrix $I_l \otimes A - \mu F \otimes (BB^{\mathsf{T}}P)$ is Hurwitz. □

If B is an identity matrix, then we have the following result.

Lemma 3.2 *Given $F \in \mathbb{R}^{l\times l}$ for any positive integer l, suppose $\underline{\lambda}_F > 0$. Then, for any $A \in \mathbb{R}^{n\times n}$, the matrix $I_l \otimes A - \mu F \otimes I_n$ is Hurwitz if $\mu > \bar{\lambda}_A\underline{\lambda}_F^{-1}$. In particular, if $\bar{\lambda}_A \leq 0$, then the matrix $I_l \otimes A - \mu F \otimes I_n$ is Hurwitz for any $\mu > 0$.*

Proof By Proposition A.2, the eigenvalues of the matrix $I_l \otimes A - \mu F \otimes I_n$ are given by

$$\{\lambda_i(A) - \mu\lambda_j(F) : i = 1, \ldots, n, j = 1, \ldots, l\}.$$

Therefore, the matrix $(I_l \otimes A) - \mu(F \otimes I_n)$ is Hurwitz if $\mu > \bar{\lambda}_A\underline{\lambda}_F^{-1}$, and is Hurwitz for any $\mu > 0$ if $\bar{\lambda}_A \leq 0$. □

We are ready to present our main results for this section.

Theorem 3.1 *Under Assumptions 2.1 and 3.1, let $P > 0$ satisfy $PA + A^{\mathsf{T}}P - PBB^{\mathsf{T}}P + I_n = 0$, and $\delta = \min_{1\leq i\leq N}\{\Re(\lambda_i(\mathscr{L}))\colon \lambda_i(\mathscr{L}) \neq 0\}$. Then, the leaderless consensus problem of system (3.5) is solvable by a distributed control law of the form (3.6) with $K = \mu B^{\mathsf{T}}P$ and $\mu \geq \frac{1}{2}\delta^{-1}$.*

Proof Let $x = \mathrm{col}(x_1, \ldots, x_N)$. Then, the closed-loop system under control law (3.6) is given by

$$\dot{x} = (I_N \otimes A - \mu\mathscr{L} \otimes (BB^{\mathsf{T}}P))x.$$

By Remark 2.1, under Assumption 2.1, $\mathscr{L}$ has exactly one zero eigenvalue and all other eigenvalues have positive real parts. Then the real Jordan canonical form of $\mathscr{L}$ is as follows:

$$J = \left[\begin{array}{c|c} 0 & 0 \cdots 0 \\ \hline 0 & \\ \vdots & \Lambda \\ 0 & \end{array}\right]$$

where $\Lambda \in \mathbb{R}^{(N-1)\times(N-1)}$ and all of its eigenvalues have positive real parts. Let $r \in \mathbb{R}^N$ be the left eigenvector of $\mathscr{L}$ corresponding to the zero eigenvalue such that $r^\top 1_N = 1$. Then there exist $Y \in \mathbb{R}^{N\times(N-1)}$ and $W \in \mathbb{R}^{(N-1)\times N}$ such that

$$V = \begin{bmatrix} 1_N & Y \end{bmatrix}, \ V^{-1} = \begin{bmatrix} r^\top \\ W \end{bmatrix}$$

and $V^{-1}\mathscr{L}V = J$.

Let $\xi = (V^{-1} \otimes I_n)x$. Then ξ is governed by the following equation:

$$\begin{aligned}\dot{\xi} &= (I_N \otimes A - \mu(V^{-1}\mathscr{L}V) \otimes (BB^\top P))\xi \\ &= \begin{bmatrix} A & 0_{n\times(N-1)n} \\ 0_{(N-1)n\times n} & I_{N-1} \otimes A - \mu\Lambda \otimes (BB^\top P) \end{bmatrix}\xi.\end{aligned}$$

Since $\mu\Re(\lambda_i(\Lambda)) \geq \mu\delta \geq 1/2$, by Lemma 3.1 with $l = N - 1$, the matrix $(I_{N-1} \otimes A - \mu\Lambda \otimes (BB^\top P)$ is Hurwitz. Therefore,

$$\lim_{t\to\infty} \left(\xi(t) - \begin{bmatrix} e^{At} & 0_{n\times(N-1)n} \\ 0_{(N-1)n\times n} & 0_{(N-1)n\times(N-1)n} \end{bmatrix}\xi(0)\right) = 0.$$

Since

$$\begin{aligned}&\left(\xi(t) - \begin{bmatrix} e^{At} & 0_{n\times(N-1)n} \\ 0_{(N-1)n\times n} & 0_{(N-1)n\times(N-1)n} \end{bmatrix}\xi(0)\right) \\ &= (V^{-1} \otimes I_n)\left(x(t) - (V \otimes I_n)\begin{bmatrix} e^{At} & 0_{n\times(N-1)n} \\ 0_{(N-1)n\times n} & 0_{(N-1)n\times(N-1)n} \end{bmatrix}(V^{-1} \otimes I_n)x(0)\right) \\ &= (V^{-1} \otimes I_n)\left(x(t) - ((1_N r^\top) \otimes e^{At})x(0)\right) \\ &= (V^{-1} \otimes I_n)\left(x(t) - 1_N \otimes ((r^\top \otimes e^{At})x(0))\right),\end{aligned}$$

we have

$$\lim_{t\to\infty} \left(x(t) - 1_N \otimes ((r^\top \otimes e^{At})x(0))\right) = 0$$

which completes the proof. □

Next, we study the leader-following consensus problem. Let $\bar{x}_i = x_i - x_0$ and $\bar{x} = \text{col}(\bar{x}_1, \ldots, \bar{x}_N)$. Then, it can be verified that, under the control law (3.8), $\bar{x}$ is governed by

$$\dot{\bar{x}} = (I_N \otimes A - H \otimes (BK))\bar{x} \tag{3.9}$$

where H is as defined in (2.5). We call (3.9) the error system. It is clear that the control law (3.8) solves the leader-following consensus problem of system (3.5) if and only if K is such that the origin of (3.9) is asymptotically stable. Thus, by Lemma 3.1, we obtain the following result.

Theorem 3.2 *Under Assumptions 2.3 and 3.1, let $P > 0$ satisfy $PA + A^{\mathsf{T}}P - PBB^{\mathsf{T}}P + I_n = 0$. Then, for $\mu \geq \frac{1}{2}\underline{\lambda}_H^{-1}$, the origin of the following system:*

$$\dot{\bar{x}} = (I_N \otimes A - \mu H \otimes (BB^{\mathsf{T}}P))\bar{x} \tag{3.10}$$

is asymptotically stable. As a result, the leader-following consensus problem of systems (3.5) and (3.7) is solvable by a distributed control law of the form (3.8) with $K = \mu B^{\mathsf{T}}P$ and $\mu \geq \frac{1}{2}\underline{\lambda}_H^{-1}$.

Proof By Lemma 2.11, under Assumption 3.1, there exists a unique $P > 0$ satisfying $PA + A^{\mathsf{T}}P - PBB^{\mathsf{T}}P + I_n = 0$. Under the control law (3.8) with $K = \mu B^{\mathsf{T}}P$, the error system (3.9) is given by (3.10). Under Assumption 2.3, $\underline{\lambda}_H > 0$. Thus, with $l = N$ and $F = H$, all the conditions in Lemma 3.1 are satisfied. Thus, by Lemma 3.1, with $\mu \geq \frac{1}{2}\underline{\lambda}_H^{-1}$, the matrix $(I_N \otimes A - \mu H \otimes (BB^{\mathsf{T}}P))$ is Hurwitz, i.e., the origin of (3.10) is asymptotically stable. □

For the special case where $B = I_n$, we have the following result upon using Lemma 3.2.

Theorem 3.3 *Under Assumption 2.3, for any $\mu > \bar{\lambda}_A\underline{\lambda}_H^{-1}$, the origin of the following system:*

$$\dot{\bar{x}} = (I_N \otimes A - \mu H \otimes I_n)\bar{x} \tag{3.11}$$

is asymptotically stable. As a result, the leader-following consensus problem of systems (3.5) and (3.7) with $B = I_n$ is solvable by a distributed control law of the form (3.8) with $K = \mu I_n$ and $\mu > \bar{\lambda}_A\underline{\lambda}_H^{-1}$.

Proof Under the control law (3.8) with $K = \mu I_n$, the error system (3.9) with $B = I_n$ is given by (3.11). Under Assumption 2.3, $\underline{\lambda}_H > 0$. Thus, by Lemma 3.2, the matrix $(I_N \otimes A - \mu H \otimes I_n)$ with $\mu > \bar{\lambda}_A\underline{\lambda}_H^{-1}$ is Hurwitz, i.e., the origin of (3.11) is asymptotically stable. □

Example 3.1 Let us first consider the leaderless consensus problem of system (3.5) with $N = 6$ and

$$A = \begin{bmatrix} 0 & 1 & 1 \\ -1 & 1 & 0 \\ 0 & 0 & 0 \end{bmatrix}, \quad B = \begin{bmatrix} 0 \\ 1 \\ 1 \end{bmatrix}. \tag{3.12}$$

The static communication graph $\mathcal{G}$ is illustrated in Fig. 3.3. Since $\mathcal{G}$ contains a spanning tree with node 1 as a root, Assumption 2.1 is satisfied. Solving the Riccati equation $PA + A^{\mathsf{T}}P - PBB^{\mathsf{T}}P + I_n = 0$ gives

$$P = \begin{bmatrix} 11.1902 & -7.3204 & 11.2753 \\ -7.3204 & 7.1682 & -8.0022 \\ 11.2753 & -8.0022 & 12.8551 \end{bmatrix}.$$

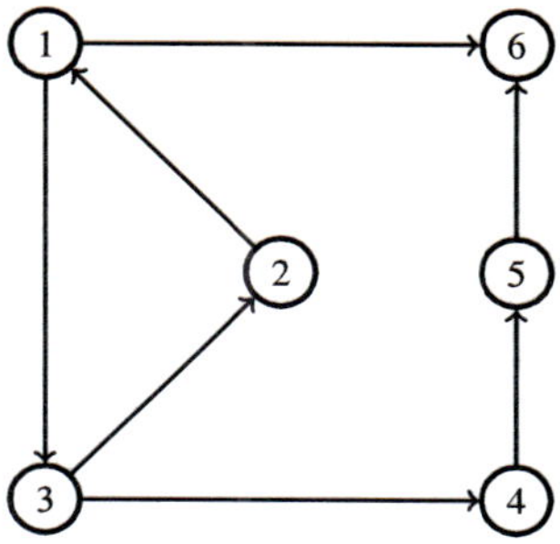

Fig. 3.3 The static communication graph $\mathscr{G}$

A direct inspection on Fig. 3.3 gives the following Laplacian matrix:

$$\mathscr{L} = \begin{bmatrix} 1 & -1 & 0 & 0 & 0 & 0 \\ 0 & 1 & -1 & 0 & 0 & 0 \\ -1 & 0 & 1 & 0 & 0 & 0 \\ 0 & 0 & -1 & 1 & 0 & 0 \\ 0 & 0 & 0 & -1 & 1 & 0 \\ -1 & 0 & 0 & 0 & -1 & 2 \end{bmatrix} \tag{3.13}$$

with

$$\lambda(\mathscr{L}) = \{2.0000, 1.0000, 1.0000, 0, 1.5000 + 0.8660\imath, 1.5000 - 0.8660\imath\}.$$

The left eigenvector corresponding to the zero eigenvalue is

$$r = [1/3, 1/3, 1/3, 0, 0, 0]^{\mathsf{T}}.$$

Since $\delta = 1$, we can set $\mu = 3$ which satisfies $\mu \geq \frac{1}{2}\delta^{-1} = 0.5$. Then, by Theorem 3.1, the control gain is given by

$$K = \mu B^{\mathsf{T}} P = \begin{bmatrix} 11.8645 & -2.5020 & 14.5587 \end{bmatrix}. \tag{3.14}$$

Simulation results on the performance of the closed-loop system under the control law (3.6) are shown in Fig. 3.4, where $x_c(t) \triangleq (r^{\mathsf{T}} \otimes e^{At})x(0)$. It can be seen that all states of the agents converge to $x_c(t)$ asymptotically.

Next, consider the leader-following consensus problem of the follower system (3.5) and the leader system (3.7) with $N = 6$, where A and B are the same as those in (3.12). The static communication graph $\bar{\mathscr{G}}$ is illustrated in Fig. 3.5, which contains a spanning tree with the node 0 as the root and hence Assumption 2.3 is satisfied. Simple inspection on Fig. 3.5 gives the same Laplacian matrix $\mathscr{L}$ as described in (3.13) and

$$\Delta = \mathscr{D}(1, 0, 1, 0, 0, 0).$$

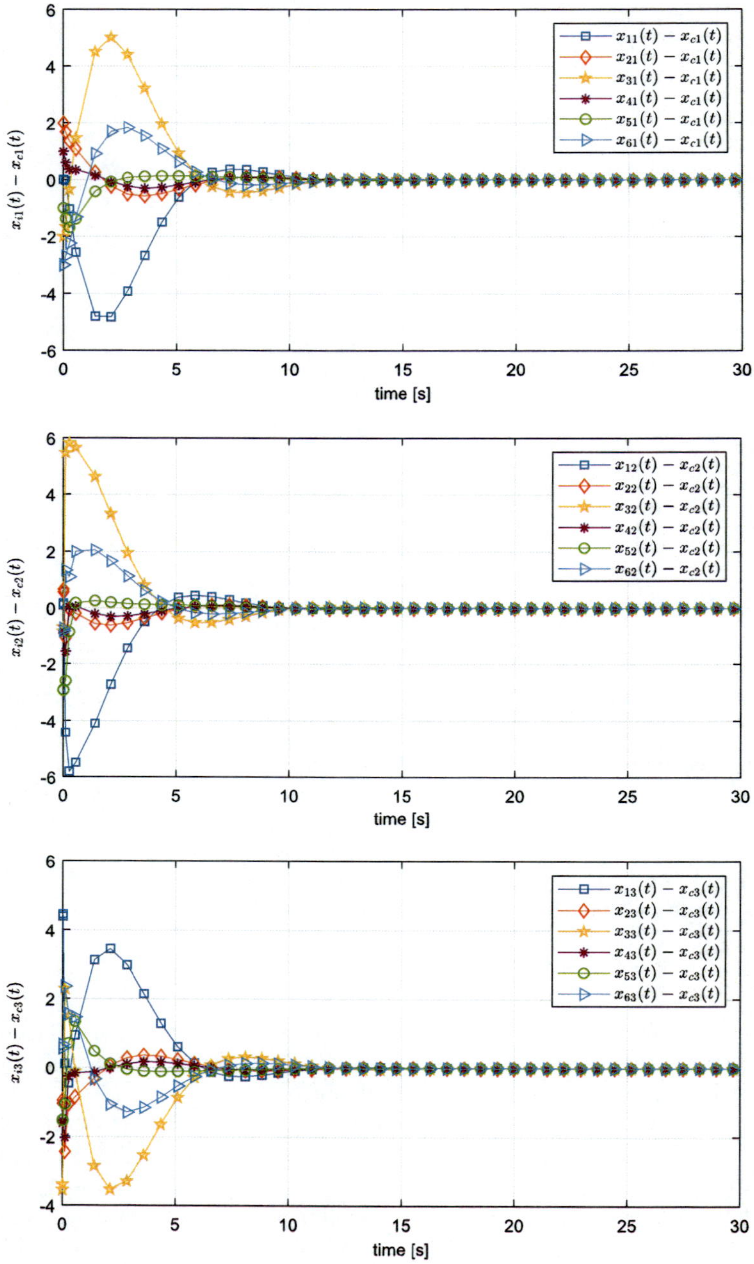

Fig. 3.4 The response of $x_{ij}(t) - x_{cj}(t)$ under the control law (3.6)

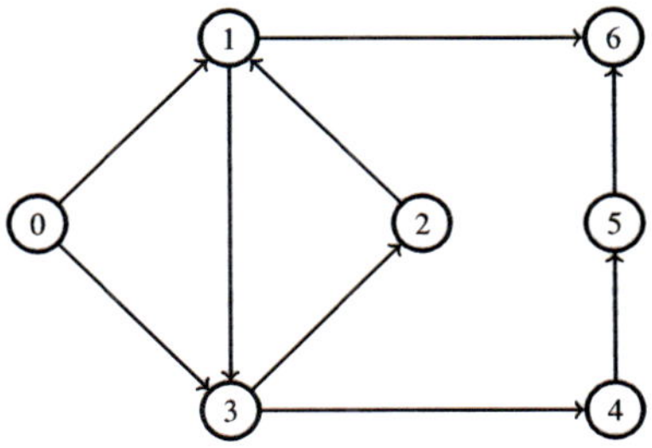

Fig. 3.5 The static communication graph $\bar{\mathscr{G}}$

Therefore,

$$H = \begin{bmatrix} 2 & -1 & 0 & 0 & 0 & 0 \\ 0 & 1 & -1 & 0 & 0 & 0 \\ -1 & 0 & 2 & 0 & 0 & 0 \\ 0 & 0 & -1 & 1 & 0 & 0 \\ 0 & 0 & 0 & -1 & 1 & 0 \\ -1 & 0 & 0 & 0 & -1 & 2 \end{bmatrix}.$$

Since

$$\lambda(H) = \{2.0000, 1.0000, 1.0000, 0.5344, 2.2328 + 0.7926\imath, 2.2328 - 0.7926\imath\},$$

with $\underline{\lambda}_H = 0.5344$, we can also choose $\mu = 3$ which satisfies $\mu \geq \frac{1}{2}\underline{\lambda}_H^{-1} = 0.9356$. Then, by Theorem 3.2, the control gain is given by

$$K = \mu B^{\mathsf{T}} P = \begin{bmatrix} 11.8645 & -2.5020 & 14.5587 \end{bmatrix}.$$

Simulation results on the performance of the closed-loop system under the control law (3.8) are shown in Fig. 3.6. Again it can be seen that all states of the agents converge to $x_0(t)$.

3.4 The Two Consensus Problems over Switching Networks

In this section, we will further study the leaderless and leader-following consensus problems over switching networks. In this case, the two consensus problems will come down to the stabilization problem of a linear time-varying system of the following form:

$$\dot{x} = \left(I_N \otimes A - F_{\sigma(t)} \otimes (BK)\right) x \tag{3.15}$$

where $\sigma(t)$ is a piecewise constant switching signal with dwell time τ, $A \in \mathbb{R}^{n\times n}$, $B \in \mathbb{R}^{n\times m}$, and $F_{\sigma(t)} \in \mathbb{R}^{N\times N}$ are given, and $K \in \mathbb{R}^{m\times n}$ is to be designed.

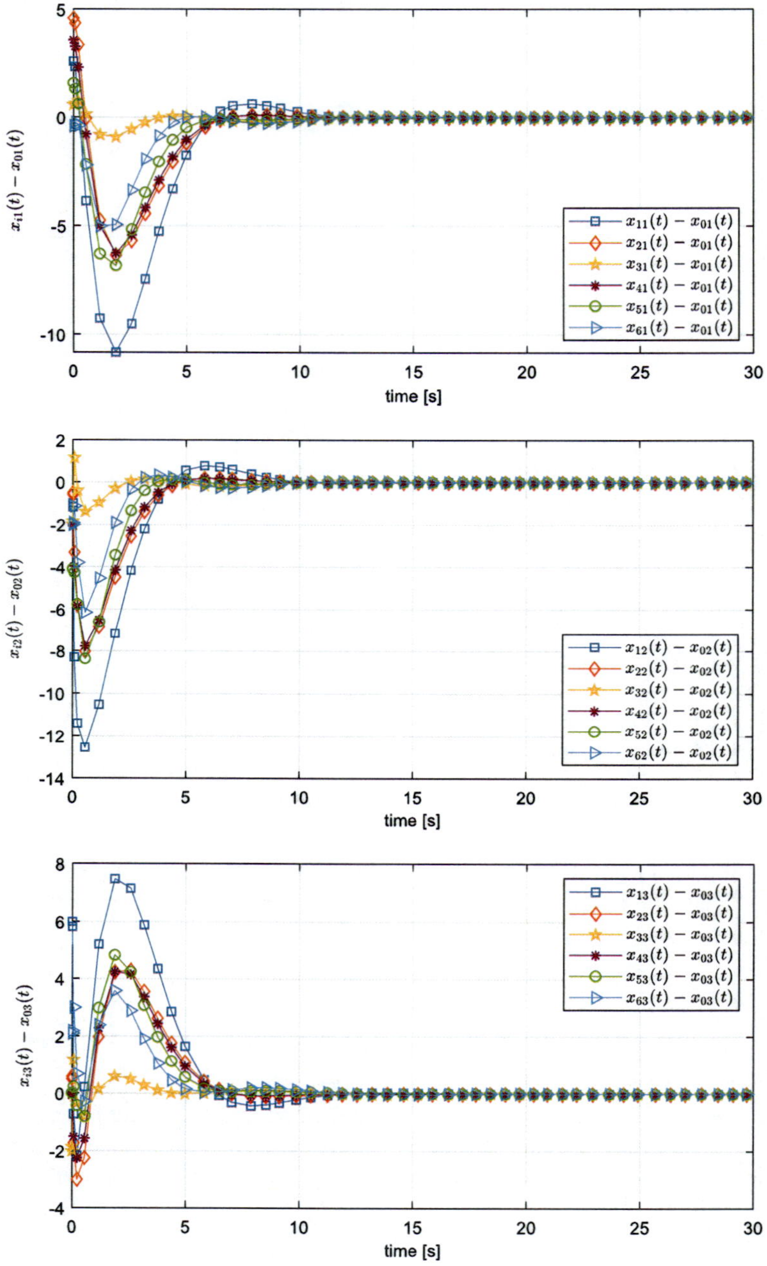

Fig. 3.6 The response of $x_{ij}(t) - x_{0j}(t)$ under the control law (3.8)

For the static network case, $F_{\sigma(t)}$ is a constant matrix. Due to Lemma 3.1, the stabilization problem of (3.15) is a simultaneous eigenvalue placement problem. Nevertheless, for the switching network case, the stabilization problem of (3.15) is more complex. We need to establish some stability results for the linear time-varying system (3.15). For this purpose, we first state two conditions on (3.15) as follows:

Condition (i) There exists a subsequence $\{i_k : k = 0, 1, 2, \dots\}$ of $\{i : i = 0, 1, 2, \dots\}$ with $t_{i_{k+1}} - t_{i_k} < \nu$ for some $\nu > 0$ and all $k = 0, 1, 2, \dots$ such that $x(t_{i_k})$ is orthogonal to the null space of the matrix $\left(\sum_{q=i_k}^{i_{k+1}-1} F_{\sigma(t_q)}\right) \otimes I_n$.

Condition (ii) There exists a subsequence $\{i_k : k = 0, 1, 2, \dots\}$ of $\{i : i = 0, 1, 2, \dots\}$ with $t_{i_{k+1}} - t_{i_k} < \nu$ for some $\nu > 0$ and all $k = 0, 1, 2, \dots$ such that the matrix $\sum_{q=i_k}^{i_{k+1}-1} F_{\sigma(t_q)}$ is nonsingular.

Lemma 3.3 *Consider the following system:*

$$\dot{\xi} = \left(I_N \otimes \bar{A} - F_{\sigma(t)} \otimes (\bar{B}\bar{K})\right)\xi \tag{3.16}$$

where $\bar{A} \in \mathbb{R}^{n\times n}$, $\bar{B} \in \mathbb{R}^{n\times m}$, $\bar{K} \in \mathbb{R}^{m\times n}$, and $F_{\sigma(t)}$ is symmetric and positive semidefinite for any $t \geq 0$. Suppose $(\bar{A}, \bar{B})$ is controllable and $\bar{A}$ is anti-symmetric. Letting $\bar{K} = \mu\bar{B}^{\mathsf{T}}$ with $\mu > 0$ gives

$$\dot{\xi} = \left(I_N \otimes \bar{A} - \mu F_{\sigma(t)} \otimes (\bar{B}\bar{B}^{\mathsf{T}})\right)\xi. \tag{3.17}$$

Then, Condition (i) with $x(t) = \xi(t)$ implies

$$\lim_{t\to\infty} \xi(t) = 0 \tag{3.18}$$

and Condition (ii) with $x(t) = \xi(t)$ implies the origin of system (3.17) is asymptotically stable.

Proof Let

$$V(\xi(t)) = \frac{1}{2}\xi^{\mathsf{T}}(t)\xi(t).$$

Then, the derivative of $V(\xi(t))$ along system (3.17) exists on every interval $[t_i, t_{i+1})$, $i = 0, 1, 2, \dots$, and is given by, upon using $\bar{A}^{\mathsf{T}} + \bar{A} = 0$,

$$\begin{aligned}\dot{V}(\xi(t))\,|_{3.17} &= \frac{1}{2}\xi^{\mathsf{T}}(t)(I_N \otimes \bar{A}^{\mathsf{T}} - \mu F_{\sigma(t)} \otimes (\bar{B}\bar{B}^{\mathsf{T}}) + I_N \otimes \bar{A} - \mu F_{\sigma(t)} \otimes (\bar{B}\bar{B}^{\mathsf{T}}))\xi(t) \\ &= -\mu\xi^{\mathsf{T}}(t)(F_{\sigma(t)} \otimes (\bar{B}\bar{B}^{\mathsf{T}}))\xi(t) \leq 0.\end{aligned} \tag{3.19}$$

So $V(\xi(t)) \leq V(\xi(0))$, i.e., $\|\xi(t)\| \leq \|\xi(0)\|$ for any $t \geq 0$. Thus, system (3.17) is stable. Since $\sigma(t) \in \mathcal{P}$ and $\mathcal{P}$ is a finite set, $\|F_{\sigma(t)}\|$ is uniformly bounded over $[0, \infty)$. Then, by (3.17), $\|\dot{\xi}(t)\|$ is uniformly bounded over $[0, \infty)$. So $\ddot{V}(\xi(t))$ is uniformly bounded over $[0, \infty)$. Therefore, by Corollary 2.6, $\lim_{t\to\infty} \dot{V}(\xi(t)) = 0$. Thus, by (3.19),

$$\lim_{t\to\infty} \xi^{\mathsf{T}}(t)(F_{\sigma(t)} \otimes (\bar{B}\bar{B}^{\mathsf{T}}))\xi(t) = 0$$

which in turn implies

$$\lim_{t\to\infty} (F^{\frac{1}{2}}_{\sigma(t)} \otimes \bar{B}^{\mathsf{T}})\xi(t) = 0. \tag{3.20}$$

Since $\|(F^{\frac{1}{2}}_{\sigma(t)} \otimes I_n)\|$ is uniformly bounded over $[0, \infty)$, (3.20) implies

$$\lim_{t\to\infty} (F_{\sigma(t)} \otimes \bar{B}^{\mathsf{T}})\xi(t) = 0. \tag{3.21}$$

We will show that (3.21) implies (3.18) by the following two steps:

Step-1: We first show that (3.21) implies

$$\lim_{t\to\infty} (F_{\sigma(t)} \otimes I_n)\xi(t) = 0. \tag{3.22}$$

In fact, (3.21) is equivalent to

$$\lim_{t\to\infty} (I_N \otimes \bar{B}^{\mathsf{T}})(F_{\sigma(t)} \otimes I_n)\xi(t) = 0. \tag{3.23}$$

Let $\eta(t) = (F_{\sigma(t)} \otimes I_n)\xi(t)$. Then (3.23) becomes

$$\lim_{t\to\infty} (I_N \otimes \bar{B}^{\mathsf{T}})\eta(t) = 0. \tag{3.24}$$

Since $\xi(t)$ is uniformly bounded over $[0, \infty)$, $\eta(t)$ is also uniformly bounded over $[0, \infty)$, and the derivative of $\eta(t)$ that exists on every interval $[t_i, t_{i+1})$, $i = 0, 1, 2, \ldots$, is given by

$$\begin{aligned}\dot{\eta} &= (F_{\sigma(t)} \otimes I_n)(I_N \otimes \bar{A} - \mu F_{\sigma(t)} \otimes (\bar{B}\bar{B}^{\mathsf{T}}))\xi \\ &= (I_N \otimes \bar{A} - \mu F_{\sigma(t)} \otimes (\bar{B}\bar{B}^{\mathsf{T}}))(F_{\sigma(t)} \otimes I_n)\xi \\ &= (I_N \otimes \bar{A} - \mu F_{\sigma(t)} \otimes (\bar{B}\bar{B}^{\mathsf{T}}))\eta.\end{aligned} \tag{3.25}$$

It can be seen that system (3.25) is in the form of (2.27) with $\psi(t) = \eta(t)$, $M = I_N \otimes \bar{A}$, and $Q(t) = \mu F_{\sigma(t)} \otimes (\bar{B}\bar{B}^{\mathsf{T}}) = \mu(I_N \otimes \bar{B})(F_{\sigma(t)} \otimes \bar{B}^{\mathsf{T}})$, which is uniformly bounded over $[0, \infty)$. Let $G = I_N \otimes \bar{B}^{\mathsf{T}}$. Then (3.21) and (3.24) imply $\lim_{t\to\infty} Q(t)\eta(t) = 0$, and $\lim_{t\to\infty} G\eta(t) = 0$, respectively. By Lemma 2.13, we have

$$\lim_{t\to\infty} (I_N \otimes \bar{B}^{\mathsf{T}})(I_N \otimes \bar{A})^k\eta(t) = 0, \ k = 0, 1, 2, \ldots. \tag{3.26}$$

Since $(\bar{A}, \bar{B})$ is controllable and $\bar{A} + \bar{A}^{\mathsf{T}} = 0$, the pair $(\bar{B}^{\mathsf{T}}, \bar{A})$ is observable. Thus, $(I_N \otimes \bar{B}^{\mathsf{T}}, I_N \otimes \bar{A})$ is also observable. By (3.26), we can obtain $\lim_{t\to\infty} \eta(t) = 0$, i.e., (3.22) holds.

Step-2: We now show that (3.22) implies (3.18). For any $t \geq 0$, and any finite $T_0 \geq 0$,

$$\xi(t+T_0) = e^{(I_N\otimes\bar{A})T_0}\xi(t) + \Lambda(t,T_0) \tag{3.27}$$

where

$$\Lambda(t,T_0) = \int_t^{t+T_0} e^{(I_N\otimes\bar{A})(t+T_0-s)}(-\mu F_{\sigma(s)}\otimes(\bar{B}\bar{B}^{\mathsf{T}}))\xi(s)ds.$$

Let $\Upsilon = \max_{t\in[0,T_0]}\|e^{(I_N\otimes\bar{A})t}\|$. Then Υ is finite since $\|e^{(I_N\otimes\bar{A})t}\|$ is continuous on $[0, T_0]$. Thus,

$$\begin{aligned}\|\Lambda(t,T_0)\| &\leq \int_t^{t+T_0}\|e^{(I_N\otimes\bar{A})(t+T_0-s)}\|\cdot\|\mu F_{\sigma(s)}\otimes(\bar{B}\bar{B}^{\mathsf{T}})\xi(s)\|ds\\ &\leq \Upsilon\|I_N\otimes\bar{B}\|\int_t^{t+T_0}\|\mu(F_{\sigma(s)}\otimes\bar{B}^{\mathsf{T}})\xi(s)\|ds.\end{aligned}$$

By (3.21) and Lemma 2.14,

$$\lim_{t\to\infty}\Lambda(t,T_0) = 0. \tag{3.28}$$

Since the product of $I_N\otimes\bar{A}$ and $F_{\sigma(t+T_0)}\otimes I_n$ commutes, so does the product of $e^{(I_N\otimes\bar{A})T_0}$ and $F_{\sigma(t+T_0)}\otimes I_n$. Then, by (3.27), (3.28), and the fact that $\|F_{\sigma(t+T_0)}\otimes I_n\|$ is uniformly bounded over $[0,\infty)$, we have

$$\begin{aligned}\lim_{t\to\infty} e^{(I_N\otimes\bar{A})T_0}(F_{\sigma(t+T_0)}\otimes I_n)\xi(t) &= \lim_{t\to\infty}(F_{\sigma(t+T_0)}\otimes I_n)e^{(I_N\otimes\bar{A})T_0}\xi(t)\\ &= \lim_{t\to\infty}(F_{\sigma(t+T_0)}\otimes I_n)(\xi(t+T_0)-\Lambda(t,T_0))\\ &= \lim_{t\to\infty}(\eta(t+T_0)-(F_{\sigma(t+T_0)}\otimes I_n)\Lambda(t,T_0))\\ &= 0.\end{aligned}$$

Since $e^{(I_N\otimes\bar{A})T_0}$ is nonsingular, we have

$$\lim_{t\to\infty}(F_{\sigma(t+T_0)}\otimes I_n)\xi(t) = 0. \tag{3.29}$$

Letting $t = t_{i_k}$ and $t+T_0 = t_{i_k+j}$ in (3.29) gives

$$\lim_{k\to\infty}(F_{\sigma(t_{i_k+j})}\otimes I_n)\xi(t_{i_k}) = 0 \tag{3.30}$$

where $j = 0, 1, \ldots, (i_{k+1}-i_k-1)$, respectively. Let $J_k = \left(\sum_{q=i_k}^{i_{k+1}-1} F_{\sigma(t_q)}\right)\otimes I_n$ and $\zeta_k = J_k\xi(t_{i_k})$. Then (3.30) leads to

$$\lim_{k\to\infty}\zeta_k = 0. \tag{3.31}$$

Under Condition (i), $\xi(t_{i_k})$ is orthogonal to the null space of J_k. Let $J_k^\dagger$ denote the Moore–Penrose inverse of J_k. Applying Lemma A.1 with $A = J_k$ and $x = \xi(t_{i_k})$, we have

$$\xi(t_{i_k}) = J_k^\dagger J_k \xi(t_{i_k}) = J_k^\dagger \zeta_k. \tag{3.32}$$

Since $(i_{k+1} - i_k - 1) \leq \lfloor \frac{\nu}{\tau} \rfloor + 1$, and $\mathscr{P}$ is a finite set, the set $\{J_k^\dagger\colon\ k = 1, 2, \ldots\}$ contains only finitely many distinct matrices. Thus, there exists a finite positive number λ such that, for all $k = 1, 2, \ldots$, $\|J_k^\dagger\| \leq \lambda$. Then (3.32) implies

$$\|\xi(t_{i_k})\| \leq \lambda \|\zeta_k\|. \tag{3.33}$$

Therefore, (3.31) and (3.33) imply

$$\lim_{k\to\infty} \xi(t_{i_k}) = 0. \tag{3.34}$$

By (3.19), $\|\xi(t)\|$ is non-increasing over $[0, \infty)$. This fact together with (3.34) concludes $\lim_{t\to\infty} \xi(t) = 0$.

Under Condition (ii), J_k is nonsingular. Hence, the Moore–Penrose inverse of J_k becomes the inverse of J_k. We still have (3.34), and hence, any solution of system (3.17) will approach the origin asymptotically. Therefore, the origin of system (3.17) is asymptotically stable. □

Lemma 3.3 only guarantees the asymptotic stability of system (3.17) under Condition (ii). In some applications such as the establishment of the distributed observer to be studied in the next chapter, it is desired that system (3.17) is exponentially stable under Condition (ii). In fact, it is possible to establish the exponential stability of system (3.17) by using the so-called generalized Krasovskii–LaSalle theorem. Here, we will only state the conclusion and leave the proof of the following lemma to Sect. A.4.

Lemma 3.4 *Consider the system* (3.17) *where* $(\bar{A}, \bar{B})$ *is controllable and* $\bar{A}$ *is anti-symmetric. Then, Condition (ii) implies the origin of system* (3.17) *is exponentially stable.*

Remark 3.2 The matrix A is said to be neutrally stable if all the eigenvalues of A are semi-simple with zero real parts. If A is neutrally stable, then there exists a nonsingular matrix T such that $TAT^{-1} = \bar{A}$ where $\bar{A}$ is anti-symmetric. Let $\xi = (I_N \otimes T)x$. Then (3.15) is transformed to (3.16) with $\bar{B} = TB$ and $\bar{K} = KT^{-1}$. It is clear that the controllability of (A, B) implies the controllability of $(\bar{A}, \bar{B})$. Suppose that A is neutrally stable and (A, B) is controllable. Then, by Lemma 3.3, under the control gain $K = \bar{K}T = \mu\bar{B}^\mathsf{T} T = \mu B^\mathsf{T} T^\mathsf{T} T = \mu B^\mathsf{T} P$ where $P = T^\mathsf{T} T$, Condition (i) implies that the solution of (3.15) will approach the origin asymptotically, and Condition (ii) implies that the origin of the system (3.15) is asymptotically stable. In fact, by Lemma 3.4, Condition (ii) implies the origin of the system (3.15) is exponentially stable. Since $\bar{A} + \bar{A}^\mathsf{T} = 0$, i.e., $TAT^{-1} + T^{-\mathsf{T}} A^\mathsf{T} T^\mathsf{T} = 0$, we have

$PA + A^\top P = 0$. Thus, P is the positive definite and symmetric matrix satisfying $PA + A^\top P = 0$.

Both Lemmas 3.3 and 3.4 require the neutral stability of A and the controllability of the pair (A, B). In what follows, we show that the neutral stability of A can be relaxed to the marginal stability of A, i.e., all the eigenvalues of A have non-positive real parts and those eigenvalues with zero real parts are semi-simple, and the controllability of (A, B) can be relaxed to the stabilizability of (A, B).

Assumption 3.2 A is marginally stable, and (A, B) is stabilizable.

Theorem 3.4 *Consider system (3.15) where $F_{\sigma(t)}$ is symmetric and positive semidefinite for any $t \geq 0$. Under Assumption 3.2, there exists a matrix K such that Condition (i) implies*

$$\lim_{t\to\infty} x(t) = 0$$

and Condition (ii) implies the origin of system (3.15) is exponentially stable.

Proof Since A is marginally stable, there is a nonsingular matrix T_1 such that $T_1 A T_1^{-1} = \begin{bmatrix} A_u & 0 \\ 0 & A_s \end{bmatrix}$ and $T_1 B = \begin{bmatrix} B_u \\ B_s \end{bmatrix}$ where $A_u \in \mathbb{R}^{n_u \times n_u}$ is neutrally stable, $A_s \in \mathbb{R}^{(n-n_u)\times(n-n_u)}$ is a Hurwitz matrix, $B_u \in \mathbb{R}^{n_u \times m}$, and $B_s \in \mathbb{R}^{(n-n_u)\times m}$. Since (A, B) is stabilizable, (A_u, B_u) is also stabilizable, which implies the controllability of (A_u, B_u) as A_u is neutrally stable.

Let $z = (I_N \otimes T_1)x$. Then it can be verified that

$$\dot{z} = \left(I_N \otimes \begin{bmatrix} A_u & 0 \\ 0 & A_s \end{bmatrix} - F_{\sigma(t)} \otimes \begin{bmatrix} B_u K T_1^{-1} \\ B_s K T_1^{-1} \end{bmatrix}\right) z.$$

Partition z as $z = \mathrm{col}(z_{1u}, z_{1s}, \ldots, z_{Nu}, z_{Ns})$ where, for $i = 1, \ldots, N$, $z_{iu} \in \mathbb{R}^{n_u}$, $z_{is} \in \mathbb{R}^{(n-n_u)}$. Let $\bar{x} = \mathrm{col}(z_{1u}, z_{2u}, \ldots, z_{Nu}, z_{1s}, z_{2s}, \ldots, z_{Ns})$. Then there exists a nonsingular matrix $T_2 \in \mathbb{R}^{Nn \times Nn}$ such that $\bar{x} = T_2 z$.

Now, let $T = T_2(I_N \otimes T_1)$, and $K = [K_1, 0_{m\times(n-n_u)}]T_1$ with $K_1 \in \mathbb{R}^{m \times n_u}$. Then, it can be verified that $\bar{x} = Tx$ is governed by the following equation:

$$\dot{\bar{x}} = \begin{bmatrix} I_N \otimes A_u - F_{\sigma(t)} \otimes B_u K_1 & 0 \\ -F_{\sigma(t)} \otimes B_s K_1 & I_N \otimes A_s \end{bmatrix} \bar{x}. \tag{3.35}$$

Let $\bar{x}_1 = \mathrm{col}(z_{1u}, z_{2u}, \ldots, z_{Nu})$ and $\bar{x}_2 = \mathrm{col}(z_{1s}, z_{2s}, \ldots, z_{Ns})$. Partition $\bar{x}$ as $\bar{x} = \mathrm{col}(\bar{x}_1, \bar{x}_2)$. Then (3.35) can be expanded as follows:

$$\dot{\bar{x}}_1 = (I_N \otimes A_u - F_{\sigma(t)} \otimes B_u K_1)\bar{x}_1 \tag{3.36a}$$
$$\dot{\bar{x}}_2 = -(F_{\sigma(t)} \otimes B_s K_1)\bar{x}_1 + (I_N \otimes A_s)\bar{x}_2. \tag{3.36b}$$

Since A_u is neutrally stable and (A_u, B_u) is controllable, by Remark 3.2, under the control gain $K_1 = \mu B_u^\top P$ where $\mu > 0$ and P is the positive definite and symmetric matrix satisfying $PA_u + A_u^{\mathsf{T}}P = 0$, Condition (i) with $x = \bar{x}_1$ implies the solution of (3.36a) will approach the origin asymptotically, and Condition (ii) implies the system (3.36a) is exponentially stable. Since $I_N \otimes A_s$ is Hurwitz, by Lemma 2.5, $\bar{x}_2$ will approach the origin asymptotically under Condition (i) and exponentially under Condition (ii). Thus, under the gain matrix $K = \mu[B_u^{\mathsf{T}}P, 0_{m\times(n-n_u)}]T_1$, Condition (i) implies the solution of (3.15) will approach the origin asymptotically, and Condition (ii) implies the origin of system (3.15) is exponentially stable. □

We are now ready to obtain the results for the two consensus problems over switching graphs.

Theorem 3.5 *Under Assumptions 2.4, 2.5, and 3.2, the leaderless consensus problem of system (3.5) is solvable by a distributed state feedback control law of the form (3.6).*

Proof Under the distributed state feedback control law (3.6), the closed-loop system of agent i is

$$\dot{x}_i = Ax_i + BK\sum_{j=1}^{N} a_{ij}(t)(x_j - x_i). \tag{3.37}$$

Let

$$x_c(t) = \frac{x_1(t) + x_2(t) + \cdots + x_N(t)}{N}.$$

Then $x_c(t)$ is called the center of all agents at time t. Since the communication graph is undirected, by direct calculation, we have

$$\dot{x}_c = \frac{\sum_{i=1}^{N}\dot{x}_i}{N} = A\left(\frac{\sum_{i=1}^{N}x_i}{N}\right) = Ax_c. \tag{3.38}$$

Decompose $x_i(t)$ into

$$x_i(t) = x_c(t) + w_i(t),\ i = 1,\ldots,N. \tag{3.39}$$

Then (3.39) can be put in the following compact form:

$$x(t) = 1_N \otimes x_c(t) + w(t) \tag{3.40}$$

where $x(t) = \mathrm{col}(x_1(t), x_2(t),\ldots,x_N(t))$ and $w(t)=\mathrm{col}(w_1(t), w_2(t),\ldots,w_N(t))$. The vector $w(t)$ is called the (group) disagreement vector. Using (3.37), (3.38), and (3.40) shows that the disagreement vector $w(t)$ satisfies the following system:

$$\dot{w} = \left(I_N \otimes A - \mathcal{L}_{\sigma(t)} \otimes (BK)\right)w$$

which has the form of system (3.15) with $F_{\sigma(t)} = \mathcal{L}_{\sigma(t)}$. Under Assumption 2.5, $\mathcal{L}_{\sigma(t)}$ is symmetric. Since

$$\sum_{i=1}^{N} w_i(t) = \sum_{i=1}^{N} x_i(t) - N x_c(t) = 0,$$

$w(t)$ is orthogonal to span$\{1_N \otimes I_n\}$ for any $t \geq 0$. By Remark 2.4, under Assumption 2.4, the null space of $\sum_{q=i_k}^{i_{k+1}-1} \mathcal{L}_{\sigma(t_q)}$ is span$\{1_N\}$. So the null space of $\left(\sum_{q=i_k}^{i_{k+1}-1} \mathcal{L}_{\sigma(t_q)}\right) \otimes I_n$ is span$\{1_N \otimes I_n\}$. Thus, $w(t_{i_k})$ is orthogonal to the null space of $\left(\sum_{q=i_k}^{i_{k+1}-1} \mathcal{L}_{\sigma(t_q)}\right) \otimes I_n$. Then, by Theorem 3.4, there exists a gain matrix K such that $\lim_{t\to\infty} w(t) = 0$. Thus, by (3.39), all the states $x_i(t)$ asymptotically converge to $x_c(t)$.

From the proof of Theorem 3.4, $K = \mu[B_u^\top P, 0]T_1$ with $\mu > 0$. In particular, if A is neutrally stable, then $K = \mu B^\top P$ where $P > 0$ is such that $PA + A^\top P = 0$. □

For the leader-following consensus problem, we have the following result.

Theorem 3.6 *Under Assumptions 2.5, 2.6, and 3.2, the leader-following consensus problem of the follower system (3.5) and the leader system (3.7) is solvable by a distributed state feedback control law of the form (3.8).*

Proof Under the distributed state feedback control law (3.8), the closed-loop system of agent i is

$$\dot{x}_i = A x_i + BK \left(\sum_{j=0}^{N} a_{ij}(t)(x_j - x_i) \right). \tag{3.41}$$

Let $\xi_i = x_i - x_0$, $\xi = \text{col}(\xi_1, \ldots, \xi_N)$, i.e., $\xi(t) = x(t) - 1_N \otimes x_0(t)$, where $x = \text{col}(x_1, \ldots, x_N)$. Then by (3.7) and (3.41), the overall closed-loop system with $\xi(t)$ as the state can be written in the following form:

$$\dot{\xi} = \left(I_N \otimes A - H_{\sigma(t)} \otimes (BK) \right) \xi, \tag{3.42}$$

where $H_{\sigma(t)}$ is as defined in (2.10). It is noted that (3.42) has the form of system (3.15) with $F_{\sigma(t)} = H_{\sigma(t)}$. Then, by Lemma 2.3, under Assumptions 2.5 and 2.6, $H_{\sigma(t)}$ satisfies Condition (ii). Thus, by Theorem 3.4, there exists a gain matrix K such that the origin of system (3.42) is exponentially stable. Hence, all the state $x_i(t)$ exponentially converge to $x_0(t)$.

Again, from the proof of Theorem 3.4, $K = \mu[B_u^\top P, 0]T_1$ with $\mu > 0$. In particular, if A is neutrally stable, then $K = \mu B^\top P$ where $P > 0$ is such that $PA + A^\top P = 0$. □

Remark 3.3 Suppose A is neutrally stable. Then there exists a positive definite matrix $P \in \mathbb{R}^{n \times n}$ satisfying $PA + A^\top P = 0$. Let $C = B^\top P$. If we attach an output equation $y_i = C x_i$, $i = 0, 1, \ldots, N$, to systems (3.5) and (3.7), then (3.5) and (3.7) together with the output equation $y_i = C x_i$, $i = 0, 1, \ldots, N$, are passive, and the observability of $(B^{\mathsf{T}}P, A)$ is equivalent to the observability of (C, A). Thus, for such systems, by Theorem 3.5, the leaderless consensus problem can be solved by the following static output feedback control law:

$$u_i = \mu \sum_{j=1}^{N} a_{ij}(t)(y_j - y_i).$$

Similarly, by Theorem 3.6, the leader-following consensus problem can be solved by the following static output feedback control law:

$$u_i = \mu \sum_{j=0}^{N} a_{ij}(t)(y_j - y_i).$$

In both Theorems 3.5 and 3.6, we require the matrix A be marginally stable and the switching communication graph $\mathcal{G}_{\sigma(t)}$ be undirected. If $B = I_n$, it is possible to remove the assumption that the switching communication graph $\mathcal{G}_{\sigma(t)}$ is undirected, and to handle the matrix A with unstable modes.

Theorem 3.7 *Consider system (3.15) where $B = I_n$, $\bar{\lambda}_A \leq 0$, and $F_{\sigma(t)} = \mathcal{L}_{\sigma(t)}$. Under Assumption 2.4, the leaderless consensus problem of system (3.5) is solvable by the distributed state feedback control law (3.6) with $K = \mu I_n$ and $\mu > 0$.*

Proof Under the control law (3.6) with $K = \mu I_n$, the closed-loop system is

$$\dot{x} = (I_N \otimes A - \mu \mathcal{L}_{\sigma(t)} \otimes I_n)x.$$

Let $\varphi = (I_N \otimes e^{-At})x$, and $\varphi_i = e^{-At}x_i$, $i = 1, 2, \ldots, N$. Then

$$\begin{aligned}
\dot{\varphi} &= -(I_N \otimes Ae^{-At})x + (I_N \otimes e^{-At})\dot{x} \\
&= -(I_N \otimes Ae^{-At})x + (I_N \otimes e^{-At})(I_N \otimes A - \mu \mathcal{L}_{\sigma(t)} \otimes I_n)x \\
&= (I_N \otimes e^{-At})(-\mu \mathcal{L}_{\sigma(t)} \otimes I_n)x \\
&= -\mu(\mathcal{L}_{\sigma(t)} \otimes I_n)(I_N \otimes e^{-At})x \\
&= -\mu(\mathcal{L}_{\sigma(t)} \otimes I_n)\varphi.
\end{aligned}$$

By Corollary 2.1, for $i = 1, 2, \ldots, N$, all $\varphi_i(t)$ converge to the same vector denoted by φ_0 exponentially as $t \to \infty$. Therefore, for $i = 1, 2, \ldots, N$, all $x_i(t)$ converge to the same time function $e^{At}\varphi_0$ exponentially as $t \to \infty$. □

Theorem 3.8 *Consider system (3.15) where $B = I_n$ and $\bar{\lambda}_A \leq 0$. Under Assumption 2.6, the leader-following consensus problem of the follower system (3.5) and the leader system (3.7) is solvable by the distributed state feedback control law (3.8) with $K = \mu I_n$ and $\mu > 0$.*

Proof From the proof of Theorem 3.6, it suffices to show that the system (3.42) with $B = I_n$ and $K = \mu I_n$ is exponentially stable. For this purpose, let $\varphi = (I_N \otimes e^{-At})\xi$. Then

$$\begin{aligned}\dot{\varphi} &= -(I_N \otimes Ae^{-At})\xi + (I_N \otimes e^{-At})\dot{\xi} \\ &= -(I_N \otimes Ae^{-At})\xi + (I_N \otimes e^{-At})(I_N \otimes A - \mu H_{\sigma(t)} \otimes I_n)\xi \\ &= (I_N \otimes e^{-At})(-\mu H_{\sigma(t)} \otimes I_n)\xi \\ &= -\mu(H_{\sigma(t)} \otimes I_n)(I_N \otimes e^{-At})\xi \\ &= -\mu(H_{\sigma(t)} \otimes I_n)\varphi.\end{aligned}$$

By Corollary 2.3, $\varphi(t) \to 0$ exponentially, i.e., there exist $\alpha_1, \lambda_1 > 0$ such that

$$\|\varphi(t)\| \leq \alpha_1 \|\varphi(0)\| e^{-\lambda_1 t} = \alpha_1 \|\xi(0)\| e^{-\lambda_1 t}.$$

Moreover, since A has no eigenvalues with positive real parts, there exists a polynomial $Q(t)$ such that

$$\|(I_N \otimes e^{At})\| \leq Q(t).$$

Then

$$\|\xi(t)\| \leq \|(I_N \otimes e^{At})\| \cdot \|\varphi(t)\| \leq \alpha_1 \|\xi(0)\| Q(t) e^{-\lambda_1 t}. \tag{3.43}$$

Therefore, the proof is completed by invoking Lemma 2.6. □

Remark 3.4 From (3.43), it can be seen that the assumption $\bar{\lambda}_A \leq 0$ can be weakened to $\bar{\lambda}_A < \lambda_1$.

Example 3.2 Let us first consider the leaderless consensus problem of system (3.5) with $N = 4$ and

$$A = \begin{bmatrix} \frac{1}{2} & \frac{1}{2} & -\frac{1}{2} \\ -\frac{1}{2} & -\frac{1}{2} & -\frac{1}{2} \\ 1 & 1 & 0 \end{bmatrix}, \quad B = \begin{bmatrix} 0.3 \\ 0.3 \\ 0.3 \end{bmatrix}. \tag{3.44}$$

The switching communication graph $\mathcal{G}_{\sigma(t)}$ is defined by the following piecewise constant switching signal:

$$\sigma(t) = \begin{cases} 1 & \text{if } 2s \leq t < 2s + 0.5 \\ 2 & \text{if } 2s + 0.5 \leq t < 2s + 1 \\ 3 & \text{if } 2s + 1 \leq t < 2s + 1.5 \\ 4 & \text{if } 2s + 1.5 \leq t < 2s + 2 \end{cases} \tag{3.45}$$

where $s = 0, 1, 2, \ldots$, and the four graphs $\mathcal{G}_i$, $i = 1, 2, 3, 4$, as illustrated in Fig. 3.7, are all undirected. Note that none of $\mathcal{G}_i$ is connected. However, let $\{t_k = 0.5k\colon k = 0, 1, 2, \ldots\}$. Then $\bigcup_{i=k}^{k+3} \mathcal{G}_{\sigma(t_i)}$ is connected for any $k = 0, 1, 2, \ldots$. Therefore, Assumptions 2.4 and 2.5 are satisfied. It can be verified that Assumption 3.2 is also

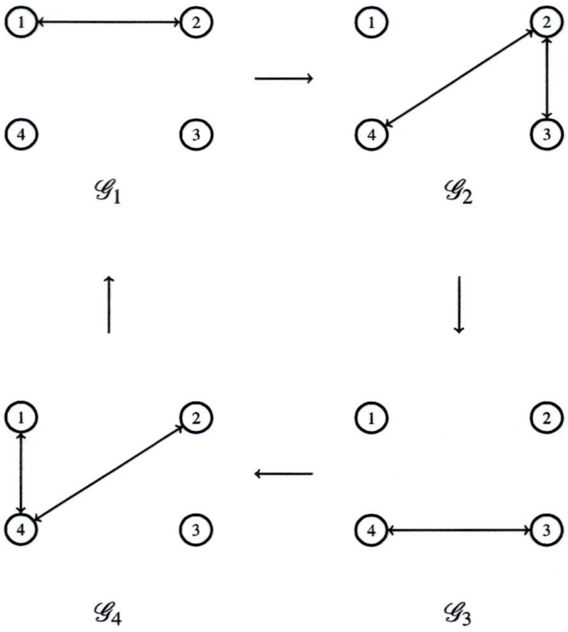

Fig. 3.7 The switching communication graph $\mathcal{G}_{\sigma(t)}$ with $\mathcal{P} = \{1, 2, 3, 4\}$

satisfied. In particular, since the eigenvalues of A are $\{0, \pm\imath\}$, A is neutrally stable. Solving $PA + A^{\mathsf{T}}P = 0$ gives

$$P = \begin{bmatrix} 3 & 1 & -1 \\ 1 & 3 & 1 \\ -1 & 1 & 3 \end{bmatrix}.$$

Then, by Theorem 3.5, with $\mu = 1$, the control gain is given by

$$K = B^{\mathsf{T}}P = \begin{bmatrix} 0.9 & 1.5 & 0.9 \end{bmatrix}. \tag{3.46}$$

Simulation results on the performance of the closed-loop system under the control law (3.6) are shown in Fig. 3.8. It can be seen that all states of the agents converge to their center.

Next, consider the leader-following consensus problem of the follower system (3.5) and the leader system (3.7) with $N = 4$, where A and B are the same as those in (3.44). The switching communication graph $\bar{\mathcal{G}}_{\sigma(t)}$ is defined by the same piecewise constant switching signal (3.45) and the four graphs $\bar{\mathcal{G}}_i$, $i = 1, 2, 3, 4$, are illustrated in Fig. 3.9 with node 0 associated with the leader. It can be seen that, for $i = 1, 2, 3, 4$, the subgraphs $\mathcal{G}_i$ of the graphs $\bar{\mathcal{G}}_i$ are the same as those in Fig. 3.7, and

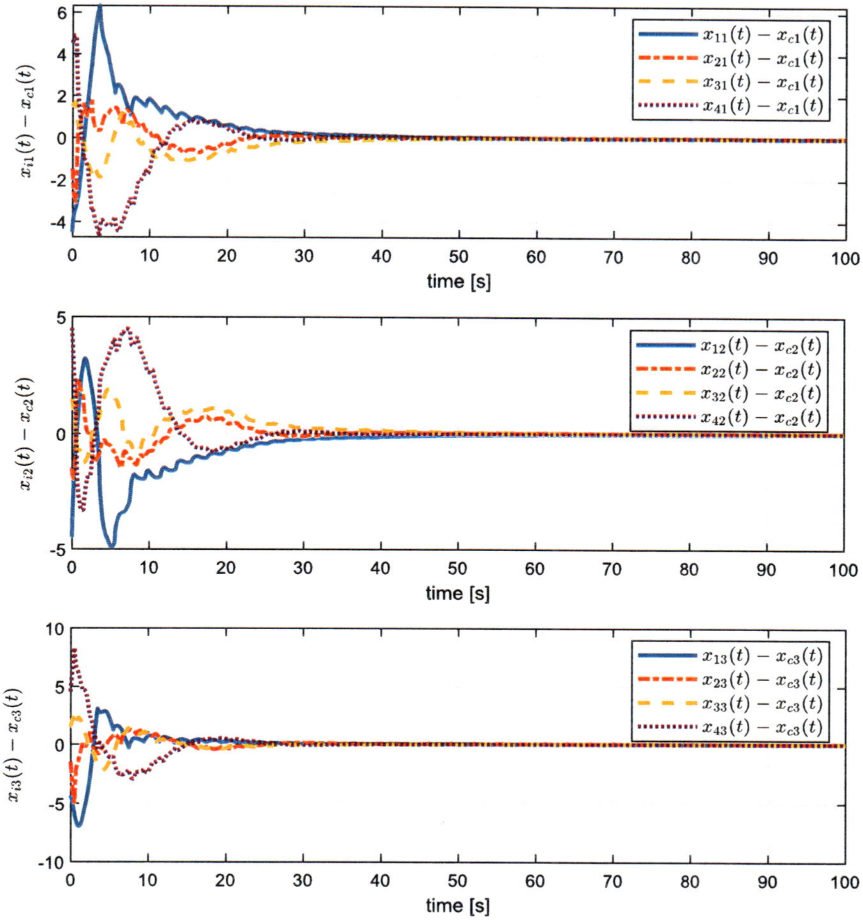

Fig. 3.8 The response of $x_{ij}(t) - x_{cj}(t)$ under the control law (3.6)

$$\begin{aligned} \Delta_1 &= \mathscr{D}(1, 0, 0, 0), \ \ \Delta_2 = \mathscr{D}(0, 0, 0, 0) \\ \Delta_3 &= \mathscr{D}(0, 0, 1, 0), \ \ \Delta_4 = \mathscr{D}(0, 0, 0, 0). \end{aligned}$$

By direct calculation, the system matrix in (3.42) is not Hurwitz for all $t \geq 0$. However, it can be verified that every node in $\bigcup_{i=k}^{k+3} \bar{\mathscr{G}}_{\sigma(t_i)}$ is reachable from the node 0 for any $k = 0, 1, 2, \ldots$. Therefore, Assumptions 2.5 and 2.6 are satisfied. Since A is neutrally stable, by Theorem 3.6, with $\mu = 1$, the control gain is given by

$$K = B^{\mathsf{T}} P = \begin{bmatrix} 0.9 \ 1.5 \ 0.9 \end{bmatrix}.$$

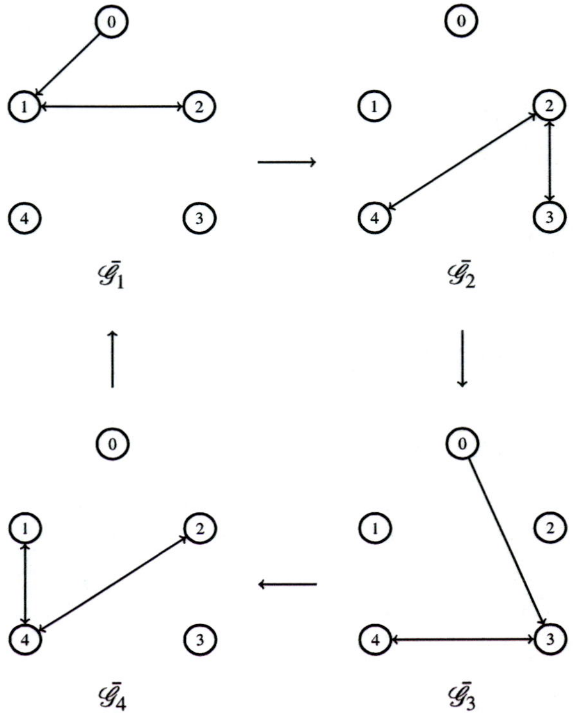

Fig. 3.9 The switching communication graph $\bar{\mathscr{G}}_{\sigma(t)}$ with $\mathscr{P} = \{1, 2, 3, 4\}$

Figure 3.10 shows the performance of the closed-loop system under the control law (3.8). It can be seen that the states of all agents converge to the state of their leader.

3.5 A Dual Result to Theorem 3.4

Lemma 3.3 plays the key role in solving the two consensus problems. In order to deal with the problem of the existence of the distributed observer to be studied in the next chapter, we need to consider a problem dual to the stabilization problem of the system (3.15), i.e., the stabilization problem of the following linear time-varying system:

$$\dot{x} = \big(I_N \otimes A - F_{\sigma(t)} \otimes (LC)\big)\, x \tag{3.47}$$

where $\sigma(t)$ is a piecewise constant switching signal with dwell time τ, $A \in \mathbb{R}^{n\times n}$, $C \in \mathbb{R}^{m\times n}$, and $F_{\sigma(t)} \in \mathbb{R}^{N\times N}$ are given, and $L \in \mathbb{R}^{n\times m}$ is to be designed.

We make the following assumption:

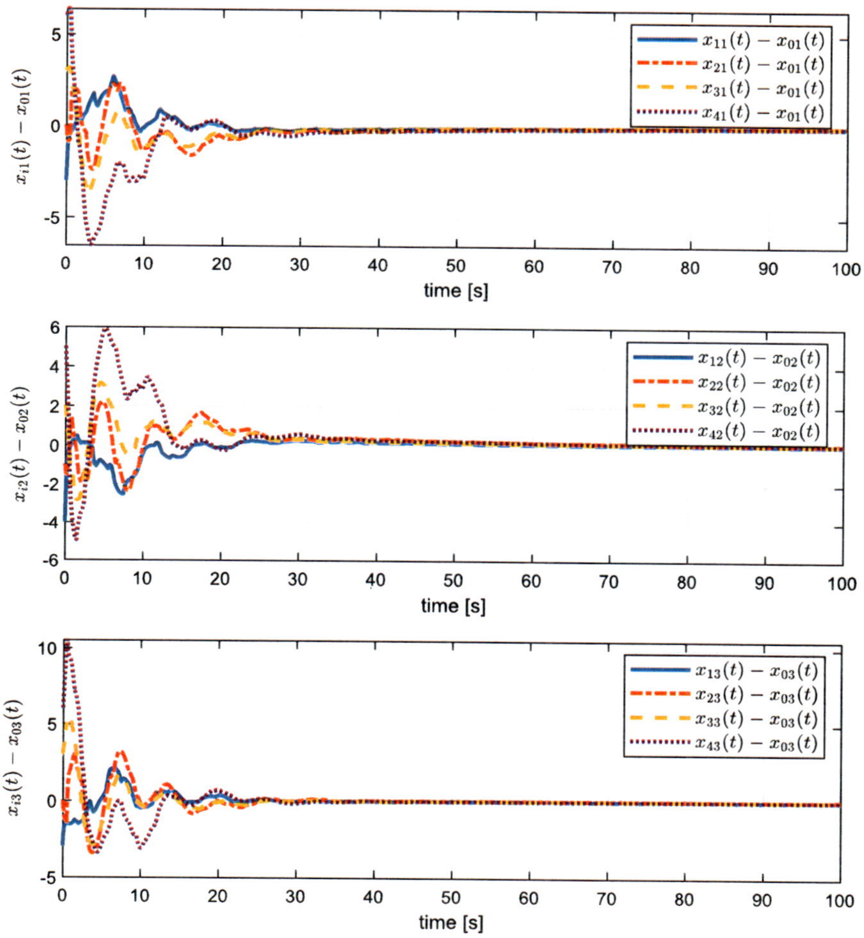

Fig. 3.10 The response of $x_{ij}(t) - x_{0j}(t)$ under the control law (3.8)

Assumption 3.3 *A is marginally stable, and (C, A) is detectable.*

Let us first consider a special case of (3.47) as follows:

$$\dot{\xi} = \left(I_N \otimes \bar{A} - F_{\sigma(t)} \otimes (\bar{L}\bar{C})\right)\xi \tag{3.48}$$

where $\bar{A} \in \mathbb{R}^{n\times n}$ is anti-symmetric, $\bar{C} \in \mathbb{R}^{m\times n}$, $\bar{L} \in \mathbb{R}^{n\times m}$, $F_{\sigma(t)} \in \mathbb{R}^{N\times N}$, and $(\bar{C}, \bar{A})$ is observable. Let $\bar{L} = \mu\bar{C}^{\mathsf{T}}$. Then, we have

$$\dot{\xi} = \left(I_N \otimes \bar{A} - \mu F_{\sigma(t)} \otimes (\bar{C}^{\mathsf{T}}\bar{C})\right)\xi. \tag{3.49}$$

It can be seen that (3.49) is in the same form as (3.17) with $\bar{C} = \bar{B}^{\mathsf{T}}$. Since the observability of $(\bar{C}, \bar{A})$ is equivalent to the controllability of $(\bar{A}^{\mathsf{T}}, \bar{C}^{\mathsf{T}})$, which is equivalent to the controllability of $(\bar{A}, \bar{C}^{\mathsf{T}})$ as $\bar{A}$ is anti-symmetric, we can directly apply Lemma 3.4 to (3.49) to obtain the following result.

Lemma 3.5 *Consider system* (3.49). *Suppose* $(\bar{C}, \bar{A})$ *is observable and* $\bar{A}$ *is anti-symmetric. Then, for any* $\mu > 0$*, Condition (ii) with* $x(t) = \xi(t)$ *implies the origin of the system* (3.49) *is exponentially stable.*

Remark 3.5 Similar to Remark 3.2, if A is neutrally stable, then there exists a nonsingular matrix T such that $TAT^{-1} = \bar{A}$ where $\bar{A}$ is anti-symmetric. Let $\xi = (I_N \otimes T)x$. Then (3.47) is transformed to (3.48) with $\bar{C} = CT^{-1}$ and $\bar{L} = TL$. Then, $L = T^{-1}\bar{L} = \mu T^{-1}\bar{C}^{\mathsf{T}} = \mu T^{-1}T^{-\mathsf{T}}C^{\mathsf{T}} = \mu PC^{\mathsf{T}}$ where $P = T^{-1}T^{-\mathsf{T}}$. Also, the observability of (C, A) is equivalent to the observability of $(\bar{C}, \bar{A})$. Moreover, the observability of $(\bar{C}, \bar{A})$ is equivalent to the controllability of $(\bar{A}^{\mathsf{T}}, \bar{C}^{\mathsf{T}})$, and thus is equivalent to the controllability of $(\bar{A}, \bar{C}^{\mathsf{T}})$ as $\bar{A}$ is anti-symmetric. Since $\bar{A} + \bar{A}^{\mathsf{T}} = 0$, i.e., $TAT^{-1} + T^{-\mathsf{T}}A^{\mathsf{T}}T^{\mathsf{T}} = 0$, we have $PA^{\mathsf{T}} + AP = 0$. As a result, we have the following corollary of Lemma 3.5.

Corollary 3.1 *Consider system* (3.47). *Suppose* (C, A) *is observable and* A *is neutrally stable. Let* P *be the positive definite and symmetric matrix satisfying* $PA^{\mathsf{T}} + AP = 0$*, and* $L = \mu PC^{\mathsf{T}}$ *with* $\mu > 0$*. Then, Condition (ii) implies the origin of system* (3.47) *is exponentially stable.*

Theorem 3.9 *Consider system* (3.47) *where* $F_{\sigma(t)}$ *is symmetric and positive semidefinite for any* $t \geq 0$*. Under Assumption 3.3, there exists a matrix* L *such that Condition (ii) implies the origin of the system* (3.47) *is exponentially stable.*

Proof Since A is marginally stable, there is a nonsingular matrix T_1 such that $T_1AT_1^{-1} = \begin{bmatrix} A_u & 0 \\ 0 & A_s \end{bmatrix}$ and $CT_1^{-1} = \begin{bmatrix} C_u & C_s \end{bmatrix}$ where $A_u \in \mathbb{R}^{n_u \times n_u}$ is neutrally stable, $A_s \in \mathbb{R}^{(n-n_u)\times(n-n_u)}$ is a Hurwitz matrix, $C_u \in \mathbb{R}^{m\times n_u}$, and $C_s \in \mathbb{R}^{m\times(n-n_u)}$. Since (C, A) is detectable, (C_u, A_u) is also detectable, which implies the observability of (C_u, A_u) as A_u is neutrally stable.

Let $z = (I_N \otimes T_1)x$. Then it can be verified that

$$\dot{z} = \left(I_N \otimes \begin{bmatrix} A_u & 0 \\ 0 & A_s \end{bmatrix} - F_{\sigma(t)} \otimes \begin{bmatrix} T_1LC_u & T_1LC_s \end{bmatrix}\right) z.$$

Partition z as $z = \mathrm{col}(z_{1u}, z_{1s}, \dots, z_{Nu}, z_{Ns})$ where, for $i = 1, \dots, N$, $z_{iu} \in \mathbb{R}^{n_u}$, $z_{is} \in \mathbb{R}^{(n-n_u)}$. Let $\bar{x} = \mathrm{col}(z_{1u}, z_{2u}, \dots, z_{Nu}, z_{1s}, z_{2s}, \dots, z_{Ns})$. Then there exists a nonsingular matrix $T_2 \in \mathbb{R}^{Nn\times Nn}$ such that $\bar{x} = T_2z$.

Now, let $T = T_2(I_N \otimes T_1)$, and $L = T_1^{-1}\begin{bmatrix} L_1 \\ 0_{(n-n_u)\times m} \end{bmatrix}$ with $L_1 \in \mathbb{R}^{n_u\times m}$. Then, it can be verified that $\bar{x} = Tx$ is governed by the following equation:

$$\dot{\bar{x}} = \begin{bmatrix} I_N \otimes A_u - F_{\sigma(t)} \otimes L_1C_u & -F_{\sigma(t)} \otimes L_1C_s \\ 0 & I_N \otimes A_s \end{bmatrix} \bar{x}. \tag{3.50}$$

Partition $\bar{x}$ as $\bar{x} = \text{col}(\bar{x}_1, \bar{x}_2)$ with $\bar{x}_1 \in \mathbb{R}^{n_u N}$ and $\bar{x}_2 \in \mathbb{R}^{(n-n_u)N}$. Then (3.50) can be expanded as follows:

$$\dot{\bar{x}}_1 = (I_N \otimes A_u - F_{\sigma(t)} \otimes L_1 C_u)\bar{x}_1 - (F_{\sigma(t)} \otimes L_1 C_s)\bar{x}_2 \tag{3.51a}$$

$$\dot{\bar{x}}_2 = (I_N \otimes A_s)\bar{x}_2. \tag{3.51b}$$

Since A_u is neutrally stable and (C_u, A_u) is observable, by Corollary 3.1, under the gain $L_1 = \mu P C_u^\top$ where $\mu > 0$ and P is the positive definite and symmetric matrix satisfying $PA_u^\top + A_u P = 0$, Condition (ii) implies the system $\dot{\bar{x}}_1 = (I_N \otimes A_u - F_{\sigma(t)} \otimes L_1 C_u)\bar{x}_1$ is exponentially stable. This fact together with the Hurwitzness of $I_N \otimes A_s$ implies that system (3.51) is exponentially stable. Thus, under the gain matrix $L = \mu T_1^{-1} \begin{bmatrix} PC_u^\top \\ 0_{(n-n_u)\times m} \end{bmatrix}$, Condition (ii) implies that the origin of system (3.47) is exponentially stable. □

3.6 Notes and References

The two consensus problems for linear multi-agent systems have received intensive attention over the past two decades (see the survey paper [1] and the books [2, 3]). Consensus problems over the static networks can be converted into the simultaneous stabilization problem. The consensus problem for single integrator and double integrator systems were first studied in, say, [4–6]. For general linear systems, the leaderless consensus problem was solved by LQR-based design in [7], which lays the foundation of the approach in Sect. 3.3. This technique can be extended to some classes of switching networks that are connected at every time instant or is frequently connected with time period T [8].

The case of jointly connected switching networks is more difficult than the static network case since the network can be disconnected at all time instants and the closed-loop system is time-varying. Reference [9] first handled the single integrator discrete-time multi-agent systems over undirected jointly connected networks. Such analysis was later extended to the case of directed networks in [6], and general linear systems with the input matrix in some special form in [10]. Some other attempts are based on Barbalat's lemma [11] or LaSalle's invariance principle [12]. The materials in Sect. 3.4 are based on generalized Barbalat's lemma given in [13]. Its discrete-time counterpart can be found in [14]. There are some other methods in this direction, such as non-smooth analysis [15], max – min Lyapunov function [16], generalized LaSalle-type theorems [17], and so on.

This chapter only focused on the consensus problem of continuous-time systems. The study on the consensus problem of general discrete-time systems over static communication graphs can be found in, say, [18–21], and over switching communication graphs can be found in, say, [14, 22–26].

Various scenarios of the cyclic pursuit problem were studied in, say, [27–30]. The numerical example in Sect. 3.1 is a simplified version of the cyclic pursuit problem from [28, 30].

Theorem 3.4 assumes that the network is undirected and all the eigenvalues of the matrix A have zero or negative real parts. In fact, the undirected assumption on the network was weakened in [31], and, when the switching signal is periodic, the restriction on the eigenvalues of A can be removed [32].

References

1. Olfati-Saber R, Fax JA, Murray RM (2007) Consensus and cooperation in networked multi-agent systems. Proc IEEE 95(1):215–223
2. Qu Z (2008) Cooperative control of dynamical systems: applications to autonomous vehicles. Springer, London
3. Ren W, Beard RW (2008) Distributed consensus in multi-vehicle cooperative control: theory and applications. Springer, London
4. Hu J, Hong Y (2007) Leader-following coordination of multi-agent systems with coupling time delays. Phys A: Stat Mech Appl 374(2):853–863
5. Olfati-Saber R, Murray RM (2004) Consensus problems in networks of agents with switching topology and time-delays. IEEE Trans Autom Control 49(9):1520–1533
6. Ren W, Beard RW (2005) Consensus seeking in multiagent systems under dynamically changing interaction topologies. IEEE Trans Autom Control 50(5):655–661
7. Tuna SE (2008) LQR-based coupling gain for synchronization of linear systems. arXiv:0801.3390
8. Wang J, Cheng D, Hu X (2008) Consensus of multi-agent linear dynamic systems. Asian J Cortrol 10(2):144–155
9. Jadbabaie A, Lin J, Morse AS (2003) Coordination of groups of mobile autonomous agents using nearest neighbor rules. IEEE Trans Autom Control 48(6):988–1001
10. Su S, Lin Z (2016) Distributed consensus control of multi-agent systems with higher order agent dynamics and dynamically changing directed interaction topologies. IEEE Trans Autom Control 61(2):515–519
11. Ni W, Cheng D (2010) Leader-following consensus of multi-agent systems under fixed and switching topologies. Syst & Control Lett 59(3):209–217
12. Cheng D, Wang J, Hu X (2008) An extension of LaSalle's invariance principle and its application to multi-agent consensus. IEEE Trans Autom Control 53(7):885–890
13. Su Y, Huang J (2012) Stability of a class of linear switching systems with applications to two consensus problems. IEEE Trans Autom Control 57(6):1420–1430
14. Su Y, Huang J (2012) Two consensus problems for discrete-time multi-agent systems with switching network topology. Automatica 48(9):1988–1997
15. Lin Z (2005) Coupled dynamic systems: from structure towards stability and stabilizability, PhD dissertation, University of Toronto, Toronto, Canada
16. Moreau L (2004) Stability of continuous-time distributed consensus algorithms. In: Proceedings of the 41st IEEE conference on decision and control, pp 3998–4003
17. Lee T, Tan Y, Su Y, Mareels I (2021) Invariance principles and observability in switched systems with an application in consensus. IEEE Trans Autom Control 66(11): 5128–5143
18. Tuna SE (2008) Synchronizing linear systems via partial-state coupling. Automatica 44:2179–2184
19. You K, Xie L (2011) Network topology and communication data rate for consensusability of discrete-time multi-agent systems. IEEE Trans Autom Control 56(10):2262–2275

20. Hengster-Movric K, You K, Lewis FL, Xie L (2013) Synchronization of discrete-time multi-agent systems on graphs using Riccati design. Automatica 49:414–423
21. Liu J, Huang J (2019) A spectral property of a graph matrix and its application to the leader-following consensus problem of discrete-time multi-agent systems. IEEE Trans Autom Control 64(6):2583–2589
22. Qin J, Gao H, Yu C (2014) On discrete-time convergence for general linear multi-agent systems under dynamic topology. IEEE Trans Autom Control 59(4):1054–1059
23. Huang J (2017) The consensus for discrete-time linear multi-agent systems under directed switching networks. IEEE Trans Autom Control 62(8):4086–4092
24. Lee T, Xia W, Su Y, Huang J (2018) Exponential consensus of discrete-time systems based on a novel Krasovskii–LaSalle theorem under directed switching networks. Automatica 97:189–199
25. Liu J, Huang J (2021) Discrete-time leader-following consensus over switching digraphs with general system modes. IEEE Trans Autom Control 66(3):1238–1245
26. Liu T, Huang J (2021) Discrete-time distributed observers over jointly connected switching networks and an application. IEEE Trans Autom Control 66(4):1918–1924
27. Klamkin MS, Newman DJ (1971) Cyclic pursuit or the three bugs problem. Amer Math Mon 78(6):631–639
28. Bruckstein AM, Cohen N, Efrat A (1991) Ants, crickets and frogs in cyclic pursuit. Center for intelligent systems. Technical Report #9105, Technion-Israel Institute of Technology, Haifa, Israel
29. Richardson TJ (2001) Non-mutual captures in cyclic pursuit. Ann Math Artif Intell 31:127–146
30. Marshall JA, Broucke ME, Francis BA (2004) Formations of vehicles in cyclic pursuit. IEEE Trans Autom Control 49(11):1963–1974
31. Liu T, Lee T, Huang J (2019) An exponential stability result for a class of linear switched systems and its application. In: IEEE conference on decision and control, pp 1551–1556
32. He C, Huang J (2022) Adaptive distributed observer for general linear leader systems over periodic switching digraphs. Automatica 137(3):110021

Chapter 4
The Distributed Observer Approach

4.1 Introduction

In Chap. 3, we studied both the leaderless consensus problem and the leader-following consensus problem for linear homogeneous multi-agent systems by distributed control laws. The control laws used in Chap. 3 are static feedback control laws in the diffusive form. In practice, multi-agent systems can be much more complex than the systems encountered in Chap. 3. The complexity is characterized by one or more attributes such as nonlinearity, model uncertainty, external disturbances, and heterogeneity. Also, the tasks of control may be more demanding than mere state consensus. Some of the typical control tasks are asymptotic tracking of the outputs of all subsystems to a class of reference inputs, disturbance rejection, formation, etc. The complexity of both system dynamics and control tasks entails the development of more advanced cooperative control techniques.

One of the systematic approaches for synthesizing a distributed control law for complex multi-agent systems is called the distributed observer approach, which aims at handling a variety of cooperative control problems involving both a leader system and a follower system. The distributed observer is a distributed dynamic compensator capable of estimating and transmitting the leader's information to every follower subsystem over the communication network. The distributed observer has experienced three phases of developments. In the first phase, the distributed observer only estimates and transmits the leader's signal to every follower subsystem assuming every follower knows the dynamics of the leader. In a more practical setting, not every follower can access the dynamics of the leader. Thus, in the second phase, the distributed observer is further rendered the capability of estimating and transmitting not only the leader's signal but also the dynamics of the leader to every follower subsystem provided that the leader's children know the information of the leader. Such a dynamic compensator is called adaptive distributed observer for a known leader system. The third phase focuses on such leaders whose dynamics contain uncertain parameters. Thus, none of the follower subsystems knows the exact dynamics of the

H. Cai et al., *Cooperative Control of Multi-agent Systems*, Advances in Industrial Control,
https://doi.org/10.1007/978-3-030-98377-2_4

leader. Such a dynamic compensator is called adaptive distributed observer for an uncertain leader. All these three types of distributed observers will be detailed in this chapter.

4.2 A Framework for Synthesizing Distributed Control Laws

In this section, we will give an overall introduction to the distributed observer approach. Consider a leader system of the form (1.3), which is repeated below for convenience:

$$\dot{v}_0 = f_0(v_0, w_0) \tag{4.1a}$$

$$y_{m0} = h_{m0}(v_0, w_0) \tag{4.1b}$$

where $v_0 \in \mathbb{R}^q$ and $y_{m0} \in \mathbb{R}^{p_{m0}}$ are the state and the measurement output of the leader system, respectively, and $w_0 \in \mathbb{R}^{n_{w0}}$ represents a constant unknown vector.

For $i = 1, \ldots, N$, the dynamics of the ith subsystem of the follower system are described by

$$\dot{x}_i = f_i(x_i, u_i, v_0, w) \tag{4.2a}$$

$$e_i = h_i(x_i, u_i, v_0, w) \tag{4.2b}$$

$$y_{mi} = h_{mi}(x_i, u_i, v_0, w) \tag{4.2c}$$

where $x_i \in \mathbb{R}^{n_i}$, $u_i \in \mathbb{R}^{m_i}$, $e_i \in \mathbb{R}^{p_i}$, $y_{mi} \in \mathbb{R}^{p_{mi}}$, and $w \in \mathbb{R}^{n_w}$ are the system state, control input, regulated output, measurement output, and constant unknown vector, respectively. It is noted that (4.2) is composed of (1.4) and (1.21).

As in Chap. 3, the leader system and the follower system together are viewed as a multi-agent system with $N+1$ subsystems whose communication topology is described by a switching communication graph $\bar{\mathscr{G}}_{\sigma(t)} = (\bar{\mathscr{V}}, \bar{\mathscr{E}}_{\sigma(t)})$ with $\bar{\mathscr{V}} = \{0, 1, \ldots, N\}$ and $\bar{\mathscr{E}}_{\sigma(t)} \subseteq \bar{\mathscr{V}} \times \bar{\mathscr{V}}$ for all $t \geq 0$. Here, the node 0 is associated with the leader system (4.1) and the node i, $i = 1, \ldots, N$, is associated with the ith follower subsystem of (4.2). For $i = 0, 1, \ldots, N$, $j = 1, \ldots, N$, $i \neq j$, $(i, j) \in \bar{\mathscr{E}}_{\sigma(t)}$ if and only if follower j can receive the information from follower i or the leader system at time instant t. Since the leader system (4.1) does not have an input, there is no such edges as $(i, 0)$, $i = 1, \ldots, N$. For $i = 1, \ldots, N$, define $\bar{\mathscr{N}}_i(t) = \{j : (j, i) \in \bar{\mathscr{E}}(t)\}$ and we call $\bar{\mathscr{N}}_i(t)$ the neighbor set of the node i at time t. Let $\bar{\mathscr{A}}_{\sigma(t)} = [a_{ij}(t)]_{i,j=0}^N$ denote the weighted adjacency matrix of $\bar{\mathscr{G}}_{\sigma(t)}$. Moreover, let $\mathscr{G}_{\sigma(t)} = (\mathscr{V}, \mathscr{E}_{\sigma(t)})$ be a subgraph of $\bar{\mathscr{G}}_{\sigma(t)}$ where $\mathscr{V} = \{1, \ldots, N\}$ and $\mathscr{E}_{\sigma(t)} = \bar{\mathscr{E}}_{\sigma(t)} \bigcap (\mathscr{V} \times \mathscr{V})$. Let $\mathscr{L}_{\sigma(t)}$ be the Laplacian of $\mathscr{G}_{\sigma(t)}$ and $H_{\sigma(t)} = \mathscr{L}_{\sigma(t)} + \mathscr{D}(a_{10}(t), \ldots, a_{N0}(t))$. If the switching communication graph becomes static, then $\bar{\mathscr{G}}_{\sigma(t)}$, $\mathscr{G}_{\sigma(t)}$, and $H_{\sigma(t)}$ reduce to $\bar{\mathscr{G}}$, $\mathscr{G}$, and H, respectively.

As described in Sect. 1.2.3, a distributed dynamic feedback control law is given as follows:

$$u_i = k_i(\xi_i, \xi_j, y_{mi}, y_{mj}, j \in \bar{\mathcal{N}}_i(t)) \tag{4.3a}$$

$$\dot{\xi}_i = \psi_i(\xi_i, \xi_j, y_{mi}, y_{mj}, j \in \bar{\mathcal{N}}_i(t)) \tag{4.3b}$$

where k_i and ψ_i are some globally defined smooth functions to be designed. It can be seen that, for $i = 1, \ldots, N$, at any time instant $t \geq 0$, $u_i(t)$ depends on $\xi_j(t)$, $y_{mj}(t)$, $j \neq i$, if and only if the node j is a neighbor of the node i at time t. Thus, the control law (4.3) satisfies the communication constraints imposed by the neighbor sets $\bar{\mathcal{N}}_i(t)$, $i = 1, \ldots, N$.

We consider the following cooperative control problem:

Problem 4.1 Given systems (4.1) and (4.2) and the switching communication graph $\bar{\mathcal{G}}_{\sigma(t)}$, find a distributed control law of the form (4.3) such that, for any initial condition, the solution of the closed-loop system exists for all $t \geq 0$ and satisfies $\lim_{t\to\infty} e_i(t) = 0$, $i = 1, \ldots, N$.

Remark 4.1 The leader-following consensus problem studied in Chap. 3 can be viewed as a special case of Problem 4.1 with $v_0 = x_0$, $e_i = x_i - v_0$, $i = 1, \ldots, N$, and $y_{mi} = x_i$, $i = 0, 1, \ldots, N$, and various static control laws used in Chap. 3 are also the special cases of the control law (4.3) where the dimensions of ξ_i, $i = 1, \ldots, N$, are all equal to zero.

As aforementioned, the complexity of both the system dynamics and the control tasks makes Problem 4.1 more complicated and challenging than the two consensus problems studied in Chap. 3. In what follows, we will outline a general framework for synthesizing a distributed control law of the form (4.3) to solve Problem 4.1, i.e., the distributed observer approach.

If $N = 1$, then Problem 4.1 reduces to the classical output regulation problem, which will be more precisely described in Chap. 7 for linear systems. Typically, the problem can be solved by a control law of the following form:

$$u_1 = k_1(z_1, y_{m1}, y_{m0}) \tag{4.4a}$$

$$\dot{z}_1 = g_1(z_1, y_{m1}, y_{m0}). \tag{4.4b}$$

If $N > 1$, and, for all $i = 1, \ldots, N$ and all $t \geq 0$, $\bar{\mathcal{N}}_i(t) = \{0\}$, or what is the same, the leader is the only neighbor of each subsystem of (4.2) for all $t \geq 0$, then every follower is able to access the information of the leader for all $t \geq 0$. In this case, by designing the control law of the form (4.4) for each subsystem of (4.2), we can obtain an overall control law as follows:

$$u_i = k_i(z_i, y_{mi}, y_{m0}) \tag{4.5a}$$

$$\dot{z}_i = g_i(z_i, y_{mi}, y_{m0}), \quad i = 1, \ldots, N. \tag{4.5b}$$

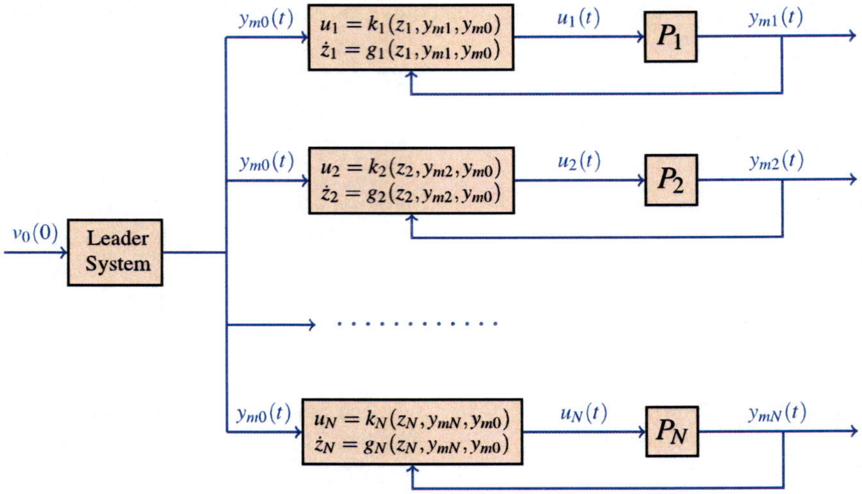

Fig. 4.1 Purely decentralized control scheme

Since, for $i = 1, \ldots, N$, u_i does not have to rely on the information of other follower subsystems, we call (4.5) a purely decentralized control law, which leads to the control scheme illustrated in Fig. 4.1. In general, since not all the followers can take y_{m0} for feedback control for all time t due to the communication constraints imposed by $\bar{\mathcal{G}}_{\sigma(t)}$, one has to take advantage of information sharing among the different subsystems, thus leading to the distributed control law of the form (4.3). For complex systems, designing a distributed control law of the form (4.3) is much more challenging than designing the purely decentralized control law (4.5). In what follows, we will describe a framework for synthesizing a distributed control law of the form (4.3) based on the purely decentralized control law of the form (4.5) and a so-called distributed observer described below.

For simplicity, here we only focus on the class of linear leader systems of the form (1.13), which is repeated below for convenience:

$$\dot{v}_0 = S_0 v_0 \tag{4.6a}$$

$$y_{m0} = W_0 v_0 \tag{4.6b}$$

where $S_0 \in \mathbb{R}^{q \times q}$ and $W_0 \in \mathbb{R}^{p_{m0} \times q}$ are known constant matrices.

Given the switching communication graph $\bar{\mathcal{G}}_{\sigma(t)}$ and the leader system, consider a distributed dynamic compensator of the following form:

$$\dot{v}_i = \phi_i(W_0 v_i, W_0 v_j, j \in \bar{\mathcal{N}}_i(t)), \quad i = 1, \ldots, N. \tag{4.7}$$

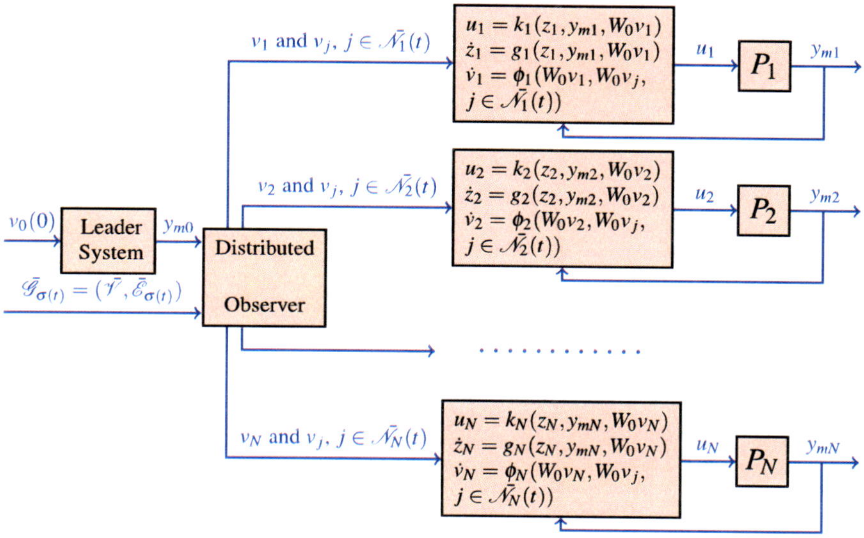

Fig. 4.2 Distributed observer-based control law

If, for any initial condition, the solution of system (4.6) and system (4.7) satisfies, for $i = 1, \ldots, N$, $\lim_{t\to\infty}(v_i(t) - v_0(t)) = 0$, then the compensator (4.7) is called a distributed observer for the linear leader system (4.6). It is noted that (4.7) is driven by the measurement output y_{m0} of the leader system since $W_0 v_0 = y_{m0}$.

Suppose we have had a purely decentralized control law (4.5) and a distributed observer (4.7). Then composing (4.5) and (4.7) yields the following control law:

$$u_i = k_i(z_i, y_{mi}, W_0 v_i) \tag{4.8a}$$

$$\dot{z}_i = g_i(z_i, y_{mi}, W_0 v_i) \tag{4.8b}$$

$$\dot{v}_i = \phi_i(W_0 v_i, W_0 v_j, j \in \bar{\mathcal{N}}_i(t)) \tag{4.8c}$$

which is in the form of (4.3) with $\xi_i = \text{col}(z_i, v_i)$ and $\psi_i = \text{col}(g_i, \phi_i)$. We call a control law of the form (4.8) a distributed observer-based control law, which leads to the control scheme illustrated in Fig. 4.2.

Two issues need to be addressed. First, for a given leader system (4.6) and a switching communication graph $\bar{\mathcal{G}}_{\sigma(t)}$, whether a distributed observer for the leader system exists. Second, even if a distributed observer for the leader system (4.6) exists, one cannot take for granted that the distributed observer-based control law (4.8) solves Problem 4.1 since, after all, (4.5) and (4.8) are two different control laws. If the solvability of Problem 4.1 by a control law of the form (4.5) implies the solvability of Problem 4.1 by a control law of the form (4.8), then we say the control law (4.8) satisfies the certainty equivalence principle.

Clearly, the degree of the difficulty of solving Problem 4.1 not only depends on the dynamics of the systems (4.2) and (4.6), but also the property of the switching communication graph $\bar{\mathscr{G}}_{\sigma(t)}$. As in Chap. 3, the mildest assumption on the switching communication graph $\bar{\mathscr{G}}_{\sigma(t)}$ is as follows:

Assumption 4.1 There exists a subsequence $\{i_k : k = 0, 1, 2, \dots\}$ of $\{i : i = 0, 1, 2, \dots\}$ with $t_{i_{k+1}} - t_{i_k} < \nu$ for some $\nu > 0$ and all $k = 0, 1, 2, \dots$ such that the union graph $\bar{\mathscr{G}}([t_{i_k}, t_{i_{k+1}}))$ contains a spanning tree with the node 0 as the root.

The static communication graph is a special case of the switching communication graph when $\rho = 1$. For this special case, Assumption 4.1 reduces to the following one.

Assumption 4.2 $\bar{\mathscr{G}}$ contains a spanning tree with the node 0 as the root.

In some cases, we also need the following condition.

Assumption 4.3 $\mathscr{G}_{\sigma(t)}$ is undirected for any $t \geq 0$.

Assumptions for the linear leader system (4.6) are listed as follows.

Assumption 4.4 (W_0, S_0) is detectable.

Assumption 4.5 The matrix S_0 has no eigenvalues with positive real parts.

Assumption 4.6 The matrix S_0 is marginally stable, i.e., there exists a positive definite matrix P_0 satisfying $P_0 S_0^{\mathrm{T}} + S_0 P_0 \leq 0$.

Remark 4.2 Since the concept of the distributed observer for the leader system includes the conventional Luenberger observer as a special case when $N = 1$, Assumption 4.4 is a necessary condition for the existence of a distributed observer for the leader system. Assumptions 4.5 and 4.6 impose various conditions on the location of the eigenvalues of the matrix S_0 and will be used to invoke various results in Chap. 3.

4.3 Distributed Observer for a Known Leader System

In this section, we consider the leader system (4.6) and assume that the system matrices S_0 and W_0 are known by every follower for all $t \geq 0$. In this case, the problem of finding a distributed observer for the leader system is closely related to the leader-following consensus problem studied in Chap. 3. In fact, the leader-following consensus problem of the leader system (3.7) and the follower system (3.5) is to design a distributed control law of the form (3.8) repeated below:

$$u_i = K\left(\sum_{j=0}^{N} a_{ij}(t)(x_j - x_i)\right), \quad i = 1, \dots, N \tag{4.9}$$

where $K \in \mathbb{R}^{m\times n}$, such that, for $i = 1, \ldots, N$, $\lim_{t\to\infty}(x_i(t) - x_0(t)) = 0$.

It is known from Chap. 3 that, under the control law (4.9), $\bar{x} = x - 1_N \otimes x_0$ with $x = \mathrm{col}(x_1, \ldots, x_N)$ is governed by the following system:

$$\dot{\bar{x}} = (I_N \otimes A - H_{\sigma(t)} \otimes (BK))\bar{x}. \tag{4.10}$$

Thus, the control law (4.9) solves the leader-following consensus problem for (3.5) and (3.7) if and only if there exists some K such that the origin of system (4.10) is asymptotically stable.

This fact motivates the following distributed observer candidate:

$$\dot{v}_i = S_0 v_i + L_0 \sum_{j=0}^{N} a_{ij}(t) W_0(v_j - v_i), \ i = 1, \ldots, N \tag{4.11}$$

where $v_i \in \mathbb{R}^q$, and $L_0 \in \mathbb{R}^{q\times p_{m0}}$ is a constant gain matrix to be designed.

For $i = 1, \ldots, N$, let $\tilde{v}_i = v_i - v_0$, $\tilde{v} = \mathrm{col}(\tilde{v}_1, \ldots, \tilde{v}_N)$. Then we have

$$\dot{\tilde{v}} = (I_N \otimes S_0 - H_{\sigma(t)} \otimes (L_0 W_0))\tilde{v}. \tag{4.12}$$

We call system (4.12) the error system. Clearly, the dynamic compensator (4.11) is a distributed observer for the leader system if and only if the origin of the error system (4.12) is asymptotically stable.

The error system (4.12) is in the form of (3.47) with $\tilde{v} = x$, $q = n$, $S_0 = A$, $H_{\sigma(t)} = F_{\sigma(t)}$, $L_0 = L$, and $W_0 = C$. Thus, we can directly invoke Theorem 3.9 to obtain the following result:

Theorem 4.1 *Given system (4.6) and a switching communication graph $\bar{\mathcal{G}}_{\sigma(t)}$, under Assumptions 4.1, 4.3, 4.4, and 4.6, there exists a constant matrix $L_0 \in \mathbb{R}^{q\times p_{m0}}$ such that the origin of the error system (4.12) is exponentially stable, that is, for any $v_i(0) \in \mathbb{R}^q$, $i = 0, 1, \ldots, N$, the solution of (4.11) exists for all $t \geq 0$ and satisfies*

$$\lim_{t\to\infty} \tilde{v}_i(t) = 0, \ i = 1, \ldots, N$$

exponentially.

Remark 4.3 The observer gain L_0 can be calculated according to the proof of Theorem 3.9 as follows. If S_0 is neutrally stable, let $P_0 > 0$ satisfy $P_0 S_0^{\mathrm{T}} + S_0 P_0 = 0$. Then $L_0 = \mu_v P_0 W_0^{\mathrm{T}}$ with $\mu_v > 0$. If S_0 is not neutrally stable, then let T_1 be such that $T_1 S_0 T_1^{-1} = \begin{bmatrix} S_{0u} & 0 \\ 0 & S_{0s} \end{bmatrix}$ and $W_0 T_1^{-1} = \begin{bmatrix} W_{0u} & W_{0s} \end{bmatrix}$ where $S_{0u} \in \mathbb{R}^{q_u\times q_u}$ is neutrally stable, $S_{0s} \in \mathbb{R}^{(q-q_u)\times(q-q_u)}$ is a Hurwitz matrix, $W_{0u} \in \mathbb{R}^{p_{m0}\times q_u}$, and $W_{0s} \in \mathbb{R}^{p_{m0}\times(q-q_u)}$. Since (W_0, S_0) is detectable, (W_{0u}, S_{0u}) is observable as S_{0u} is neutrally stable.

Let $z = (I_N \otimes T_1)\tilde{v}$ and partition z as $z = \mathrm{col}(z_{1u}, z_{1s}, \ldots, z_{Nu}, z_{Ns})$ where, for $i = 1, \ldots, N$, $z_{iu} \in \mathbb{R}^{q_u}$ and $z_{is} \in \mathbb{R}^{(q-q_u)}$. Let $\bar{x} =$

$\text{col}(z_{1u}, z_{2u}, \ldots, z_{Nu}, z_{1s}, z_{2s}, \ldots, z_{Ns})$. Then there exists a nonsingular matrix $T_2 \in \mathbb{R}^{Nq\times Nq}$ such that $\bar{x} = T_2 z$.

Since S_{0u} is neutrally stable, there exists positive definite and symmetric matrix P_0 satisfying $P_0 S_{0u}^{\mathsf{T}} + S_{0u} P_0 = 0$. Let $T = T_2(I_N \otimes T_1)$. Then $L_0 = \mu_v T_1^{-1} \begin{bmatrix} P_0 W_{0u}^{\mathsf{T}} \\ 0_{(q-q_u)\times p_{m0}} \end{bmatrix}$ with $\mu_v > 0$.

It is noted that (4.11) is driven by the measurement output y_{m0} of the leader system since $W_0 v_0 = y_{m0}$. If $y_{m0} = v_0$, then (4.11) reduces to the following form:

$$\dot{v}_i = S_0 v_i + \mu_v \sum_{j=0}^{N} a_{ij}(t)(v_j - v_i), i = 1, \ldots, N \tag{4.13}$$

and the error system (4.12) reduces to the following one:

$$\dot{\tilde{v}} = (I_N \otimes S_0 - \mu_v H_{\sigma(t)} \otimes I_q)\tilde{v}. \tag{4.14}$$

Applying Theorem 3.8 to (4.14) gives the following result:

Theorem 4.2 *Given system (4.6) with $W_0 = I_q$ and a switching communication graph $\bar{\mathcal{G}}_{\sigma(t)}$, under Assumptions 4.1 and 4.5, for any $v_i(0) \in \mathbb{R}^q$, $i = 0, 1, \ldots, N$, and any $\mu_v > 0$, the solution of (4.13) exists for all $t \geq 0$ and satisfies*

$$\lim_{t\to\infty} \tilde{v}_i(t) = 0, \ i = 1, \ldots, N$$

exponentially.

Next, we consider the special case where the switching communication graph $\bar{\mathcal{G}}_{\sigma(t)}$ becomes a static one $\bar{\mathcal{G}}$. Then, the distributed observer candidate (4.11) reduces to the following one:

$$\dot{v}_i = S_0 v_i + L_0 \sum_{j=0}^{N} a_{ij} W_0 (v_j - v_i), \ i = 1, \ldots, N \tag{4.15}$$

and the error system (4.12) reduces to the following one:

$$\dot{\tilde{v}} = (I_N \otimes S_0 - H \otimes (L_0 W_0))\tilde{v}. \tag{4.16}$$

Since the matrix $(I_N \otimes S_0 - H \otimes (L_0 W_0))$ is Hurwitz if and only if the matrix $(I_N \otimes S_0^{\mathsf{T}} - H^{\mathsf{T}} \otimes (W_0^{\mathsf{T}} L_0^{\mathsf{T}}))$ is Hurwitz, by Lemma 3.1 with $l = N$, $A = S_0^{\mathsf{T}}$, $F = H^{\mathsf{T}}$, $B = W_0^{\mathsf{T}}$, and $\mu B^{\mathsf{T}} P = L_0^{\mathsf{T}}$, we have the following result:

Theorem 4.3 *Given system (4.6) and a static communication graph $\bar{\mathscr{G}}$, under Assumptions 4.2 and 4.4, let $P_0 > 0$ be the unique solution of $P_0 S_0^{\mathsf{T}} + S_0 P_0 - P_0 W_0^{\mathsf{T}} W_0 P_0 + I_q = 0$. Then, for any $v_i(0) \in \mathbb{R}^q$, $i = 0, 1, \ldots, N$, the solution of (4.15) with $L_0 = \mu_v P_0 W_0^{\mathsf{T}}$ and $\mu_v \geq \frac{1}{2}\underline{\lambda}_H^{-1}$ exists for all $t \geq 0$ and satisfies*

$$\lim_{t\to\infty} \tilde{v}_i(t) = 0, \ i = 1, \ldots, N$$

exponentially.

If $y_{m0} = v_0$, i.e., $W_0 = I_q$, then, by letting $L_0 = \mu_v I_q$, the distributed observer (4.15) reduces to the following form:

$$\dot{v}_i = S_0 v_i + \mu_v \sum_{j=0}^{N} a_{ij}(v_j - v_i), \ i = 1, \ldots, N \tag{4.17}$$

and the error system (4.16) reduces to the following one:

$$\dot{\tilde{v}} = (I_N \otimes S_0 - \mu_v(H \otimes I_q)\tilde{v}. \tag{4.18}$$

By Theorem 3.3, we have the following result:

Theorem 4.4 *Given system (4.6) with $W_0 = I_q$ and a static communication graph $\bar{\mathscr{G}}$, under Assumption 4.2, for any $v_i(0) \in \mathbb{R}^q$, $i = 0, 1, \ldots, N$, the solution of (4.17) with $\mu_v > \bar{\lambda}_{S_0}\underline{\lambda}_H^{-1}$ exists for all $t \geq 0$ and satisfies*

$$\lim_{t\to\infty} \tilde{v}_i(t) = 0, \ i = 1, \ldots, N$$

exponentially.

4.4 Adaptive Distributed Observer for a Known Leader System

As pointed out in Sect. 4.1, in a more practical setting, not every follower can access the system matrices S_0 and W_0 of the leader for all $t \geq 0$. Also, as described in Sect. 1.3, in addition to the measurement output y_{m0}, the leader system may also provide a reference signal y_0 of the form (1.23), which, for convenience, is repeated as follows:

$$y_0 = F_0 v_0 \tag{4.19}$$

where $F_0 \in \mathbb{R}^{p\times q}$ is a constant matrix and may not be accessed by the control of every follower subsystem. Thus, in this section, we further consider designing a distributed observer for the leader system (4.6) and (4.19) by only assuming that the children of the leader system know the matrices S_0, W_0, and F_0. In this case, the distributed observer introduced in the last section is not feasible. We need to further introduce the so-called adaptive distributed observer for the leader system (4.6) and (4.19), which not only estimates the state of the leader but also the three matrices S_0, W_0, and F_0. We will consider four scenarios.

4.4.1 Case 1

First, we consider the case where the switching communication graph $\bar{\mathcal{G}}_{\sigma(t)}$ is a static one $\bar{\mathcal{G}}$. Under Assumption 4.4, let P_0 be the solution of $P_0 S_0^{\mathsf{T}} + S_0 P_0 - P_0 W_0^{\mathsf{T}} W_0 P_0 + I_q = 0$ and $L_0 = P_0 W_0^{\mathsf{T}}$. Then, we propose the following adaptive distributed observer candidate:

$$\dot{S}_i = \mu_S \sum_{j=0}^{N} a_{ij}(S_j - S_i) \tag{4.20a}$$

$$\dot{F}_i = \mu_F \sum_{j=0}^{N} a_{ij}(F_j - F_i) \tag{4.20b}$$

$$\dot{W}_i = \mu_W \sum_{j=0}^{N} a_{ij}(W_j - W_i) \tag{4.20c}$$

$$\dot{L}_i = \mu_L \sum_{j=0}^{N} a_{ij}(L_j - L_i) \tag{4.20d}$$

$$\dot{v}_i = S_i v_i + \mu_v L_i \sum_{j=0}^{N} a_{ij}(W_j v_j - W_i v_i) \tag{4.20e}$$

$$\hat{y}_i = F_i v_i \tag{4.20f}$$

where, for $i = 1, \ldots, N$, the quantities S_i, F_i, W_i, L_i, and $\hat{y}_i$ are the estimates of S_0, F_0, W_0, L_0, and y_0, respectively, by the ith subsystem of (4.20).

For $i = 1, \ldots, N$, let $\tilde{S}_i = S_i - S_0$, $\tilde{F}_i = F_i - F_0$, $\tilde{W}_i = W_i - W_0$, $\tilde{L}_i = L_i - L_0$, $\tilde{v}_i = v_i - v_0$, and $\tilde{y}_i = \hat{y}_i - y_0 = F_i v_i - F_0 v_0 = F_i \tilde{v}_i + \tilde{F}_i v_0$. Then, we have

$$
\begin{aligned}
& L_i \sum_{j=0}^{N} a_{ij}(W_j v_j - W_i v_i) \\
&= L_i \sum_{j=0}^{N} a_{ij} \left((\tilde{W}_j + W_0) v_j - (\tilde{W}_i + W_0) v_i \right) \\
&= L_i W_0 \sum_{j=0}^{N} a_{ij}(v_j - v_i) + L_i \sum_{j=0}^{N} a_{ij}(\tilde{W}_j v_j - \tilde{W}_i v_i) \\
&= L_i W_0 \sum_{j=0}^{N} a_{ij}(\tilde{v}_j - \tilde{v}_i) + L_i \sum_{j=0}^{N} a_{ij} \left(\tilde{W}_j(\tilde{v}_j + v_0) - \tilde{W}_i(\tilde{v}_i + v_0) \right) \\
&= L_0 W_0 \sum_{j=0}^{N} a_{ij}(\tilde{v}_j - \tilde{v}_i) + \tilde{L}_i W_0 \sum_{j=0}^{N} a_{ij}(\tilde{v}_j - \tilde{v}_i) \\
&\quad + L_i \sum_{j=0}^{N} a_{ij}(\tilde{W}_j \tilde{v}_j - \tilde{W}_i \tilde{v}_i) + L_i \sum_{j=0}^{N} a_{ij}(\tilde{W}_j - \tilde{W}_i) v_0.
\end{aligned}
$$

Thus,

$$
\begin{aligned}
\dot{\tilde{v}}_i &= S_i v_i + \mu_v L_i \sum_{j=0}^{N} a_{ij}(W_j v_j - W_i v_i) - S_0 v_0 \\
&= S_0 v_i - S_0 v_0 + S_i v_i - S_0 v_i + \mu_v L_i \sum_{j=0}^{N} a_{ij}(W_j v_j - W_i v_i) \\
&= S_0 \tilde{v}_i + \tilde{S}_i v_i + \mu_v L_i \sum_{j=0}^{N} a_{ij}(W_j v_j - W_i v_i) \\
&= S_0 \tilde{v}_i + \tilde{S}_i \tilde{v}_i + \tilde{S}_i v_0 + \mu_v L_0 W_0 \sum_{j=0}^{N} a_{ij}(\tilde{v}_j - \tilde{v}_i) + \mu_v \tilde{L}_i W_0 \sum_{j=0}^{N} a_{ij}(\tilde{v}_j - \tilde{v}_i) \\
&\quad + \mu_v L_i \sum_{j=0}^{N} a_{ij}(\tilde{W}_j \tilde{v}_j - \tilde{W}_i \tilde{v}_i) + \mu_v L_i \sum_{j=0}^{N} a_{ij}(\tilde{W}_j - \tilde{W}_i) v_0 \\
&= S_0 \tilde{v}_i + \mu_v L_0 W_0 \sum_{j=0}^{N} a_{ij}(\tilde{v}_j - \tilde{v}_i) \\
&\quad + \tilde{S}_i \tilde{v}_i + \mu_v \tilde{L}_i W_0 \sum_{j=0}^{N} a_{ij}(\tilde{v}_j - \tilde{v}_i) + \mu_v L_i \sum_{j=0}^{N} a_{ij}(\tilde{W}_j \tilde{v}_j - \tilde{W}_i \tilde{v}_i) \\
&\quad + \tilde{S}_i v_0 + \mu_v L_i \sum_{j=0}^{N} a_{ij}(\tilde{W}_j - \tilde{W}_i) v_0.
\end{aligned}
$$

Furthermore, let $v = \text{col}(v_1, \ldots, v_N)$, $\tilde{v} = \text{col}(\tilde{v}_1, \ldots, \tilde{v}_N) = v - 1_N \otimes v_0$, $\tilde{S} = \text{col}(\tilde{S}_1, \ldots, \tilde{S}_N)$, $\tilde{F} = \text{col}(\tilde{F}_1, \ldots, \tilde{F}_N)$, $\tilde{W} = \text{col}(\tilde{W}_1, \ldots, \tilde{W}_N)$, $\tilde{L} = \text{col}(\tilde{L}_1, \ldots, \tilde{L}_N)$, $\tilde{S}_d = \mathscr{D}(\tilde{S}_1, \ldots, \tilde{S}_N)$, $\tilde{W}_d = \mathscr{D}(\tilde{W}_1, \ldots, \tilde{W}_N)$, $\tilde{L}_d = \mathscr{D}(\tilde{L}_1, \ldots, \tilde{L}_N)$, and $L_d = \mathscr{D}(L_1, \ldots, L_N)$. Then, it is ready to verify that

$$\begin{bmatrix} L_0 W_0 \sum_{j=0}^{N} a_{1j}(\tilde{v}_j - \tilde{v}_1) \\ \vdots \\ L_0 W_0 \sum_{j=0}^{N} a_{Nj}(\tilde{v}_j - \tilde{v}_N) \end{bmatrix} = -(H \otimes (L_0 W_0))\tilde{v}$$

$$\begin{bmatrix} \tilde{L}_1 W_0 \sum_{j=0}^{N} a_{1j}(\tilde{v}_j - \tilde{v}_1) \\ \vdots \\ \tilde{L}_N W_0 \sum_{j=0}^{N} a_{Nj}(\tilde{v}_j - \tilde{v}_N) \end{bmatrix} = -\tilde{L}_d(H \otimes W_0)\tilde{v}$$

$$\begin{bmatrix} L_1 \sum_{j=0}^{N} a_{1j}(\tilde{W}_j \tilde{v}_j - \tilde{W}_1 \tilde{v}_1) \\ \vdots \\ L_N \sum_{j=0}^{N} a_{Nj}(\tilde{W}_j \tilde{v}_j - \tilde{W}_N \tilde{v}_N) \end{bmatrix} = -L_d(H \otimes I_{p_{m0}})\tilde{W}_d \tilde{v}$$

$$\begin{bmatrix} L_1 \sum_{j=0}^{N} a_{1j}(\tilde{W}_j - \tilde{W}_1) v_0 \\ \vdots \\ L_N \sum_{j=0}^{N} a_{Nj}(\tilde{W}_j - \tilde{W}_N) v_0 \end{bmatrix} = -L_d(H \otimes I_{p_{m0}})\tilde{W}_d(1_N \otimes v_0).$$

Thus, we obtain the following error system:

$$\dot{\tilde{S}} = -\mu_S \left(H \otimes I_q\right) \tilde{S} \tag{4.21a}$$

$$\dot{\tilde{F}} = -\mu_F \left(H \otimes I_p\right) \tilde{F} \tag{4.21b}$$

$$\dot{\tilde{W}} = -\mu_W \left(H \otimes I_{p_{m0}}\right) \tilde{W} \tag{4.21c}$$

$$\dot{\tilde{L}} = -\mu_L \left(H \otimes I_q\right) \tilde{L} \tag{4.21d}$$

$$\begin{aligned} \dot{\tilde{v}} = & (I_N \otimes S_0 - \mu_v(H \otimes (L_0 W_0)))\tilde{v} \\ & + (\tilde{S}_d - \mu_v \tilde{L}_d(H \otimes W_0) - \mu_v L_d(H \otimes I_{p_{m0}})\tilde{W}_d)\tilde{v} \\ & + (\tilde{S}_d - \mu_v L_d(H \otimes I_{p_{m0}})\tilde{W}_d)(1_N \otimes v_0). \end{aligned} \tag{4.21e}$$

Let

$$\begin{aligned} A &= I_N \otimes S_0 - \mu_v(H \otimes (L_0 W_0)) \\ M(t) &= \tilde{S}_d - \mu_v \tilde{L}_d(H \otimes W_0) - \mu_v L_d(H \otimes I_{p_{m0}})\tilde{W}_d \\ \psi(t) &= \tilde{S}_d - \mu_v L_d(H \otimes I_{p_{m0}})\tilde{W}_d. \end{aligned}$$

Then, system (4.21e) can be put into the following form:

$$\dot{\tilde{v}} = A\tilde{v} + M(t)\tilde{v} + \psi(t)(1_N \otimes v_0). \tag{4.22}$$

Now, we are ready to present the following result:

Theorem 4.5 *Given system (4.6) with the reference output (4.19) and a static communication graph $\bar{\mathcal{G}}$, under Assumptions 4.2 and 4.4, if $\mu_S, \mu_F, \mu_W > \bar{\lambda}_{S_0}\underline{\lambda}_H^{-1}$, $\mu_v > \frac{1}{2}\underline{\lambda}_H^{-1}$, then, for any $\mu_L > 0$ and any initial condition, the solution of system (4.20) exists for all $t \geq 0$ and satisfies*

$$\lim_{t\to\infty} \tilde{S}(t) = 0$$
$$\lim_{t\to\infty} \tilde{F}(t) = 0$$
$$\lim_{t\to\infty} \tilde{W}(t) = 0$$
$$\lim_{t\to\infty} \tilde{L}(t) = 0$$
$$\lim_{t\to\infty} \tilde{v}(t) = 0$$

exponentially. As a result,

$$\lim_{t\to\infty} \tilde{y}_i(t) = 0, \ i = 1, \ldots, N$$

exponentially.

Proof Note that (4.22) is in the form (2.16) with $x = \tilde{v}$, $\epsilon = 1$, and $F(t) = \psi(t)(1_N \otimes v_0)$. Under Assumption 4.2, by Lemma 2.1, $-H$ is Hurwitz. Therefore, $\tilde{S}(t)$, $\tilde{F}(t)$, $\tilde{W}(t)$, $\tilde{L}(t)$ and thus $\tilde{S}_d(t)$, $\tilde{W}_d(t)$, $\tilde{L}_d(t)$ tend to zero exponentially as t tends to infinity. Thus, $\|\tilde{S}_d(t)\| \leq f_1(t)e^{-\mu_S\underline{\lambda}_H t}$ and $\|\tilde{W}_d(t)\| \leq f_2(t)e^{-\mu_W\underline{\lambda}_H t}$ for some polynomial functions $f_1(t)$ and $f_2(t)$. By Theorem 3.2, under Assumptions 4.2 and 4.4, with $L_0 = P_0 W_0^\mathsf{T}$, A is Hurwitz if $\mu_v > \frac{1}{2}\underline{\lambda}_H^{-1}$. Also, $M(t)$ tends to zero exponentially since $\tilde{S}(t)$, $\tilde{W}(t)$, $\tilde{L}(t)$ and hence $\tilde{S}_d(t)$, $\tilde{W}_d(t)$, $\tilde{L}_d(t)$ tend to zero exponentially.

Note that

$$\|1_N \otimes v_0(t)\| \leq \|I_N \otimes e^{S_0 t}\| \cdot \|1_N \otimes v_0(0)\| \leq f_3(t)e^{\bar{\lambda}_{S_0} t}\|1_N \otimes v_0(0)\|$$

for some polynomial function $f_3(t)$.

Since $\mu_S\underline{\lambda}_H, \mu_W\underline{\lambda}_H > \bar{\lambda}_{S_0}$, there exist $\alpha_1 > 0$ and $\min\{\mu_S\underline{\lambda}_H, \mu_W\underline{\lambda}_H\} > \lambda_1 > \bar{\lambda}_{S_0}$ such that

$$\|\psi(t)\| \leq \alpha_1 e^{-\lambda_1 t}.$$

Then, by Lemma 2.6, for some $\alpha_2 > 0$ and $0 < \lambda_2 < (\lambda_1 - \bar{\lambda}_{S_0})$,

$$\begin{aligned}
&\|\psi(t)(1_N \otimes v_0(t))\| \\
&\leq \alpha_1 \|1_N \otimes v_0(0)\| f_3(t) e^{-(\lambda_1-\bar{\lambda}_{S_0})t} \\
&< \alpha_2 \|1_N \otimes v_0(0)\| e^{-\lambda_2 t}.
\end{aligned}$$

Therefore, by Lemma 2.7, $\lim_{t\to\infty} \tilde{v}(t) = 0$ exponentially. Noting that $\tilde{y}_i = F_i\tilde{v}_i + \tilde{F}_i v_0$ and $\mu_F \underline{\lambda}_H > \bar{\lambda}_{S_0}$, again by Lemma 2.6, $\tilde{F}_i(t)v_0(t)$ decays to zero exponentially, which further implies that $\lim_{t\to\infty} \tilde{y}_i(t) = 0$ exponentially. □

4.4.2 Case 2

Next, we consider the case where the switching communication graph $\bar{\mathcal{G}}_{\sigma(t)}$ is again a static one $\bar{\mathcal{G}}$ and $y_{m0} = v_0$. Since W_0 is an identity matrix, there is no need to introduce $L_i, i = 0, 1, \ldots, N$. Thus, the adaptive distributed observer (4.20) reduces to the following form:

$$\dot{S}_i = \mu_S \sum_{j=0}^{N} a_{ij}(S_j - S_i) \tag{4.23a}$$

$$\dot{F}_i = \mu_F \sum_{j=0}^{N} a_{ij}(F_j - F_i) \tag{4.23b}$$

$$\dot{v}_i = S_i v_i + \mu_v \sum_{j=0}^{N} a_{ij}(v_j - v_i) \tag{4.23c}$$

$$\hat{y}_i = F_i v_i. \tag{4.23d}$$

Simple calculation gives

$$\begin{aligned}
\dot{\tilde{v}}_i &= S_i v_i + \mu_v \sum_{j=0}^{N} a_{ij}(v_j - v_i) - S_0 v_0 \\
&= S_0 \tilde{v}_i + \tilde{S}_i \tilde{v}_i + \tilde{S}_i v_0 + \mu_v \sum_{j=0}^{N} a_{ij}(\tilde{v}_j - \tilde{v}_i).
\end{aligned}$$

As a result, the error system (4.21) reduces to the following form:

$$\dot{\tilde{S}} = -\mu_S \left(H \otimes I_q\right) \tilde{S} \tag{4.24a}$$

$$\dot{\tilde{F}} = -\mu_F \left(H \otimes I_p\right) \tilde{F} \tag{4.24b}$$

$$\dot{\tilde{v}} = \left(I_N \otimes S_0 - \mu_v (H \otimes I_q)\right) \tilde{v} + \tilde{S}_d \tilde{v} + \tilde{S}_d (1_N \otimes v_0). \tag{4.24c}$$

Let

$$\begin{aligned} A &= \big(I_N \otimes S_0 - \mu_v (H \otimes I_q)\big) \\ M(t) &= \tilde{S}_d \\ \psi(t) &= \tilde{S}_d. \end{aligned}$$

Then, system (4.24c) can be put into the following form:

$$\dot{\tilde{v}} = A\tilde{v} + M(t)\tilde{v} + \psi(t)(1_N \otimes v_0). \tag{4.25}$$

We have the following result:

Theorem 4.6 *Given system (4.6) with the reference output (4.19) and a static communication graph $\bar{\mathcal{G}}$, under Assumption 4.2, if $\mu_S, \mu_F, \mu_v > \bar{\lambda}_{S_0}\underline{\lambda}_H^{-1}$, then, for any initial condition, the solution of system (4.23) exists for all $t \geq 0$ and satisfies*

$$\begin{aligned} \lim_{t\to\infty} \tilde{S}(t) &= 0 \\ \lim_{t\to\infty} \tilde{F}(t) &= 0 \\ \lim_{t\to\infty} \tilde{v}(t) &= 0 \end{aligned}$$

exponentially. As a result,

$$\lim_{t\to\infty} \tilde{y}_i(t) = 0, \ i = 1, \ldots, N$$

exponentially.

Proof Note that (4.25) is in the form (2.16) with $x = \tilde{v}$, $\epsilon = 1$, and $F(t) = \psi(t)(1_N \otimes v_0)$. Under Assumption 4.2, by Lemma 2.1, $-H$ is Hurwitz. Therefore, $\tilde{F}(t)$, $\tilde{S}(t)$, and thus $\tilde{S}_d(t)$ tend to zero exponentially as t tends to infinity, and thus, $\|\tilde{S}_d(t)\| \leq f_1(t)e^{-\mu_S\underline{\lambda}_H t}$ and $\|\tilde{F}(t)\| \leq f_2(t)e^{-\mu_F\underline{\lambda}_H t}$ for some polynomial functions $f_1(t)$ and $f_2(t)$. Similar to the proof of Theorem 4.5, since $\mu_S\underline{\lambda}_H > \bar{\lambda}_{S_0}$, $\psi(t)(1_N \otimes v_0)$ decays to zero exponentially. Moreover, since $\mu_v\underline{\lambda}_H > \bar{\lambda}_{S_0}$, by Lemma 3.2, A is Hurwitz. As a result, by Lemma 2.7, $\lim_{t\to\infty} \tilde{v}_i(t) = 0$ exponentially. Noting that $\tilde{y}_i = F_i\tilde{v}_i + \tilde{F}_i v_0$ and $\mu_F\underline{\lambda}_H > \bar{\lambda}_{S_0}$, again by Lemma 2.6, $\tilde{F}_i v_0$ decays to zero exponentially, which further implies that $\lim_{t\to\infty} \tilde{y}_i(t) = 0$ exponentially. □

4.4.3 Case 3

We now turn to the case where the communication graph $\bar{\mathcal{G}}_{\sigma(t)}$ is switching. By Theorem 4.1, under Assumptions 4.1, 4.3, 4.4, and 4.6, there exists a constant matrix

$L_0 \in \mathbb{R}^{q\times p_{m0}}$ such that the origin of the error system (4.12) is exponentially stable. Let the adaptive distributed observer candidate be designed as follows:

$$\dot{S}_i = \mu_S \sum_{j=0}^{N} a_{ij}(t)(S_j - S_i) \tag{4.26a}$$

$$\dot{F}_i = \mu_F \sum_{j=0}^{N} a_{ij}(t)(F_j - F_i) \tag{4.26b}$$

$$\dot{W}_i = \mu_W \sum_{j=0}^{N} a_{ij}(t)(W_j - W_i) \tag{4.26c}$$

$$\dot{L}_i = \mu_L \sum_{j=0}^{N} a_{ij}(t)(L_j - L_i) \tag{4.26d}$$

$$\dot{v}_i = S_i v_i + \mu_v L_i \sum_{j=0}^{N} a_{ij}(t)(W_j v_j - W_i v_i) \tag{4.26e}$$

$$\hat{y}_i = F_i v_i. \tag{4.26f}$$

Similar to the derivation of (4.21), we have the following error system:

$$\dot{\tilde{S}} = -\mu_S \left(H_{\sigma(t)} \otimes I_q\right) \tilde{S} \tag{4.27a}$$

$$\dot{\tilde{F}} = -\mu_F \left(H_{\sigma(t)} \otimes I_p\right) \tilde{F} \tag{4.27b}$$

$$\dot{\tilde{W}} = -\mu_W \left(H_{\sigma(t)} \otimes I_{p_{m0}}\right) \tilde{W} \tag{4.27c}$$

$$\dot{\tilde{L}} = -\mu_L \left(H_{\sigma(t)} \otimes I_q\right) \tilde{L} \tag{4.27d}$$

$$\begin{aligned} \dot{\tilde{v}} &= (I_N \otimes S_0 - \mu_v(H_{\sigma(t)} \otimes (L_0 W_0)))\tilde{v} \\ &\quad + (\tilde{S}_d - \mu_v \tilde{L}_d(H_{\sigma(t)} \otimes W_0) - \mu_v L_d(H_{\sigma(t)} \otimes I_{p_{m0}})\tilde{W}_d)\tilde{v} \\ &\quad + (\tilde{S}_d - \mu_v L_d(H_{\sigma(t)} \otimes I_{p_{m0}})\tilde{W}_d)(1_N \otimes v_0). \end{aligned} \tag{4.27e}$$

Let

$$\begin{aligned} A(t) &= I_N \otimes S_0 - \mu_v(H_{\sigma(t)} \otimes (L_0 W_0)) \\ M(t) &= \tilde{S}_d - \mu_v \tilde{L}_d(H_{\sigma(t)} \otimes W_0) - \mu_v L_d(H_{\sigma(t)} \otimes I_{p_{m0}})\tilde{W}_d \\ \psi(t) &= \tilde{S}_d - \mu_v L_d(H_{\sigma(t)} \otimes I_{p_{m0}})\tilde{W}_d. \end{aligned}$$

Then, system (4.27e) can be put into the following form:

$$\dot{\tilde{v}} = A(t)\tilde{v} + M(t)\tilde{v} + \psi(t)(1_N \otimes v_0). \tag{4.28}$$

Now, we are ready to present the following result:

Theorem 4.7 *Given system (4.6) with the reference output (4.19) and a switching communication graph $\bar{\mathcal{G}}_{\sigma(t)}$, under Assumptions 4.1, 4.3, 4.4, and 4.6, there exists a constant matrix $L_0 \in \mathbb{R}^{q \times p_{m0}}$, such that, for any $\mu_S, \mu_F, \mu_W, \mu_L, \mu_v > 0$, and any initial condition, the solution of system (4.26) exists for all $t \geq 0$ and is such that*

$$\lim_{t\to\infty} \tilde{S}(t) = 0$$
$$\lim_{t\to\infty} \tilde{F}(t) = 0$$
$$\lim_{t\to\infty} \tilde{W}(t) = 0$$
$$\lim_{t\to\infty} \tilde{L}(t) = 0$$
$$\lim_{t\to\infty} \tilde{v}(t) = 0$$

exponentially. As a result,

$$\lim_{t\to\infty} \tilde{y}_i(t) = 0, \ i = 1, \ldots, N$$

exponentially.

Proof Note that (4.28) is in the form (2.17) with $x = \tilde{v}$ and $F(t) = \psi(t)(1_N \otimes v_0)$. By Corollary 2.3, under Assumption 4.1, for any $\mu_S, \mu_F, \mu_W, \mu_L > 0$, systems (4.27a) to (4.27d) are all exponentially stable. Thus, $\tilde{S}(t)$, $\tilde{F}(t)$, $\tilde{W}(t)$, and $\tilde{L}(t)$ tend to zero exponentially, which implies $M(t)$ and $\psi(t)$ tend to zero exponentially, and, under the additional Assumption 4.6, similar to the proof of Theorem 4.5, $\psi(t)(1_N \otimes v_0(t))$ tends to zero exponentially, too. By Theorem 4.1, under Assumptions 4.1, 4.3, 4.4, and 4.6, there exists a constant matrix $L_0 \in \mathbb{R}^{q \times p_{m0}}$, which is described in Remark 4.3, such that the following system:

$$\dot{\tilde{v}} = A(t)\tilde{v}$$

is exponentially stable. Thus, by Lemma 2.8, $\lim_{t\to\infty} \tilde{v}(t) = 0$ exponentially. As a result, $\lim_{t\to\infty} \tilde{y}_i(t) = 0$ exponentially. □

4.4.4 Case 4

In this case, we consider the special case where $y_{m0} = v_0$. As in Case 2, since W_0 is an identity matrix, there is no need to introduce L_i, $i = 0, 1, \ldots, N$. Thus, the adaptive distributed observer (4.26) reduces to the following form:

$$\dot{S}_i = \mu_S \sum_{j=0}^{N} a_{ij}(t)(S_j - S_i) \tag{4.29a}$$

$$\dot{F}_i = \mu_F \sum_{j=0}^{N} a_{ij}(t)(F_j - F_i) \tag{4.29b}$$

$$\dot{v}_i = S_i v_i + \mu_v \sum_{j=0}^{N} a_{ij}(t)(v_j - v_i) \tag{4.29c}$$

$$\hat{y}_i = F_i v_i \tag{4.29d}$$

and the error system (4.27) reduces to the following form:

$$\dot{\tilde{S}} = -\mu_S \left(H_{\sigma(t)} \otimes I_q\right) \tilde{S} \tag{4.30a}$$

$$\dot{\tilde{F}} = -\mu_F \left(H_{\sigma(t)} \otimes I_p\right) \tilde{F} \tag{4.30b}$$

$$\dot{\tilde{v}} = \left(I_N \otimes S_0 - \mu_v(H_{\sigma(t)} \otimes I_q)\right) \tilde{v} + \tilde{S}_d \tilde{v} + \tilde{S}_d(1_N \otimes v_0). \tag{4.30c}$$

Let $A(t) = I_N \otimes S_0 - \mu_v(H_{\sigma(t)} \otimes I_q)$ and $M(t) = \psi(t) = \tilde{S}_d(t)$. Then, system (4.30) can also be put in the form of (4.28). We have the following result:

Theorem 4.8 *Given system (4.6) with the reference output (4.19) and a switching communication graph $\bar{\mathcal{G}}_{\sigma(t)}$, under Assumptions 4.1 and 4.5, for any $\mu_S, \mu_F, \mu_v > 0$, and any initial condition, the solution of system (4.29) exists for all $t \geq 0$ and satisfies*

$$\lim_{t\to\infty} \tilde{S}(t) = 0$$
$$\lim_{t\to\infty} \tilde{F}(t) = 0$$
$$\lim_{t\to\infty} \tilde{v}(t) = 0$$

exponentially. As a result,

$$\lim_{t\to\infty} \tilde{y}_i(t) = 0, \ i = 1, \ldots, N$$

exponentially.

Proof From the proof of Theorem 4.7, it is clear that, under Assumption 4.1, for any $\mu_S, \mu_F > 0$, $\lim_{t\to\infty} \tilde{S}(t) = 0$ and $\lim_{t\to\infty} \tilde{F}(t) = 0$ exponentially. Thus, for any $\mu_S > 0$, $\lim_{t\to\infty} \tilde{S}_d(t) = 0$ exponentially. By Theorem 4.2, under Assumptions 4.1 and 4.5, for any $\mu_v > 0$, the following system:

$$\dot{\tilde{v}} = (I_N \otimes S_0 - \mu_v(H_{\sigma(t)} \otimes I_q))\tilde{v}$$

is exponentially stable.

Under Assumption 4.5, similar to the proof of Theorem 4.5, $\lim_{t\to\infty}\psi(t)(1_N \otimes v_0(t)) = 0$ exponentially. Thus, by Lemma 2.8, $\lim_{t\to\infty}\tilde{v}(t) = 0$ exponentially, and hence, $\lim_{t\to\infty}\tilde{y}_i(t) = 0$ exponentially. □

Remark 4.4 In many cases, the role of the matrices F_0 and W_0 is to select the elements of v_0. In such cases, the elements of F_0 and W_0 are either 1 or 0, and we can assume F_0 and W_0 are known by every follower. Thus, there is no need to estimate F_0 and W_0. As a result, the adaptive distributed observer (4.20) is simplified to the following:

$$\dot{S}_i = \mu_S \sum_{j=0}^{N} a_{ij}(S_j - S_i) \tag{4.31a}$$

$$\dot{L}_i = \mu_L \sum_{j=0}^{N} a_{ij}(L_j - L_i) \tag{4.31b}$$

$$\dot{v}_i = S_i v_i + \mu_v L_i \sum_{j=0}^{N} a_{ij}(W_0 v_j - W_0 v_i) \tag{4.31c}$$

$$\hat{y}_i = F_0 v_i, \tag{4.31d}$$

the adaptive distributed observer (4.23) is simplified to the following:

$$\dot{S}_i = \mu_S \sum_{j=0}^{N} a_{ij}(S_j - S_i) \tag{4.32a}$$

$$\dot{v}_i = S_i v_i + \mu_v \sum_{j=0}^{N} a_{ij}(v_j - v_i) \tag{4.32b}$$

$$\hat{y}_i = F_0 v_i, \tag{4.32c}$$

the adaptive distributed observer (4.26) is simplified to the following:

$$\dot{S}_i = \mu_S \sum_{j=0}^{N} a_{ij}(t)(S_j - S_i) \tag{4.33a}$$

$$\dot{L}_i = \mu_L \sum_{j=0}^{N} a_{ij}(t)(L_j - L_i) \tag{4.33b}$$

$$\dot{v}_i = S_i v_i + \mu_v L_i \sum_{j=0}^{N} a_{ij}(t)(W_0 v_j - W_0 v_i) \tag{4.33c}$$

$$\hat{y}_i = F_0 v_i, \tag{4.33d}$$

and the adaptive distributed observer (4.29) is simplified to the following:

$$\dot{S}_i = \mu_S \sum_{j=0}^{N} a_{ij}(t)(S_j - S_i) \tag{4.34a}$$

$$\dot{v}_i = S_i v_i + \mu_v \sum_{j=0}^{N} a_{ij}(t)(v_j - v_i) \tag{4.34b}$$

$$\hat{y}_i = F_0 v_i. \tag{4.34c}$$

4.5 Adaptive Distributed Observer for an Uncertain Leader System

Both the distributed observer and the adaptive distributed observer assume that the dynamics of the leader system are known by at least some follower subsystems. In practice, a leader system may contain some unknown parameters, and hence, none of the followers knows the exact dynamics of the leader system. In this section, we consider the following uncertain leader system:

$$\dot{v}_0 = \mathbf{S}(\omega_0)v_0 \tag{4.35a}$$

$$y_{m0} = v_0 \tag{4.35b}$$

where $v_0 \in \mathbb{R}^q$, $\omega_0 = \text{col}(\omega_{01}, \ldots, \omega_{0l}) \in \mathbb{R}^l$, and $\mathbf{S}(\omega_0) \in \mathbb{R}^{q\times q}$. The system (4.35) is in the form of system (1.12) with $w_0 = \omega_0$ and $W_0(w_0) = I_q$.

We make the following assumptions for the static communication graph $\bar{\mathcal{G}}$ and the matrix $\mathbf{S}(\omega_0)$.

Assumption 4.7 The static communication graph $\bar{\mathcal{G}}$ contains a spanning tree with the node 0 as the root, and $\mathcal{G}$ is undirected.

Assumption 4.8 For all $\omega_0 \in \mathbb{R}^l$, the matrix $\mathbf{S}(\omega_0)$ is neutrally stable and nonsingular.

Under Assumption 4.8, without loss of generality, we can assume that

$$\mathbf{S}(\omega_0) = \mathcal{D}(\omega_{01}, \ldots, \omega_{0l}) \otimes a, \ a = \begin{bmatrix} 0 & 1 \\ -1 & 0 \end{bmatrix}$$

which indicates that $q = 2l$.

Note that a neutrally stable matrix allows semi-simple eigenvalues at the origin. But, since the eigenvalue at the origin is known and can be dealt with by the distributed observer, we exclude this case by requiring that the matrix $\mathbf{S}(\omega_0)$ be nonsingular in Assumption 4.8.

It can be seen that the state of the leader system (4.35) takes the following form:

$$
v_0(t) = \begin{bmatrix} C_1 \sin(\omega_{01} t + \phi_1) \\ C_1 \cos(\omega_{01} t + \phi_1) \\ \vdots \\ C_l \sin(\omega_{0l} t + \phi_l) \\ C_l \cos(\omega_{0l} t + \phi_l) \end{bmatrix} \tag{4.36}
$$

where, for $i = 1, \ldots, l$, $C_i = \sqrt{v_{0,2i-1}^2(0) + v_{0,2i}^2(0)}$, ϕ_i is such that $v_{0,2i-1}(0) = C_i \sin(\phi_i)$ and $v_{0,2i}(0) = C_i \cos(\phi_i)$, and they are all unknown. Reference signals in the form of (4.36) are common in practice since many periodic functions can be approximated by a trigonometric polynomial function. In what follows, we will establish an adaptive distributed observer for the uncertain leader system (4.35).

To begin with, define the following mapping: for any $x = \text{col}(x_1, \ldots, x_{2l}) \in \mathbb{R}^{2l}$, let $\phi : \mathbb{R}^{2l} \mapsto \mathbb{R}^{l \times 2l}$ be such that

$$
\phi(x) = \begin{bmatrix} -x_2 & x_1 & 0 & 0 & \cdots & 0 & 0 \\ 0 & 0 & -x_4 & x_3 & \cdots & 0 & 0 \\ \vdots & \vdots & \vdots & \vdots & \ddots & \vdots & \vdots \\ 0 & 0 & 0 & 0 & \cdots & -x_{2l} & x_{2l-1} \end{bmatrix}. \tag{4.37}
$$

Then, for $i = 1, \ldots, N$, we propose the following distributed dynamic compensator:

$$
\dot{v}_i = \mathbf{S}(\omega_i) v_i + \mu_v \sum_{j=0}^{N} a_{ij}(v_j - v_i) \tag{4.38a}
$$

$$
\dot{\omega}_i = \mu_\omega \phi \left(\sum_{j=0}^{N} a_{ij}(v_j - v_i) \right) v_i \tag{4.38b}
$$

where $v_i \in \mathbb{R}^{2l}$, $\omega_i \in \mathbb{R}^l$, and $\mu_v, \mu_\omega > 0$.

What makes (4.38) interesting is that it is independent of the unknown vector ω_0.

Before going on, we present the following proposition:

Proposition 4.1 *For any $z \in \mathbb{R}^l$ and $x, y \in \mathbb{R}^{2l}$,*

$$
\phi(x) y = -\phi(y) x \tag{4.39a}
$$

$$
x^\mathsf{T} \mathbf{S}(z) = z^\mathsf{T} \phi(x) \tag{4.39b}
$$

$$
\mathbf{S}(z) x = -\phi^\mathsf{T}(x) z. \tag{4.39c}
$$

Proof Let $z = \text{col}(z_1, \ldots, z_l)$, $x = \text{col}(x_1, \ldots, x_{2l})$, and $y = \text{col}(y_1, \ldots, y_{2l})$.

From (4.37), we have

$$\phi(x)y = \begin{bmatrix} -x_2 & x_1 & \dots & 0 & 0 \\ \vdots & \vdots & \ddots & \vdots & \vdots \\ 0 & 0 & \dots & -x_{2l} & x_{2l-1} \end{bmatrix} \begin{bmatrix} y_1 \\ \vdots \\ y_{2l} \end{bmatrix}$$
$$= \begin{bmatrix} x_1 y_2 - y_1 x_2 \\ \vdots \\ y_{2l} x_{2l-1} - x_{2l} y_{2l-1} \end{bmatrix}$$

and

$$\phi(y)x = \begin{bmatrix} -y_2 & y_1 & \dots & 0 & 0 \\ \vdots & \vdots & \ddots & \vdots & \vdots \\ 0 & 0 & \dots & -y_{2l} & y_{2l-1} \end{bmatrix} \begin{bmatrix} x_1 \\ \vdots \\ x_{2l} \end{bmatrix}$$
$$= \begin{bmatrix} y_1 x_2 - x_1 y_2 \\ \vdots \\ x_{2l} y_{2l-1} - y_{2l} x_{2l-1} \end{bmatrix}.$$

Hence, (4.39a) holds.

To show (4.39b), direct calculation gives

$$x^{\mathsf{T}} \mathbf{S}(z) = \left[-z_1 x_2, z_1 x_1, \dots, -z_l x_{2l}, z_l x_{2l-1} \right]$$

and

$$z^{\mathsf{T}} \phi(x) = [z_1, \dots, z_l] \begin{bmatrix} -x_2 & x_1 & \dots & 0 & 0 \\ \vdots & \vdots & \ddots & \vdots & \vdots \\ 0 & 0 & \dots & -x_{2l} & x_{2l-1} \end{bmatrix}$$
$$= \left[-z_1 x_2, z_1 x_1, \dots, -z_l x_{2l}, z_l x_{2l-1} \right].$$

Hence, (4.39b) holds.

To show (4.39c), using (4.39b) gives

$$\mathbf{S}(z) x = \left(x^{\mathsf{T}} \mathbf{S}^{\mathsf{T}}(z) \right)^{\mathsf{T}} = \left(x^{\mathsf{T}} \mathbf{S}(-z) \right)^{\mathsf{T}} = - \left(x^{\mathsf{T}} \mathbf{S}(z) \right)^{\mathsf{T}} = - \left(z^{\mathsf{T}} \phi(x) \right)^{\mathsf{T}} = -\phi^{\mathsf{T}}(x) z.$$

□

For $i = 1, \dots, N$, let $\tilde{v}_i = v_i - v_0$, $\tilde{\omega}_i = \omega_i - \omega_0$, and

$$e_{vi} = \sum_{j=0}^{N} a_{ij} (v_i - v_j). \tag{4.40}$$

Then, it follows from (4.38), (4.39), and (4.40) that

$$\dot{\tilde{v}}_i = \mathbf{S}(\omega_0)\tilde{v}_i - \mu_v e_{vi} + \mathbf{S}(\tilde{\omega}_i)v_i$$
$$\dot{\tilde{\omega}}_i = -\mu_\omega \phi(e_{vi})v_i = \mu_\omega \phi(v_i)e_{vi}.$$

Let $v = \mathrm{col}(v_1, \ldots, v_N)$, $\tilde{v} = \mathrm{col}(\tilde{v}_1, \ldots, \tilde{v}_N)$, $\tilde{\omega} = \mathrm{col}(\tilde{\omega}_1, \ldots, \tilde{\omega}_N)$, $e_v = \mathrm{col}(e_{v1}, \ldots, e_{vN})$, and $\mathbf{S_d}(\tilde{\omega}) = \mathscr{D}(\mathbf{S}(\tilde{\omega}_1), \ldots, \mathbf{S}(\tilde{\omega}_N))$. Then, we have

$$e_v = (H \otimes I_{2l})\tilde{v} \tag{4.41}$$

and

$$\dot{\tilde{v}} = (I_N \otimes \mathbf{S}(\omega_0) - \mu_v (H \otimes I_{2l}))\,\tilde{v} + \mathbf{S_d}(\tilde{\omega})v \tag{4.42a}$$
$$\dot{\tilde{\omega}} = \mu_\omega \phi_d(v)e_v \tag{4.42b}$$

where $\phi_d(v) = \mathscr{D}(\phi(v_1), \ldots, \phi(v_N))$.

Now we present the following result:

Theorem 4.9 *Given system (4.35) and the static communication graph $\bar{\mathscr{G}}$, under Assumptions 4.7 and 4.8, for any $v_0(0)$, $v_i(0)$, and $\omega_i(0)$, $i = 1, \ldots, N$, and any $\mu_v, \mu_\omega > 0$, the solution $v_i(t)$ and $\omega_i(t)$ of (4.38) exist and are uniformly bounded over $[0, \infty)$ and are such that $\lim_{t\to\infty} \tilde{v}_i(t) = 0$, $\lim_{t\to\infty} \dot{\tilde{\omega}}_i(t) = 0$, and $\lim_{t\to\infty} \mathbf{S_d}(\tilde{\omega}(t))\, v(t) = 0$.*

Proof Let

$$V = \frac{1}{2}\left(\tilde{v}^\mathsf{T}(H \otimes I_{2l})\,\tilde{v} + \mu_\omega^{-1}\tilde{\omega}^\mathsf{T}\tilde{\omega}\right).$$

Since, under Assumption 4.7, the matrix H is symmetric and positive definite, V is positive definite. By (4.42), we have

$$\dot{V} = \tilde{v}^\mathsf{T}(H \otimes \mathbf{S}(\omega_0))\,\tilde{v} - \mu_v \tilde{v}^\mathsf{T}\left(H^2 \otimes I_{2l}\right)\tilde{v} + \tilde{v}^\mathsf{T}(H \otimes I_{2l})\,\mathbf{S_d}(\tilde{\omega})v + \mu_\omega^{-1}\tilde{\omega}^\mathsf{T}\dot{\tilde{\omega}}.$$

Since $\mathbf{S}(\omega_0)$ is anti-symmetric and H is symmetric, $H \otimes \mathbf{S}(\omega_0)$ is anti-symmetric. Thus, we have $\tilde{v}^\mathsf{T}(H \otimes \mathbf{S}(\omega_0))\,\tilde{v} = 0$. Then $\dot{V}$ reduces to

$$\dot{V} = -\mu_v \tilde{v}^\mathsf{T}\left(H^2 \otimes I_{2l}\right)\tilde{v} + e_v^\mathsf{T}\mathbf{S_d}(\tilde{\omega})v + \mu_\omega^{-1}\tilde{\omega}^\mathsf{T}\dot{\tilde{\omega}}. \tag{4.43}$$

By (4.39b), we have, for $i = 1, \ldots, N$,

$$e_{vi}^\mathsf{T}\mathbf{S}(\tilde{\omega}_i)\,v_i = -v_i^\mathsf{T}\mathbf{S}(\tilde{\omega}_i)e_{vi} = -\tilde{\omega}_i^\mathsf{T}\phi(v_i)e_{vi}$$

which can be put in the following compact form:

$$e_v^\mathsf{T}\mathbf{S_d}(\tilde{\omega})v = -\tilde{\omega}^\mathsf{T}\phi_d(v)e_v.$$

Using the above equality in (4.43) gives

$$\begin{aligned}\dot{V} &= -\mu_v \tilde{v}^\mathsf{T}\left(H^2 \otimes I_{2l}\right)\tilde{v} - \tilde{\omega}^\mathsf{T}\phi_d(v)e_v + \mu_\omega^{-1}\tilde{\omega}^\mathsf{T}\dot{\tilde{\omega}} \\ &= -\mu_v \tilde{v}^\mathsf{T}\left(H^2 \otimes I_{2l}\right)\tilde{v} + \tilde{\omega}^\mathsf{T}\left(-\phi_d(v)e_v + \mu_\omega^{-1}\dot{\tilde{\omega}}\right).\end{aligned}$$

Using (4.42b) gives

$$\dot{V} = -\mu_v \tilde{v}^\mathsf{T}\left(H^2 \otimes I_{2l}\right)\tilde{v} \leq 0.$$

Since V is positive definite and $\dot{V}$ is negative semidefinite, $V(t)$ is uniformly bounded over $[0, \infty)$, which means $\tilde{v}(t)$ and $\tilde{\omega}(t)$ are uniformly bounded over $[0, \infty)$. From (4.42a), $\dot{\tilde{v}}(t)$ is uniformly bounded over $[0, \infty)$, which implies $\ddot{V}(t)$ is uniformly bounded over $[0, \infty)$. By Barbalat's Lemma, i.e., Lemma 2.9, we have $\lim_{t\to\infty} \dot{V}(t) = 0$, which further implies $\lim_{t\to\infty} \tilde{v}_i(t) = 0$. Thus, by (4.41), we have $\lim_{t\to\infty} e_v(t) = 0$, which together with (4.42b) yields $\lim_{t\to\infty} \dot{\tilde{\omega}}_i(t) = 0$. Finally, we have

$$\ddot{\tilde{v}} = \left(I_N \otimes \mathbf{S}\left(\omega_0\right) - \mu_v(H \otimes I_{2l})\right)\dot{\tilde{v}} + \mathbf{S_d}(\dot{\tilde{\omega}})v + \mathbf{S_d}(\tilde{\omega})\dot{v}.$$

We have shown both $\tilde{v}(t)$ and $\tilde{\omega}(t)$ are uniformly bounded over $[0, \infty)$. Since $v_0(t)$ is uniformly bounded over $[0, \infty)$ and ω_0 is constant, $v_i(t)$ and $\omega_i(t)$ are also uniformly bounded over $[0, \infty)$. As a result, by (4.38a), $\dot{v}_i(t)$ is uniformly bounded over $[0, \infty)$. From (4.42), $\dot{\tilde{v}}(t)$ and $\dot{\tilde{\omega}}(t)$ are also uniformly bounded over $[0, \infty)$. Thus, $\ddot{\tilde{v}}(t)$ is uniformly bounded over $[0, \infty)$. By Barbalat's Lemma again, we have $\lim_{t\to\infty} \dot{\tilde{v}}(t) = 0$, which together with (4.42a) implies $\lim_{t\to\infty} \mathbf{S_d}\left(\tilde{\omega}(t)\right) v(t) = 0$.

□

Since Theorem 4.9 does not guarantee $\lim_{t\to\infty} \tilde{\omega}(t) = 0$, in the following, we will show that it is possible to make $\lim_{t\to\infty} \tilde{\omega}(t) = 0$ if the signal $\phi(v(t))$ is persistently exciting.[1] For this purpose, we need to establish one more lemma as follows:

Lemma 4.1 *Suppose* $f(t), g(t) : [t_0, \infty) \mapsto \mathbb{R}^{n\times m}$ *are piecewise continuous in* t *and uniformly bounded over* $[t_0, \infty)$, *and* $\lim_{t\to\infty}(g(t) - f(t)) = 0$. *Then* $f(t)$ *is persistently exciting if and only if* $g(t)$ *is.*

Proof We first show that, if $f(t)$ is persistently exciting, then $g(t)$ is also persistently exciting. Since $f(t)$ is persistently exciting, by Definition A.3, there exist positive constants k and T_f such that

$$\int_t^{t+T_f} f(\tau)f^\mathsf{T}(\tau)d\tau \geq kI_n,\ \forall\, t \geq t_0. \tag{4.44}$$

Since $f(t)$ and $g(t)$ are uniformly bounded over $[t_0, \infty)$, and $\lim_{t\to\infty}(f(t) - g(t)) = 0$, we have

[1] See Definition A.3 for the definition of persistently exciting.

$$\lim_{t\to\infty}\left(f(t)f^{\mathsf{T}}(t)-g(t)g^{\mathsf{T}}(t)\right)=0. \tag{4.45}$$

Thus, for any given k and T_f, there exists a positive constant $T_0 \geq t_0$ such that, for all $t \geq T_0$,

$$-\frac{k}{2T_f}I_n \leq g(t)g^{\mathsf{T}}(t)-f(t)f^{\mathsf{T}}(t) \leq \frac{k}{2T_f}I_n. \tag{4.46}$$

Then, for all $t \geq T_0$, we have

$$\begin{aligned}\int_t^{t+T_f} g(\tau)g^{\mathsf{T}}(\tau)d\tau &\geq \int_t^{t+T_f}\left(f(\tau)f^{\mathsf{T}}(\tau)-\frac{k}{2T_f}I_n\right)d\tau\\ &= \int_t^{t+T_f} f(\tau)f^{\mathsf{T}}(\tau)d\tau-\frac{k}{2}I_n\\ &\geq \frac{k}{2}I_n\end{aligned} \tag{4.47}$$

where the last inequality follows from (4.44).

Let $T_g = T_0 + T_f$. Then, for all $t \geq t_0$, we have

$$\begin{aligned}&\int_t^{t+T_g} g(\tau)g^{\mathsf{T}}(\tau)d\tau-\frac{k}{2}I_n\\ &=\int_t^{t+T_0} g(\tau)g^{\mathsf{T}}(\tau)d\tau+\int_{t+T_0}^{t+T_0+T_f} g(\tau)g^{\mathsf{T}}(\tau)d\tau-\frac{k}{2}I_n.\end{aligned} \tag{4.48}$$

Since $g(t)g^{\mathsf{T}}(t)$ is symmetric and positive semidefinite for any $t \geq t_0$, $\int_t^{t+T_0} g(\tau)\cdot g^{\mathsf{T}}(\tau)d\tau$ is symmetric and positive semidefinite for any $T_0 \geq 0$ and $t \geq t_0$. Then, using (4.47) in (4.48) gives

$$\int_t^{t+T_g} g(\tau)g^{\mathsf{T}}(\tau)d\tau \geq \frac{k}{2}I_n,\ \forall\, t \geq t_0. \tag{4.49}$$

Thus, $g(t)$ is persistently exciting. Interchanging the functions $f(t)$ and $g(t)$ in the above steps shows that $f(t)$ is persistently exciting if $g(t)$ is. □

Theorem 4.10 *Under Assumptions 4.7 and 4.8, if* $col(v_{0,2i-1}(0), v_{0,2i}(0)) \neq 0$ *for all* $i = 1, \ldots, l$, *then, for any* $\mu_v, \mu_\omega > 0$, $\lim_{t\to\infty}\tilde{\omega}(t)=0$ *and* $\lim_{t\to\infty}\tilde{v}(t)=0$ *exponentially.*

Proof First note that, by Eq. (4.41) and Proposition 4.1, we can rewrite Eqs. (4.42a) and (4.42b) into the following form:

$$\dot{\tilde{v}} = \left(I_N \otimes \mathbf{S}(\omega_0) - \mu_v(H \otimes I_{2l})\right)\tilde{v} - \phi_d^{\mathsf{T}}(v)\tilde{\omega} \tag{4.50a}$$

$$\dot{\tilde{\omega}} = \mu_\omega \phi_d(v)(H \otimes I_{2l})\tilde{v}. \tag{4.50b}$$

Let $x = \tilde{v}$ and $z = \tilde{\omega}$. Then (4.50) is in the following form:

$$\dot{x} = Ax + \Omega^{\mathsf{T}}(t)z \tag{4.51a}$$
$$\dot{z} = -\Lambda\Omega(t)Px \tag{4.51b}$$

with

$$A = (I_N \otimes \mathbf{S}(\omega_0) - \mu_v(H \otimes I_{2l})), \ \Omega(t) = -\phi_d(v(t)), \ \Lambda = \mu_\omega I_{Nl}, \ P = H \otimes I_{2l}.$$

By Lemma A.5, the origin of system (4.51) is exponentially stable if
(i) $A \in \mathbb{R}^{n\times n}$ is Hurwitz, and $P \in \mathbb{R}^{n\times n}$ is a symmetric and positive definite matrix satisfying

$$A^{\mathsf{T}}P + PA = -Q$$

for some symmetric and positive definite matrix Q, and
(ii) $\|\Omega(t)\|$ and $\|\dot{\Omega}(t)\|$ are uniformly bounded over $[0, \infty)$, and $\Omega(t)$ is persistently exciting.

We will verify that (4.50) satisfies the above two conditions. Since H is symmetric and positive definite, and all the eigenvalues of $\mathbf{S}(\omega_0)$ have zero real parts, by Proposition A.2, A is Hurwitz for any $\mu_v > 0$. Next, it can be verified that

$$A^{\mathsf{T}}P + PA = -2\mu_v\left(H^2 \otimes I_{2l}\right) = -Q < 0.$$

It remains to verify that $\|\phi_d(v(t))\|$ and $\|\phi_d(\dot{v}(t))\|$ are uniformly bounded over $[0, \infty)$ and $\phi_d(v(t))$ is persistently exciting. Since $v_0(t)$ and $\dot{v}_0(t)$ are uniformly bounded over $[0, \infty)$, by Theorem 4.9, $v_i(t)$ and $\dot{v}_i(t)$ are also uniformly bounded over $[0, \infty)$. Thus, $\|\phi_d(v(t))\|$ and $\|\phi_d(\dot{v}(t))\|$ are uniformly bounded over $[0, \infty)$.

By Theorem 4.9, we have $\lim_{t\to\infty}(v_i(t) - v_0(t)) = 0$, which implies

$$\lim_{t\to\infty}(\phi(v_i(t)) - \phi(v_0(t))) = 0.$$

Thus, by Lemma 4.1, we only need to show that $\phi(v_0(t))$ is persistently exciting. In fact, from (4.36), we have

$$\begin{aligned}\phi(v_0(t))\,\phi^{\mathsf{T}}(v_0(t)) &= \begin{bmatrix} v_{01}(0)^2 + v_{02}(0)^2 & & \\ & \ddots & \\ & & v_{0,2l-1}(0)^2 + v_{0,2l}(0)^2 \end{bmatrix} \\ &= \begin{bmatrix} C_1^2 & & \\ & \ddots & \\ & & C_l^2 \end{bmatrix}.\end{aligned}$$

Since, for $i = 1, \ldots, l$, $\text{col}(v_{0,2i-1}(0), v_{0,2i}(0)) \neq 0$, $\phi(v_0(t))\phi^{\mathsf{T}}(v_0(t))$ is a constant positive definite matrix. Thus, $\phi(v_0(t))$ is persistently exciting. Then the proof is completed by resorting to Lemma A.5. □

4.6 Numerical Examples

In this section, we will illustrate the design methods for various distributed observers for a leader system of the form (4.6) with $N = 4$ using numerical examples.

Both static and switching communication graphs will be considered. The static communication graphs are shown in Fig. 4.3, and it can be seen that $\bar{\mathscr{G}}_a$ satisfies Assumption 4.2 and $\bar{\mathscr{G}}_b$ satisfies Assumption 4.7. For $\bar{\mathscr{G}}_a$, it can be calculated that

$$H = \begin{bmatrix} 2 & 0 & -1 & 0 \\ 0 & 1 & 0 & -1 \\ -1 & -1 & 2 & 0 \\ 0 & 0 & -1 & 1 \end{bmatrix}$$

and $\underline{\lambda}_H = 0.1332$. Thus, $\underline{\lambda}_H^{-1} = 7.5075$.

Figure 4.4 shows a switching communication graph $\bar{\mathscr{G}}_{\sigma(t)}$ with the switching signal defined as follows:

$$\sigma(t) = \begin{cases} 1 \text{ if } 2s \leq t < 2s + 0.5 \\ 2 \text{ if } 2s + 0.5 \leq t < 2s + 1 \\ 3 \text{ if } 2s + 1 \leq t < 2s + 1.5 \\ 4 \text{ if } 2s + 1.5 \leq t < 2s + 2 \end{cases}$$

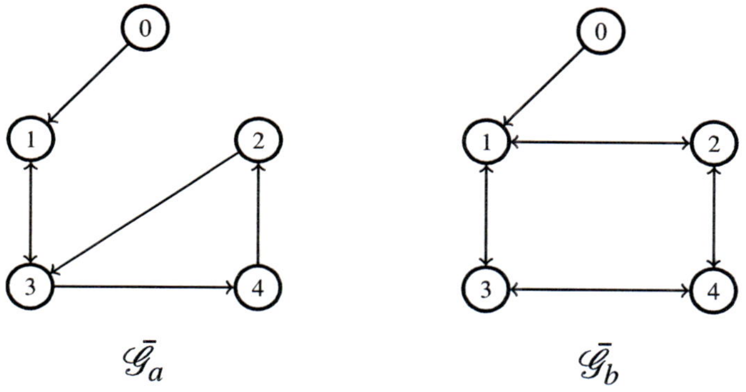

Fig. 4.3 Two static communication graphs, where $\bar{\mathscr{G}}_a$ satisfies Assumption 4.2 and $\bar{\mathscr{G}}_b$ satisfies Assumption 4.7

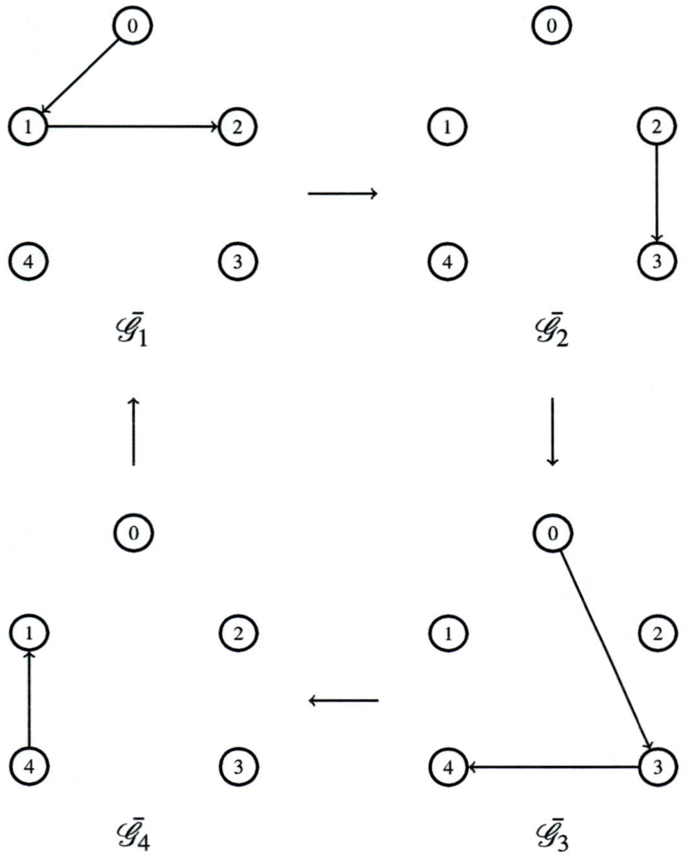

Fig. 4.4 The switching communication graph $\bar{\mathcal{G}}_{\sigma(t)}$ satisfying Assumption 4.1

where $s = 0, 1, 2, \ldots$. It is noted that none of the four graphs $\bar{\mathcal{G}}_i$, $i = 1, 2, 3, 4$, in Fig. 4.4 is connected. However, let $\{t_k = 0.5k : k = 0, 1, 2, \ldots\}$. Then, $\bigcup_{i=k}^{k+3} \bar{\mathcal{G}}_{\sigma(t_i)}$ contains a spanning tree with the node 0 as the root for any $k = 0, 1, 2, \ldots$. Therefore, $\bar{\mathcal{G}}_{\sigma(t)}$ satisfies Assumption 4.1.

Figure 4.5 shows a switching communication graph $\bar{\mathcal{G}}_{\sigma(t)}$ with the same switching signal as in Fig. 4.4. Similarly, it can be verified that $\bar{\mathcal{G}}_{\sigma(t)}$ satisfies both Assumptions 4.1 and 4.3.

Example 4.1 *(Distributed observer (4.15) over the static graph $\bar{\mathcal{G}}_a$ illustrated in Fig. 4.3)* Suppose the leader system is given in the form of (4.6) with

$$S_0 = \begin{bmatrix} 0.1 & 2 & 3 \\ 0 & 0 & 4 \\ 0 & -4 & -0.5 \end{bmatrix}, \quad W_0 = \begin{bmatrix} 1 & 1 & 1 \end{bmatrix}.$$

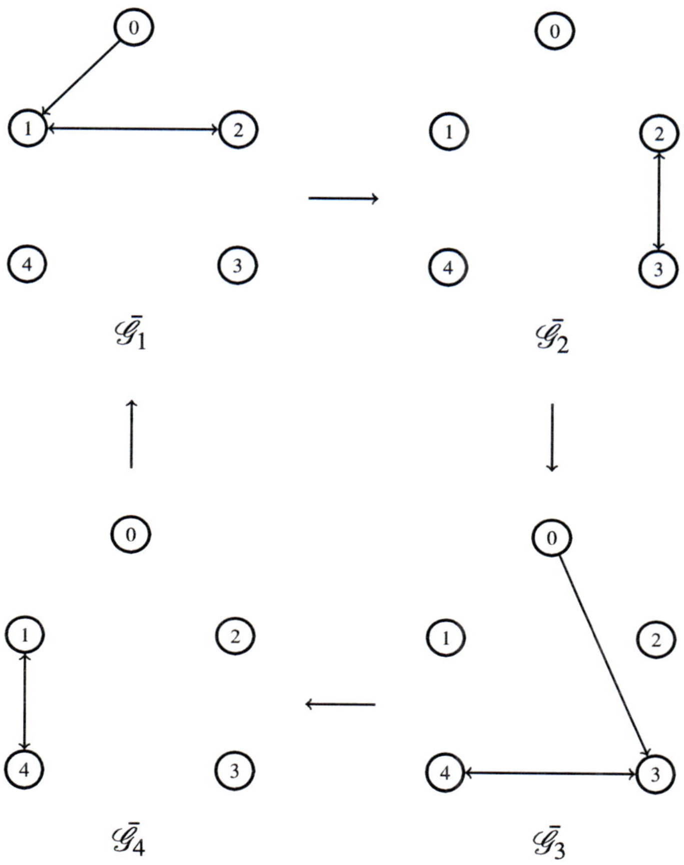

Fig. 4.5 The switching communication graph $\bar{\mathcal{G}}_{\sigma(t)}$ satisfying both Assumptions 4.1 and 4.3

It can be verified that Assumption 4.4 is satisfied. Thus, a distributed observer of the form (4.15) can be synthesized by Theorem 4.3. Solving the following Riccati equation:

$$P_0 S_0^\top + S_0 P_0 - P_0 W_0^\top W_0 P_0 + I_q = 0$$

gives

$$P_0 = \begin{bmatrix} 1.29 & 0.25 & 0.04 \\ 0.25 & 0.61 & -0.04 \\ 0.04 & -0.04 & 0.76 \end{bmatrix}.$$

Let $L_0 = \mu_v P_0 W_0^\top$ where $\mu_v = 10$ which is greater than $\frac{1}{2}\lambda_H^{-1} = 3.7538$. With $v_0(0) = \mathrm{col}(0, 10, 20)$ and $v_i(0) = 0$, $i = 1, 2, 3, 4$, the performance of the dis-

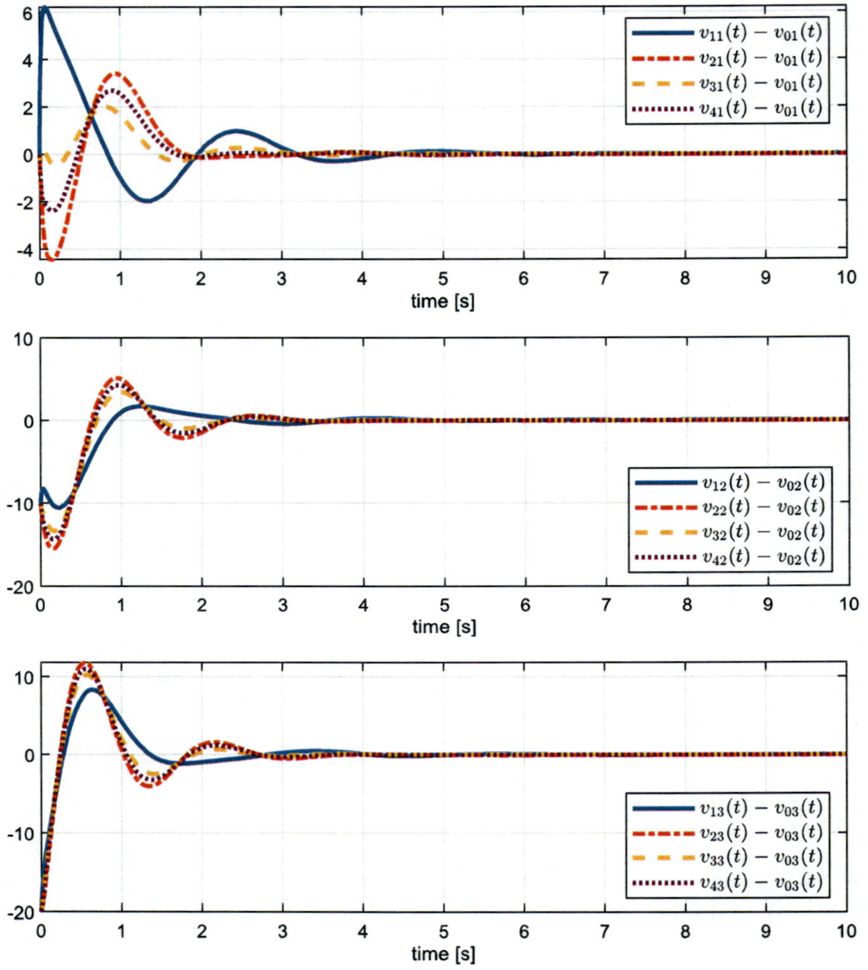

Fig. 4.6 The response of $v_{ij}(t) - v_{0j}(t)$ of the distributed observer (4.15)

tributed observer (4.15) is shown in Fig. 4.6. It can be seen that all the estimation errors converge to zero.

Example 4.2 (*Distributed observer (4.17) over the static graph $\bar{\mathscr{G}}_a$ illustrated in Fig. 4.3*) Suppose the leader system is given in the form of (4.6) with

$$S_0 = \begin{bmatrix} 0.1 & 2 & 3 \\ 0 & 0 & 4 \\ 0 & -4 & -0.5 \end{bmatrix}, \quad W_0 = I_3.$$

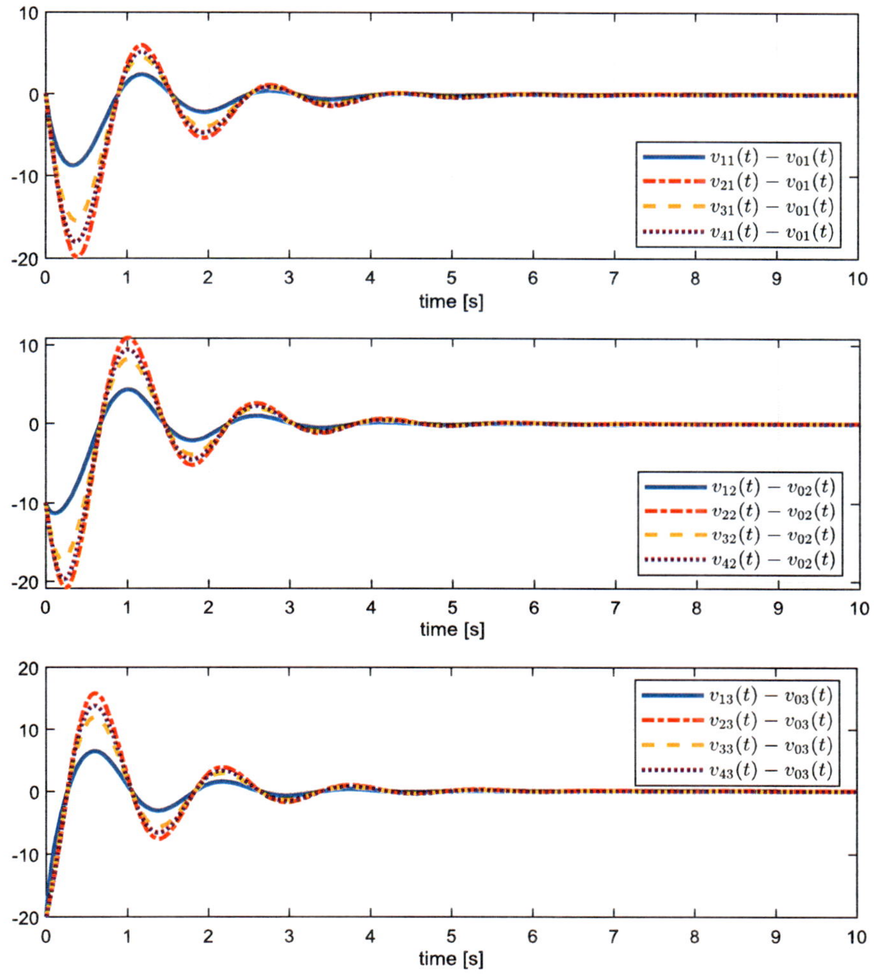

Fig. 4.7 The response of $v_{ij}(t) - v_{0j}(t)$ of the distributed observer (4.17)

Since $\bar{\lambda}_{S_0} = 0.1$, by Theorem 4.4, it suffices to choose $\mu_v = 5$ which is greater than $\bar{\lambda}_{S_0}\underline{\lambda}_H^{-1} = 0.75$. With $v_0(0) = \text{col}(0, 10, 20)$ and $v_i(0) = 0, i = 1, 2, 3, 4$, the performance of the distributed observer (4.17) is shown in Fig. 4.7. Again, all the estimation errors tend to zero asymptotically.

Example 4.3 (*Distributed observer (4.11) over the switching graph $\bar{\mathscr{G}}_{\sigma(t)}$ illustrated in Fig. 4.5*) Suppose the leader system is given in the form of (4.6) with

$$S_0 = \begin{bmatrix} 0 & 0 & 0 \\ 0 & 0 & 2 \\ 0 & -2 & 0 \end{bmatrix}, \quad W_0 = \begin{bmatrix} 1 & 1 & 1 \end{bmatrix}.$$

It can be verified that Assumptions 4.4 and 4.6 are satisfied. Thus, a distributed observer of the form (4.11) can be synthesized by Theorem 4.1. Let $P_0 = I_3$ which satisfies $P_0 S_0^{\mathrm{T}} + S_0 P_0 = 0$. Let $L_0 = \mu_v P_0 W_0^{\mathrm{T}}$ with $\mu_v = 1 > 0$. With $v_0(0) = \mathrm{col}(5, 10, 20)$ and $v_i(0) = 0$, $i = 1, 2, 3, 4$, the performance of the distributed observer (4.11) is shown in Fig. 4.8. As expected, all the estimation errors tend to zero asymptotically.

Example 4.4 (*Distributed observer (4.13) over the switching graph $\bar{\mathcal{G}}_{\sigma(t)}$ illustrated in Fig. 4.4*) Suppose the leader system is given in the form of (4.6) with

$$S_0 = \begin{bmatrix} 0 & 0.1 & 0.1 \\ 0 & 0 & 0.1 \\ 0 & 0 & 0 \end{bmatrix}, \quad W_0 = I_3.$$

It can be verified that the leader system satisfies Assumption 4.5. Thus, a distributed observer of the form (4.13) can be synthesized by Theorem 4.2. Let $\mu_v = 10 > 0$. With $v_0(0) = \mathrm{col}(0.3, 0.2, 0.1)$ and $v_i(0) = 0, i = 1, 2, 3, 4$, the performance of the distributed observer (4.13) is shown in Fig. 4.9. It can be seen again that all the estimation errors tend to zero asymptotically.

Example 4.5 (*Adaptive distributed observer (4.20) over the static graph $\bar{\mathcal{G}}_a$ illustrated in Fig. 4.3*) Suppose the leader system is the same as in Example 4.1 with $F_0 = I_3$. Thus, Assumption 4.4 is satisfied. By Theorem 4.5, we can obtain an adaptive distributed observer of the form (4.20). Since the matrix $F_0 = I_3$, there is no need to estimate the matrix F_0. Let $L_0 = \mu_L P_0 W_0^{\mathrm{T}}$ where P_0 is the same as in Example 4.1, and $\mu_L = 10 > 0$. Let $\mu_S = \mu_W = 10$ which are greater than $\bar{\lambda}_{S_0} \underline{\lambda}_H^{-1} = 0.75$ and $\mu_v = 30$ which are greater than $\frac{1}{2} \underline{\lambda}_H^{-1} = 3.7538$. With $v_0(0) = \mathrm{col}(0, 10, 20)$ and the initial state of all other variables being zero, the performance of the adaptive distributed observer (4.20) is shown in Fig. 4.10. As expected, all the estimation errors tend to zero asymptotically.

Example 4.6 (*Adaptive distributed observer (4.23) over the static graph $\bar{\mathcal{G}}_a$ illustrated in Fig. 4.3*) Suppose the leader system is the same as in Example 4.2 with $F_0 = I_3$. Then by Theorem 4.6, we can obtain an adaptive distributed observer of the form (4.23). Again, there is no need to estimate F_0. Let $\mu_S = \mu_v = 10$, both of which are greater than $\bar{\lambda}_{S_0} \underline{\lambda}_H^{-1} = 0.75$. With $v_0(0) = \mathrm{col}(0, 10, 20)$ and $S_i(0) = 0$, $v_i(0) = 0, i = 1, 2, 3, 4$, the performance of the adaptive distributed observer (4.23) is shown in Fig. 4.11. Again, all the estimation errors tend to zero asymptotically.

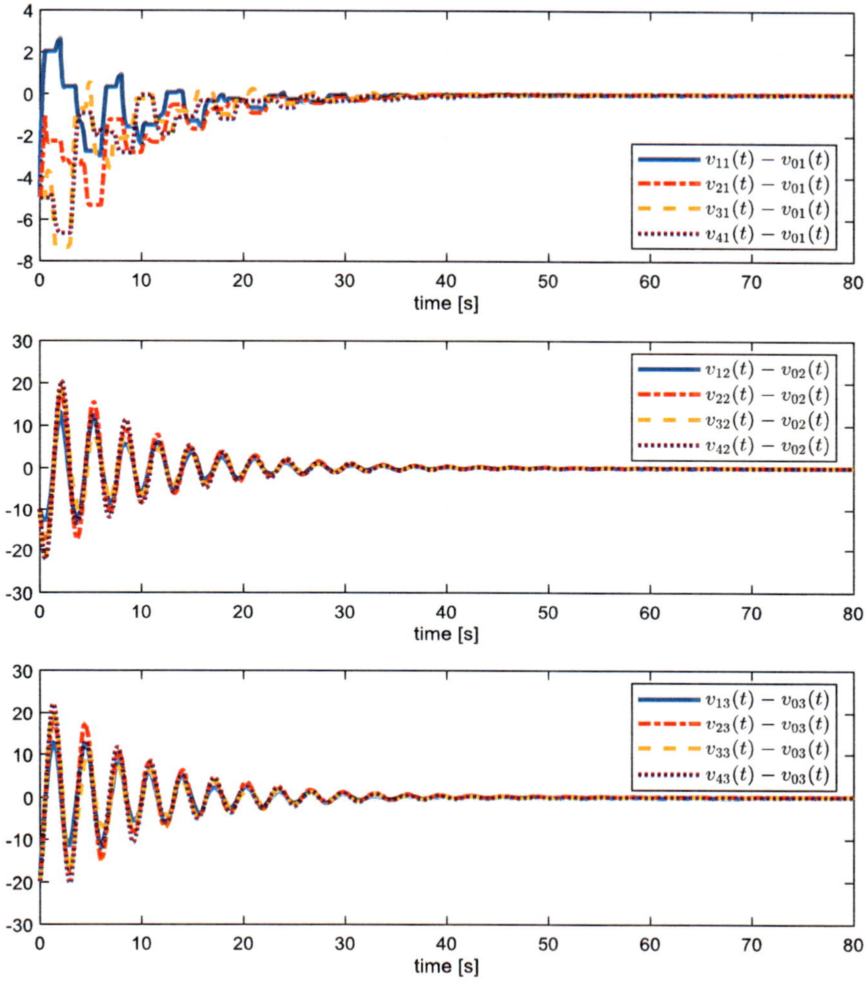

Fig. 4.8 The response of $v_{ij}(t) - v_{0j}(t)$ of the distributed observer (4.11)

Example 4.7 (*Adaptive distributed observer* (4.29) *over the switching graph* $\bar{\mathscr{G}}_{\sigma(t)}$ *illustrated in Fig.* 4.4) Suppose the leader system is the same as in Example 4.4 with $F_0 = I_3$, and thus Assumption 4.5 is satisfied. By Theorem 4.8, we can obtain an adaptive distributed observer of the form (4.29). Again, there is no need to estimate F_0. Let $\mu_S = \mu_v = 10 > 0$. With $v_0(0) = \text{col}(0.3, 0.2, 0.1)$, $S_i(0) = 0$, and $v_i(0) = 0$, $i = 1, 2, 3, 4$, the performance of the adaptive distributed observer (4.29) is shown in Fig. 4.12. Again, it can be seen that all the estimation errors tend to zero asymptotically.

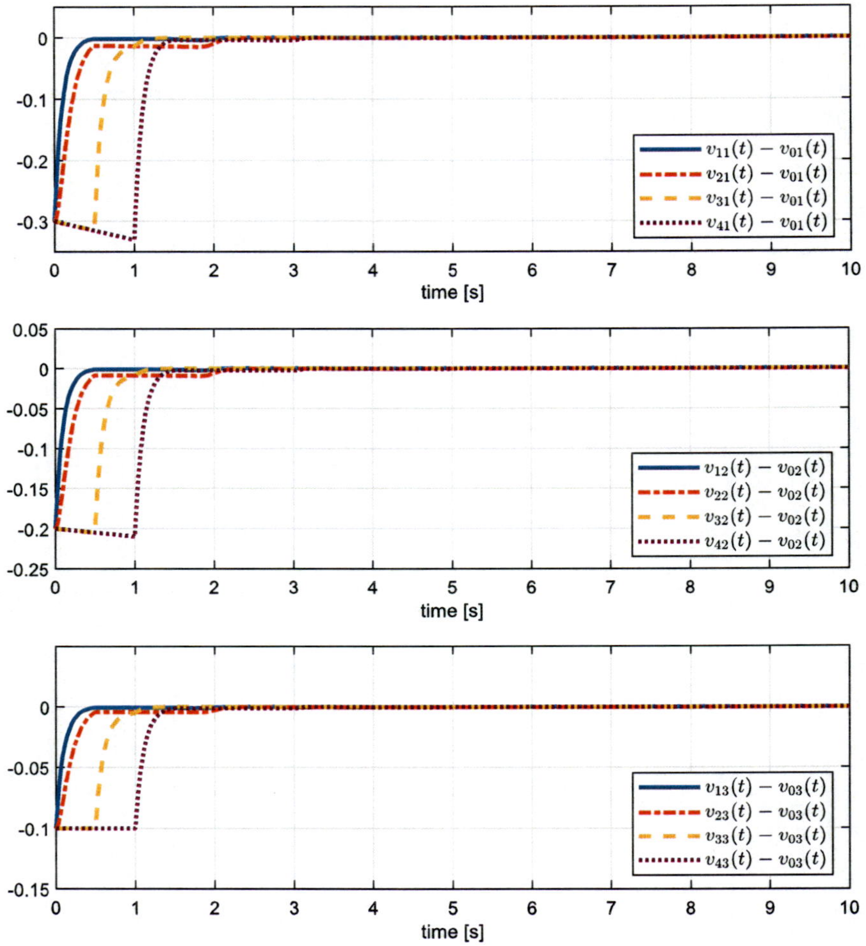

Fig. 4.9 The response of $v_{ij}(t) - v_{0j}(t)$ of the distributed observer (4.13)

Example 4.8 (*Adaptive distributed observer (4.38) for an uncertain leader system over the static graph $\bar{\mathcal{G}}_b$ illustrated in Fig. 4.3*) Suppose the leader system is uncertain and is given in the form of (4.35) with

$$\mathbf{S}(\omega_0) = \mathcal{D}(\omega_{01}a, \omega_{02}a, \omega_{03}a), \ a = \begin{bmatrix} 0 & 1 \\ -1 & 0 \end{bmatrix} \tag{4.52}$$

where $\omega_{0i} > 0$, $i = 1, 2, 3$. Thus, (4.52) satisfies Assumption 4.8. By Theorem 4.9, we can obtain an adaptive distributed observer of the form (4.38). Let $\mu_v = 80 > 0$ and $\mu_\omega = 80 > 0$. Suppose $\omega_0 = \text{col}(1, 2, 10)$. With $v_0(0) = \text{col}(0, 1, 0, 1, 0, 1)$,

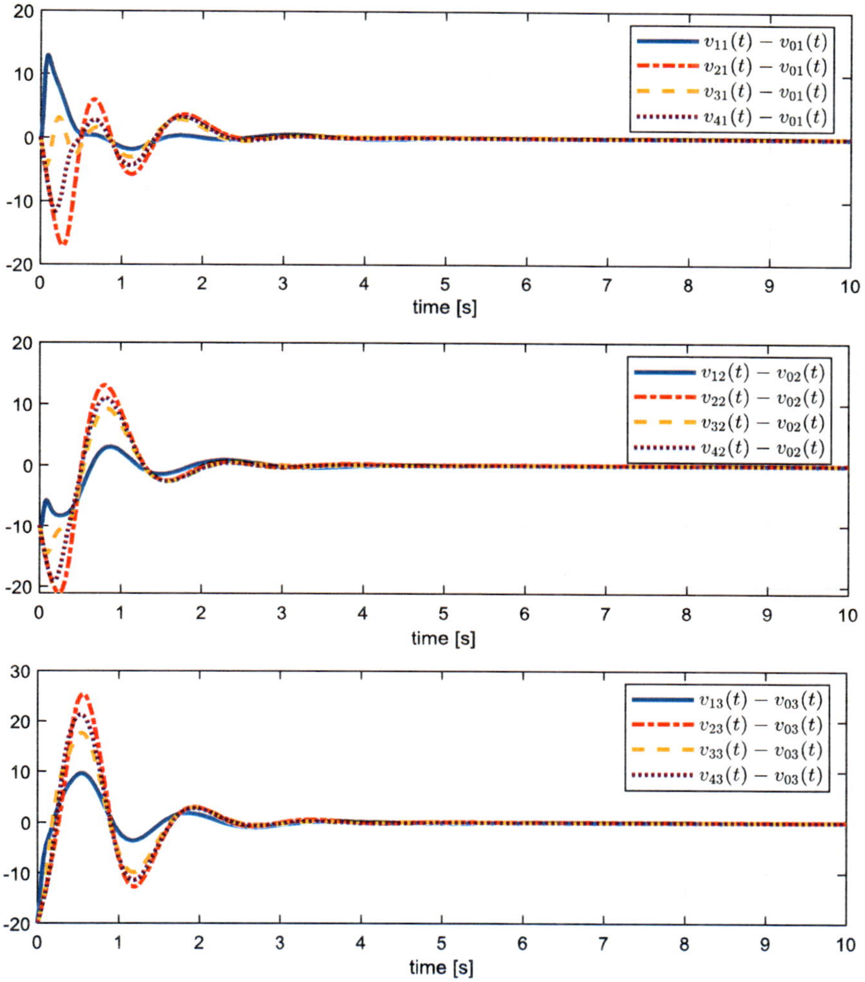

Fig. 4.10 The response of $v_{ij}(t) - v_{0j}(t)$ of the adaptive distributed observer (4.20)

$\omega_i(0) = 0$, and $v_i(0) = 0$, $i = 1, 2, 3, 4$, the estimation errors of the adaptive distributed observer are shown in Figs. 4.13 and 4.14, which all tend to zero asymptotically.

It can be verified that the initial state $v_0(0) = \text{col}(0, 1, 0, 1, 0, 1)$ satisfies the condition of Theorem 4.10. Thus, the estimated parameters $\omega_i(t)$, $i = 1, 2, 3, 4$, will all converge to the actual values of the unknown parameters. Figure 4.15 confirms Theorem 4.10.

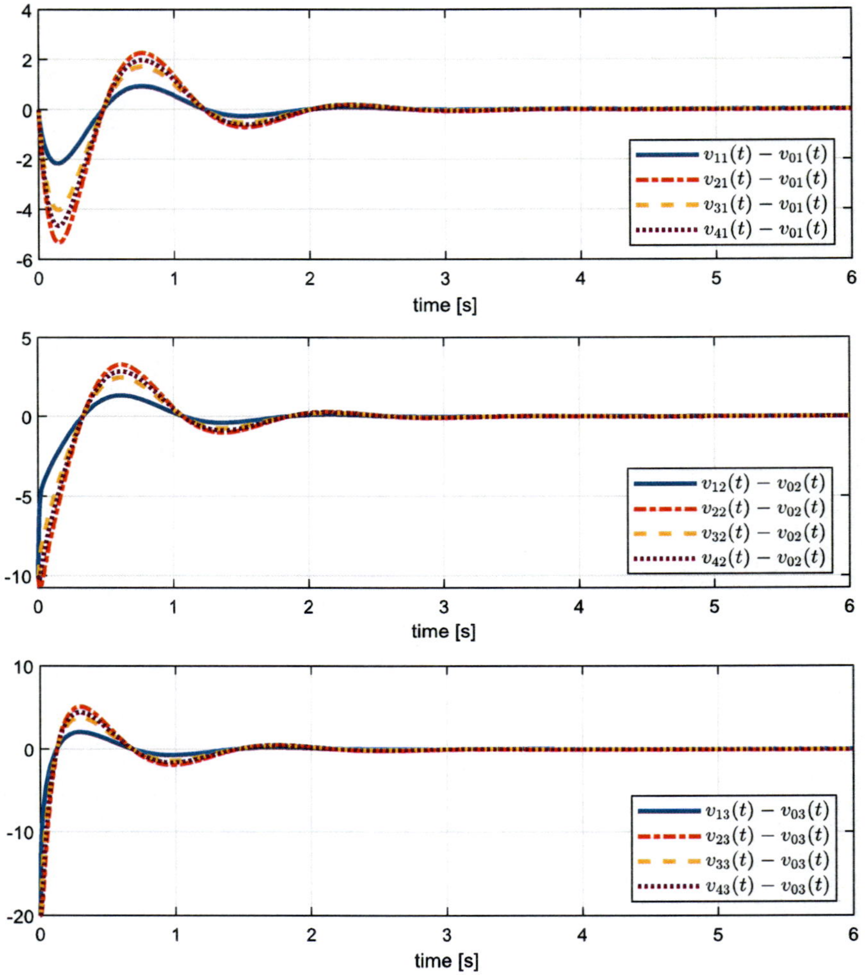

Fig. 4.11 The response of $v_{ij}(t) - v_{0j}(t)$ of the adaptive distributed observer (4.23)

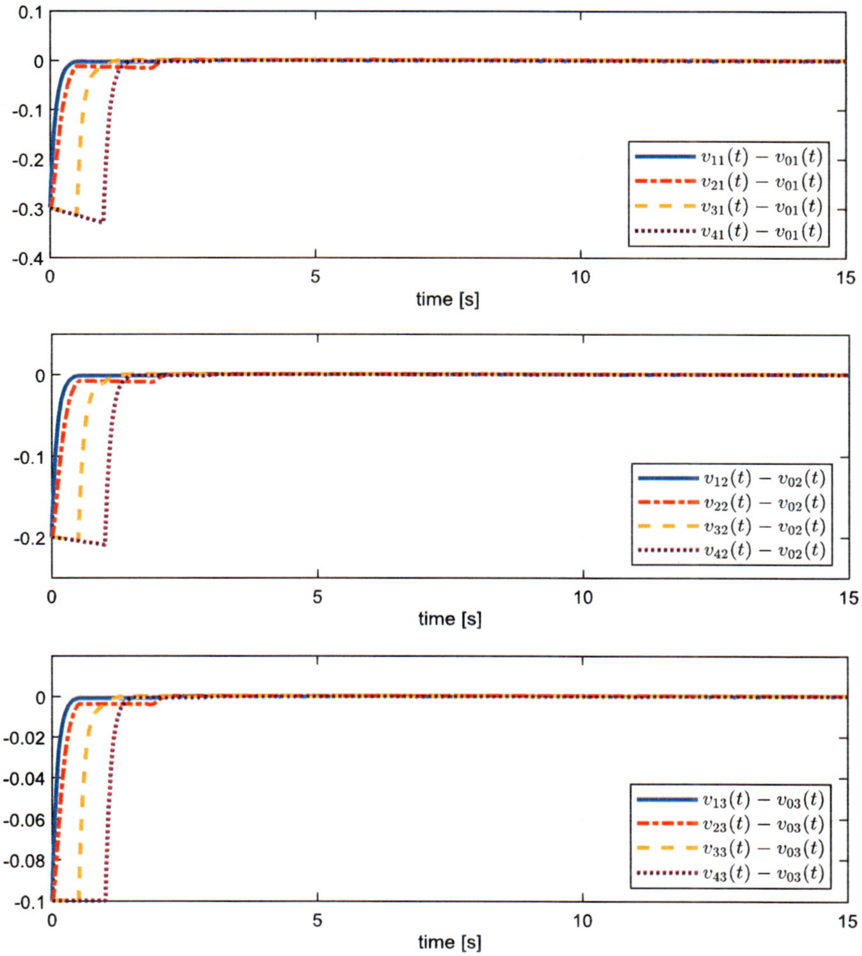

Fig. 4.12 The response of $v_{ij}(t) - v_{0j}(t)$ of the adaptive distributed observer (4.29)

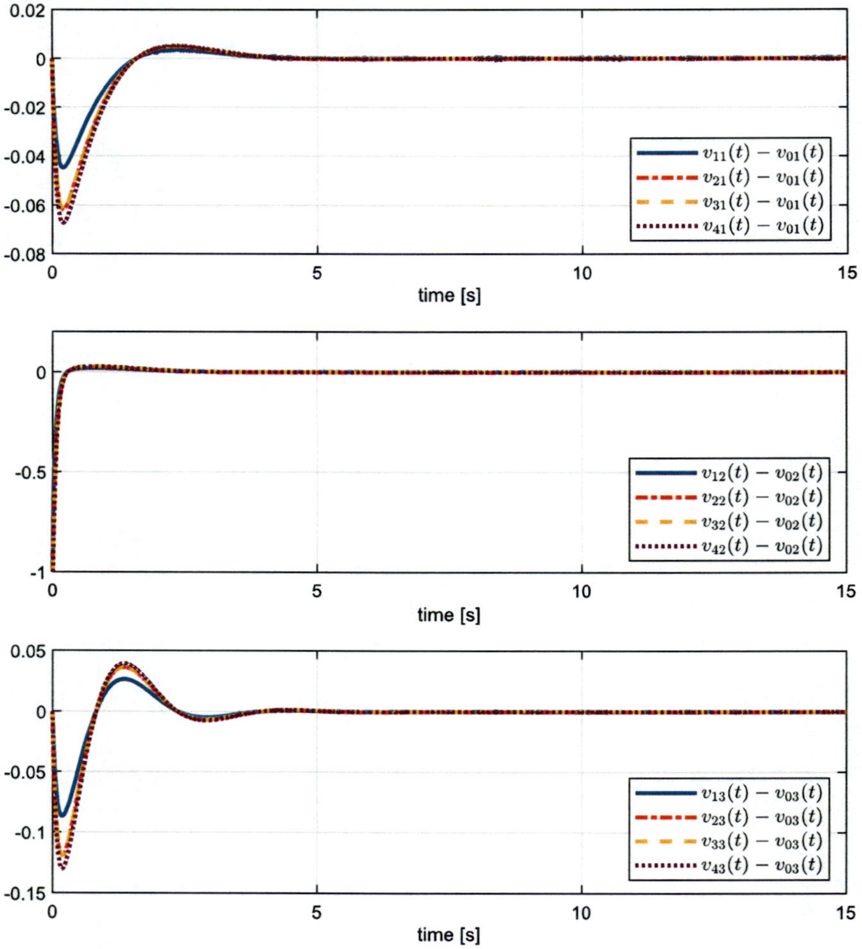

Fig. 4.13 The response of $v_{ij}(t) - v_{0j}(t)$ of the adaptive distributed observer (4.38)

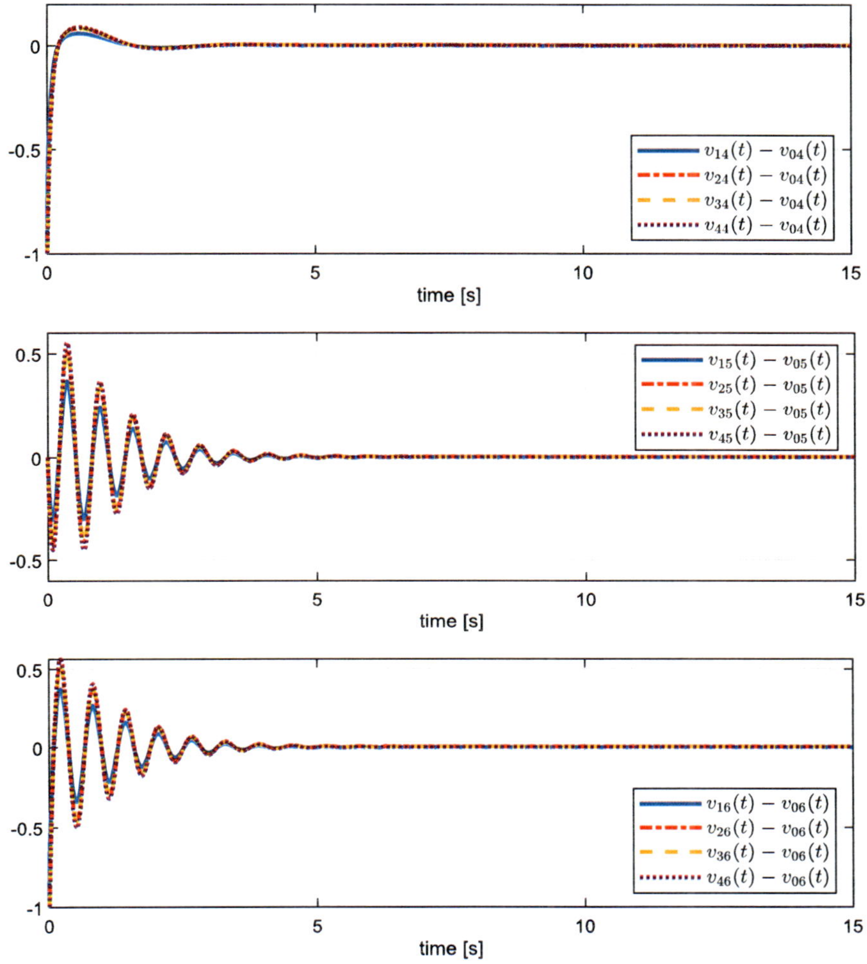

Fig. 4.14 The response of $v_{ij}(t) - v_{0j}(t)$ of the adaptive distributed observer (4.38)

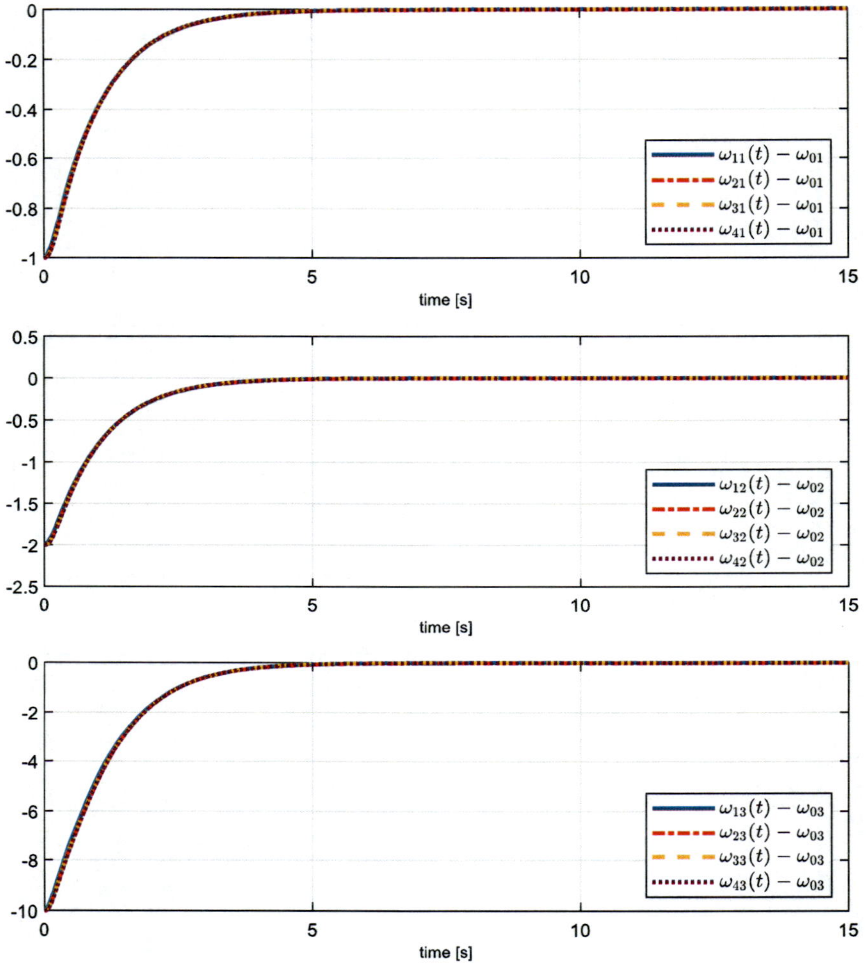

Fig. 4.15 The response of $\omega_{ij}(t) - \omega_{0j}$ of the adaptive distributed observer (4.38)

4.7 Notes and References

The distributed observer of the form (4.11) with $L_0 = \mu I_q$ for the leader system (4.6) with $W_0 = I$ was first proposed in [1] over static networks and was then extended to switching networks in [2]. Reference [3] considered the case with the general W_0. Reference [4] considered the case where each agent only has access to part of the output of the leader system. Finite-time convergence was considered for a special leader system in the chained integrator form in [5]. The case of leader systems with bounded input was studied in [6].

The adaptive distributed observer over switching networks for a known leader system was first proposed in [7] under the assumption that the matrix S_0 is neutrally stable and the communication graph is undirected. The undirected communication graph assumption was removed in [8], and the neutral stability assumption of the matrix S_0 was relaxed to the assumption that S_0 can have any eigenvalue with non-positive real part in [9]. The adaptive distributed observer over static networks was established for any matrix S_0 in [10]. There are also some other extensions regarding the adaptive distributed observers. For example, reference [11] designed an adaptive distributed observer for the case of multiple leaders. In [12], the adaptive distributed observer was modified for bipartite communication networks. Adaptive distributed observers with adaptive gains were proposed in [13].

The adaptive distributed observer for an uncertain leader system was first proposed in [14]. However, [14] did not consider the convergence of the estimation of the parameters of the matrix S_0 to the parameters of the matrix S_0. Reference [15] further proposed the adaptive distributed observer for the uncertain leader system (4.35) of the form (4.38) and showed that, as long as v_0 is persistently exciting, the estimated unknown entries of the matrix $S_0(\omega)$ will asymptotically tend to the actual value of the unknown entries of the matrix $S_0(\omega)$. Reference [16] studied the cooperative output regulation problem subject to an uncertain leader system and also gave the preliminary version of Lemma 4.1. References [17, 18] further considered the adaptive distributed observer for an uncertain leader with an unknown output over directed acyclic static graphs and directed acyclic switching graphs, respectively. Reference [18] also gave the current version of Lemma 4.1. Other results relevant to the adaptive distributed observer for the uncertain leader system can be found in [19].

The distributed observer for discrete-time leader systems was studied over static networks in [20] and over switching networks in [21, 22]. The adaptive distributed observer for discrete-time leader systems was studied over switching networks in [22, 23].

References

1. Su Y, Huang J (2012) Cooperative output regulation of linear multi-agent systems. IEEE Trans Autom Control 57(4):1062–1066
2. Su Y, Huang J (2012) Cooperative output regulation with application to multi-agent consensus under switching network. IEEE Trans Syst Man Cybern Part B: Cybern 42(3):864–875
3. Huang J (2016) Certainty equivalence, separation principle, and cooperative output regulation of multi-agent systems by the distributed observer approach. In: Control of complex systems: theory and applications, vol 14, pp 421–449
4. Chen K, Wang J, Zhao Z, Lai G, Lyu Y (2022) Output consensus of heterogeneous multiagent systems: a distributed observer-based approach. IEEE Trans Syst Man Cybern: Syst 52(1): 370-376
5. Zuo Z, Defoort M, Tian B, Ding Z (2020) Distributed consensus observer for multiagent systems with high-order integrator dynamics. IEEE Trans Autom Control 65(4):1771–1778

6. Hua Y, Dong X, Hu G, Li Q, Ren Z (2019) Distributed time-varying output formation tracking for heterogeneous linear multiagent systems with a nonautonomous leader of unknown input. IEEE Trans Autom Control 64(10):4292–4299
7. Cai H, Huang J (2016) The leader-following consensus for multiple uncertain Euler-Lagrange systems with an adaptive distributed observer. IEEE Trans Autom Control 61(10):3152–3157
8. Liu W, Huang J (2017) Adaptive leader-following consensus for a class of higher-order nonlinear multi-agent systems with directed switching networks. Automatica 79:84–92
9. Liu T, Huang J (2019) Leader-following consensus with disturbance rejection for uncertain Euler-Lagrange systems over switching networks. Int J Robust Nonlinear Control 29(18):6638–6656
10. Cai H, Lewis FL, Hu G, Huang J (2017) The adaptive distributed observer approach to the cooperative output regulation of linear multi-agent systems. Automatica 75:299–305
11. Liang H, Zhou Y, Ma H, Zhou Q (2019) Adaptive distributed observer approach for cooperative containment control of nonidentical networks. IEEE Trans Syst Man Cybern: Syst 49(2):299–307
12. Zhang H, Zhou Y, Liu Y, Sun J (online) Cooperative bipartite containment control for multiagent systems based on adaptive distributed observer. IEEE Trans Cybern. https://doi.org/10.1109/TCYB.2020.3031933
13. Dong Y, Xu S, Hu X (2017) Coordinated control with multiple dynamic leaders for uncertain Lagrangian systems via self-tuning adaptive distributed observer. Int J Robust Nonlinear Control 27(16):2708–2721
14. Modares H, Nageshrao SP, Lopes GAD, Babuška, R, Lewis FL (2016) Optimal model-free output synchronization of heterogeneous systems using off-policy reinforcement learning. Automatica 71:334–341
15. Wang S, Huang J (2018) Adaptive leader-following consensus for multiple Euler-Lagrange systems with an uncetain leader. IEEE Trans Neural Netw Learn Syst 30(7):2188–2196
16. Wang S, Huang J (2019) Cooperative output regulation of linear multi-agent systems subject to an uncertain leader system. Int J Control 94(4):952–960
17. Wang S, Huang J (2022) Adaptive distributed observer for an uncertain leader with an unknown output over directed acyclic graphs. Int J Control. https://doi.org/10.1080/00207179.2020.1766117
18. He C, Huang J (2022) Adaptive distributed observer for an uncertain leader over acyclic switching digraphs. Int J Robust Nonlinear Control (to appear)
19. Wu Y, Lu R, Shi P, Su H, Wu ZG (2017) Adaptive output synchronization of heterogeneous network with an uncertain leader. Automatica 76:183–192
20. Yan Y, Huang J (2016) Cooperative output regulation of discrete-time linear time-delay multi-agent systems. IET Control Theory & Appl 10(16):2019–2026
21. Yan Y, Huang J (2017) Cooperative output regulation of discrete-time linear time-delay multi-agent systems under switching networks. Neurocomputing 241:108–114
22. Liu T, Huang J (2021) Discrete-time distributed observers over jointly connected switching networks and an application. IEEE Trans Autom Control 66(4):1918–1924
23. Huang J (2017) The cooperative output regulation problem of discrete-time linear multi-agent systems by the adaptive distributed observer. IEEE Trans Autom Control 62(4):1979–1984

Chapter 5
Leader-Following Consensus of Multiple Euler–Lagrange Systems

Euler–Lagrange systems can describe the motion behaviors of many practical systems such as robot manipulators and rigid-body systems. In this chapter, we will study the leader-following consensus problem of multiple uncertain Euler–Lagrange systems. This chapter is organized as follows. Section 5.1 derives the Euler–Lagrange equation and describes some properties of Euler–Lagrange systems. Section 5.2 treats the asymptotic tracking problem for a single uncertain Euler–Lagrange system for three control scenarios. Section 5.3 studies how to solve the asymptotic tracking problem for a single uncertain Euler–Lagrange system via various estimation-based control laws. On the basis of Sects. 5.2 and 5.3, we further present the solution of the leader-following consensus problem of multiple uncertain Euler–Lagrange systems via the three types of distributed observers in Sect. 5.4. In Sect. 5.5, we use some numerical examples to illustrate our designs.

5.1 Euler–Lagrange Systems

The mathematical model of a system with n degrees of freedom can be derived using the following Euler–Lagrange equation:

$$\frac{d}{dt}\left(\frac{\partial L(q,\dot{q})}{\partial \dot{q}}\right) - \frac{\partial L(q,\dot{q})}{\partial q} = u \tag{5.1}$$

where $q, u \in \mathbb{R}^n$ are the generalized coordinate and force vector, respectively; $L(q,\dot{q}) \in \mathbb{R}$ is the Lagrangian function, which is defined by

$$L(q,\dot{q}) = T(q,\dot{q}) - V(q)$$

H. Cai et al., *Cooperative Control of Multi-agent Systems*, Advances in Industrial Control,
https://doi.org/10.1007/978-3-030-98377-2_5

with $T(q,\dot q)$ and $V(q)\in\mathbb{R}$ denoting system kinetic and potential energy, respectively.

Let the system kinetic energy be given by

$$T(q,\dot q)=\frac{1}{2}\dot q^{\mathsf T}M(q)\dot q$$

where $M(q)\in\mathbb{R}^{n\times n}$ is the positive definite generalized inertial matrix, and $V(q)$ represents the gravitational potential energy. Then,

$$L(q,\dot q)=\frac{1}{2}\dot q^{\mathsf T}M(q)\dot q-V(q)$$

and

$$\frac{\partial L(q,\dot q)}{\partial \dot q}=\frac{1}{2}M(q)\dot q+\frac{1}{2}M(q)^{\mathsf T}\dot q=M(q)\dot q.$$

As a result,

$$\frac{d}{dt}\left(\frac{\partial L(q,\dot q)}{\partial \dot q}\right)=\dot M(q)\dot q+M(q)\ddot q. \tag{5.2}$$

Moreover, we have

$$\frac{\partial L(q,\dot q)}{\partial q}=\frac{1}{2}\frac{\partial(\dot q^{\mathsf T}M(q)\dot q)}{\partial q}-\frac{\partial V(q)}{\partial q}. \tag{5.3}$$

Substituting (5.2) and (5.3) into (5.1) gives

$$\dot M(q)\dot q+M(q)\ddot q-\frac{1}{2}\frac{\partial(\dot q^{\mathsf T}M(q)\dot q)}{\partial q}+\frac{\partial V(q)}{\partial q}=u.$$

Let

$$C(q,\dot q)\dot q=\dot M(q)\dot q-\frac{1}{2}\frac{\partial(\dot q^{\mathsf T}M(q)\dot q)}{\partial q}$$

with $C(q,\dot q)\in\mathbb{R}^{n\times n}$, and $G(q)=\partial V(q)/\partial q$. Then we have

$$M(q)\ddot q+C(q,\dot q)\dot q+G(q)=u \tag{5.4}$$

where $C(q,\dot q)\dot q\in\mathbb{R}^n$ represents the Coriolis and centripetal force vector and $G(q)\in\mathbb{R}^n$ denotes the gravity force vector.

Some properties of system (5.4) are listed as follows.

Property 5.1 For all $q\in\mathbb{R}^n$, $M(q)\geq k_{\underline{m}}I_n$ for some $k_{\underline{m}}>0$.

Remark 5.1 By Property 5.1, $k_{\underline{m}}M(q)^{-1}\leq I_n$, and hence $M(q)^{-1}$ is bounded for all $q\in\mathbb{R}^n$.

The decomposition $C(q,\dot{q})\dot{q}$ is not unique. By adopting the Christoffel symbols, i.e.,

$$C_{ij} = \frac{1}{2}\sum_{k=1}^{n} \frac{\partial M_{ij}}{\partial q_k}\dot{q}_k + \frac{1}{2}\sum_{k=1}^{n}\left(\frac{\partial M_{ik}}{\partial q_j} - \frac{\partial M_{jk}}{\partial q_i}\right)\dot{q}_k$$

where C_{ij} and M_{ij} denote the elements of $C(q,\dot{q})$ and $M(q)$ on the ith row and jth column, respectively, the following property holds.

Property 5.2 The matrix $\dot{M}(q) - 2C(q,\dot{q})$ is anti-symmetric.

Property 5.3

$$M(q)x + C(q,\dot{q})y + G(q) = Y(q,\dot{q},x,y)\Theta, \ \forall x, y \in \mathbb{R}^n$$

where $Y(q,\dot{q},x,y) \in \mathbb{R}^{n\times l}$ is a known regression matrix and $\Theta \in \mathbb{R}^l$ is a nonzero constant vector consisting of the uncertain parameters. Moreover, if $q(t)$, $\dot{q}(t)$, $x(t)$, and $y(t)$ are uniformly bounded over $[0,\infty)$, then $C(q(t),\dot{q}(t))$ and $Y(q(t),\dot{q}(t), x(t), y(t))$ are also uniformly bounded over $[0,\infty)$.

5.2 Tracking Control of a Single Euler–Lagrange System

Let $q_0(t)$ denote the desired generalized position vector, and is assumed to be twice differentiable over $t \geq 0$. The tracking control problem of system (5.4) is described as follows.

Problem 5.1 Given system (5.4), design a control law of the following form:

$$\begin{aligned} u &= f(q,\dot{q},q_0,\dot{q}_0,\ddot{q}_0,z) \\ \dot{z} &= g(q,\dot{q},q_0,\dot{q}_0,\ddot{q}_0) \end{aligned}$$

such that

$$\lim_{t\to\infty}(q(t)-q_0(t)) = 0, \ \lim_{t\to\infty}(\dot{q}(t)-\dot{q}_0(t)) = 0.$$

The solvability of Problem 5.1 has much to do with the property of $q_0(t)$. We will consider two cases for $q_0(t)$ as follows.

Assumption 5.1 $\dot{q}_0(t)$ and $\ddot{q}_0(t)$ are uniformly bounded over $[0,\infty)$.

Assumption 5.2 $q_0(t)$ is uniformly bounded over $[0,\infty)$.

It is noted that Assumption 5.1 allows $q_0(t)$ to be unbounded. An example is $q_0(t) = t$. In case $q_0(t)$ is unbounded, we need the following assumption to guarantee the solvability of Problem 5.1.

Assumption 5.3 If $\dot{q}(t)$, $x(t)$, and $y(t)$ are uniformly bounded over $[0, \infty)$, then both $C(q(t), \dot{q}(t))$ and $Y(q(t), \dot{q}(t), x(t), y(t))$ are uniformly bounded over $[0, \infty)$.

To introduce our specific control law, let

$$s = \dot{q} - \zeta \tag{5.5}$$

where

$$\zeta = \dot{q}_0 - \alpha(q - q_0) \tag{5.6}$$

with $\alpha > 0$.

Let $Y = Y(q, \dot{q}, \dot{\zeta}, \zeta)$. Then our control law is given as follows:

$$u = -Ks + Y\hat{\Theta} \tag{5.7a}$$

$$\dot{\hat{\Theta}} = -\Lambda^{-1}Y^{\mathsf{T}}s \tag{5.7b}$$

where $K \in \mathbb{R}^{n\times n}$, $\Lambda \in \mathbb{R}^{l\times l}$ are positive definite gain matrices.

Remark 5.2 The control law (5.7) is a type of adaptive control law in which (5.7b) provides the estimate $\hat{\Theta}$ of the unknown parameter vector Θ.

Theorem 5.1 *Given system* (5.4), *Problem* 5.1 *is solvable by the control law* (5.7) *for both of the following two cases:*

1. *Assumptions* 5.1 *and* 5.2 *are satisfied;*
2. *Assumptions* 5.1 *and* 5.3 *are satisfied.*

Proof Substituting (5.7a) into (5.4) gives

$$M(q)\ddot{q} + C(q, \dot{q})\dot{q} + G(q) = -Ks + Y\hat{\Theta} \tag{5.8}$$

and subtracting $Y\Theta$ on both sides of (5.8) gives

$$M(q)\ddot{q} + C(q, \dot{q})\dot{q} + G(q) - Y\Theta = -Ks + Y\hat{\Theta} - Y\Theta.$$

Using Property 5.3 gives

$$M(q)\ddot{q} + C(q, \dot{q})\dot{q} + G(q) - M(q)\dot{\zeta} - C(q, \dot{q})\zeta - G(q) = -Ks + Y\tilde{\Theta}$$

where $\tilde{\Theta} = \hat{\Theta} - \Theta$. Then, we have

$$M(q)\dot{s} + C(q, \dot{q})s + Ks = Y\tilde{\Theta}. \tag{5.9}$$

Let

$$V = \frac{1}{2}(s^{\mathsf{T}}M(q)s + \tilde{\Theta}^{\mathsf{T}}\Lambda\tilde{\Theta}).$$

Noting $s^{\mathsf{T}}(\dot{M}(q) - 2C(q, \dot{q}))s \equiv 0$ since $\dot{M}(q) - 2C(q, \dot{q})$ is anti-symmetric gives

$$\begin{aligned}
\dot{V} &= s^{\mathsf{T}} M(q)\dot{s} + \frac{1}{2} s^{\mathsf{T}} \dot{M}(q) s + \tilde{\Theta}^{\mathsf{T}} \Lambda \dot{\tilde{\Theta}} \\
&= s^{\mathsf{T}}(-C(q, \dot{q})s - Ks + Y\tilde{\Theta}) + \frac{1}{2} s^{\mathsf{T}} \dot{M}(q) s + \tilde{\Theta}^{\mathsf{T}} \Lambda \dot{\tilde{\Theta}} \\
&= -s^{\mathsf{T}} K s + s^{\mathsf{T}} Y \tilde{\Theta} - \tilde{\Theta}^{\mathsf{T}} \Lambda \Lambda^{-1} Y^{\mathsf{T}} s \\
&= -s^{\mathsf{T}} K s \leq 0.
\end{aligned} \tag{5.10}$$

Since $M(q)$ is positive definite by Property 5.1, $V(t)$ is lower bounded for all $t \geq 0$. By (5.10), $\dot{V} \leq 0$, which implies that $\lim_{t\to\infty} V(t)$ exists and is finite, and thus $s(t)$ and $\tilde{\Theta}(t)$ are uniformly bounded over $[0, \infty)$.

Next, we will show that, for both cases in Theorem 5.1, $\dot{s}(t)$ is uniformly bounded over $[0, \infty)$. For this purpose, note that from (5.5), we have

$$\dot{s} = \ddot{q} - \dot{\zeta}.$$

By (5.6),

$$\dot{\zeta} = \ddot{q}_0 - \alpha(\dot{q} - \dot{q}_0). \tag{5.11}$$

Combining (5.5) and (5.6) gives

$$\dot{q} + \alpha q = \dot{q}_0 + \alpha q_0 + s \tag{5.12}$$

or

$$(\dot{q} - \dot{q}_0) + \alpha(q - q_0) = s. \tag{5.13}$$

- Case 1: under Assumptions 5.1 and 5.2, by (5.12), $q(t)$ and $\dot{q}(t)$ are uniformly bounded over $[0, \infty)$ by viewing (5.12) as a stable first-order system in q with a bounded input. Therefore, $\zeta(t)$ and $\dot{\zeta}(t)$ are also uniformly bounded over $[0, \infty)$, which in turn implies that $Y(t)$ is uniformly bounded over $[0, \infty)$ by Property 5.3. Then, by (5.9), Remark 5.1, and Property 5.3 again, $\dot{s}(t)$ is uniformly bounded over $[0, \infty)$.
- Case 2: by (5.13), $q(t) - q_0(t)$ and $\dot{q}(t) - \dot{q}_0(t)$ are uniformly bounded over $[0, \infty)$ by viewing (5.13) as a stable first-order system in $q - q_0$ with a bounded input. Also, since $\dot{q}_0(t)$ is uniformly bounded over $[0, \infty)$, $\dot{q}(t)$ is uniformly bounded over $[0, \infty)$. Under Assumption 5.1, $\zeta(t)$ and $\dot{\zeta}(t)$ are also uniformly bounded over $[0, \infty)$, which in turn, under Assumption 5.3, implies that $Y(t)$ is uniformly bounded over $[0, \infty)$. Then, by (5.9), Remark 5.1, and Assumption 5.3, $\dot{s}(t)$ is uniformly bounded over $[0, \infty)$.

Since $\dot{s}(t)$ is uniformly bounded over $[0, \infty)$, so is $\ddot{V}(t)$. Then, by Barbalat's lemma, i.e., Lemma 2.9, $\lim_{t\to\infty} s(t) = 0$. Thus, by (5.13),

$$\dot{q} - \dot{q}_0 = -\alpha(q - q_0) + s$$

and the proof is complete by invoking Lemma 2.5. □

5.3 Estimation-Based Control

In the last section, the control law makes use of the accurate reference signals q_0, $\dot{q}_0$, and $\ddot{q}_0$ to solve Problem 5.1. In this section, we further consider using the estimated reference signals to solve the tracking problem. The results of this section will lay a foundation for dealing with the leader-following consensus problem of multiple Euler–Lagrange systems.

Given a time sequence $\{t_k : k = 0, 1, 2, \ldots\}$ with $t_0 = 0$ and a dwell time $\tau > 0$, suppose $q_0^p(t)$ and $q_0^v(t)$ are differentiable over each time interval $[t_k, t_{k+1})$. Also, let

$$\tilde{q}_0^p(t) \triangleq q_0^p(t) - q_0(t)$$
$$\tilde{q}_0^v(t) \triangleq q_0^v(t) - \dot{q}_0(t).$$

Then, it is assumed that

$$\lim_{t\to\infty} \tilde{q}_0^p(t) = 0$$
$$\lim_{t\to\infty} \tilde{q}_0^v(t) = 0$$

and

$$\lim_{t\to\infty} (\dot{q}_0^p(t) - \dot{q}_0(t)) = 0$$
$$\lim_{t\to\infty} (\dot{q}_0^v(t) - \ddot{q}_0(t)) = 0.$$

As a result, let $\Delta q_0^v(t) = q_0^v(t) - \dot{q}_0^p(t)$. Then $\lim_{t\to\infty} \Delta q_0^v(t) = \lim_{t\to\infty}(q_0^v(t) - \dot{q}_0(t)) + \lim_{t\to\infty}(\dot{q}_0(t) - \dot{q}_0^p(t)) = 0$.

Problem 5.2 Given system (5.4), design a control law u of the following form:

$$u = f(q, \dot{q}, q_0^p, \dot{q}_0^p, q_0^v, \dot{q}_0^v, z)$$
$$\dot{z} = g(q, \dot{q}, q_0^p, \dot{q}_0^p, q_0^v, \dot{q}_0^v)$$

such that

$$\lim_{t\to\infty} (q(t) - q_0(t)) = 0, \quad \lim_{t\to\infty} (\dot{q}(t) - \dot{q}_0(t)) = 0.$$

Let

$$s = \dot{q} - \zeta \tag{5.14}$$

where

$$\zeta = q_0^v - \alpha(q - q_0^p) \tag{5.15}$$

with $\alpha > 0$. Thus,

$$\dot{\zeta} = \dot{q}_0^v - \alpha(\dot{q} - \dot{q}_0^p). \tag{5.16}$$

It can be seen that (5.15) is obtained from (5.6) by replacing q_0 with q_0^p and $\dot{q}_0$ with q_0^v, and (5.16) is obtained from (5.11) by replacing $\dot{q}_0$ with $\dot{q}_0^p$, and $\ddot{q}_0$ with $\dot{q}_0^v$.

Consider the following control law:

$$u = -Ks + Y(q, \dot{q}, \dot{\zeta}, \zeta)\hat{\Theta} \tag{5.17a}$$

$$\dot{\hat{\Theta}} = -\Lambda^{-1}Y^{\mathsf{T}}(q, \dot{q}, \dot{\zeta}, \zeta)s \tag{5.17b}$$

where $K \in \mathbb{R}^{n\times n}$ and $\Lambda \in \mathbb{R}^{l\times l}$ are positive definite gain matrices. Since (5.17) is obtained from (5.7) by replacing q_0, $\dot{q}_0$, $\ddot{q}_0$ with their estimates, it is called an estimation-based control law.

Theorem 5.2 *Given system* (5.4)*, Problem* 5.2 *is solvable by the control law* (5.17) *for both of the following two cases:*

1. *Assumptions* 5.1 *and* 5.2 *are satisfied;*
2. *Assumptions* 5.1 *and* 5.3 *are satisfied.*

Proof Substituting (5.17a) into (5.4) gives

$$M(q)\ddot{q} + C(q, \dot{q})\dot{q} + G(q) = -Ks + Y\hat{\Theta} \tag{5.18}$$

and subtracting $Y\Theta$ on both sides of (5.18) gives

$$M(q)\ddot{q} + C(q, \dot{q})\dot{q} + G(q) - Y\Theta = -Ks + Y\hat{\Theta} - Y\Theta.$$

Using Property 5.3 gives

$$M(q)\ddot{q} + C(q, \dot{q})\dot{q} + G(q) - M(q)\dot{\zeta} - C(q, \dot{q})\zeta - G(q) = -Ks + Y\tilde{\Theta}$$

where $\tilde{\Theta} = \hat{\Theta} - \Theta$. Then, we have

$$M(q)\dot{s} + C(q, \dot{q})s + Ks = Y\tilde{\Theta}. \tag{5.19}$$

Let

$$V = \frac{1}{2}(s^{\mathsf{T}}M(q)s + \tilde{\Theta}^{\mathsf{T}}\Lambda\tilde{\Theta}).$$

Noting $s^{\mathsf{T}}(\dot{M}(q) - 2C(q, \dot{q}))s \equiv 0$ since $\dot{M}(q) - 2C(q, \dot{q})$ is anti-symmetric gives

$$\begin{aligned} \dot{V} &= s^{\mathsf{T}} M(q)\dot{s} + \frac{1}{2} s^{\mathsf{T}} \dot{M}(q) s + \tilde{\Theta}^{\mathsf{T}} \Lambda \dot{\tilde{\Theta}} \\ &= s^{\mathsf{T}}(-C(q, \dot{q})s - Ks + Y\tilde{\Theta}) + \frac{1}{2} s^{\mathsf{T}} \dot{M}(q)s + \tilde{\Theta}^{\mathsf{T}} \Lambda \dot{\tilde{\Theta}} \\ &= -s^{\mathsf{T}} K s + s^{\mathsf{T}} Y \tilde{\Theta} - \tilde{\Theta}^{\mathsf{T}} \Lambda \Lambda^{-1} Y^{\mathsf{T}} s \\ &= -s^{\mathsf{T}} K s \leq 0. \end{aligned} \tag{5.20}$$

Since $M(q)$ is positive definite by Property 5.1, $V(t)$ is lower bounded for all $t \geq 0$. By (5.20), $\dot{V} \leq 0$, which implies that $s(t)$ and $\tilde{\Theta}(t)$ are uniformly bounded over $[0, \infty)$.

From (5.20), we have $\ddot{V}(t) = -2s^{\mathsf{T}} K \dot{s}$ on each time interval $[t_k, t_{k+1})$. Since $\dot{s}(t)$ is not continuous over $[0, \infty)$, instead of Barbalat's Lemma (Lemma 2.9), we have to apply Corollary 2.6 to conclude $\lim_{t \to \infty} \dot{V}(t) \to 0$. Thus, we need to show that, for both cases in Theorem 5.2, $\dot{s}(t)$ is uniformly bounded over $[0, \infty)$. For this purpose, note that from (5.14), we have

$$\dot{s} = \ddot{q} - \dot{\zeta}.$$

By (5.15),

$$\dot{\zeta} = \dot{q}_0^v - \alpha(\dot{q} - \dot{q}_0^p).$$

Combining (5.14) and (5.15) gives

$$\dot{q} + \alpha q = q_0^v + \alpha q_0^p + s \tag{5.21}$$

or

$$(\dot{q} - \dot{q}_0^p) + \alpha(q - q_0^p) = s + q_0^v - \dot{q}_0^p = s + \Delta q_0^v. \tag{5.22}$$

- Case 1: under Assumptions 5.1 and 5.2, by (5.21), $q(t)$ and $\dot{q}(t)$ are uniformly bounded over $[0, \infty)$ by viewing (5.21) as a stable first-order system in q with a bounded input. Therefore, $\zeta(t)$ and $\dot{\zeta}(t)$ are also uniformly bounded over $[0, \infty)$, which in turn implies that $Y(t)$ is uniformly bounded over $[0, \infty)$ by Property 5.3. Then, by (5.19), Remark 5.1, and Property 5.3 again, $\dot{s}(t)$ is uniformly bounded over $[0, \infty)$.
- Case 2: by (5.22), $q(t) - q_0^p(t)$ and $\dot{q}(t) - \dot{q}_0^p(t)$ are uniformly bounded over $[0, \infty)$ by viewing (5.22) as a stable first-order system in $q - q_0^p$ with a bounded input. Also, since $\dot{q}_0^p(t)$ is uniformly bounded over $[0, \infty)$, $\dot{q}(t)$ is uniformly bounded over $[0, \infty)$. Under Assumption 5.1, $\zeta(t)$ and $\dot{\zeta}(t)$ are also uniformly bounded over $[0, \infty)$, which together with Assumption 5.3 implies that $Y(t)$ is uniformly bounded over $[0, \infty)$. Then, by (5.19), Remark 5.1, and Assumption 5.3, $\dot{s}(t)$ is uniformly bounded over $[0, \infty)$.

Since both $s(t)$ and $\dot{s}(t)$ are uniformly bounded over $[0, \infty)$, there exists a positive number γ such that

$$\sup_{t_k \le t < t_{k+1}, k=0,1,2,\ldots} |\ddot{V}(t)| \le \gamma.$$

Therefore, by Corollary 2.5, $\lim_{t\to\infty} \dot{V}(t) \to 0$ and thus $\lim_{t\to\infty} s(t) \to 0$. Then, by (5.22) and Lemma 2.5, $\lim_{t\to\infty}(q(t) - q_0^p(t)) = 0$ and $\lim_{t\to\infty}(\dot{q}(t) - \dot{q}_0^p(t)) = 0$. Then, the proof is completed by noting that $q - q_0 = q - q_0^p + q_0^p - q_0$ and $\dot{q} - \dot{q}_0 = \dot{q} - \dot{q}_0^p + \dot{q}_0^p - \dot{q}_0$. □

5.4 Leader-Following Consensus of Multiple Euler–Lagrange Systems

Consider N Euler–Lagrange systems described by the following dynamic equations:

$$M_i(q_i)\ddot{q}_i + C_i(q_i, \dot{q}_i)\dot{q}_i + G_i(q_i) = u_i, \ i = 1, \ldots, N \tag{5.23}$$

where $q_i, u_i \in \mathbb{R}^n$. Suppose each subsystem of (5.23) satisfies Properties 5.1–5.3 with $M(q), C(q, \dot{q}), G(q), Y(q, \dot{q}, x, y)$, and Θ being replaced by $M_i(q_i), C_i(q_i, \dot{q}_i)$, $G_i(q_i)$, $Y_i(q_i, \dot{q}_i, x_i, y_i)$, and Θ_i, respectively.

Let $q_0 \in \mathbb{R}^n$ and $\dot{q}_0 \in \mathbb{R}^n$ denote the desired generalized position vector, and the desired generalized velocity vector, respectively. Tentatively, we assume, for each $i = 1, \ldots, N$, u_i can access q_0, $\dot{q}_0$, and $\ddot{q}_0$. Let

$$\begin{aligned}\zeta_i &= \dot{q}_0 - \alpha_i(q_i - q_0)\\ s_i &= \dot{q}_i - \zeta_i\end{aligned}$$

with $\alpha_i > 0$.

Then, a straightforward extension of the control law (5.7) gives the purely decentralized control law as follows:

$$\begin{aligned}u_i &= -K_i s_i + Y_i(q_i, \dot{q}_i, \dot{\zeta}_i, \zeta_i)\hat{\Theta}_i\\ \dot{\hat{\Theta}}_i &= -\Lambda_i^{-1} Y_i^{\mathsf{T}}(q_i, \dot{q}_i, \dot{\zeta}_i, \zeta_i)s_i\end{aligned}$$

where $K_i \in \mathbb{R}^{n\times n}$, $\Lambda_i \in \mathbb{R}^{l_i\times l_i}$ are positive definite gain matrices.

Nevertheless, it is unrealistic to assume q_0, $\dot{q}_0$, and $\ddot{q}_0$ are accessible by every control u_i, $i = 1, \ldots, N$. To obtain a distributed control law, we will utilize the distributed observer. For this purpose, assume q_0 is generated by the following linear autonomous system:

$$\dot{v}_0 = S_0 v_0 \tag{5.24a}$$

$$q_0 = F_0 v_0 \tag{5.24b}$$

$$y_{m0} = W_0 v_0 \tag{5.24c}$$

where $v_0 \in \mathbb{R}^m$, $S_0 \in \mathbb{R}^{m\times m}$, $F_0 \in \mathbb{R}^{n\times m}$, and $W_0 \in \mathbb{R}^{m_0\times m}$ are known constant matrices. As a result of (5.24), we have $\dot{q}_0 = F_0 S_0 v_0$ and $\ddot{q}_0 = F_0 S_0^2 v_0$. System (5.24) is the combination of (4.6) and (4.19) with the reference output $y_0 = q_0$.

We treat the system composed of (5.23) and (5.24) as a multi-agent system of $(N+1)$ agents with (5.24) as the leader and the N subsystems of (5.23) as N followers. Given systems (5.23) and (5.24), and a piecewise constant switching signal $\sigma(t)$ with a dwell time τ, we can define a switching communication graph $\bar{\mathscr{G}}_{\sigma(t)} = (\bar{\mathscr{V}}, \bar{\mathscr{E}}_{\sigma(t)})$ with $\bar{\mathscr{V}} = \{0, 1, \ldots, N\}$ and $\bar{\mathscr{E}}_{\sigma(t)} \subseteq \bar{\mathscr{V}} \times \bar{\mathscr{V}}$ for all $t \geq 0$. Here, the node 0 is associated with the leader system (5.24) and the node i, $i = 1, \ldots, N$, is associated with the ith subsystem of system (5.23), and, for $i = 0, 1, \ldots, N$, $j = 1, \ldots, N$, $i \neq j$, $(i, j) \in \bar{\mathscr{E}}_{\sigma(t)}$ if and only if u_j can use the information of agent i for control at time instant t. Let $\bar{\mathscr{N}}_i(t) = \{j : (j, i) \in \bar{\mathscr{E}}_{\sigma(t)}\}$ denote the neighbor set of agent i at time t. Let $\mathscr{G}_{\sigma(t)} = (\mathscr{V}, \mathscr{E}_{\sigma(t)})$ denote the subgraph of $\bar{\mathscr{G}}_{\sigma(t)}$ where $\mathscr{V} = \{1, \ldots, N\}$, $\mathscr{E}_{\sigma(t)} \subseteq \mathscr{V} \times \mathscr{V}$ is obtained from $\bar{\mathscr{E}}_{\sigma(t)}$ by removing all edges between the node 0 and the nodes in $\mathscr{V}$.

With $\bar{\mathscr{G}}_{\sigma(t)}$ thus defined, the leader-following consensus problem of multiple Euler–Lagrange systems is described as follows.

Problem 5.3 Given systems (5.23), (5.24) and a switching communication graph $\bar{\mathscr{G}}_{\sigma(t)}$, design a distributed control law in the following form:

$$u_i = f_i(q_i, \dot{q}_i, \varphi_i, \varphi_j,\ j \in \bar{\mathscr{N}}_i(t)) \tag{5.25a}$$

$$\dot{\varphi}_i = g_i(\varphi_i, \varphi_j,\ j \in \bar{\mathscr{N}}_i(t)) \tag{5.25b}$$

where φ_0 represents some known quantities of the leader to be specified, such that, for $i = 1, \ldots, N$, for all initial conditions $q_i(0)$, $\dot{q}_i(0)$, $\varphi_i(0)$ and $v_0(0)$, $q_i(t)$, $\dot{q}_i(t)$, $\varphi_i(t)$ exist for all $t \geq 0$ and satisfy

$$\lim_{t\to\infty}(q_i(t) - q_0(t)) = 0, \ \lim_{t\to\infty}(\dot{q}_i(t) - \dot{q}_0(t)) = 0.$$

Remark 5.3 We will employ various distributed observers to synthesize the control law of the form (5.25), which depends on the specific distributed observer used. In particular, the variable φ_0 depends on the specific distributed observer used.

Corresponding to the two cases studied in Sect. 5.2, we also consider two cases for the leader system (5.24).

Assumption 5.4 The matrix S_0 is neutrally stable, that is, all of its eigenvalues are semi-simple with zero real parts.

Assumption 5.5 The matrix S_0 has no eigenvalues with positive real parts, and $\dot{q}_0$ is uniformly bounded over $[0, \infty)$.

Remark 5.4 It can be verified that the satisfaction of Assumption 5.4 implies the satisfaction of both Assumptions 5.1 and 5.2, and the satisfaction of Assumption 5.5 implies the satisfaction of Assumption 5.1 but allows $q_0(t)$ to be a ramp function.

Remark 5.5 Let $L_0 = \mu_v P_0 W_0^{\mathsf{T}}$ where $\mu_v > 0$ and $P_0 > 0$ satisfying $P_0 S_0^{\mathsf{T}} + S_0 P_0 = 0$. Then, under Assumptions 4.1, 4.3, 4.4, and 5.4, by Theorem 4.1, a distributed observer for the leader system (5.24) can be synthesized as follows:

$$\dot{v}_i = S_0 v_i + L_0 \sum_{j=0}^{N} a_{ij}(t) W_0 (v_j - v_i), \ i = 1, \ldots, N \tag{5.26}$$

which satisfies

$$\lim_{t\to\infty} (v_i(t) - v_0(t)) = 0. \tag{5.27}$$

If $W_0 = I_m$, by Theorem 4.2, under Assumptions 4.1 and 5.5, the solution of (5.26) with $L_0 = \mu_v I_m$ where $\mu_v > 0$ also satisfies (5.27).

By the general framework described in Sect. 4.2, we can obtain a distributed observer-based control law as follows:

$$u_i = -K_i s_i + Y_i(q_i, \dot{q}_i, \dot{\zeta}_i, \zeta_i)\hat{\Theta}_i \tag{5.28a}$$

$$\dot{\hat{\Theta}}_i = -\Lambda_i^{-1} Y_i^{\mathsf{T}}(q_i, \dot{q}_i, \dot{\zeta}_i, \zeta_i) s_i \tag{5.28b}$$

$$\dot{v}_i = S_0 v_i + L_0 \sum_{j=0}^{N} a_{ij}(t) W_0 (v_j - v_i) \tag{5.28c}$$

where $K_i \in \mathbb{R}^{n\times n}$, $\Lambda_i \in \mathbb{R}^{l_i \times l_i}$ are positive definite gain matrices, and

$$\begin{aligned} q_i^p &= F_0 v_i \\ q_i^v &= F_0 S_0 v_i \\ \zeta_i &= q_i^v - \alpha_i (q_i - q_i^p) \\ s_i &= \dot{q}_i - \zeta_i \end{aligned}$$

with $\alpha_i > 0$. It is noted that (5.28) is in the form (5.25) with $\varphi_i = \text{col}(\hat{\Theta}_i, v_i)$ and $\varphi_0 = y_{m0}$.

We are ready to present our first result for solving Problem 5.3 using the distributed observer (5.26).

Theorem 5.3 *Given systems (5.23), (5.24) and a switching communication graph $\bar{\mathcal{G}}_{\sigma(t)}$, under Assumption 4.1, Problem 5.3 is solvable by the control law (5.28) for both of the following two cases:*

- *Assumptions 4.3, 4.4, and 5.4 are satisfied;*
- *$W_0 = I_m$, Assumptions 5.3 and 5.5 are satisfied.*

Proof By Remark 5.5, for both of the above two cases, we have

$$\lim_{t\to\infty} (v_i(t) - v_0(t)) = 0.$$

As a result, we have

$$\lim_{t\to\infty}(q_i^p(t)-q_0(t))=0,\ \lim_{t\to\infty}(q_i^v(t)-\dot{q}_0(t))=0.$$

Moreover, $q_i^p(t)$ and $q_i^v(t)$ are differentiable over each time interval $[t_k, t_{k+1})$ and satisfy

$$\lim_{t\to\infty}(\dot{q}_i^p(t)-\dot{q}_0(t))=0,\ \lim_{t\to\infty}(\dot{q}_i^v(t)-\ddot{q}_0(t))=0.$$

Therefore, the proof is complete by Theorem 5.2. □

If the communication graph is static, then, by Theorems 4.3, 4.4 and 5.2, we can obtain the following corollary of Theorem 5.3.

Corollary 5.1 *Given systems (5.23), (5.24) and a static communication graph $\bar{\mathscr{G}}$, under Assumptions 4.2, 4.4, 5.3, and 5.5, the control law (5.28) solves Problem 5.3 with $L_0=\mu_v P_0 W_0^{\mathrm{T}}$ where $P_0>0$ is the solution of $P_0 S_0^{\mathrm{T}}+S_0 P_0-P_0 W_0^{\mathrm{T}} W_0 P_0+I_m=0$ and $\mu_v \geq \frac{1}{2}\lambda_H^{-1}$. Also, if $W_0=I_m$, then, under Assumptions 4.2, 5.3 and 5.5, the control law (5.28) solves Problem 5.3 with $L_0=\mu_v I_m$ where $\mu_v>0$.*

Next, we consider a distributed control law utilizing the adaptive distributed observer with $y_{m0}=v_0$ as follows:

$$u_i = -K_i s_i + Y_i(q_i,\dot{q}_i,\dot{\zeta}_i,\zeta_i)\hat{\Theta}_i \tag{5.29a}$$

$$\dot{\hat{\Theta}}_i = -\Lambda_i^{-1} Y_i^{\mathrm{T}}(q_i,\dot{q}_i,\dot{\zeta}_i,\zeta_i)s_i \tag{5.29b}$$

$$\dot{S}_i = \mu_S \sum_{j=0}^{N} a_{ij}(t)(S_j - S_i) \tag{5.29c}$$

$$\dot{F}_i = \mu_F \sum_{j=0}^{N} a_{ij}(t)(F_j - F_i) \tag{5.29d}$$

$$\dot{v}_i = S_i v_i + \mu_v \sum_{j=0}^{N} a_{ij}(t)(v_j - v_i) \tag{5.29e}$$

where $\mu_S, \mu_F, \mu_v > 0$, $K_i \in \mathbb{R}^{n\times n}$, $\Lambda_i \in \mathbb{R}^{l_i \times l_i}$ are positive definite gain matrices, and

$$\begin{aligned} q_i^p &= F_i v_i \\ q_i^v &= F_i S_i v_i \\ \zeta_i &= q_i^v - \alpha_i(q_i - q_i^p) \\ s_i &= \dot{q}_i - \zeta_i \end{aligned}$$

with $\alpha_i > 0$. It is noted that (5.29) is also in the form (5.25) with $\varphi_i = \mathrm{col}(\hat{\Theta}_i, \mathrm{vec}(S_i), \mathrm{vec}(F_i), v_i)$ and $\varphi_0 = \mathrm{col}(\mathrm{vec}(S_0), \mathrm{vec}(F_0), v_0)$.

Theorem 5.4 *Given systems (5.23), (5.24) and a switching communication graph $\bar{\mathcal{G}}_{\sigma(t)}$, under Assumption 4.1, Problem 5.3 is solvable by the control law (5.29) for both of the following two cases:*

1. *Assumption 5.4 is satisfied;*
2. *Assumptions 5.3 and 5.5 are satisfied.*

Proof By Theorem 4.8, under Assumptions 4.1 and 5.4 (or 5.5), for both two cases

$$\lim_{t\to\infty}(S_i(t)-S_0)=0,\ \lim_{t\to\infty}(F_i(t)-F_0)=0,\ \lim_{t\to\infty}(v_i(t)-v_0(t))=0.$$

As a result, we have

$$\lim_{t\to\infty}(q_i^p(t)-q_0(t))=0,\ \lim_{t\to\infty}(q_i^v(t)-\dot{q}_0(t))=0.$$

Moreover, $q_i^p(t)$ and $q_i^v(t)$ are differentiable over each time interval $[t_k, t_{k+1})$, and satisfy, noting $\lim_{t\to\infty}\dot{S}_i(t)=0$, $\lim_{t\to\infty}\dot{F}_i(t)=0$, and $\lim_{t\to\infty}\sum_{j=0}^{N}a_{ij}(t)(v_j(t)-v_i(t))=0$,

$$\begin{aligned}
&\lim_{t\to\infty}(\dot{q}_i^p(t)-\dot{q}_0(t))\\
&=\lim_{t\to\infty}(\dot{F}_i(t)v_i(t)+F_i(t)\dot{v}_i(t)-F_0S_0v_0(t))\\
&=\lim_{t\to\infty}(F_i(t)S_i(t)v_i(t)-F_0S_0v_0(t))=0
\end{aligned}$$

and

$$\begin{aligned}
&\lim_{t\to\infty}(\dot{q}_i^v(t)-\ddot{q}_0(t))\\
&=\lim_{t\to\infty}(\dot{F}_i(t)S_i(t)v_i(t)+F_i(t)(\dot{S}_i(t)v_i(t)+S_i(t)\dot{v}_i(t))-F_0S_0^2v_0(t)))\\
&=\lim_{t\to\infty}(F_i(t)S_i^2(t)v_i(t)-F_0S_0^2v_0(t)))=0.
\end{aligned}$$

Therefore, the proof is complete by Theorem 5.2. □

Remark 5.6 In Theorem 5.4, it is also possible to use Theorem 4.7 to handle the case where y_{m0} is not equal to v_0. Details are skipped.

If the communication graph is static, then, by Theorems 4.6 and 5.2, we can obtain the following corollary of Theorem 5.4.

Corollary 5.2 *Given systems (5.23), (5.24) and a static communication graph $\bar{\mathcal{G}}$, under Assumptions 4.2, 5.3, and 5.5, the control law (5.29) solves Problem 5.3.*

Finally, we consider the case where the leader system contains uncertain parameters as described in (4.35). In this case, we will solve Problem 5.3 using the adaptive distributed observer (4.38) for the following uncertain leader system:

$$\dot{v}_0 = \mathbf{S}(\omega_0)v_0 \tag{5.30a}$$
$$q_0 = F_0 v_0 \tag{5.30b}$$
$$y_{m0} = v_0. \tag{5.30c}$$

For convenience, we assume that the control of every follower subsystem can access F_0. In this case, the distributed control law takes the following form:

$$u_i = -K_i s_i + Y_i(q_i, \dot{q}_i, \dot{\zeta}_i, \zeta_i)\hat{\Theta}_i \tag{5.31a}$$
$$\dot{\hat{\Theta}}_i = -\Lambda_i^{-1} Y_i^{\mathsf{T}}(q_i, \dot{q}_i, \dot{\zeta}_i, \zeta_i)s_i \tag{5.31b}$$
$$\dot{v}_i = \mathbf{S}(\omega_i)v_i + \mu_v \sum_{j=0}^{N} a_{ij}(v_j - v_i) \tag{5.31c}$$
$$\dot{\omega}_i = \mu_\omega \phi\left(\sum_{j=0}^{N} a_{ij}(v_j - v_i)\right) v_i \tag{5.31d}$$

where $\mu_v, \mu_\omega > 0$, $K_i \in \mathbb{R}^{n\times n}$, $\Lambda_i \in \mathbb{R}^{l_i \times l_i}$ are positive definite gain matrices, and

$$\begin{aligned} q_i^p &= F_0 v_i \\ q_i^v &= F_0 \mathbf{S}(\omega_i)v_i \\ \zeta_i &= q_i^v - \alpha_i(q_i - q_i^p) \\ s_i &= \dot{q}_i - \zeta_i \end{aligned}$$

with $\alpha_i > 0$.

Theorem 5.5 *Given systems* (5.23), (5.30) *and a static communication graph* $\bar{\mathscr{G}}$, *under Assumptions* 4.7 *and* 4.8, *Problem* 5.3 *is solvable by the control law* (5.31).

Proof By Theorem 4.9, under Assumptions 4.7 and 4.8, both $v_i(t)$ and $\omega_i(t)$ are uniformly bounded over $[0, \infty)$ and satisfy

$$\lim_{t\to\infty}(v_i(t) - v_0(t)) = 0,\ \lim_{t\to\infty}\dot{\omega}_i(t) = 0,\ \lim_{t\to\infty}\mathbf{S}(\omega_i(t) - \omega_0)v_i(t) = 0.$$

Then, we have

$$\begin{aligned} &\lim_{t\to\infty}(\mathbf{S}(\omega_i(t))v_i(t) - \mathbf{S}(\omega_0)v_0(t)) \\ =&\lim_{t\to\infty}(\mathbf{S}(\omega_i(t))v_i(t) - \mathbf{S}(\omega_0)v_i(t) + \mathbf{S}(\omega_0)v_i(t) - \mathbf{S}(\omega_0)v_0(t)) \\ =&\lim_{t\to\infty}(\mathbf{S}(\omega_i(t) - \omega_0)v_i(t) + \mathbf{S}(\omega_0)(v_i(t) - v_0(t))) \\ =&\,0 \end{aligned} \tag{5.32}$$

and

$$
\begin{aligned}
&\lim_{t\to\infty}(\mathbf{S}(\omega_i(t))^2 v_i(t) - \mathbf{S}(\omega_0)^2 v_0(t)) \\
&= \lim_{t\to\infty}(\mathbf{S}(\omega_i(t))^2 v_i(t) - \mathbf{S}(\omega_0)^2 v_i(t) + \mathbf{S}(\omega_0)^2 v_i(t) - \mathbf{S}(\omega_0)^2 v_0(t)) \\
&= \lim_{t\to\infty}(\mathbf{S}(\omega_i(t)+\omega_0)\mathbf{S}(\omega_i(t)-\omega_0)v_i + \mathbf{S}(\omega_0)^2(v_i(t)-v_0(t))) \\
&= 0.
\end{aligned}
\tag{5.33}
$$

Noting that

$$q_i^p - q_0 = F_0 v_i - F_0 v_0,\ q_i^v - \dot{q}_0 = F_0\mathbf{S}(\omega_i)v_i - F_0\mathbf{S}(\omega_0)v_0$$

gives

$$\lim_{t\to\infty}(q_i^p(t) - q_0(t)) = 0,\ \lim_{t\to\infty}(q_i^v(t) - \dot{q}_0(t)) = 0.$$

Since $\lim_{t\to\infty}\dot{\omega}_i(t) = 0$ and $v_i(t)$ is uniformly bounded over $[0,\infty)$, we have

$$\lim_{t\to\infty} F_0\mathbf{S}(\dot{\omega}_i(t))v_i(t) = 0.$$

Also, since $q_i^p(t)$ and $q_i^v(t)$ are differentiable for all $t \ge 0$, we have

$$\dot{q}_i^p - \dot{q}_0 = F_0\dot{v}_i - F_0\dot{v}_0 = F_0(\mathbf{S}(\omega_i)v_i - \mathbf{S}(\omega_0)v_0) + F_0\mu_v\sum_{j=0}^{N}a_{ij}(v_j - v_i) \tag{5.34}$$

and

$$\dot{q}_i^v - \ddot{q}_0 = F_0\mathbf{S}(\dot{\omega}_i)v_i + F_0\mathbf{S}^2(\omega_i)v_i + \mu_v F_0\mathbf{S}(\omega_i)\sum_{j=0}^{N}a_{ij}(v_j - v_i) - F_0\mathbf{S}^2(\omega_0)v_0. \tag{5.35}$$

Thus, by noting $\lim_{t\to\infty}\sum_{j=0}^{N}a_{ij}(v_j(t) - v_i(t)) = 0$, (5.32) and (5.34) imply

$$\lim_{t\to\infty}(\dot{q}_i^p(t) - \dot{q}_0(t)) = 0$$

and (5.33) and (5.35) imply

$$\lim_{t\to\infty}(\dot{q}_i^v(t) - \ddot{q}_0(t)) = 0.$$

Therefore, the proof is complete by Theorem 5.2. □

5.5 Numerical Examples

In this section, we illustrate our design methods by numerical examples.

Example 5.1 q_0 **and** $\dot{q}_0$ **are uniformly bounded** In this case, as shown in Fig. 5.1, we consider a group of four three-link cylindrical robot arms whose motion behaviors are described by

$$M_i(q_i)\ddot{q}_i + C_i(q_i, \dot{q}_i)\dot{q}_i + G_i(q_i) = u_i$$

where $q_i = \mathrm{col}(\theta_i, h_i, r_i)$ and

$$M_i(q_i) = \begin{bmatrix} J_i + m_{i2}r_i^2 & 0 & 0 \\ 0 & m_{i1} + m_{i2} & 0 \\ 0 & 0 & m_{i2} \end{bmatrix}$$

$$C_i(q_i, \dot{q}_i) = \begin{bmatrix} m_{i2}r_i\dot{r}_i & 0 & m_{i2}r_i\dot{\theta}_i \\ 0 & 0 & 0 \\ -m_{i2}r_i\dot{\theta}_i & 0 & 0 \end{bmatrix}$$

$$G_i(q_i) = \begin{bmatrix} 0 \\ (m_{i1} + m_{i2})g \\ 0 \end{bmatrix}$$

where for the ith follower, J_i is the moment of inertia of the base link; m_{i1} and m_{i2} are the masses of the vertical link and the horizontal link, respectively. We assume that J_i, m_{i1}, and m_{i2} are unknown. The actual values of the unknown parameters are $J_i = (1 + 0.1i)$ kg·m^2, $m_{i1} = (2 + 0.1i)$ kg, $m_{i2} = (3 + 0.1i)$ kg for $i = 1, 2, 3, 4$.

Let $x = \mathrm{col}(x_1, x_2, x_3)$ and $y = \mathrm{col}(y_1, y_2, y_3)$. Then

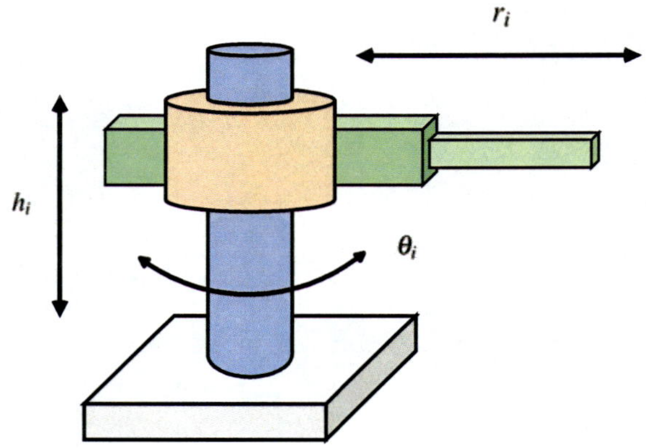

Fig. 5.1 Three-link cylindrical robot arm

$$\begin{aligned}
&M_i(q_i)x + C_i(q_i, \dot{q}_i)y + G_i(q_i) \\
&= \begin{bmatrix} (J_i + m_{i2}r_i^2)x_1 + m_{i2}r_i\dot{r}_i y_1 + m_{i2}r_i\dot{\theta}_i y_3 \\ (m_{i1} + m_{i2})x_2 + (m_{i1} + m_{i2})g \\ m_{i2}x_3 - m_{i2}r_i\dot{\theta}_i y_1 \end{bmatrix} \\
&= \begin{bmatrix} x_1 & 0 & r_i^2 x_1 + r_i\dot{r}_i y_1 + r_i\dot{\theta}_i y_3 \\ 0 & x_2 + g & x_2 + g \\ 0 & 0 & x_3 - r_i\dot{\theta}_i y_1 \end{bmatrix} \cdot \begin{bmatrix} J_i \\ m_{i1} \\ m_{i2} \end{bmatrix} \\
&\triangleq Y(q_i, \dot{q}_i, x, y)\Theta_i
\end{aligned}$$

where

$$\begin{aligned}
Y(q_i, \dot{q}_i, x, y) &= \begin{bmatrix} x_1 & 0 & r_i^2 x_1 + r_i\dot{r}_i y_1 + r_i\dot{\theta}_i y_3 \\ 0 & x_2 + g & x_2 + g \\ 0 & 0 & x_3 - r_i\dot{\theta}_i y_1 \end{bmatrix} \\
\Theta_i &= \begin{bmatrix} J_i \\ m_{i1} \\ m_{i2} \end{bmatrix}.
\end{aligned}$$

The leader system of the form (5.24) is defined by the following three matrices:

$$S_0 = \begin{bmatrix} 0 & 0 & 0 \\ 0 & 0 & 4 \\ 0 & -4 & 0 \end{bmatrix}, \quad F_0 = \begin{bmatrix} \pi/6 & 0 & 0 \\ 0 & 0.1 & 0 \\ 0 & 0 & 0.4 \end{bmatrix}, \quad W_0 = I_3.$$

Clearly, Assumption 5.4 is satisfied.

The communication graph is shown in Fig. 5.2, where the switching signal is defined as follows:

$$\sigma(t) = \begin{cases} 1 \text{ if } 0.4s \le t < 0.4s + 0.1 \\ 2 \text{ if } 0.4s + 0.1 \le t < 0.4s + 0.2 \\ 3 \text{ if } 0.4s + 0.2 \le t < 0.4s + 0.3 \\ 4 \text{ if } 0.4s + 0.3 \le t < 0.4s + 0.4 \end{cases}$$

where $s = 0, 1, 2, \ldots$. It can be seen that Assumption 4.1 is satisfied.

By Theorem 5.3, we can synthesize a distributed control law of the form (5.28) utilizing the distributed observer. Since $W_0 = I_3$, we choose $L_0 = \mu_v I_3$ with $\mu_v = 20$. Other design parameters are $\alpha_i = 10$, $K_i = 10I_3$, and $\Lambda_i = 0.2I_3$.

By Theorem 5.4, we can also synthesize a distributed control law of the form (5.29) utilizing the adaptive distributed observer. Various design parameters are selected to be $\mu_S = 20$, $\mu_F = 20$, $\mu_v = 20$, $\alpha_i = 10$, $K_i = 10I_3$, and $\Lambda_i = 0.2I_3$.

To evaluate the performance of the two control laws, let the desired generalized position vector $q_0(t) = \text{col}(\pi/6, 0.1\sin 4t, 0.4\cos 4t)$, which can be generated by the leader system with $v_0(0) = \text{col}(1, 0, 1)$.

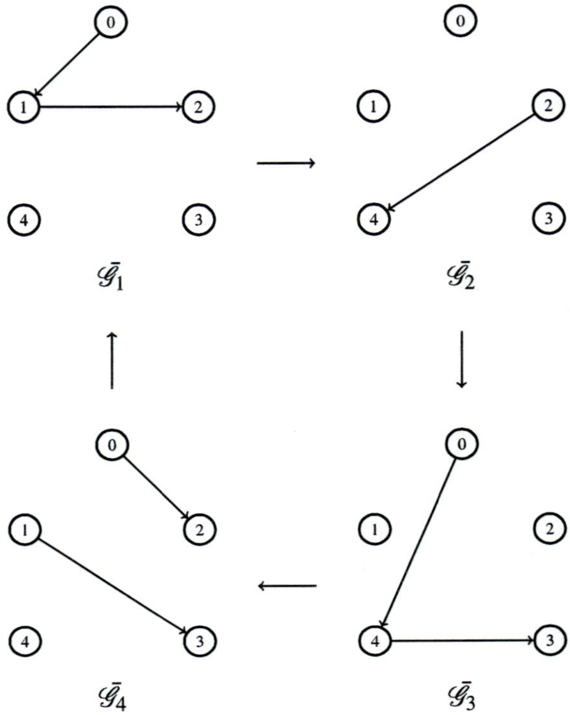

Fig. 5.2 The switching communication graph $\bar{\mathcal{G}}_{\sigma(t)}$ satisfying Assumption 4.1

Let us first evaluate the control law (5.28) with the initial values being given by $q_i(0) = 0$, $\dot{q}_i(0) = 0$, $v_i(0) = 0$, $\hat{\Theta}_i(0) = 0$. Figure 5.3 shows the tracking performance of the position and velocity for each link. It is observed that all the tracking errors tend to zero asymptotically.

Next, we evaluate the control law (5.29) with the initial values being given by $q_i(0) = 0$, $\dot{q}_i(0) = 0$, $v_i(0) = 0$, $S_i(0) = 0$, $F_i(0) = 0$, $\hat{\Theta}_i(0) = 0$. Figure 5.4 shows the tracking performance of the position and velocity for each link. Again, all the tracking errors tend to zero asymptotically.

Example 5.2 q_0 **is a ramp function** In this case, as shown by Fig. 5.5, we consider a group of four two-link revolute robot arms whose motion behaviors are described by

$$M_i(q_i)\ddot{q}_i + C_i(q_i, \dot{q}_i)\dot{q}_i + G_i(q_i) = u_i$$

where $q_i = \text{col}(\theta_{i1}, \theta_{i2})$ and

Fig. 5.3 The response of $q_{ij}(t) - q_{0j}(t)$ and $\dot{q}_{ij}(t) - \dot{q}_{0j}(t)$ under the control law (5.28)

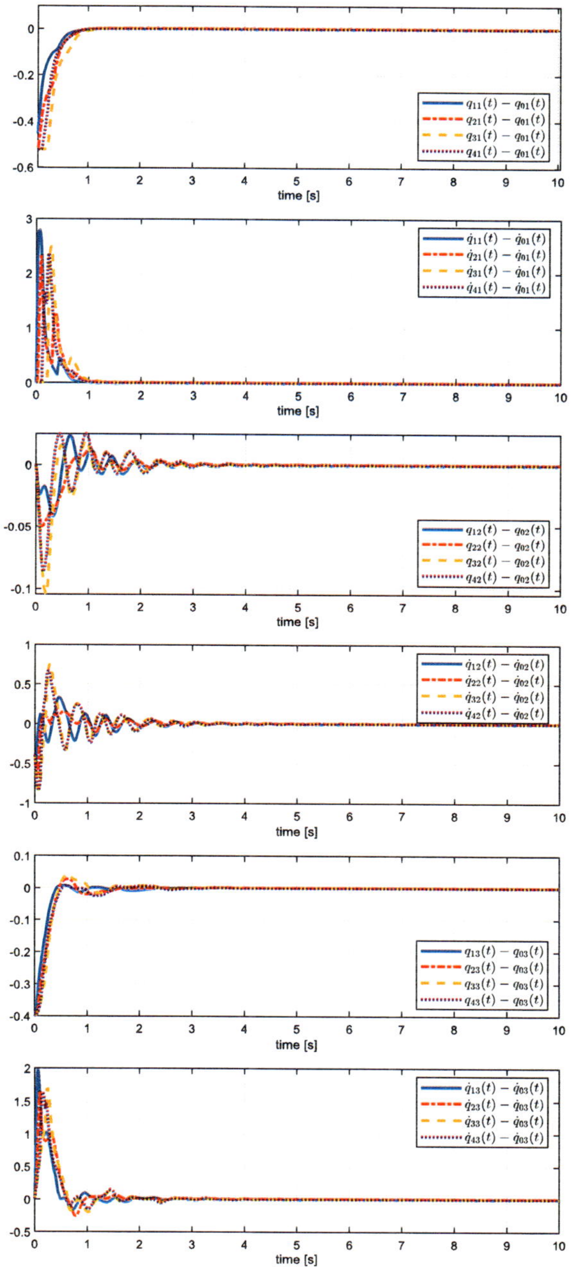

Fig. 5.4 The response of $q_{ij}(t) - q_{0j}(t)$ and $\dot{q}_{ij}(t) - \dot{q}_{0j}(t)$ under the control law (5.29)

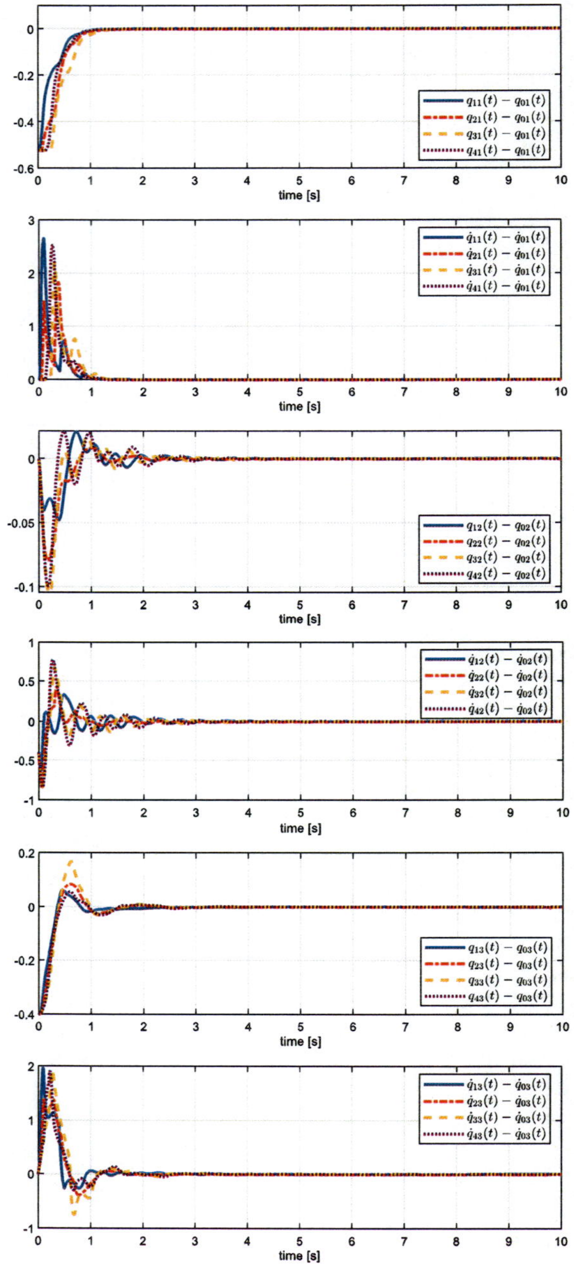

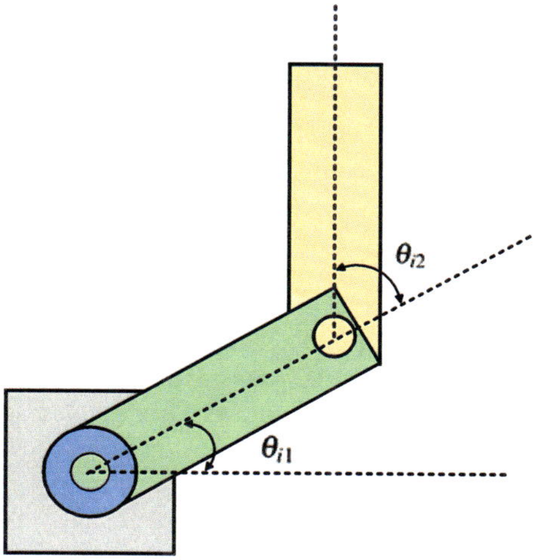

Fig. 5.5 Two-link revolute robot arm

$$M_i(q_i) = \begin{bmatrix} a_{i1} + a_{i2} + 2a_{i3}\cos\theta_{i2} & a_{i2} + a_{i3}\cos\theta_{i2} \\ a_{i2} + a_{i3}\cos\theta_{i2} & a_{i2} \end{bmatrix}$$
$$C_i(q_i, \dot{q}_i) = \begin{bmatrix} -a_{i3}(\sin\theta_{i2})\dot{\theta}_{i2} & -a_{i3}(\sin\theta_{i2})(\dot{\theta}_{i1} + \dot{\theta}_{i2}) \\ a_{i3}(\sin\theta_{i2})\dot{\theta}_{i1} & 0 \end{bmatrix}$$
$$G_i(q_i) = \begin{bmatrix} a_{i4}g\cos\theta_{i1} + a_{i5}g\cos(\theta_{i1} + \theta_{i2}) \\ a_{i5}g\cos(\theta_{i1} + \theta_{i2}) \end{bmatrix}$$

where $\Theta_i = \text{col}(a_{i1}, a_{i2}, a_{i3}, a_{i4}, a_{i5})$ are the unknown system parameters. The actual values of Θ_i are as follows:

$$\begin{aligned} \Theta_1 &= \text{col}(0.64, 1.10, 0.08, 0.64, 0.32) \\ \Theta_2 &= \text{col}(0.76, 1.17, 0.14, 0.93, 0.44) \\ \Theta_3 &= \text{col}(0.91, 1.26, 0.22, 1.27, 0.58) \\ \Theta_4 &= \text{col}(1.10, 1.36, 0.32, 1.67, 0.73). \end{aligned}$$

Let $x = \text{col}(x_1, x_2)$ and $y = \text{col}(y_1, y_2)$. Then

$$\begin{aligned} & M_i(q_i)x + C_i(q_i, \dot{q}_i)y + G_i(q_i) \\ &= \left[\begin{array}{c} (a_{i1} + a_{i2} + 2a_{i3}\cos\theta_{i2})x_1 + (a_{i2} + a_{i3}\cos\theta_{i2})x_2 \\ -a_{i3}(\sin\theta_{i2})\dot{\theta}_{i2}y_1 - a_{i3}(\sin\theta_{i2})(\dot{\theta}_{i1} + \dot{\theta}_{i2})y_2 \\ +a_{i4}g\cos\theta_{i1} + a_{i5}g\cos(\theta_{i1} + \theta_{i2}) \\ \hline (a_{i2} + a_{i3}\cos\theta_{i2})x_1 + a_{i2}x_2 \\ +a_{i3}(\sin\theta_{i2})\dot{\theta}_{i1}y_1 + a_{i5}g\cos(\theta_{i1} + \theta_{i2}) \end{array}\right] \\ &= Y(q_i, \dot{q}_i, x, y)\Theta_i \end{aligned}$$

with

$$\begin{aligned}
&Y_{11} = x_1, \; Y_{12} = x_1 + x_2 \\
&Y_{13} = 2(\cos\theta_{i2})x_1 + (\cos\theta_{i2})x_2 - (\sin\theta_{i2})\dot{\theta}_{i2}y_1 - (\sin\theta_{i2})(\dot{\theta}_{i1} + \dot{\theta}_{i2})y_2 \\
&Y_{14} = g\cos\theta_{i1}, \; Y_{15} = g\cos(\theta_{i1} + \theta_{i2}), \; Y_{21} = 0 \\
&Y_{22} = x_1 + x_2, \; Y_{23} = (\cos\theta_{i2})x_1 + (\sin\theta_{i2})\dot{\theta}_{i1}y_1 \\
&Y_{24} = 0, \; Y_{25} = g\cos(\theta_{i1} + \theta_{i2})
\end{aligned}$$

where Y_{mn} is the element of Y in the mth row and nth column. Since in both $C(q_i, \dot{q}_i)$ and $Y(q_i, \dot{q}_i, x, y)$, q_i are in the sinusoidal form, Assumption 5.3 is satisfied.

The leader system of the form (5.24) is defined by the following three matrices:

$$S_0 = \begin{bmatrix} 0 & 1 \\ 0 & 0 \end{bmatrix}, \; F_0 = \begin{bmatrix} 0.1 & 1 \\ 0.5 & 2 \end{bmatrix}, \; W_0 = I_2.$$

Clearly, Assumption 5.5 is satisfied.

We adopt the same communication graph as used in Example 5.1.

By Theorem 5.3, we can synthesize a distributed control law of the form (5.28) utilizing the distributed observer. Since $W_0 = I_2$, we choose $L_0 = \mu_v I_2$ with $\mu_v = 10$. Other design parameters are $\alpha_i = 10$, $K_i = 20I_2$, and $\Lambda_i = 0.2I_5$.

By Theorem 5.4, we can also synthesize a distributed control law of the form (5.29) utilizing the adaptive distributed observer. The control gains are selected to be $\mu_S = 10$, $\mu_F = 10$, $\mu_v = 10$, $\alpha_i = 10$, $K_i = 20I_2$, and $\Lambda_i = 0.2I_2$.

To evaluate the performance of the two control laws, let the desired generalized position vector $q_0(t) = \text{col}(0.1t + 1, 0.5t + 2)$, which can be generated by the leader system with $v_0(0) = \text{col}(0, 1)$.

To examine the system performance under the control law (5.28), let the initial values be given by $q_i(0) = 0, \dot{q}_i(0) = 0, v_i(0) = 0, \hat{\Theta}_i(0) = 0$. Figure 5.6 shows the tracking performance of the position and velocity for each link.

Next, to examine the system performance under the control law (5.29), let the initial values be given by $q_i(0) = 0$, $\dot{q}_i(0) = 0$, $v_i(0) = 0$, $S_i(0) = 0$, $F_i(0) = 0$, $\hat{\Theta}_i(0) = 0$. Figure 5.7 shows the tracking performance of the position and velocity for each link.

Example 5.3 Uncertain leader system We consider an uncertain leader system of the form (5.30) with

$$\mathbf{S}(\omega_0) = \begin{bmatrix} 0 & \omega_0 \\ -\omega_0 & 0 \end{bmatrix}, \; F_0 = I_2, W_0 = I_2, \tag{5.36}$$

where $\omega_0 > 0$ is unknown. Clearly, Assumption 4.8 is satisfied. With $v_0(0) = \text{col}(0, 1)$, the leader system (5.36) will generate a reference signal $q_0(t) = \text{col}(\sin\omega_0 t, \cos\omega_0 t)$ with unknown frequency ω_0. We consider the same Euler–Lagrange system

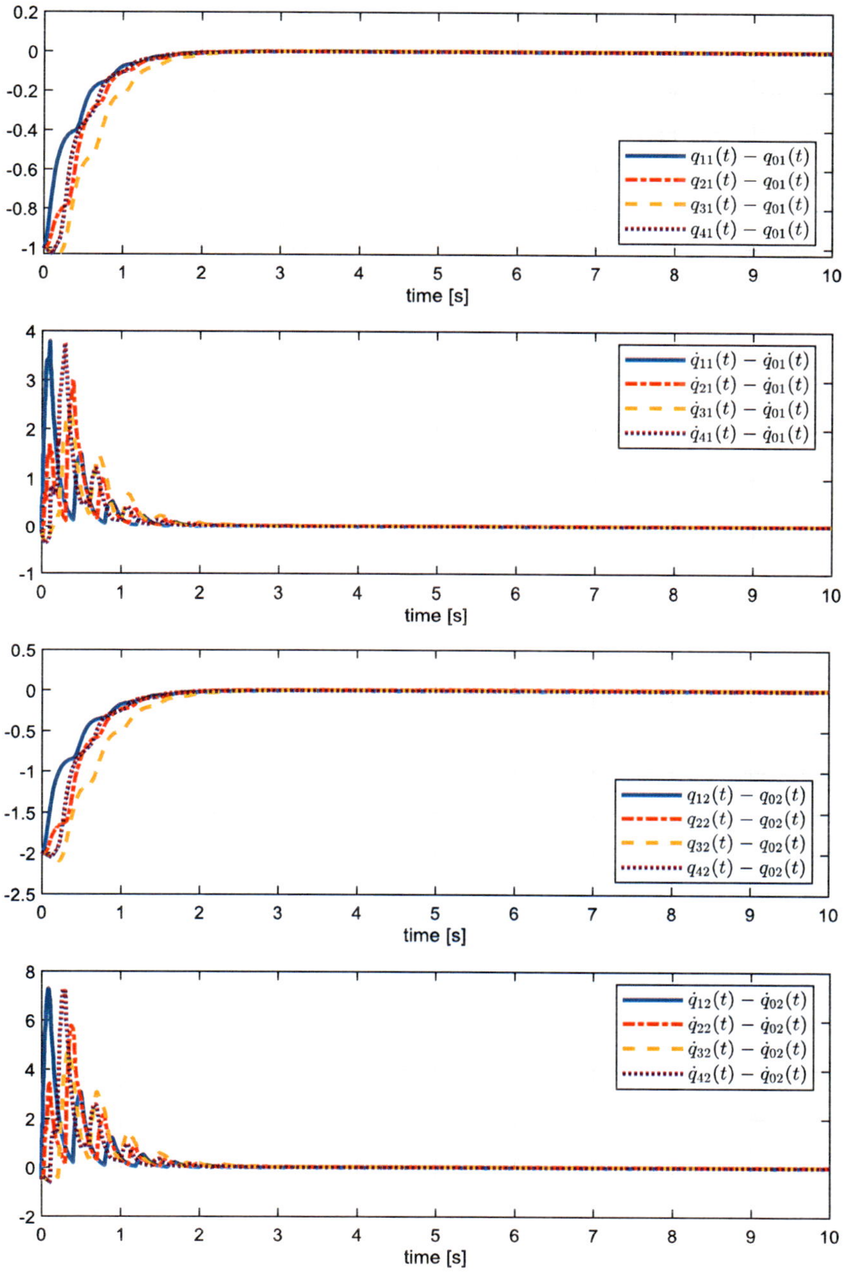

Fig. 5.6 The response of $q_{ij}(t) - q_{0j}(t)$ and $\dot{q}_{ij}(t) - \dot{q}_{0j}(t)$ under the control law (5.28)

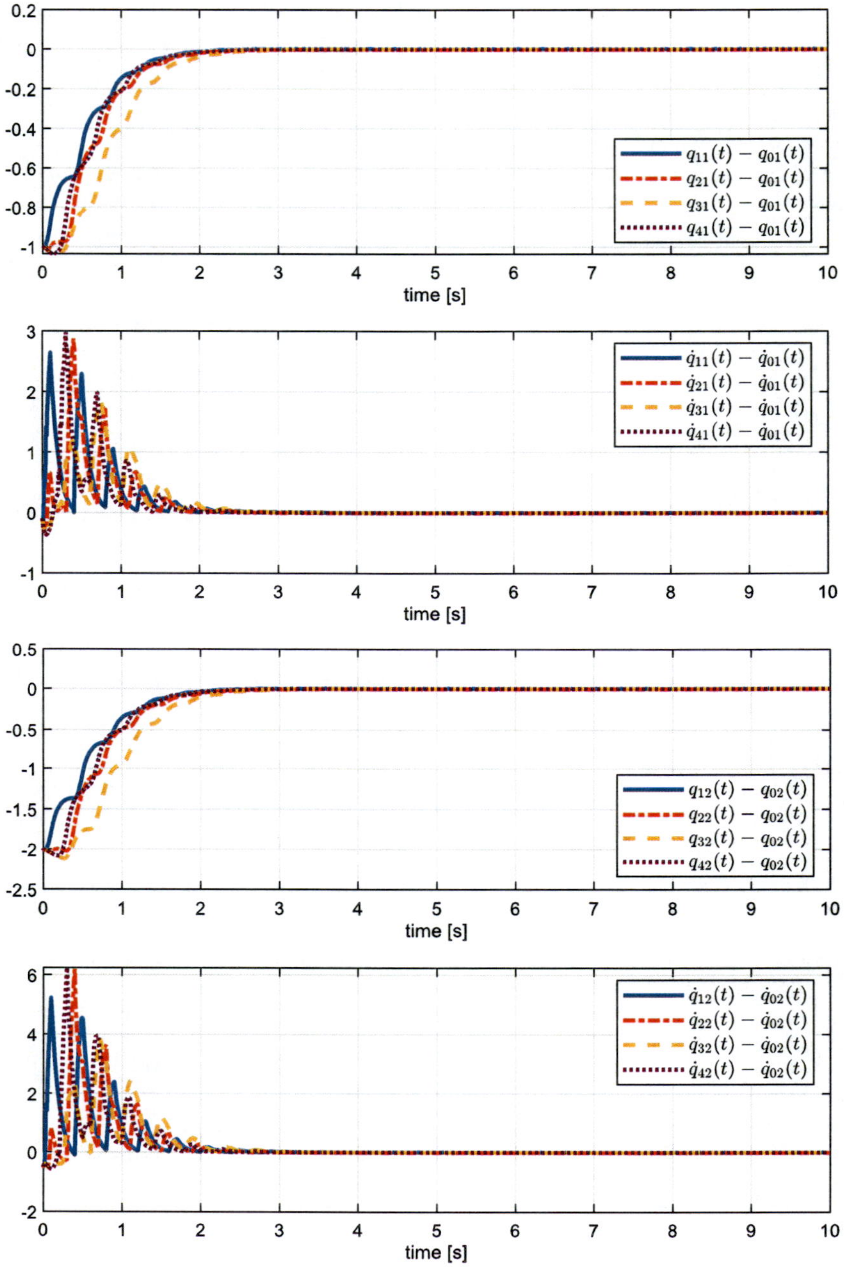

Fig. 5.7 The response of $q_{ij}(t) - q_{0j}(t)$ and $\dot{q}_{ij}(t) - \dot{q}_{0j}(t)$ under the control law (5.29)

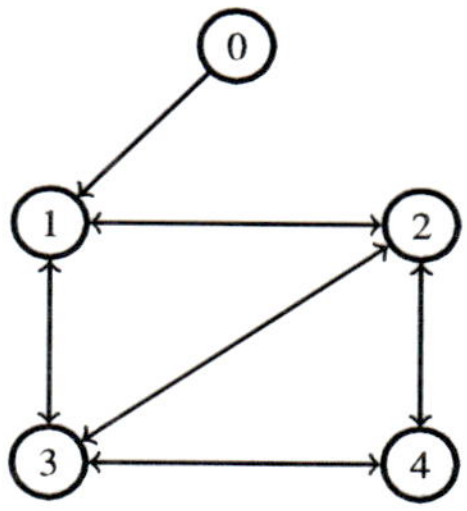

Fig. 5.8 Static communication graph $\bar{\mathscr{G}}$ satisfying Assumption 4.7

as in Example 5.2. The communication graph is shown in Fig. 5.8, which satisfies Assumption 4.7.

Thus, by Theorem 5.5, a distributed controller of the form (5.31) can be synthesized. The control gains are selected to be $\mu_v = 10$, $\mu_\omega = 10$, $\alpha_i = 10$, $K_i = 20I_2$, and $\Lambda_i = 0.2I_5$. The initial values are given by $q_i(0) = 0$, $\dot{q}_i(0) = 0$, $v_i(0) = 0$, $\omega_i(0) = 0$, $\hat{\Theta}_i(0) = 0$. Figure 5.9 shows the tracking performance of the position and velocity for each link and the convergence of the estimated frequency of the reference signal to the actual value $\omega_0 = 2$.

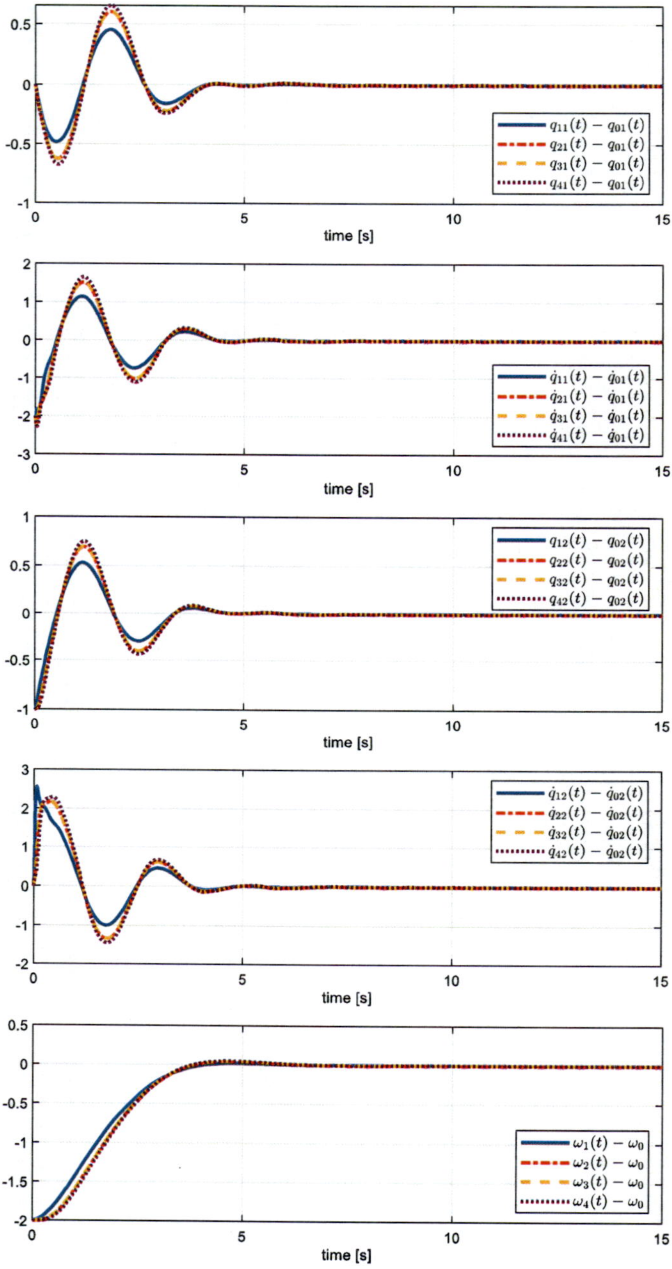

Fig. 5.9 The response of $q_{ij}(t) - q_{0j}(t)$, $\dot{q}_{ij}(t) - \dot{q}_{0j}(t)$, and $\omega_i(t) - \omega_0$ under the control law (5.31)

5.6 Notes and References

Fundamental properties of Euler–Lagrange systems can be found in, say, [1, 2]. Early papers on the trajectory tracking control of Euler–Lagrange systems by the adaptive sliding mode approach can be found in [3–5], which were later documented in [1]. For the leader-following consensus problem of multiple Euler–Lagrange systems, reference [6] employed the neural networks approximation method to deal with the nonlinear system dynamics and achieved uniformly ultimately bounded tracking. An observer-like structure was proposed in [7] to estimate the leader's state which facilitates the design of a position feedback controller. Note that all the above results were obtained over static communication networks. The leader-following consensus problem of multiple Euler–Lagrange systems over switching communication networks was first solved by [8] by the distributed observer approach. Later, for the special case where all the eigenvalues of S_0 are semi-simple with zero real parts, reference [9] further extended the results of [8] by proposing the adaptive distributed observer approach, and reference [10] extended the results of [9] by allowing S_0 to have eigenvalues with non-positive real parts. Reference [11] considered the case of uncertain leader systems. Besides the leader-following consensus problem, reference [12] considered concurrent synchronization of Euler–Lagrange systems by the contraction analysis. References [13, 14] considered the effect of time delay of communication networks for the leaderless consensus problem of multiple Euler–Lagrange systems. Note that both [12, 14] also considered the distributed tracking problem in the sense that the reference signal is known by all the followers.

The detailed models for the three-link cylindrical robot arm and the two-link revolute robot arm in Sect. 5.5 can be found, say, from [2].

The leader-following consensus problem of a class of multiple robot manipulators over switching communication networks was studied by the position feedback control utilizing the adaptive distributed observer in [15], and the work was further extended to more general multiple Euler–Lagrange systems in [16]. The leader-following consensus with disturbance rejection problem of multiple Euler–Lagrange systems was studied in [10] where the disturbances are multi-tone sinusoidal signals, and in [17] where the disturbances are any bounded signals.

References

1. Slotine J-JE, Li W (1991) Applied nonlinear control. Prentice-Hall, Englewood Cliffs
2. Lewis FL, Dawson MD, Abdallah TC (2004) Robot manipulator control: theory and practice, 2nd edn. Marcel Dekker, New York
3. Slotine J-JE, Li W (1987) On the adaptive control of robot manipulators. Int J Robot Res 6(3):49–59
4. Slotine J-JE, Li W (1988) Adaptive manipulator control: a case study. IEEE Trans Autom Control 33(11):995–1003
5. Slotine J-JE, Li W (1989) Composite adaptive control of robot manipulators. Automatica 2(4):509–519

6. Chen G, Lewis FL (2011) Distributed adaptive tracking control for synchronization of unknown networked Lagrangian systems. IEEE Trans Syst Man Cybern B Cybern 41(3):805–816
7. Yang Q, Fang H, Chen J, Jiang Z, Cao M (2017) Distributed global output-feedback control for a class of Euler–Lagrange systems. IEEE Trans Autom Control 62(9):4855–4861
8. Cai H, Huang J (2014) Leader-following consensus of multiple uncertain Euler–Lagrange systems under switching network topology. Int J Gen Syst 43(3–4):294–304
9. Cai H, Huang J (2016) The leader-following consensus for multiple uncertain Euler–Lagrange systems with an adaptive distributed observer. IEEE Trans Autom Control 61(10):3152–3157
10. Liu T, Huang J (2019) Leader-following consensus with disturbance rejection for uncertain Euler–Lagrange systems over switching networks. Int J Robust Nonlinear Control 29(18):6638–6656
11. Wang S, Huang J (2019) Adaptive leader-following consensus for multiple Euler–Lagrange systems with an uncertain leader system. IEEE Trans Neural Netw Learn Syst 30(7):2188–2196
12. Chung S, Slotine J-JE (2009) Cooperative robot control and concurrent synchronization of Lagrangian systems. IEEE Trans Rob 25(3):686–700
13. Abdessameud A, Polushin IG, Tayebi A (2014) Synchronization of Lagrangian systems with irregular communication delays. IEEE Trans Autom Control 59(1):187–193
14. Nuño E, Ortega R, Basañez L, Hill D (2011) Synchronization of networks of nonidentical Euler–Lagrange systems with uncertain parameters and communication delays. IEEE Trans Autom Control 56(4):935–941
15. He C, Huang J (2020) Leader-following consensus for a class of multiple robot manipulators over switching networks by distributed position feedback control. IEEE Trans Autom Control 65(2):890–896
16. He C, Huang J (2021) Leader-following consensus for multiple Euler–Lagrange systems by distributed position feedback control. IEEE Trans Autom Control 66(11):5561–5568
17. Wang T, Huang J (2021) Leader-following consensus of multiple uncertain Euler-Lagrange systems subject to unknown disturbances over switching networks. In: The 40th Chinese Control Conference, July 26–28, 2021, Shanghai, China

Chapter 6
Leader-Following Consensus of Multiple Rigid-Body Systems

This chapter studies the leader-following consensus problem of multiple rigid-body systems. The attitude parametrization, kinematics, and dynamics of a rigid-body system are first given in Sect. 6.1. Then, in Sect. 6.2, we treat three cases of the standard attitude control problem for a single rigid-body system. Section 6.3 studies how to deal with the three cases of the attitude control problem for a single rigid-body system via various estimation-based control laws. On the basis of Sect. 6.3, we present, in Sect. 6.4, the solvability of the three cases of the leader-following consensus problem of multiple rigid-body systems for three scenarios by various distributed observer-based control laws. In Sect. 6.5, we use some numerical examples to illustrate our designs.

6.1 Attitude Parametrization, Kinematics, and Dynamics

6.1.1 Rotation Matrix

Consider the rotation of a rigid-body with respect to an inertial frame. As shown by Fig. 6.1, the inertial frame is denoted by $\mathscr{I}$, and the rigid-body is endowed with a body frame $\mathscr{F}$ whose origin coincides with the center of mass of the rigid-body. Since we only consider the rotation of the rigid-body, we assume that the origins of $\mathscr{I}$ and $\mathscr{F}$ are identical. Let $\{\mathbf{i}, \mathbf{j}, \mathbf{k}\}$ and $\{\mathbf{i}', \mathbf{j}', \mathbf{k}'\}$ be the orthonormal and dextral basis of $\mathscr{I}$ and $\mathscr{F}$, respectively. Then the rotation matrix $C_{FI} = [c_{ij}] \in \mathbb{R}^{3\times 3}$ is given as

$$C_{FI} = \begin{bmatrix} \mathbf{i}' \cdot \mathbf{i} & \mathbf{i}' \cdot \mathbf{j} & \mathbf{i}' \cdot \mathbf{k} \\ \mathbf{j}' \cdot \mathbf{i} & \mathbf{j}' \cdot \mathbf{j} & \mathbf{j}' \cdot \mathbf{k} \\ \mathbf{k}' \cdot \mathbf{i} & \mathbf{k}' \cdot \mathbf{j} & \mathbf{k}' \cdot \mathbf{k} \end{bmatrix}$$

H. Cai et al., *Cooperative Control of Multi-agent Systems*, Advances in Industrial Control,
https://doi.org/10.1007/978-3-030-98377-2_6

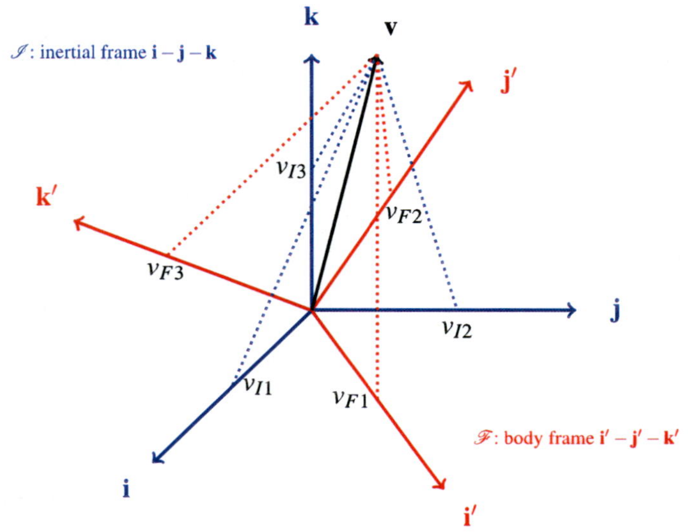

Fig. 6.1 Inertial frame and body frame

which satisfies $C_{FI}^{-1} = C_{FI}^{\mathrm{T}}$, and $\det(C_{FI}) = 1$. Since the elements of C_{FI} are inner products of two unit direction vectors, which are the cosine functions of the angles between the two unit direction vectors, the rotation matrix is also called direction cosine matrix. Let $C_{IF} = C_{FI}^{-1}$. Then, for any vector $\mathbf{v}$, letting $v_I = \mathrm{col}(v_{I1}, v_{I2}, v_{I3})$ and $v_F = \mathrm{col}(v_{F1}, v_{F2}, v_{F3})$ be the coordinates of $\mathbf{v}$ expressed in $\mathscr{I}$ and $\mathscr{F}$, respectively, gives

$$v_F = C_{FI} v_I \quad \text{and} \quad v_I = C_{FI}^{-1} v_F = C_{IF} v_F.$$

The rotation matrix C_{FI} defines the attitude of the rigid-body with respect to the inertial frame $\mathscr{I}$. However, C_{FI} has nine parameters c_{ij}, $i, j = 1, 2, 3$, which is redundant as a mathematical description for attitude due to the following constraints:

$$\begin{aligned}
\mathbf{i} \cdot \mathbf{j} = \mathbf{j} \cdot \mathbf{k} = \mathbf{k} \cdot \mathbf{i} = 0 \\
\mathbf{i} \cdot \mathbf{i} = \mathbf{j} \cdot \mathbf{j} = \mathbf{k} \cdot \mathbf{k} = 1 \\
\mathbf{i}' \cdot \mathbf{j}' = \mathbf{j}' \cdot \mathbf{k}' = \mathbf{k}' \cdot \mathbf{i}' = 0 \\
\mathbf{i}' \cdot \mathbf{i}' = \mathbf{j}' \cdot \mathbf{j}' = \mathbf{k}' \cdot \mathbf{k}' = 1.
\end{aligned}$$

To simplify the mathematical description for attitude, some other parameterizations of attitude with less parameters have been developed, among which the most widely used are Euler angles (3 parameters), Rodrigues parameters (3 parameters), modified Rodrigues parameters (3 parameters), and unit quaternion (4 parameters). In contrast to other three-parameter parameterizations, the unit quaternion is globally defined, computationally efficient, and free from kinematic singularity. Therefore, in this chapter, the unit quaternion is chosen as the mathematical description for attitude.

6.1.2 *Quaternion and Unit Quaternion*

A quaternion q is defined by

$$q = \begin{bmatrix} \hat{q} \\ \bar{q} \end{bmatrix}$$

where $\hat{q} \in \mathbb{R}^3$ and $\bar{q} \in \mathbb{R}$ are called the vector part and the scalar part of q, respectively. Let $\mathbb{Q}$ denote the set of all quaternion.

Given $q \in \mathbb{Q}$, the conjugate of q is defined as follows:

$$q^* = \begin{bmatrix} -\hat{q} \\ \bar{q} \end{bmatrix}.$$

Obviously, $\|q\| = \|q^*\|$. Moreover, if $\|q\| \neq 0$, the inverse of q is defined as follows:

$$q^{-1} = \frac{q^*}{\|q\|^2}.$$

Quaternion addition and scalar multiplication are defined in the same way as for vectors in $\mathbb{R}^4$. For $q_i, q_j \in \mathbb{Q}$, the quaternion multiplication is defined as follows:

$$q_i \odot q_j = \begin{bmatrix} \bar{q}_i \hat{q}_j + \bar{q}_j \hat{q}_i + \hat{q}_i^{\times} \hat{q}_j \\ \bar{q}_i \bar{q}_j - \hat{q}_i^{\mathsf{T}} \hat{q}_j \end{bmatrix}.$$

Due to the term of $\hat{q}_i^{\times} \hat{q}_j$, quaternion multiplication is not commutative, but is distributive and associative, i.e., for any $q_a, q_b, q_c \in \mathbb{Q}$,

$$q_a \odot (q_b + q_c) = q_a \odot q_b + q_a \odot q_c$$

and

$$q_a \odot (q_b \odot q_c) = (q_a \odot q_b) \odot q_c.$$

Moreover, for any $q_a, q_b \in \mathbb{Q}$,

$$(q_a \odot q_b)^* = q_b^* \odot q_a^*.$$

Let $q_I = \mathrm{col}(0, 0, 0, 1)$. Then, for any $q \in \mathbb{Q}$,

$$q \odot q_I = q_I \odot q = q$$

and, for any nonzero $q \in \mathbb{Q}$,

$$q \odot q^{-1} = q^{-1} \odot q = q_I.$$

Hence, q_I is called the quaternion identity.

If $q \in \mathbb{Q}$ and $\|q\| = 1$, then q is called a unit quaternion. Let $\mathbb{Q}_u$ denote the set of all unit quaternion. For $q \in \mathbb{Q}_u$, $q^{-1} = q^*/\|q\|^2 = q^*$.

More identities involving quaternion operations are provided in Sect. A.5.

6.1.3 Attitude Parameterized by Unit Quaternion

The Euler rotation theorem reveals that the general displacement of a rigid-body with one point fixed in an inertial frame is a rotation about an axis through that point. Suppose the fixed point coincides with both the origins of the body frame $\mathscr{F}$ and the inertial frame $\mathscr{I}$. Since the rotation matrix C_{FI} is orthonormal, its spectrum is given by

$$\lambda(C_{FI}) = \{1, e^{\imath\phi}, e^{-\imath\phi}\}, \ \phi \in \mathbb{R}$$

which indicates the existence of a vector $\mathbf{a}$ satisfying

$$C_{FI}\mathbf{a} = 1\mathbf{a} = \mathbf{a}, \ \mathbf{a}^{\mathsf{T}}\mathbf{a} = 1. \tag{6.1}$$

As shown by Fig. 6.2, physically, Eq. (6.1) means $\mathbf{a}$ is invariant during the rotation described by C_{FI}, and the body frame $\mathscr{F}$ can be viewed as the result of rotating $\mathscr{I}$ about the axis $\mathbf{a}$ by an angle ϕ. Let $\mathrm{tr}(C_{FI})$ denote the trace of C_{FI}, then it follows that

$$\mathrm{tr}(C_{FI}) = 1 + e^{\imath\phi} + e^{-\imath\phi} = 1 + 2\cos\phi.$$

Let a_I and a_F be the coordinates of $\mathbf{a}$ expressed in $\mathscr{I}$ and $\mathscr{F}$, respectively. By (6.1), a_I and a_F are identical. Let $a = a_I = a_F$ and

$$C(a, \phi) = \cos\phi I_3 + (1 - \cos\phi)aa^{\mathsf{T}} - \sin\phi a^{\times}.$$

We will show $C(a, \phi) = C_{FI}$. In fact, using Identities A.1 and A.3 in Sect. A.5 gives

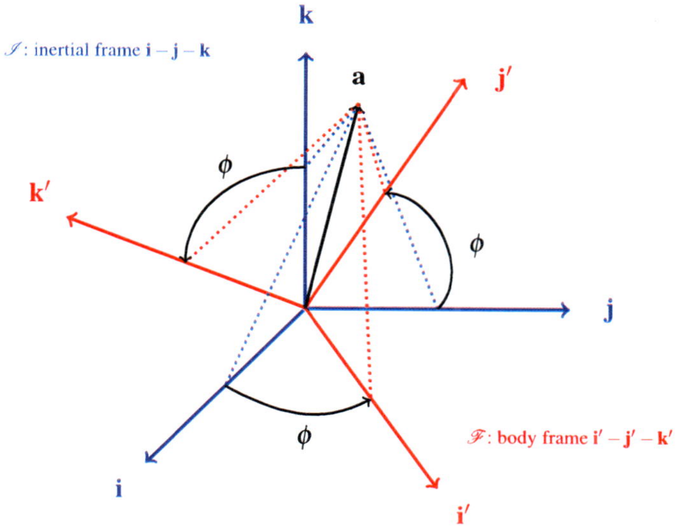

Fig. 6.2 Inertial frame, body frame, rotation axis, and rotation angle

$$\begin{aligned}
& C(a,\phi)^{\mathsf{T}} C(a,\phi) \\
&= (\cos\phi I_3 + (1-\cos\phi)aa^{\mathsf{T}} - \sin\phi a^{\times})^{\mathsf{T}}(\cos\phi I_3 + (1-\cos\phi)aa^{\mathsf{T}} - \sin\phi a^{\times}) \\
&= (\cos\phi)^2 I_3 + \cos\phi((1-\cos\phi)aa^{\mathsf{T}} - \sin\phi a^{\times}) \\
&\quad + (1-\cos\phi)\cos\phi aa^{\mathsf{T}} + (1-\cos\phi)^2 aa^{\mathsf{T}}aa^{\mathsf{T}} - (1-\cos\phi)\sin\phi aa^{\mathsf{T}}a^{\times} \\
&\quad + \cos\phi\sin\phi a^{\times} + \sin\phi(1-\cos\phi)a^{\times}aa^{\mathsf{T}} - (\sin\phi)^2 a^{\times}a^{\times} \\
&= (\cos\phi)^2 I_3 + (2\cos\phi - 2(\cos\phi)^2 + 1 - 2\cos\phi + (\cos\phi)^2)aa^{\mathsf{T}} \\
&\quad - (1-(\cos\phi)^2)(aa^{\mathsf{T}} - I_3) \\
&= I_3.
\end{aligned}$$

Therefore, $C(a,\phi)$ is an orthonormal matrix and thus $\det(C(a,\phi)) = \pm 1$. Since $\det(C(a,0)) = 1$, the continuity of $\det(C(a,\phi))$ in ϕ implies $\det(C(a,\phi)) = 1$. Hence $C(a,\phi)$ is a rotation matrix. Next, we have

$$C(a,\phi)a = \cos\phi a + (1-\cos\phi)aa^{\mathsf{T}}a - \sin\phi a^{\times}a = a$$

and

$$\operatorname{tr}(C(a,\phi)) = 3\cos\phi + 1 - \cos\phi = 1 + 2\cos\phi.$$

As a result, $C(a,\phi) = C_{FI}$.

Define the following quaternion:

$$q = \begin{bmatrix} \hat{q} \\ \bar{q} \end{bmatrix} = \begin{bmatrix} \sin\frac{\phi}{2}a \\ \cos\frac{\phi}{2} \end{bmatrix}. \tag{6.2}$$

Then $q \in \mathbb{Q}_u$ is the unit quaternion parametrization of the attitude of $\mathscr{F}$ with respect to $\mathscr{I}$. Since ϕ and $\phi' = \phi + 2\pi$ correspond to the same attitude, we have

$$-q = \begin{bmatrix} -\hat{q} \\ -\bar{q} \end{bmatrix} = \begin{bmatrix} -\sin\frac{\phi}{2}a \\ -\cos\frac{\phi}{2} \end{bmatrix} = \begin{bmatrix} \sin\frac{\phi+2\pi}{2}a \\ \cos\frac{\phi+2\pi}{2} \end{bmatrix} = \begin{bmatrix} \sin\frac{\phi'}{2}a \\ \cos\frac{\phi'}{2} \end{bmatrix}.$$

Thus, $-q$ represents the same attitude as q does.

Remark 6.1 By (6.2), if $\mathscr{I}$ and $\mathscr{F}$ coincide, then $\phi = k \cdot 2\pi$, $k = 0, \pm 1, \pm 2, \ldots$, and thus $\hat{q} = a\sin(k\pi) = 0$. On the other hand, if $\hat{q} = 0$, since $a \neq 0$, then $\phi = k \cdot 2\pi$, $k = 0, \pm 1, \pm 2, \ldots$, which indicates $\mathscr{I}$ and $\mathscr{F}$ coincide. Thus, $\mathscr{I}$ and $\mathscr{F}$ coincide if and only if $\hat{q} = 0$.

The rotation matrix C_{FI} is related to the unit quaternion q in the following way:

$$\begin{aligned} C_{FI} &= \cos\phi I_3 + (1 - \cos\phi)aa^{\mathsf{T}} - \sin\phi a^{\times} \\ &= (\bar{q}^2 - \hat{q}^{\mathsf{T}}\hat{q})I_3 + 2\hat{q}\hat{q}^{\mathsf{T}} - 2\bar{q}\hat{q}^{\times} = \mathbf{C}(q). \end{aligned} \tag{6.3}$$

On the other hand, by (6.3),

$$C_{IF} = C_{FI}^{\mathsf{T}} = (\bar{q}^2 - \hat{q}^{\mathsf{T}}\hat{q})I_3 + 2\hat{q}\hat{q}^{\mathsf{T}} + 2\bar{q}\hat{q}^{\times} = \mathbf{C}(q^{-1})$$

which indicates that q^{-1} represents the attitude of $\mathscr{I}$ with respect to $\mathscr{F}$. Moreover, $\mathbf{C}(\pm q_I) = I_3$.

By Identity A.8 in Sect. A.5, for any $q \in \mathbb{Q}_u$ and $x \in \mathbb{R}^3$,

$$q^* \odot \mathbf{Q}(x) \odot q = \mathbf{Q}(\mathbf{C}(q)x)$$

where the operation $q^* \odot \mathbf{Q}(x) \odot q$ is called the adjoint operation. For a vector $\mathbf{v}$ whose coordinates in two different frames $\mathscr{F}_1$ and $\mathscr{F}_2$ are denoted by $v_1, v_2 \in \mathbb{R}^3$, respectively,

$$v_2 = \mathbf{C}(q_{21})v_1.$$

Equivalently, we have

$$\mathbf{Q}(v_2) = q_{21}^* \odot \mathbf{Q}(v_1) \odot q_{21}$$

where $q_{21} \in \mathbb{Q}_u$ describes the attitude of $\mathscr{F}_2$ with respect to $\mathscr{F}_1$. In this sense, the adjoint operation provides an alternative way for vector coordinate transformation.

Another merit of using unit quaternion for attitude parametrization is its simplicity in describing successive rotations. Let $q_1 \in \mathbb{Q}_u$ denote the attitude of $\mathscr{F}_b$ with respect to $\mathscr{F}_a$ and let $q_2 \in \mathbb{Q}_u$ denote the attitude of $\mathscr{F}_c$ with respect to $\mathscr{F}_b$. Then

$$q_3 = q_1 \odot q_2 \tag{6.4}$$

represents the attitude of $\mathscr{F}_c$ with respect to $\mathscr{F}_a$.

6.1.4 Unit Quaternion-Based Attitude Kinematics and Dynamics

Suppose a rigid-body is rotating with respect to the inertial frame. Let $q \in \mathbb{Q}_u$ denote the attitude of the body frame $\mathscr{F}$ with respect to the inertial frame $\mathscr{I}$ and $\boldsymbol{\omega}$ be the angular velocity vector of $\mathscr{F}$ with respect to $\mathscr{I}$. Let $\omega \in \mathbb{R}^3$ be the component of $\boldsymbol{\omega}$ expressed in either $\mathscr{F}$ or $\mathscr{I}$, which is the same. Then the kinematic and dynamic equations of the rigid-body are given by

$$\dot{q} = \frac{1}{2} q \odot \mathbf{Q}(\omega) \tag{6.5a}$$

$$J\dot{\omega} = -\omega^{\times} J\omega + u \tag{6.5b}$$

where $J \in \mathbb{R}^{3\times 3}$ is the component of the moments of inertia $\mathbf{J}$ which is positive definite, and $u \in \mathbb{R}^3$ is the component of control torque $\mathbf{u}$, both of which are expressed in $\mathscr{F}$.

The following lemmas will be used in subsequent sections.

Lemma 6.1 *For any differentiable time function $q(t) \in \mathbb{Q}$, and any piecewise continuous time function $x(t) \in \mathbb{R}^3$, the solution of the following system:*

$$\dot{q} = \frac{1}{2} q \odot \mathbf{Q}(x)$$

is such that $\|q(t)\| = \|q(0)\|$, $\forall t \geq 0$.

Proof Let $V = q^{\mathsf{T}} q$. Then, it follows from Identity A.9 in Sect. A.5 that, for all $t \geq 0$,

$$\dot{V} = q^{\mathsf{T}}(q \odot \mathbf{Q}(x)) = 0$$

which completes the proof. □

Lemma 6.2 *Consider the following system:*

$$\dot{q} = \frac{1}{2} q \odot \mathbf{Q}(\omega) + d$$

where $q(t) \in \mathbb{Q}$, and $\omega(t) \in \mathbb{R}^3$ and $d(t) \in \mathbb{Q}$ are both piecewise continuous in t and uniformly bounded over $[0, \infty)$. Suppose $\lim_{t\to\infty} \|q(t)\| = 1$ and $\lim_{t\to\infty} d(t) = 0$. Let $\Omega(t) = \omega(t) + \kappa\hat{q}(t)$ with κ being any positive constant. If $\lim_{t\to\infty} \Omega(t) = 0$, then $\lim_{t\to\infty} \hat{q}(t) = 0$, and hence $\lim_{t\to\infty} \omega(t) = 0$.

Proof Define a set $\mathscr{B}(\epsilon) = \{x \in \mathbb{R}^3 : \|x\| \leq \epsilon\}$. We need to prove, without loss of generality, that, for any $0 < \epsilon < 1$, there exists $T \geq 0$ such that

$$\hat{q}(t) \in \mathscr{B}(\epsilon), \ \forall t \geq T. \tag{6.6}$$

Let $V = q^{\mathsf{T}} q$. Then $\lim_{t\to\infty} V(t) = 1$. Since $q(t)$ is continuous and $\lim_{t\to\infty} \|q(t)\| = 1$, there exists $M > 0$ such that $\|q(t)\| \leq M$ for all $t \geq 0$. Then, by Identity A.9 in Sect. A.5, we have, for all $t \geq 0$,

$$\dot{V} = q^{\mathsf{T}}(q \odot \mathbf{Q}(\omega)) + 2q^{\mathsf{T}} d = 2q^{\mathsf{T}} d \leq 2M\|d\|.$$

Since $\lim_{t\to\infty} d(t) = 0$ and $\lim_{t\to\infty} \Omega(t) = 0$, there exists $T_1 > 0$ such that, for all $t \geq T_1$,

$$\dot{V}(t) < \frac{1}{16}\kappa\epsilon^2\sqrt{1 - \frac{3}{8}\epsilon^2} \tag{6.7a}$$

$$1 - \frac{1}{8}\epsilon^2 < V(t) < 1 + \frac{1}{8}\epsilon^2 \tag{6.7b}$$

$$\|\Omega(t)\| < \frac{1}{4}\kappa\epsilon \tag{6.7c}$$

$$|\bar{d}(t)| < \frac{1}{32}\kappa\epsilon^2. \tag{6.7d}$$

If there exists no $T_2 \geq T_1$ such that $\hat{q}(T_2) \notin \mathscr{B}(\epsilon)$, then (6.6) holds with $T = T_1$ and the proof is complete. Thus, in what follows, we assume that there exists $T_2 \geq T_1$ such that $\hat{q}(T_2) \notin \mathscr{B}(\epsilon)$. We will show that this assumption will lead to the following two claims:

1. There exists $T > T_2$ such that $\hat{q}(T) \in \mathscr{B}(\epsilon/2)$ and $\hat{q}(t) \notin \mathscr{B}(\epsilon/2), \forall T_2 \leq t < T$.
2. $\hat{q}(t) \in \mathscr{B}(\epsilon/2), \forall t \geq T$.

Clearly, the verifications of these two claims complete the proof.

Proof of Claim (1):

We first prove that, for all $t \geq T_1$,

$$\|\hat{q}(t)\| \geq \frac{1}{2}\epsilon \Rightarrow \dot{\bar{q}}(t) > \frac{1}{32}\kappa\epsilon^2 > 0. \tag{6.8}$$

We have

$$\dot{\bar{q}} = -\frac{1}{2}\hat{q}^{\mathsf{T}}(\Omega - \kappa\hat{q}) + \bar{d}.$$

Then

$$\begin{aligned}\dot{\bar{q}} &= \frac{1}{2}\kappa\hat{q}^{\mathsf{T}}\hat{q} - \frac{1}{2}\hat{q}^{\mathsf{T}}\Omega + \bar{d} > \frac{1}{4}\kappa\|\hat{q}\|\epsilon - \frac{1}{8}\kappa\|\hat{q}\|\epsilon - |\bar{d}| \\ &> \frac{1}{8}\kappa\|\hat{q}\|\epsilon - \frac{1}{32}\kappa\epsilon^2 \geq \frac{1}{16}\kappa\epsilon^2 - \frac{1}{32}\kappa\epsilon^2 = \frac{1}{32}\kappa\epsilon^2 > 0.\end{aligned}$$

If Claim (1) doesn't hold, i.e., $\|\hat{q}(t)\| > \epsilon/2$, then by (6.8), $\dot{\bar{q}}(t) > (\kappa\epsilon^2)/32 > 0$, which will contradict the fact that $q(t)$ is uniformly bounded over $[0, \infty)$. Therefore, Claim (1) holds.

Proof of Claim (2):

We first prove that, for all $t \geq T_1$,

$$\|\hat{q}(t)\| = \frac{1}{2}\epsilon \text{ and } \bar{q}(t) > 0 \Rightarrow \frac{d}{dt}(\|\hat{q}(t)\|^2) < 0. \tag{6.9}$$

By (6.8), we have $\dot{\bar{q}}(t) > (\kappa\epsilon^2)/32$, and by (6.7b), we have $\bar{q}(t) > 0$ which implies

$$\bar{q}(t) > \sqrt{1 - \frac{1}{8}\epsilon^2 - \frac{1}{4}\epsilon^2} = \sqrt{1 - \frac{3}{8}\epsilon^2}.$$

Therefore

$$\frac{d}{dt}(|\bar{q}(t)|^2) = 2\bar{q}(t)\dot{\bar{q}}(t) > \frac{1}{16}\kappa\epsilon^2\sqrt{1 - \frac{3}{8}\epsilon^2}. \tag{6.10}$$

Then (6.9) follows from (6.7a) and (6.10).

Next, we prove

$$\frac{d}{dt}(\|\hat{q}(t)\|^2)|_{t=T} < 0. \tag{6.11}$$

Since $\|\hat{q}(T)\| = \epsilon/2$, by (6.9), we only need to prove $\bar{q}(T) > 0$. For this purpose, note that from the proof of Claim (1), we have $\dot{\bar{q}}(t) > 0$ for $T_2 \leq t \leq T$ since $\|\hat{q}(t)\| \geq \epsilon/2$ for $T_2 \leq t \leq T$. Hence $\bar{q}(T) > \bar{q}(T_2)$. Since $\hat{q}(T_2) \notin \mathscr{B}(\epsilon)$, we have $\|\hat{q}(T_2)\| - \|\hat{q}(T)\| > \epsilon/2$. Thus,

$$\begin{aligned} \|\hat{q}(T_2)\|^2 - \|\hat{q}(T)\|^2 &= (\|\hat{q}(T_2)\| + \|\hat{q}(T)\|)(\|\hat{q}(T_2)\| - \|\hat{q}(T)\|) \\ &> \|\hat{q}(T_2)\| \cdot \frac{1}{2}\epsilon > \frac{1}{2}\epsilon^2. \end{aligned} \tag{6.12}$$

Since (6.7b) holds for all $t \geq T_1$, $V(T) - V(T_2) > -\epsilon^2/4$, i.e.,

$$\|\hat{q}(T)\|^2 + |\bar{q}(T)|^2 - \|\hat{q}(T_2)\|^2 - |\bar{q}(T_2)|^2 > -\frac{1}{4}\epsilon^2. \tag{6.13}$$

By (6.12) and (6.13), we have

$$|\bar{q}(T)|^2 - |\bar{q}(T_2)|^2 > \|\hat{q}(T_2)\|^2 - \|\hat{q}(T)\|^2 - \frac{1}{4}\epsilon^2 > \frac{1}{2}\epsilon^2 - \frac{1}{4}\epsilon^2 = \frac{1}{4}\epsilon^2 > 0.$$

Therefore, $|\bar{q}(T)| > |\bar{q}(T_2)|$, which together with the fact that $\bar{q}(T) > \bar{q}(T_2)$ implies $\bar{q}(T) > 0$.

Now suppose Claim (2) is not true. Then, from (6.11), there exists $T_3 > T$ such that $\|\hat{q}(t)\| < \epsilon/2$ for $T < t < T_3$, $\|\hat{q}(T_3)\| = \epsilon/2$, and

$$\frac{d}{dt}(\|\hat{q}(t)\|^2)|_{t=T_3} > 0.$$

By (6.7b), we have

$$|\bar{q}(t)| > \sqrt{1 - \frac{3}{8}\epsilon^2}, \ \forall T \le t \le T_3.$$

However, since $\bar{q}(T) > 0$, by the continuity of $\bar{q}(t)$, we have $\bar{q}(t) > 0$ for $T \le t \le T_3$, which together with $\|\hat{q}(T_3)\| = \epsilon/2$ implies, by (6.9), that

$$\frac{d}{dt}(\|\hat{q}(t)\|^2)|_{t=T_3} < 0.$$

The contradiction shows that Claim (2) must be true. □

6.2 Tracking Control of a Single Rigid-Body System

Let $q_0 \in \mathbb{Q}_u$ denote the attitude of some reference frame $\mathscr{R}$ with respect to the inertial frame $\mathscr{I}$, and $\boldsymbol{\omega_0}$ be the angular velocity vector of $\mathscr{R}$ with respect to $\mathscr{I}$. Let $\omega_0 \in \mathbb{R}^3$ be the component of $\boldsymbol{\omega_0}$ expressed in $\mathscr{R}$. Then the kinematic equation for the reference frame is given by

$$\dot{q}_0 = \frac{1}{2} q_0 \odot \mathbf{Q}(\omega_0). \tag{6.14}$$

Let

$$q_e = q_0^{-1} \odot q \tag{6.15a}$$
$$\omega_e = \omega - \mathbf{C}(q_e)\omega_0 \tag{6.15b}$$

where $q_e \in \mathbb{Q}_u$, $\omega_e \in \mathbb{R}^3$. Then, q_e denotes the attitude of the body frame $\mathscr{F}$ with respect to the reference frame $\mathscr{R}$ and $\mathbf{C}(q_e) = C_{FR}$ is the rotation matrix describing the attitude of the rigid-body with respect to the reference frame. By Remark 6.1, the attitude of the body frame $\mathscr{F}$ coincides with the reference frame $\mathscr{R}$ if and only if $\hat{q}_e = 0$. If $\hat{q}_e = 0$, then $q_e = q_I$ or $q_e = -q_I$. In either case, $\mathbf{C}(q_e) = I_3$. Therefore, if $\hat{q}_e = 0$, the angular velocity of the body frame will be the same as that of the reference frame if and only if $\omega_e = 0$. In this sense, q_e and ω_e are called the attitude and angular velocity tracking errors, respectively.

By Identity A.10 in Sect. A.5, the error dynamics are given by

$$\dot{q}_e = \frac{1}{2} q_e \odot \mathbf{Q}(\omega_e) \tag{6.16a}$$

$$J\dot{\omega}_e = -\omega^\times J\omega + J\left(\omega_e^\times \mathbf{C}(q_e)\omega_0 - \mathbf{C}(q_e)\dot{\omega}_0\right) + u. \tag{6.16b}$$

We will consider two classes of control laws as follows:

1. State Feedback Control:

$$u = f(q, \omega, q_0, \omega_0, \dot{\omega}_0, z) \tag{6.17a}$$

$$\dot{z} = g(q, \omega, q_0, \omega_0, \dot{\omega}_0, z) \tag{6.17b}$$

   where f and g are some globally defined smooth functions.
2. Attitude Feedback Control:

$$u = f(q, q_0, \omega_0, \dot{\omega}_0, \rho) \tag{6.18a}$$

$$\dot{\rho} = h(\rho, q, q_0) \tag{6.18b}$$

   where $\rho \in \mathbb{Q}$, and f and h are some globally defined smooth functions.

Remark 6.2 The control law (6.17) contains the special case where the dimension of z is zero. In this case, the control law (6.17) is called a static state feedback control law. The control law (6.18) is independent of the angular velocity ω, and is thus called the attitude feedback control law. The attitude feedback control law (6.18) contains an additional dynamic compensator governing ρ whose specific form is given in (6.26).

Now, the tracking control problem for a single rigid-body system is described as follows.

Problem 6.1 Given systems (6.5) and (6.14), design a state feedback control law of the form (6.17) such that, for any $q(0), q_0(0) \in \mathbb{Q}_u$, any $\omega(0)$, and any $z(0)$, or an attitude feedback control law of the form (6.18) such that, for any $q(0), q_0(0) \in \mathbb{Q}_u$, any $\omega(0)$, and any $\rho(0) \neq 0$, the trajectory of the closed-loop system exists for all $t \geq 0$ and is uniformly bounded over $[0, \infty)$, and

$$\lim_{t\to\infty} \hat{q}_e(t) = 0, \quad \lim_{t\to\infty} \omega_e(t) = 0.$$

Remark 6.3 It will be seen shortly that the solution $\rho(t)$ of (6.26) with $\rho(0) = 0$ is zero for all $t \geq 0$. In this case, it will not have any effect on the attitude feedback control law (6.18). That is why, in the statement of Problem 6.1, we require $\rho(0) \neq 0$.

As we require that the trajectory of the closed-loop system be uniformly bounded over $[0, \infty)$ for any initial state, to make Problem 6.1 well-posed, we need the following assumption.

Assumption 6.1 All $\omega_0(t)$, $\dot{\omega}_0(t)$, and $\ddot{\omega}_0(t)$ exist and are uniformly bounded over $[0, \infty)$.

We will apply Lemma 6.2 to obtain the solution of Problem 6.1. For this purpose, we first perform on (6.16) the following transformation:

$$\Omega_e = \omega_e + \kappa \hat{q}_e \tag{6.19}$$

where $\kappa > 0$.

As a result, (6.16) is transformed to the following form:

$$\dot{q}_e = \frac{1}{2} q_e \odot \mathbf{Q}(\omega_e) \tag{6.20a}$$

$$\begin{aligned} J\dot{\Omega}_e &= J\dot{\omega}_e + \kappa J \dot{\hat{q}}_e \\ &= -\omega^{\times} J\omega + J\left(\omega_e^{\times} \mathbf{C}(q_e)\omega_0 - \mathbf{C}(q_e)\dot{\omega}_0 + \frac{\kappa}{2}(\hat{q}_e^{\times} + \bar{q}_e I_3)\omega_e\right) + u. \end{aligned} \tag{6.20b}$$

Remark 6.4 By Lemma 6.2, if $\lim_{t\to\infty} \|q_e(t)\| = 1$, and the control u is such that $\lim_{t\to\infty} \Omega_e(t) = 0$, then $\lim_{t\to\infty} \hat{q}_e(t) = 0$, and hence $\lim_{t\to\infty} \omega_e(t) = 0$. By Lemma 6.1, $\|q_0(0)\| = 1$ and $\|q(0)\| = 1$ imply $\|q_0(t)\| = 1$ and $\|q(t)\| = 1$, respectively, $\forall t \geq 0$, which further imply $\|q_e(t)\| = 1$, $\forall t \geq 0$. Thus, Problem 6.1 reduces to the stabilization problem for the subsystem (6.20b). In what follows, we will design three types of control laws to stabilize (6.20b) according to the following three cases.

6.2.1 Case 1: J Is Known and ω, q Are Available

Theorem 6.1 *Under Assumption 6.1, Problem 6.1 is solvable by the following static feedback control law:*

$$u = \omega^{\times} J\omega - J\left(\omega_e^{\times} \mathbf{C}(q_e)\omega_0 - \mathbf{C}(q_e)\dot{\omega}_0 + \frac{\kappa}{2}(\hat{q}_e^{\times} + \bar{q}_e I_3)\omega_e\right) - k\Omega_e \tag{6.21}$$

where $k > 0$.

Proof Substituting (6.21) into (6.20b) gives

$$\dot{\Omega}_e = -kJ^{-1}\Omega_e.$$

Since J is positive definite and $k > 0$, $-kJ^{-1}$ is Hurwitz. Thus, $\lim_{t\to\infty} \Omega_e(t) = 0$. By Remark 6.4, $\|q_e(t)\| = 1$, $\forall t \geq 0$. Applying Lemma 6.2 to system (6.20a) and noting (6.19) gives $\lim_{t\to\infty} \hat{q}_e(t) = 0$, and $\lim_{t\to\infty} \omega_e(t) = 0$. The proof is thus completed. □

6.2.2 Case 2: J Is Uncertain and ω, q Are Available

Let us first define the following operator. For $x = \mathrm{col}(x_1, x_2, x_3) \in \mathbb{R}^3$, $L(x)$ is defined as

$$L(x) = \begin{bmatrix} x_1 & 0 & 0 & 0 & x_3 & x_2 \\ 0 & x_2 & 0 & x_3 & 0 & x_1 \\ 0 & 0 & x_3 & x_2 & x_1 & 0 \end{bmatrix}. \tag{6.22}$$

Let

$$J = \begin{bmatrix} J_{11} & J_{12} & J_{13} \\ J_{12} & J_{22} & J_{23} \\ J_{13} & J_{23} & J_{33} \end{bmatrix}$$

and

$$\Theta = \mathrm{col}\,(J_{11}, J_{22}, J_{33}, J_{23}, J_{13}, J_{12}).$$

Then it can be verified that, for any $x \in \mathbb{R}^3$, $Jx = L(x)\Theta$. Therefore, (6.20b) can be rewritten as follows:

$$J\dot{\Omega}_e = \chi_e \Theta + u \tag{6.23}$$

where

$$\chi_e = -\omega^{\times} L(\omega) + L\left(\omega_e^{\times}\mathbf{C}(q_e)\omega_0 - \mathbf{C}(q_e)\dot{\omega}_0 + \frac{\kappa}{2}(\hat{q}_e^{\times} + \bar{q}_e I_3)\omega_e\right).$$

Theorem 6.2 *Under Assumption 6.1, Problem 6.1 is solvable by the following dynamic feedback control law:*

$$u = -\chi_e \hat{\Theta} - k\Omega_e \tag{6.24a}$$

$$\dot{\hat{\Theta}} = \Lambda^{-1}\chi_e^{\mathsf{T}}\Omega_e \tag{6.24b}$$

where Λ is some positive definite gain matrix and $k > 0$.

Proof Let $\tilde{\Theta} = \hat{\Theta} - \Theta$. Then, substituting (6.24a) into (6.23) gives

$$J\dot{\Omega}_e = -\chi_e \tilde{\Theta} - k\Omega_e. \tag{6.25}$$

Let

$$V = \frac{1}{2}(\Omega_e^{\mathsf{T}} J \Omega_e + \tilde{\Theta}^{\mathsf{T}} \Lambda \tilde{\Theta}).$$

Then, along the trajectories of (6.24b) and (6.25),

$$\dot{V} = \Omega_e^{\mathsf{T}}(-\chi_e\tilde{\Theta} - k\Omega_e) + \tilde{\Theta}^{\mathsf{T}}\Lambda\Lambda^{-1}\chi_e^{\mathsf{T}}\Omega_e = -k\|\Omega_e\|^2 \leq 0.$$

Therefore, both $\Omega_e(t)$ and $\tilde{\Theta}(t)$ are uniformly bounded over $[0, \infty)$. Under Assumption 6.1, $\omega_e(t)$ and $\omega(t)$ are also uniformly bounded over $[0, \infty)$. Therefore, $\chi_e(t)$ is uniformly bounded over $[0, \infty)$, and so is $\dot{\Omega}_e(t)$. As a result, $\ddot{V}(t)$ is uniformly bounded over $[0, \infty)$, and hence $\dot{V}(t)$ is uniformly continuous. By Barbalat's lemma, i.e., Lemma 2.9, $\lim_{t\to\infty} \dot{V}(t) = 0$, and thus $\lim_{t\to\infty} \Omega_e(t) = 0$. By Remark 6.4, $\|q_e(t)\| = 1, \forall t \geq 0$. Therefore, by Lemma 6.2, $\lim_{t\to\infty} \hat{q}_e(t) = 0$, and $\lim_{t\to\infty} \omega_e(t) = 0$. □

Theorem 6.2 does not say anything about the convergence of $\hat{\Theta}$. The following result presents the sufficient condition to guarantee the convergence of $\hat{\Theta}$ to Θ.

Theorem 6.3 *Let* $\pi_0 = \omega_0^{\mathsf{T}} L(\dot{\omega}_0)$. *Under Assumption 6.1, if* $\pi_0^{\mathsf{T}}(t)$ *is persistently exciting, then* $\lim_{t\to\infty} \tilde{\Theta}(t) = 0$.

Proof By (6.25),

$$J\ddot{\Omega}_e = -\dot{\chi}_e\tilde{\Theta} - \chi_e\dot{\tilde{\Theta}} - k\dot{\Omega}_e.$$

By Theorem 6.2, all $\Omega_e(t)$, $\tilde{\Theta}(t)$, $\omega_e(t)$, $\omega(t)$, $\chi_e(t)$, and $\dot{\Omega}_e(t)$ are uniformly bounded over $[0, \infty)$. Then under Assumption 6.1, both $\dot{\omega}_e(t)$ and $\dot{\omega}(t)$ are uniformly bounded over $[0, \infty)$, and thus $\dot{\chi}_e(t)$ is also uniformly bounded over $[0, \infty)$. Therefore, $\ddot{\Omega}_e(t)$ is uniformly bounded over $[0, \infty)$, and hence $\dot{\Omega}_e$ is uniformly continuous. Since $\lim_{t\to\infty} \Omega_e(t) = 0$, by Barbalat's lemma, i.e., Lemma 2.9, $\lim_{t\to\infty} \dot{\Omega}_e(t) = 0$, which together with (6.25) implies

$$\lim_{t\to\infty} \chi_e(t)\tilde{\Theta}(t) = 0.$$

By Theorem 6.2, $\lim_{t\to\infty} \mathbf{C}(q_e(t)) = I_3$. Therefore

$$\lim_{t\to\infty} (\chi_e(t) + \omega_0(t)^{\times} L(\omega_0(t)) + L(\dot{\omega}_0(t))) = 0.$$

Thus,

$$\lim_{t\to\infty} (\omega_0(t)^{\times} L(\omega_0(t)) + L(\dot{\omega}_0(t)))\tilde{\Theta}(t) = 0.$$

Since $\omega_0(t)$ is uniformly bounded over $[0, \infty)$, we have

$$\lim_{t\to\infty} \omega_0^{\mathsf{T}}(t)(\omega_0(t)^{\times} L(\omega_0(t)) + L(\dot{\omega}_0(t)))\tilde{\Theta}(t) = 0$$

which together with Identity A.1 in Sect. A.5 gives

$$\lim_{t\to\infty} \omega_0^{\mathsf{T}}(t) L(\dot{\omega}_0(t))\tilde{\Theta}(t) = \lim_{t\to\infty} \pi_0(t)\tilde{\Theta}(t) = 0.$$

By Theorem 6.2, $\lim_{t\to\infty} \dot{\tilde{\Theta}}(t) = 0$. This fact together with the assumption that $\pi_0^\mathsf{T}(t)$ is persistently exciting implies, by Lemma A.4, $\lim_{t\to\infty} \tilde{\Theta}(t) = 0$. □

6.2.3 Case 3: *J* Is Known and Only *q* Is Available

In this case, we will consider solving Problem 6.1 by an attitude feedback control law. For this purpose, let $\rho \in \mathbb{Q}$ be governed by the following equation:

$$\dot{\rho} = \frac{1}{2}\rho \odot \mathbf{Q}(\beta) \tag{6.26}$$

where $\beta \in \mathbb{R}^3$ is defined as follows:

$$\alpha = \rho^* \odot q_e \tag{6.27a}$$

$$\beta = \Gamma\hat{\alpha} \tag{6.27b}$$

with $\Gamma \in \mathbb{R}^{3\times 3}$ being some positive definite gain matrix.

It is noted that (6.27a) defines a nonsingular transformation such that $\rho^* = \alpha \odot q_e^{-1}$, and it can be verified that $\alpha \in \mathbb{Q}$ satisfies

$$\begin{aligned}
\dot{\alpha} &= \dot{\rho}^* \odot q_e + \rho^* \odot \dot{q}_e \\
&= \frac{1}{2}(\mathbf{Q}(\beta)^* \odot \rho^* \odot q_e + \rho^* \odot q_e \odot \mathbf{Q}(\omega_e)) \\
&= \frac{1}{2}(\mathbf{Q}(-\beta) \odot \alpha + \alpha \odot \mathbf{Q}(\omega_e)) \\
&= \frac{1}{2}\begin{bmatrix} \bar{\alpha}(\omega_e - \beta) + \hat{\alpha}^\times(\omega_e + \beta) \\ \hat{\alpha}^\mathsf{T}(\beta - \omega_e) \end{bmatrix}.
\end{aligned} \tag{6.28}$$

Theorem 6.4 *Under Assumption 6.1, Problem 6.1 is solvable by the following attitude feedback control law:*

$$u = -k_1\hat{q}_e - k_2\hat{\alpha} + J\mathbf{C}(q_e)\dot{\omega}_0 + \big(\mathbf{C}(q_e)\omega_0\big)^\times J\mathbf{C}(q_e)\omega_0 \tag{6.29a}$$

$$\dot{\rho} = \frac{1}{2}\rho \odot \mathbf{Q}(\beta) \tag{6.29b}$$

where $k_1, k_2 > 0$.

Proof Substituting (6.29a) into (6.16b) gives

$$
\begin{aligned}
J\dot{\omega}_e =& -(\omega_e + \mathbf{C}(q_e)\omega_0)^\times J(\omega_e + \mathbf{C}(q_e)\omega_0) + J\left(\omega_e^\times \mathbf{C}(q_e)\omega_0 - \mathbf{C}(q_e)\dot{\omega}_0\right)\\
& - k_1\hat{q}_e - k_2\hat{\alpha} + J\mathbf{C}(q_e)\dot{\omega}_0 + \left(\mathbf{C}(q_e)\omega_0\right)^\times J\mathbf{C}(q_e)\omega_0\\
=& -\omega_e^\times J\omega_e - \omega_e^\times J\mathbf{C}(q_e)\omega_0 - (\mathbf{C}(q_e)\omega_0)^\times J\omega_e + J\omega_e^\times \mathbf{C}(q_e)\omega_0 - k_1\hat{q}_e - k_2\hat{\alpha}\\
=& -\omega_e^\times J\omega_e - \omega_e^\times J\mathbf{C}(q_e)\omega_0 - (\mathbf{C}(q_e)\omega_0)^\times J\omega_e - J(\mathbf{C}(q_e)\omega_0)^\times\omega_e - k_1\hat{q}_e - k_2\hat{\alpha}.
\end{aligned} \tag{6.30}
$$

Put the system consisting of (6.20a), (6.28), and (6.30) as follows:

$$
\dot{q}_e = \frac{1}{2}q_e \odot \mathbf{Q}(\omega_e) \tag{6.31a}
$$

$$
J\dot{\omega}_e = -\omega_e^\times J\omega_e - \omega_e^\times J\mathbf{C}(q_e)\omega_0 - (\mathbf{C}(q_e)\omega_0)^\times J\omega_e - J(\mathbf{C}(q_e)\omega_0)^\times\omega_e - k_1\hat{q}_e - k_2\hat{\alpha} \tag{6.31b}
$$

$$
\dot{\alpha} = \frac{1}{2}\begin{bmatrix} \bar{\alpha}(\omega_e - \beta) + \hat{\alpha}^\times(\omega_e + \beta) \\ \hat{\alpha}^\mathsf{T}(\beta - \omega_e) \end{bmatrix}. \tag{6.31c}
$$

Let

$$
V = \frac{1}{2}\left(\frac{1}{2}\omega_e^\mathsf{T} J\omega_e + k_1\|q_e - q_I\|^2 + k_2\|\alpha - q_I\|^2\right).
$$

Then, noting that $(\mathbf{C}(q_e)\omega_0)^\times J + J(\mathbf{C}(q_e)\omega_0)^\times$ is anti-symmetric gives the derivative of V along the trajectory of (6.31) as follows:

$$
\begin{aligned}
\dot{V} =& \frac{1}{2}\omega_e^\mathsf{T}(-k_1\hat{q}_e - k_2\hat{\alpha}) + \frac{1}{2}k_1\hat{q}_e^\mathsf{T}(\bar{q}_e I_3 + \hat{q}_e^\times)\omega_e + k_1(\bar{q}_e - 1)(-\frac{1}{2}\hat{q}_e^\mathsf{T}\omega_e)\\
& + k_2\hat{\alpha}^\mathsf{T}(\frac{1}{2}\bar{\alpha}(\omega_e - \beta) + \frac{1}{2}\hat{\alpha}^\times(\omega_e + \beta)) + k_2(\bar{\alpha} - 1)(\frac{1}{2}\hat{\alpha}^\mathsf{T}(\beta - \omega_e)) \\
=& -\frac{1}{2}k_2\hat{\alpha}^\mathsf{T}\beta = -\frac{1}{2}k_2\hat{\alpha}^\mathsf{T}\Gamma\hat{\alpha} \leq 0
\end{aligned} \tag{6.32}
$$

where we have used Identity A.1 in Sect. A.5. Therefore, $\omega_e(t)$, $\alpha(t)$, and hence $\beta(t)$ are all uniformly bounded over $[0, \infty)$. Thus, $\dot{\alpha}(t)$ and hence $\ddot{V}(t)$ are uniformly bounded over $[0, \infty)$. Therefore, by Barbalat's lemma, i.e., Lemma 2.9, $\lim_{t\to\infty}\dot{V}(t) = 0$, and thus $\lim_{t\to\infty}\hat{\alpha}(t) = 0$. Since $q_e \in \mathbb{Q}_u$, by Identity A.7, $\|\alpha(t)\| = \|\rho(t)\|$, and by Lemma 6.1, $\|\rho(t)\| = \|\rho(0)\|$ for all $t \geq 0$. Therefore, $\|\alpha(t)\| = \|\rho(0)\|$ for all $t \geq 0$, which together with the facts that $\alpha(t)$ is continuous and $\lim_{t\to\infty}\hat{\alpha}(t) = 0$ implies that $\lim_{t\to\infty}\alpha(t)$ exists. Since $\omega_0(t)$ is uniformly bounded over $[0, \infty)$, by (6.30), $\dot{\omega}_e(t)$ is uniformly bounded over $[0, \infty)$. As a result, $\ddot{\alpha}(t)$ is uniformly bounded over $[0, \infty)$ and thus $\lim_{t\to\infty}\dot{\alpha}(t) = 0$ by Barbalat's lemma. Since $\|\rho(0)\| \neq 0$, $\lim_{t\to\infty}\bar{\alpha}(t) \neq 0$. Therefore, since $\lim_{t\to\infty}\beta(t) = 0$, (6.28) implies that $\lim_{t\to\infty}\omega_e(t) = 0$. Since $\dot{\omega}_0(t)$ is also uniformly bounded over $[0, \infty)$, $\ddot{\omega}_e(t)$ is uniformly bounded over $[0, \infty)$, and thus $\lim_{t\to\infty}\dot{\omega}_e(t) = 0$ by Barbalat's lemma. Since both $\lim_{t\to\infty}\hat{\alpha}(t) = 0$ and $\lim_{t\to\infty}\dot{\omega}_e(t) = 0$, it follows from (6.30) that $\lim_{t\to\infty}\hat{q}_e(t) = 0$. Thus the proof is complete. □

Remark 6.5 If we view β and $\hat{\alpha}$ as the input and output of system (6.31), respectively, then, by (6.32),

$$-\frac{1}{2}k_2\hat{\alpha}^\mathsf{T}\beta = \dot{V}$$

which indicates that system (6.31) is lossless. Then, letting the input $\beta = -\Gamma\hat{\alpha}$ makes $\dot{V}$ negative semidefinite, which further leads to $\lim_{t\to\infty}\hat{\alpha}(t) = 0$ by invoking Barbalat's lemma. This interpretation gives the origin of the control law (6.29).

6.3 Estimation-Based Control

In the previous section, the control law made use of the accurate state of the reference frame to solve Problem 6.1. In this section, as in Sect. 5.3, we further consider using the estimated state of the reference frame to solve the attitude control problem. The results of this section will lay a foundation for dealing with the leader-following consensus problem of multiple rigid-body systems.

Let $\zeta_0(t), \zeta_0^d(t) \in \mathbb{R}^3$. Given a time sequence $\{t_k : k = 0, 1, 2, \ldots\}$ with $t_0 = 0$ and a dwell time $\tau > 0$, suppose $\zeta_0(t)$, $\zeta_0^d(t)$, $\dot{\zeta}_0(t)$, $\ddot{\zeta}_0(t)$, and $\dot{\zeta}_0^d(t)$ are piecewise continuous and uniformly bounded over $[0, \infty)$, and $\psi_0(t) \in \mathbb{Q}$ is governed by the following equation:

$$\dot{\psi}_0 = \frac{1}{2}\psi_0 \odot \mathbf{Q}(\zeta_0) + \Delta(t) \tag{6.33}$$

where $\Delta(t), \dot{\Delta}(t) \in \mathbb{Q}$ are continuous on each interval $[t_j, t_{j+1})$, $j = 0, 1, \ldots$, and uniformly bounded over $[0, \infty)$ with $\lim_{t\to\infty}\Delta(t) = 0$ exponentially. Moreover, assume

$$\lim_{t\to\infty}(\psi_0(t) - q_0(t)) = 0 \tag{6.34a}$$

$$\lim_{t\to\infty}(\zeta_0(t) - \omega_0(t)) = 0 \tag{6.34b}$$

$$\lim_{t\to\infty}(\zeta_0^d(t) - \dot{\omega}_0(t)) = 0 \tag{6.34c}$$

$$\lim_{t\to\infty}(\zeta_0^d(t) - \dot{\zeta}_0(t)) = 0 \tag{6.34d}$$

exponentially.

Corresponding to the two classes of control laws (6.17) and (6.18), this section considers the following two classes of control laws:

1. Estimation-Based State Feedback Control:

$$u = f(q, \omega, \psi_0, \zeta_0, \zeta_0^d, z) \tag{6.35a}$$

$$\dot{z} = g(q, \omega, \psi_0, \zeta_0, \zeta_0^d, z) \tag{6.35b}$$

   where f and g are some globally defined smooth functions.
2. Estimation-Based Attitude Feedback Control:

$$u = f(q, \psi_0, \zeta_0, \zeta_0^d, \rho) \tag{6.36a}$$
$$\dot{\rho} = h(\rho, q, \psi_0) \tag{6.36b}$$

where f and h are some globally defined smooth functions.

We now describe the problem as follows.

Problem 6.2 Given systems (6.5) and (6.14), design a control law of the form (6.35) such that, for any $q(0) \in \mathbb{Q}_u$, any $\omega(0)$, and any $z(0)$, or a control law of the form (6.36) such that, for any $q(0) \in \mathbb{Q}_u$, any $\omega(0)$, and any $\rho(0) \neq 0$, the trajectory of the closed-loop system exists for all $t \geq 0$ and is uniformly bounded over $[0, \infty)$, and

$$\lim_{t\to\infty} \hat{q}_e(t) = 0, \quad \lim_{t\to\infty} \omega_e(t) = 0.$$

Since neither of the control laws (6.35) and (6.36) can rely on q_0 and ω_0, let us first replace q_0 and ω_0 in (6.15) by their estimation ψ_0 and ζ_0 to obtain the following estimated tracking errors:

$$q_\epsilon = \psi_0^* \odot q$$
$$\omega_\epsilon = \omega - \mathbf{C}(q_\epsilon)\zeta_0.$$

Remark 6.6 Since $\|q(t)\| = 1$ for all $t \geq 0$, by Identity A.7 in Sect. A.5, $\|q_\epsilon(t)\| = \|\psi_0(t)\|$ for all $t \geq 0$. Since $\|q_0(t)\| = 1$ for all $t \geq 0$, by (6.34a), $\lim_{t\to\infty} \|\psi_0(t)\| = 1$ exponentially. Therefore, $\lim_{t\to\infty} \|q_\epsilon(t)\| = 1$ exponentially.

Lemma 6.3 *If* $\lim_{t\to\infty} \hat{q}_\epsilon(t) = 0$ *and* $\lim_{t\to\infty} \omega_\epsilon(t) = 0$, *then* $\lim_{t\to\infty} \hat{q}_e(t) = 0$ *and* $\lim_{t\to\infty} \omega_e(t) = 0$.

Proof First, note that

$$q_\epsilon - q_e = \psi_0^* \odot q - q_0^* \odot q = (\psi_0 - q_0)^* \odot q.$$

Thus, it follows from (6.34a) that $\lim_{t\to\infty}(q_\epsilon(t) - q_e(t)) = 0$, and hence $\lim_{t\to\infty} \hat{q}_e(t) = 0$.

Second, by Remark 6.6, $\lim_{t\to\infty} \|\bar{q}_\epsilon(t)\| = 1$. Therefore,

$$\lim_{t\to\infty} \mathbf{C}(q_e(t)) = \lim_{t\to\infty} \mathbf{C}(q_\epsilon(t)) = I_3.$$

Then, noting that

$$\omega_e - \omega_\epsilon = \omega - \mathbf{C}(q_e)\omega_0 - \omega + \mathbf{C}(q_\epsilon)\zeta_0 = \mathbf{C}(q_\epsilon)\zeta_0 - \mathbf{C}(q_e)\omega_0$$

and (6.34b) concludes $\lim_{t\to\infty} \omega_e(t) = 0$. □

By Identity A.10 in Sect. A.5, we have

$$\dot{q}_\epsilon = \frac{1}{2} q_\epsilon \odot \mathbf{Q}(\omega_\epsilon) + d_{q\epsilon} \tag{6.37a}$$

$$\begin{aligned} J\dot{\omega}_\epsilon &= -\omega^\times J\omega + J\omega_\epsilon^\times \mathbf{C}(q_\epsilon)\zeta_0 - J\mathbf{C}(q_\epsilon)\dot{\zeta}_0 + u - Jd_{C\epsilon}\zeta_0 \\ &= -\omega^\times J\omega + J\omega_\epsilon^\times \mathbf{C}(q_\epsilon)\zeta_0 - J\mathbf{C}(q_\epsilon)\zeta_0^d + u - d_{\omega\epsilon} \end{aligned} \tag{6.37b}$$

where

$$\begin{aligned} d_{q\varepsilon} &= \frac{1}{2}(q_\epsilon^\mathsf{T} q_\epsilon - 1)\mathbf{Q}(\zeta_0) \odot q_\epsilon + \Delta^* \odot q \\ d_{C\epsilon} &= 2\bar{q}_\epsilon \bar{d}_{q\epsilon} I_3 - 2\hat{q}_\epsilon^\mathsf{T} \hat{d}_{q\epsilon} I_3 + 2\hat{d}_{q\epsilon}\hat{q}_\epsilon^\mathsf{T} + 2\hat{q}_\epsilon \hat{d}_{q\epsilon}^\mathsf{T} - 2\bar{d}_{q\epsilon}\hat{q}_\epsilon^\times - 2\bar{q}_\epsilon \hat{d}_{q\epsilon}^\times \\ d_{\omega\epsilon} &= Jd_{C\epsilon}\zeta_0 + J\mathbf{C}(q_\epsilon)(\dot{\zeta}_0 - \zeta_0^d). \end{aligned}$$

Let

$$\Omega_\epsilon = \omega_\epsilon + \kappa \hat{q}_\epsilon \tag{6.38}$$

where $\kappa > 0$. Then q_ϵ and Ω_ϵ are governed by

$$\dot{q}_\epsilon = \frac{1}{2} q_\epsilon \odot \mathbf{Q}(\omega_\epsilon) + d_{q\epsilon} \tag{6.39a}$$

$$\begin{aligned} J\dot{\Omega}_\epsilon =& -\omega^\times J\omega + J(\omega_\epsilon^\times \mathbf{C}(q_\epsilon)\zeta_0 - \mathbf{C}(q_\epsilon)\zeta_0^d) + u - d_{\omega\epsilon} \\ &+ \frac{\kappa}{2} J(\bar{q}_\epsilon I_3 + \hat{q}_\epsilon^\times)\omega_\epsilon + \kappa J\hat{d}_{q\epsilon} \\ =& -\omega^\times J\omega + J\left(\omega_\epsilon^\times \mathbf{C}(q_\epsilon)\zeta_0 - \mathbf{C}(q_\epsilon)\zeta_0^d + \frac{\kappa}{2}(\bar{q}_\epsilon I_3 + \hat{q}_\epsilon^\times)\omega_\epsilon\right) + u - d_{\Omega\epsilon} \end{aligned} \tag{6.39b}$$

where $d_{\Omega\epsilon} = d_{\omega\epsilon} - \kappa J\hat{d}_{q\epsilon}$.

Remark 6.7 Comparing with (6.20), the right-hand sides of (6.39a) and (6.39b) contain extra terms $d_{q\epsilon}$ and $d_{\Omega\epsilon}$, respectively. Under Assumption 6.1, $\zeta_0(t)$ is uniformly bounded over $[0, \infty)$. By Remark 6.6, $d_{q\epsilon}(t)$ and hence $d_{C\epsilon}(t)$, $d_{\omega\epsilon}(t)$, $d_{\Omega\epsilon}(t)$ are all uniformly bounded over $[0, \infty)$ and decay to zero exponentially, and $\lim_{t\to\infty} \|q_\epsilon(t)\| = 1$. Thus, by Lemma 6.2, it suffices to stabilize (6.39b) by a control law of the form (6.35).

Again, we will consider three cases as in Sect. 6.2.

6.3.1 *Case 1: J Is Known and ω, q Are Available*

Replacing q_0, ω_0, and $\dot{\omega}_0$ in (6.21) with their estimation ψ_0, ζ_0, and ζ_0^d, respectively, gives the following estimation-based control law:

$$u = \omega^{\times} J\omega - J\left(\omega_{\epsilon}^{\times}\mathbf{C}(q_{\epsilon})\zeta_0 - \mathbf{C}(q_{\epsilon})\zeta_0^d + \frac{\kappa}{2}(\bar{q}_{\epsilon}I_3 + \hat{q}_{\epsilon}^{\times})\omega_{\epsilon}\right) - k\Omega_{\epsilon} \tag{6.40}$$

where $k > 0$.

Theorem 6.5 *Under Assumption 6.1, Problem 6.2 is solvable by the control law (6.40).*

Proof Substituting (6.40) into (6.39b) gives

$$\dot{\Omega}_{\epsilon} = -kJ^{-1}\Omega_{\epsilon} - J^{-1}d_{\Omega\epsilon}.$$

Since J is positive definite, $k > 0$, kJ^{-1} is positive definite and hence $-kJ^{-1}$ is Hurwitz. By Remark 6.7 and Lemma 2.5, $\lim_{t\to\infty}\Omega_{\epsilon}(t) = 0$. By Remark 6.6, $\lim_{t\to\infty}\|q_{\epsilon}(t)\| = 1$ exponentially. Thus, we can apply Lemma 6.2 to (6.39a) to conclude that $\lim_{t\to\infty}\hat{q}_{\epsilon}(t) = 0$ and $\lim_{t\to\infty}\omega_{\epsilon}(t) = 0$. The proof is then completed by resorting to Lemma 6.3. □

Remark 6.8 The term $d_{\Omega\epsilon}$ in (6.39b) can be viewed as an exponentially decaying disturbance to a stable linear system. Thus, even though the control law (6.40) does not cancel $d_{\Omega\epsilon}$ in (6.39b), it still drives $\Omega_{\epsilon}(t)$ to the origin exponentially.

6.3.2 Case 2: *J* Is Uncertain and *ω*, *q* Are Available

By adopting the same operator L as the one defined in (6.22), system (6.39b) can be rewritten as follows:

$$J\dot{\Omega}_{\epsilon} = \chi_{\epsilon}\Theta + u - d_{\Omega\epsilon} \tag{6.41}$$

where

$$\chi_{\epsilon} = -\omega^{\times}L(\omega) + L\left(\omega_{\epsilon}^{\times}\mathbf{C}(q_{\epsilon})\zeta_0 - \mathbf{C}(q_{\epsilon})\zeta_0^d + \frac{\kappa}{2}(\hat{q}_{\epsilon}^{\times} + \bar{q}_{\epsilon}I_3)\omega_{\epsilon}\right).$$

Replacing q_0, ω_0, and $\dot{\omega}_0$ in (6.24) with ψ_0, ζ_0, and ζ_0^d, respectively, gives the following estimation-based control law:

$$u = -\chi_{\epsilon}\hat{\Theta} - k\Omega_{\epsilon} \tag{6.42a}$$

$$\dot{\hat{\Theta}} = \Lambda^{-1}\chi_{\epsilon}^{\mathsf{T}}\Omega_{\epsilon} \tag{6.42b}$$

where Λ is some positive definite gain matrix and $k > 0$.

Theorem 6.6 *Under Assumption 6.1, Problem 6.2 is solvable by the control law (6.42).*

Proof Let $\tilde{\Theta} = \hat{\Theta} - \Theta$. Then, substituting (6.42a) into (6.41) gives

$$J\dot{\Omega}_\epsilon = -\chi_\epsilon \tilde{\Theta} - k\Omega_\epsilon - d_{\Omega\epsilon}. \tag{6.43}$$

Let

$$V = \frac{1}{2}(\Omega_\epsilon^\mathsf{T} J \Omega_\epsilon + \tilde{\Theta}^\mathsf{T} \Lambda \tilde{\Theta}). \tag{6.44}$$

Then

$$\begin{aligned}\dot{V} =& \Omega_\epsilon^\mathsf{T}(-\chi_\epsilon \tilde{\Theta} - k\Omega_\epsilon - d_{\Omega\epsilon}) + \tilde{\Theta}^\mathsf{T} \Lambda \Lambda^{-1} \chi_\epsilon^\mathsf{T} \Omega_\epsilon \\ =& -k\|\Omega_\epsilon\|^2 - \Omega_\epsilon^\mathsf{T} d_{\Omega\epsilon} \\ \leq& -k\|\Omega_\epsilon\|^2 + \frac{k}{2}\|\Omega_\epsilon\|^2 + \frac{1}{2k}\|d_{\Omega\epsilon}\|^2 \\ =& -\frac{k}{2}\|\Omega_\epsilon\|^2 + \frac{1}{2k}\|d_{\Omega\epsilon}\|^2.\end{aligned}$$

Let

$$W = V - \frac{1}{2k}\int_0^t \|d_{\Omega\epsilon}(\tau)\|^2 d\tau.$$

By Remark 6.7, $d_{\Omega\epsilon}(t)$ decays to zero exponentially. Thus $\int_0^\infty \|d_{\Omega\epsilon}(\tau)\|^2 d\tau$ exists and is bounded. Therefore, W is lower bounded. Since $\dot{W} = -\frac{k}{2}\|\Omega_\epsilon\|^2 \leq 0$, $\lim_{t\to\infty} W(t)$ exists and is finite, which together with (6.44) implies that $\Omega_\epsilon(t)$ and $\tilde{\Theta}(t)$ are uniformly bounded over $[0, \infty)$. Then, under Assumption 6.1, $\omega_\epsilon(t)$ and $\omega(t)$ are uniformly bounded over $[0, \infty)$. Therefore, $\chi_\epsilon(t)$ is uniformly bounded over $[0, \infty)$, which together with (6.43) implies that $\dot{\Omega}_\epsilon(t)$ is uniformly bounded over $[0, \infty)$. Hence, $\ddot{W}(t)$ is uniformly bounded over $[0, \infty)$. Thus, by Barbalat's lemma, i.e., Lemma 2.9, $\lim_{t\to\infty} \dot{W}(t) = 0$. Thus, $\lim_{t\to\infty} \Omega_\epsilon(t) = 0$. Therefore, applying Lemma 6.2 to (6.39a) and noting (6.38) gives $\lim_{t\to\infty} \hat{q}_\epsilon(t) = 0$ and $\lim_{t\to\infty} \omega_\epsilon(t) = 0$. Finally, the proof is completed by resorting to Lemma 6.3. □

The following result further addresses the issue of the convergence of $\tilde{\Theta}(t)$ to the origin.

Theorem 6.7 *Let* $\pi_0 = \omega_0^\mathsf{T} L(\dot{\omega}_0)$. *Under Assumption 6.1, if* $\pi_0^\mathsf{T}(t)$ *is persistently exciting, then* $\lim_{t\to\infty} \tilde{\Theta}(t) = 0$.

Proof First, we will show the following:

(a) $\lim_{t\to\infty} \Omega_\epsilon(t)$ exists;

(b) $\Omega_\epsilon(t)$ is twice differentiable on each time interval $[t_j, t_{j+1})$, $j = 0, 1, 2, ...$;

(c) There exists a positive constant K such that

$$\sup_{t\in[t_j, t_{j+1}), j=0,1,2,\dots} |\ddot{\Omega}_\epsilon(t)| \leq K.$$

Part (a) has been established in Theorem 6.6. Since $\dot{\chi}_\epsilon$ and $\dot{d}_{\Omega\epsilon}$ exist on each time interval $[t_j, t_{j+1})$, Part (b) is satisfied by noting that

$$J\ddot{\Omega}_\epsilon = -\dot{\chi}_\epsilon\tilde{\Theta} - \chi_\epsilon\dot{\tilde{\Theta}} - k\dot{\Omega}_\epsilon - \dot{d}_{\Omega\epsilon}.$$

From the proof of Theorem 6.6, $\dot{\Omega}_\epsilon(t)$ is uniformly bounded over $[0, \infty)$, and so are $\dot{\omega}_\epsilon(t)$ and $\dot{\omega}(t)$ under Assumption 6.1. Therefore, $\chi_\epsilon(t)$ and $\dot{\chi}_\epsilon(t)$ are both uniformly bounded over $[0, \infty)$. Moreover, it can be directly verified that $\dot{\tilde{\Theta}}(t)$ and $\dot{d}_{\Omega\epsilon}(t)$ are also uniformly bounded over $[0, \infty)$. Therefore, Part (c) is satisfied. By Corollary 2.5, $\lim_{t\to\infty}\dot{\Omega}_\epsilon(t) = 0$. The rest of the proof is similar to the proof of Theorem 6.3 and thus is omitted. □

6.3.3 Case 3: *J* Is Known and Only *q* Is Available

Replacing q_0, ω_0, and $\dot{\omega}_0$ in (6.29) with ψ_0, ζ_0, and ζ_0^d, respectively, gives the following estimation-based attitude feedback control law:

$$u = -k_1\hat{q}_\epsilon - k_2\hat{\alpha} + J\mathbf{C}(q_\epsilon)\zeta_0^d + \left(\mathbf{C}(q_\epsilon)\zeta_0\right)^\times J\mathbf{C}(q_\epsilon)\zeta_0 \tag{6.45a}$$

$$\dot{\rho} = \frac{1}{2}\rho \odot \mathbf{Q}(\beta) \tag{6.45b}$$

where $\rho \in \mathbb{Q}, \alpha = \rho^* \odot q_\epsilon, \beta = \Gamma\hat{\alpha}$ with Γ being some positive definite gain matrix, $k_1, k_2 > 0$.

Under the control law (6.45), instead of (6.28), α is governed by the following equation:

$$\begin{aligned}\dot{\alpha} &= \dot{\rho}^* \odot q_\epsilon + \rho^* \odot \dot{q}_\epsilon \\ &= \frac{1}{2}(\mathbf{Q}(\beta)^* \odot \rho^* \odot q_\epsilon + \rho^* \odot q_\epsilon \odot \mathbf{Q}(\omega_\epsilon)) + \rho^* \odot d_{q\epsilon} \\ &= \frac{1}{2}\begin{pmatrix}\bar{\alpha}(\omega_\epsilon - \beta) + \hat{\alpha}^\times(\omega_\epsilon + \beta) \\ \hat{\alpha}^\mathsf{T}(\beta - \omega_\epsilon)\end{pmatrix} + d_\alpha\end{aligned} \tag{6.46}$$

where $d_\alpha = \rho^* \odot d_{q\epsilon}$.

Remark 6.9 Comparing with (6.28), the right-hand side of (6.46) contains an extra term d_α. By Lemma 6.1, $\|\rho(t)\| = \|\rho(0)\|$ for all $t \geq 0$. By Remark 6.7, d_α will decay to zero exponentially.

Theorem 6.8 *Under Assumption 6.1, Problem 6.2 is solvable by the control law (6.45).*

Proof Substituting (6.45a) into (6.37b) gives

$$\begin{aligned}J\dot{\omega}_\epsilon =& -(\omega_\epsilon + \mathbf{C}(q_\epsilon)\zeta_0)^\times J(\omega_\epsilon + \mathbf{C}(q_\epsilon)\zeta_0) + J\left(\omega_\epsilon^\times\mathbf{C}(q_\epsilon)\zeta_0 - \mathbf{C}(q_\epsilon)\zeta_0^d\right) - d_{\omega\epsilon} \\ &- k_1\hat{q}_\epsilon - k_2\hat{\alpha} + J\mathbf{C}(q_\epsilon)\zeta_0^d + \left(\mathbf{C}(q_\epsilon)\zeta_0\right)^\times J\mathbf{C}(q_\epsilon)\zeta_0\end{aligned}$$

$$
\begin{aligned}
&= -\omega_\epsilon^\times J\omega_\epsilon - \omega_\epsilon^\times J\mathbf{C}(q_\epsilon)\zeta_0 - (\mathbf{C}(q_\epsilon)\zeta_0)^\times J\omega_\epsilon \\
&\quad + J\omega_\epsilon^\times \mathbf{C}(q_\epsilon)\zeta_0 - k_1\hat{q}_\epsilon - k_2\hat{\alpha} - d_{\omega\epsilon} \\
&= -\omega_\epsilon^\times J\omega_\epsilon - \omega_\epsilon^\times J\mathbf{C}(q_\epsilon)\zeta_0 - (\mathbf{C}(q_\epsilon)\zeta_0)^\times J\omega_\epsilon \\
&\quad - J(\mathbf{C}(q_\epsilon)\zeta_0)^\times \omega_\epsilon - k_1\hat{q}_\epsilon - k_2\hat{\alpha} - d_{\omega\epsilon}.
\end{aligned}
\tag{6.47}
$$

Let $W = \frac{1}{2}\omega_\epsilon^\mathsf{T} J\omega_\epsilon$. Noting that $(\mathbf{C}(q_\epsilon)\zeta_0)^\times J + J(\mathbf{C}(q_\epsilon)\zeta_0)^\times$ is anti-symmetric gives

$$
\dot{W} = \omega_\epsilon^\mathsf{T}(-k_1\hat{q}_\epsilon - k_2\hat{\alpha} - d_{\omega\epsilon}).
$$

Since $q_\epsilon(t)$, $\alpha(t)$, and $d_{\omega\epsilon}(t)$ are all uniformly bounded over $[0, \infty)$, we have

$$
\dot{W} \le c_1\sqrt{W}
$$

for some $c_1 > 0$. Thus, for any $W(0) > 0$, by Theorem A.3, there exists c_2 such that

$$
W(t) \le \frac{1}{4}(c_2 + c_1 t)^2
$$

which in turn implies

$$
\|\omega_\epsilon(t)\| \le c_3|c_2 + c_1 t|
$$

for some $c_3 > 0$.

Let

$$
V = \frac{1}{2}\left(\frac{1}{2}\omega_\epsilon^\mathsf{T} J\omega_\epsilon + k_1\|q_\epsilon - q_I\|^2 + k_2\|\alpha - q_I\|^2\right).
$$

Then

$$
\begin{aligned}
\dot{V} =& \frac{1}{2}\omega_\epsilon^\mathsf{T}(-k_1\hat{q}_\epsilon - k_2\hat{\alpha} - d_{\omega\epsilon}) + k_1\hat{q}_\epsilon^\mathsf{T}\left(\frac{1}{2}(\bar{q}_e I_3 + \hat{q}_\epsilon^\times)\omega_\epsilon + \hat{d}_{q\varepsilon}\right) \\
&+ k_1(\bar{q}_e - 1)\left(-\frac{1}{2}\hat{q}_\epsilon^\mathsf{T}\omega_\epsilon + \bar{d}_{q\epsilon}\right) + k_2\hat{\alpha}^\mathsf{T}\left(\frac{1}{2}\bar{\alpha}(\omega_\epsilon - \beta) + \frac{1}{2}\hat{\alpha}^\times(\omega_\epsilon + \beta) + \hat{d}_\alpha\right) \\
&+ k_2(\bar{\alpha} - 1)\left(\frac{1}{2}\hat{\alpha}^\mathsf{T}(\beta - \omega_\epsilon) + \bar{d}_\alpha\right) \\
=& -\frac{1}{2}k_2\hat{\alpha}^\mathsf{T}\Gamma\hat{\alpha} + d_V
\end{aligned}
$$

where $d_V = -\frac{1}{2}\omega_\epsilon^\mathsf{T} d_{\omega\epsilon} + k_1(q_\epsilon - q_I)^\mathsf{T} d_{q\epsilon} + k_2(\alpha - q_I)d_\alpha$. By Remarks 6.7 and 6.9, $d_{\omega\epsilon}$, $d_{q\epsilon}$, and d_α all decay to zero exponentially. Thus, by Lemma 2.6, we have $\lim_{t\to\infty} d_V(t) = 0$ exponentially. Let

$$
\bar{V} = V - \int_0^t d_V(\tau)d\tau.
$$

Since $\int_0^\infty d_V(\tau)d\tau$ exists and is bounded, $\bar{V}$ is lower bounded, and

$$\dot{\bar{V}} = -\frac{1}{2}k_2\hat{\alpha}^{\mathsf{T}}\Gamma\hat{\alpha} \leq 0.$$

Therefore, $\bar{V}(t)$ is uniformly bounded over $[0, \infty)$ and so are $\omega_\epsilon(t)$ and $\alpha(t)$. Moreover, $\ddot{\bar{V}}(t)$ is uniformly bounded over $[0, \infty)$, which implies that $\lim_{t\to\infty}\dot{\bar{V}}(t) = 0$. Therefore, $\lim_{t\to\infty}\hat{\alpha}(t) = 0$. Noting that by Identity A.7 in Sect. A.5 and Remark 6.9, $\|\alpha(t)\| = \|\rho(0)\| \cdot \|q_\epsilon(t)\|$ gives $\lim_{t\to\infty}\|\alpha(t)\| = \|\rho(0)\|$ and thus $\lim_{t\to\infty}\alpha(t)$ exists since $\alpha(t)$ is continuous. Moreover, it can be verified that $\alpha(t)$ is twice differentiable on each time interval $[t_j, t_{j+1})$. Under Assumption 6.1, $\dot{\omega}_\epsilon(t)$ is uniformly bounded over $[0, \infty)$. Then noting that $\dot{d}_{q\epsilon}(t)$ is uniformly bounded over $[0, \infty)$ concludes that $\ddot{\alpha}(t)$ is uniformly bounded over $[0, \infty)$. Therefore, by Corollary 2.5, $\lim_{t\to\infty}\dot{\alpha}(t) = 0$, which together with the facts that $\lim_{t\to\infty}\beta(t) = 0$ and $\rho(0) \neq 0$ implies $\lim_{t\to\infty}\bar{\alpha}(t) \neq 0$ and thus $\lim_{t\to\infty}\omega_\epsilon(t) = 0$. It can be verified that $\ddot{\omega}_\epsilon(t)$ exists and is uniformly bounded over $[0, \infty)$ and again resorting to Corollary 2.5 gives $\lim_{t\to\infty}\dot{\omega}_\epsilon(t) = 0$. Finally, it follows from (6.47) that $\lim_{t\to\infty}\hat{q}_\epsilon(t) = 0$. The proof is thus completed upon using Lemma 6.3. □

6.4 Leader-Following Consensus of Multiple Rigid-Body Systems

6.4.1 Problem Formulation

Consider N rigid-body systems described as follows:

$$\dot{q}_i = \frac{1}{2}q_i \odot \mathbf{Q}(\omega_i) \tag{6.48a}$$

$$J_i\dot{\omega}_i = -\omega_i^\times J_i\omega_i + u_i \tag{6.48b}$$

where, for $i = 1, \ldots, N$, $q_i \in \mathbb{Q}_u$, $\omega_i, u_i \in \mathbb{R}^3$, $J_i \in \mathbb{R}^{3\times3}$ denote the attitude, angular velocity, control input, and inertial matrix of the ith rigid-body system, respectively, all expressed in the body frame.

As in Sect. 5.4, let us tentatively assume that, for each $i = 1, \ldots, N$, the control u_i of every subsystem of (6.48) can access q_0, ω_0, and $\dot{\omega}_0$. Then, a straightforward extension of control law (6.21) gives the following purely decentralized control law:

$$u_i = \omega_i^\times J_i\omega_i - J_i\left(\omega_{ei}^\times\mathbf{C}(q_{ei})\omega_0 - \mathbf{C}(q_{ei})\dot{\omega}_0 + \frac{\kappa_i}{2}(\hat{q}_{ei}^\times + \bar{q}_{ei}I_3)\omega_{ei}\right) - k_i\Omega_{ei} \tag{6.49}$$

where, for $i = 1, \ldots, N$, $\kappa_i, k_i > 0$, and

$$
\begin{aligned}
q_{ei} &= q_0^{-1} \odot q_i \\
\omega_{ei} &= \omega_i - \mathbf{C}(q_{ei})\omega_0 \\
\Omega_{ei} &= \omega_{ei} + \kappa_i \hat{q}_{ei}.
\end{aligned}
$$

Nevertheless, it is unrealistic to assume that q_0, ω_0, and $\dot{\omega}_0$ are accessible by every control u_i, $i = 1, \ldots, N$. To obtain a distributed control law, we assume q_0 and ω_0 are generated by the following nonlinear autonomous system:

$$\dot{q}_0 = \frac{1}{2} q_0 \odot \mathbf{Q}(\omega_0) \tag{6.50a}$$

$$\dot{v}_0 = S_0 v_0 \tag{6.50b}$$

$$\omega_0 = F_0 v_0 \tag{6.50c}$$

where $q_0 \in \mathbb{Q}_u$, $v_0 \in \mathbb{R}^m$, $\omega_0 \in \mathbb{R}^3$, and $S_0 \in \mathbb{R}^{m \times m}$, $F_0 \in \mathbb{R}^{3 \times m}$ are constant matrices. As a result, $\dot{\omega}_0 = F_0 S_0 v_0$.

To guarantee the satisfaction of Assumption 6.1, we make the following assumption.

Assumption 6.2 All the eigenvalues of S_0 are semi-simple with zero real parts.

As in Chap. 5, we treat the system composed of (6.48) and (6.50) as a multi-agent system of $(N + 1)$ agents with (6.50) as the leader and the N subsystems of (6.48) as N followers. Given systems (6.48) and (6.50), and a piecewise constant switching signal $\sigma(t)$ with the dwell time τ, we can define a switching communication graph $\bar{\mathscr{G}}_{\sigma(t)} = (\bar{\mathscr{V}}, \bar{\mathscr{E}}_{\sigma(t)})$ with $\bar{\mathscr{V}} = \{0, 1, \ldots, N\}$ and $\bar{\mathscr{E}}_{\sigma(t)} \subseteq \bar{\mathscr{V}} \times \bar{\mathscr{V}}$ for all $t \geq 0$. Here, the node 0 is associated with the leader system (6.50) and the node i, $i = 1, \ldots, N$, is associated with the ith subsystem of system (6.48), and, for $i = 0, 1, \ldots, N$, $j = 1, \ldots, N, i \neq j$, $(i, j) \in \bar{\mathscr{E}}_{\sigma(t)}$ if and only if u_j can use the information of agent i for control at time instant t. Let $\bar{\mathscr{N}}_i(t) = \{j : (j, i) \in \bar{\mathscr{E}}_{\sigma(t)}\}$ denote the neighbor set of agent i at time t.

Corresponding to the two classes of control laws (6.17) and (6.18), this section considers the following two classes of distributed control laws:

1. Distributed State Feedback Control:

$$u_i = f_i(q_i, \omega_i, \varphi_i, \varphi_j, \ j \in \bar{\mathscr{N}}_i(t)) \tag{6.51a}$$

$$\dot{\varphi}_i = g_i(q_i, \omega_i, \varphi_i, \varphi_j, \ j \in \bar{\mathscr{N}}_i(t)), \quad i = 1, \ldots, N \tag{6.51b}$$

   where φ_0 consists of (q_0, v_0) and possibly some parameters of the leader system to be specified later, and, for $i = 1, \ldots, N$, f_i and g_i are some globally defined smooth functions.
2. Distributed Attitude Feedback Control:

$$u_i = f_i(q_i, \rho_i, \varphi_i, \varphi_j, \ j \in \bar{\mathcal{N}}_i(t)) \tag{6.52a}$$

$$\dot{\varphi}_i = g_i(\varphi_i, \varphi_j, \ j \in \bar{\mathcal{N}}_i(t)) \tag{6.52b}$$

$$\dot{\rho}_i = h_i(\rho_i, q_i, \varphi_i, \varphi_j, \ j \in \bar{\mathcal{N}}_i(t)), \ \ i = 1, \ldots, N \tag{6.52c}$$

where, for $i = 1, \ldots, N$, f_i, g_i, and h_i are some globally defined smooth functions.

Now, the leader-following attitude consensus problem of multiple rigid-body systems is described as follows.

Problem 6.3 Given systems (6.48), (6.50), and a switching communication graph $\bar{\mathcal{G}}_{\sigma(t)}$, find a distributed state feedback control law of the form (6.51) such that, for any $q_0(0) \in \mathbb{Q}_u$, any $v_0(0)$, and any $q_i(0) \in \mathbb{Q}_u$, any $\omega_i(0)$, and any $\varphi_i(0)$, $i = 1, \ldots, N$, or a distributed attitude feedback control law of the form (6.52) such that, for any $q_0(0) \in \mathbb{Q}_u$, any $v_0(0)$, and any $q_i(0) \in \mathbb{Q}_u$, any $\omega_i(0)$, any $\varphi_i(0)$, and any $\rho_i(0) \neq 0$, $i = 1, \ldots, N$, the trajectory of the closed-loop system exists for all $t \geq 0$ and is uniformly bounded over $[0, \infty)$, and satisfies

$$\lim_{t\to\infty} \hat{q}_{ei}(t) = 0, \ \lim_{t\to\infty} \omega_{ei}(t) = 0.$$

6.4.2 Distributed Observer Design

Let us first note that the reference system (6.50) consists of the attitude reference system described by (6.50a) with the measurement output $y_{m0} = q_0$ and the velocity reference system described by (6.50b) and (6.50c) with the measurement output $y_{m0} = v_0$. Since the velocity reference system is independent of q_0, we can separately design the distributed observer/adaptive distributed observer for the velocity reference system and the attitude reference system. From Chap. 4, the distributed observer for the velocity reference system with $y_{m0} = v_0$ is as follows:

$$\dot{v}_i = S_0 v_i + \mu_v \sum_{j=0}^{N} a_{ij}(t)(v_j - v_i) \tag{6.53}$$

where, for $i = 1, \ldots, N$, $v_i \in \mathbb{R}^m$. By Theorem 4.2, under Assumptions 4.1 and 6.2, for any $\mu_v > 0$, the solution of (6.53) is such that

$$\lim_{t\to\infty} (v_i(t) - v_0(t)) = 0 \tag{6.54}$$

exponentially.

On the other hand, from (4.29), the adaptive distributed observer for the velocity reference system with $y_{m0} = v_0$ and $y_0 = \omega_0 = F_0 v_0$ is as follows:

$$\dot{S}_i = \mu_S \sum_{j=0}^{N} a_{ij}(t)(S_j - S_i) \tag{6.55a}$$

$$\dot{F}_i = \mu_F \sum_{j=0}^{N} a_{ij}(t)(F_j - F_i) \tag{6.55b}$$

$$\dot{v}_i = S_i v_i + \mu_v \sum_{j=0}^{N} a_{ij}(t)(v_j - v_i) \tag{6.55c}$$

$$\xi_i = F_i v_i \tag{6.55d}$$

where, for $i = 1, \ldots, N$, $S_i \in \mathbb{R}^{m\times m}$, $F_i \in \mathbb{R}^{3\times m}$, and $v_i \in \mathbb{R}^m$, $\xi_i \in \mathbb{R}^3$, μ_S, μ_F, μ_v are positive constants.

Remark 6.10 For $i = 1, \ldots, N$, let $\tilde{v}_i = v_i - v_0$, $\tilde{S}_i = S_i - S_0$, $\tilde{F}_i = F_i - F_0$, and $\tilde{\xi}_i = \xi_i - \omega_0$. Then, by Theorem 4.8, under Assumptions 4.1 and 6.2, for any positive constants μ_S, μ_F, μ_v, the error variables $\tilde{S}_i(t)$, $\tilde{F}_i(t)$, $\tilde{v}_i(t)$, and $\tilde{\xi}_i(t)$ all tend to zero exponentially as t goes to infinity.

Next, we further develop a distributed observer for the attitude reference system (6.50a). For this purpose, let

$$\dot{\eta}_i = \frac{1}{2}\eta_i \odot \mathbf{Q}(\xi_i) + \mu_\eta \sum_{j=0}^{N} a_{ij}(t)(\eta_j - \eta_i) \tag{6.56}$$

where, for $i = 1, \ldots, N$, $\eta_i \in \mathbb{Q}$, $\xi_i(t) \in \mathbb{R}^3$ is piecewise continuous and uniformly bounded over $[0, \infty)$, $\mu_\eta > 0$, and $\eta_0 = q_0$.

For convenience of manipulating quaternion multiplication, define a matrix function $M(\cdot) : \mathbb{R}^3 \to \mathbb{R}^{4\times 4}$, such that for each $y = \text{col}(y_1, y_2, y_3) \in \mathbb{R}^3$,

$$M(y) = \begin{bmatrix} 0 & y_3 & -y_2 & y_1 \\ -y_3 & 0 & y_1 & y_2 \\ y_2 & -y_1 & 0 & y_3 \\ -y_1 & -y_2 & -y_3 & 0 \end{bmatrix}.$$

Then, we have the following simple result.

Proposition 6.1 *For any* $q \in \mathbb{Q}$, $y \in \mathbb{R}^3$,
(i) $q \odot \mathbf{Q}(y) = M(y)q$.
(ii) For any $q(0) \in \mathbb{Q}$ *and any piecewise continuous-time function* $y(t) \in \mathbb{R}^3$, *the solution of the following system:*

$$\dot{q} = \frac{1}{2}M(y)q$$

is such that $\|q(t)\| = \|q(0)\|$, $\forall t \geq 0$.

Part (i) of Proposition 6.1 is obtained by direct verification, and Part (ii) of Proposition 6.1 is a rephrasing of Lemma 6.1. It is noted that the matrix $M(y)$ is anti-symmetric and linear in y.

Lemma 6.4 *Under Assumptions 4.1 and 6.2, let $q_0(t)$ and $\omega_0(t)$ be any time functions generated by the leader system (6.50) with $q_0(0) \in \mathbb{Q}_u$. Suppose*

$$\lim_{t\to\infty}(\xi_i(t) - \omega_0(t)) = 0, \quad i = 1, \ldots, N$$

exponentially. Then, for any $\mu_\eta > 0$, and any initial condition $\eta_i(0) \in \mathbb{Q}$, $i = 1, \ldots, N$, the solution of (6.56) is such that $\lim_{t\to\infty}(\eta_i(t) - q_0(t)) = 0$ exponentially.

Proof Using Part (i) of Proposition 6.1, system (6.56) can be written as follows:

$$\dot{\eta}_i = \frac{1}{2}M(\xi_i)\eta_i + \mu_\eta \sum_{j=0}^{N} a_{ij}(t)(\eta_j - \eta_i). \tag{6.57}$$

For $i = 1, \ldots, N$, let $x_i = \eta_i - q_0$ and $\tilde{\xi}_i = \xi_i - \omega_0$. Then

$$\begin{aligned}\dot{x}_i &= \frac{1}{2}M(\xi_i)\eta_i - \frac{1}{2}M(\omega_0)q_0 + \mu_\eta \sum_{j=0}^{N} a_{ij}(t)(\eta_j - \eta_i)\\ &= \frac{1}{2}M(\xi_i)x_i + \mu_\eta \sum_{j=0}^{N} a_{ij}(t)(x_j - x_i) + \frac{1}{2}M(\tilde{\xi}_i)q_0.\end{aligned} \tag{6.58}$$

Letting $x = \mathrm{col}(x_1, \ldots, x_N)$ and $\xi = \mathrm{col}(\xi_1, \ldots, \xi_N)$ gives

$$\dot{x} = \mathcal{M}(\xi)x - \mu_\eta(H_{\sigma(t)} \otimes I_4)x + F(t) \tag{6.59}$$

where

$$\begin{aligned}\mathcal{M}(\xi) &= \frac{1}{2}\mathcal{D}(M(\xi_1), \ldots, M(\xi_N))\\ F(t) &= \frac{1}{2}\mathrm{col}\left(M(\tilde{\xi}_1)q_0, \ldots, M(\tilde{\xi}_N)q_0\right).\end{aligned}$$

To ascertain the exponential convergence property of system (6.59), we further put it into the following form:

$$\dot{x} = \left(\frac{1}{2}I_N \otimes M(\omega_0) - \mu_\eta(H_{\sigma(t)} \otimes I_4)\right)x + \left(\mathcal{M}(\xi) - \frac{1}{2}I_N \otimes M(\omega_0)\right)x + F(t) \tag{6.60}$$

which is in the form of (2.17) with

$$A(t) = \frac{1}{2} I_N \otimes M(\omega_0) - \mu_\eta (H_{\sigma(t)} \otimes I_4)$$
$$M(t) = \mathscr{M}(\xi) - \frac{1}{2} I_N \otimes M(\omega_0) = \frac{1}{2}\mathscr{D}(M(\tilde{\xi}_1), \ldots, M(\tilde{\xi}_N))$$

by viewing $\omega_0(t)$ as a uniformly bounded time function. By Remark 6.10, both $M(t)$ and $F(t)$ converge to zero exponentially as t tends to infinity. To make use of Lemma 2.8, it suffices to establish that the origin of the following system:

$$\dot{x} = \left(\frac{1}{2} I_N \otimes M(\omega_0) - \mu_\eta (H_{\sigma(t)} \otimes I_4)\right) x \tag{6.61}$$

is exponential stable.

For this purpose, let $\Phi(\tau, t) \in \mathbb{R}^{4\times 4}$ be the state transition matrix of the system

$$\dot{\varphi} = \frac{1}{2} M(\omega_0)\varphi$$

where $\varphi \in \mathbb{R}^4$, and define $z(t) = (I_N \otimes \Phi(0, t))x(t)$. Then, we have $z(0) = x(0)$, and along the trajectory of system (6.61),

$$\begin{aligned}
\dot{z} &= \left(I_N \otimes \frac{\partial}{\partial t}\Phi(0, t)\right) x + (I_N \otimes \Phi(0, t))\dot{x} \\
&= -\frac{1}{2} (I_N \otimes \Phi(0, t)M(\omega_0))\, x + (I_N \otimes \Phi(0, t))\Big(\frac{1}{2} I_N \otimes M(\omega_0) - \mu_\eta (H_{\sigma(t)} \otimes I_4)\Big)x \\
&= -\mu_\eta (I_N \otimes \Phi(0, t))(H_{\sigma(t)} \otimes I_4)x \\
&= -\mu_\eta (H_{\sigma(t)} \otimes I_4)(I_N \otimes \Phi(0, t))x \\
&= -\mu_\eta (H_{\sigma(t)} \otimes I_4)z.
\end{aligned} \tag{6.62}$$

By Corollary 2.3, under Assumption 4.1, for any $\mu_\eta > 0$, the origin of the linear switched system (6.62) is exponentially stable, i.e., there exist positive constants γ and λ such that

$$\|z(t)\| \le \gamma \|z(0)\| e^{-\lambda t} = \gamma \|x(0)\| e^{-\lambda t}.$$

By Part (ii) of Proposition 6.1, for any $\varphi(0)$, $\|\varphi(t)\| = \|\Phi(t, 0)\varphi(0)\| = \|\varphi(0)\|$ for all $t \ge 0$, and hence the matrix $\Phi(t, 0)$ is uniformly bounded over $[0, \infty)$. As a result,

$$\|x(t)\| = \|(I_N \otimes \Phi(t, 0))z(t)\| \le \|(I_N \otimes \Phi(t, 0))\| \cdot \|z(t)\| \le \delta \|x(0)\| e^{-\lambda t}$$

for some $\delta > 0$. Thus, we have shown that the origin of the linear switched system (6.61) is exponentially stable.

By Lemma 2.8, for any initial condition $x(0)$, the solution of system (6.60) is such that $\lim_{t\to\infty} x(t) = 0$ exponentially. Consequently, for $i = 1, \ldots, N$, $\lim_{t\to\infty}(\eta_i(t) - q_0(t)) = 0$ exponentially. □

Lemma 6.4 immediately leads to the following result.

Corollary 6.1 *Under Assumptions 4.1 and 6.2, let $q_0(t)$ and $\omega_0(t)$ be any time functions generated by the leader system (6.50) with $q_0(0) \in \mathbb{Q}_u$. Then,*
(i) for any $\eta_i(0), v_i(0)$, $i = 1, \ldots, N$, the solution of the following compensator:

$$\dot{\eta}_i = \frac{1}{2}\eta_i \odot \mathbf{Q}(F_0 v_i) + \mu_\eta \sum_{j=0}^{N} a_{ij}(t)(\eta_j - \eta_i) \tag{6.63a}$$

$$\dot{v}_i = S_0 v_i + \mu_v \sum_{j=0}^{N} a_{ij}(t)(v_j - v_i), \quad i = 1, \ldots, N \tag{6.63b}$$

where $\eta_0 = q_0$ and $\mu_\eta, \mu_v > 0$, is such that, for $i = 1, \ldots, N$,

$$\lim_{t\to\infty}(\eta_i(t) - q_0(t)) = 0, \quad \lim_{t\to\infty}(v_i(t) - v_0(t)) = 0$$

exponentially.
(ii) for any $\eta_i(0), S_i(0), F_i(0), v_i(0)$, $i = 1, \ldots, N$, the solution of the following compensator:

$$\dot{\eta}_i = \frac{1}{2}\eta_i \odot \mathbf{Q}(\xi_i) + \mu_\eta \sum_{j=0}^{N} a_{ij}(t)(\eta_j - \eta_i) \tag{6.64a}$$

$$\dot{S}_i = \mu_S \sum_{j=0}^{N} a_{ij}(t)(S_j - S_i) \tag{6.64b}$$

$$\dot{F}_i = \mu_F \sum_{j=0}^{N} a_{ij}(t)(F_j - F_i) \tag{6.64c}$$

$$\dot{v}_i = S_i v_i + \mu_v \sum_{j=0}^{N} a_{ij}(t)(v_j - v_i), \quad i = 1, \ldots, N \tag{6.64d}$$

where $\eta_0 = q_0$, $\xi_i = F_i v_i$, and $\mu_\eta, \mu_S, \mu_F, \mu_v > 0$ is such that, for $i = 1, \ldots, N$,

$$\begin{aligned}
&\lim_{t\to\infty}(\eta_i(t) - q_0(t)) = 0\\
&\lim_{t\to\infty}(S_i(t) - S_0) = 0\\
&\lim_{t\to\infty}(F_i(t) - F_0) = 0\\
&\lim_{t\to\infty}(v_i(t) - v_0(t)) = 0
\end{aligned}$$

exponentially. As a result, $\lim_{t\to\infty}(\xi_i(t) - \omega_0(t)) = 0$ *exponentially.*

That is, (6.63) is a distributed observer for the leader system (6.50), and (6.64) is an adaptive distributed observer for the leader system (6.50).

6.4.3 Solvability of Problem 6.3 for Three Cases

In what follows, we will synthesize a distributed observer-based control law by composing the distributed observer or the adaptive distributed observer with the purely decentralized control law in the way described in Sect. 4.2.

We will also consider three cases as in Sect. 6.3.

6.4.3.1 Case 1: For $i = 1, \ldots, N$, J_i Is Known and ω_i, q_i Are Available

Let us first synthesize a distributed control law by composing the distributed observer (6.63) with the purely decentralized control law (6.49). For this purpose, let

$$\xi_i = F_0 v_i \tag{6.65a}$$

$$\xi_i^d = F_0 S_0 v_i \tag{6.65b}$$

$$q_{\epsilon i} = \eta_i^* \odot q_i \tag{6.65c}$$

$$\omega_{\epsilon i} = \omega_i - \mathbf{C}(q_{\epsilon i})\xi_i \tag{6.65d}$$

$$\Omega_{\epsilon i} = \omega_{\epsilon i} + \kappa_i \hat{q}_{\epsilon i} \tag{6.65e}$$

where $\kappa_i, k_i > 0$. Then, the distributed observer-based control law is as follows:

$$u_i = \omega_i^\times J_i \omega_i - J_i \left(\omega_{\epsilon i}^\times \mathbf{C}(q_{\epsilon i})\xi_i - \mathbf{C}(q_{\epsilon i})\xi_i^d + \frac{\kappa_i}{2}(\bar{q}_{\epsilon i} I_3 + \hat{q}_{\epsilon i}^\times)\omega_{\epsilon i}\right) - k_i \Omega_{\epsilon i} \tag{6.66a}$$

$$\dot{\eta}_i = \frac{1}{2}\eta_i \odot \mathbf{Q}(F_0 v_i) + \mu_\eta \sum_{j=0}^{N} a_{ij}(t)(\eta_j - \eta_i) \tag{6.66b}$$

$$\dot{v}_i = S_0 v_i + \mu_v \sum_{j=0}^{N} a_{ij}(t)(v_j - v_i) \tag{6.66c}$$

where $\eta_0 = q_0$, $\mu_v, \mu_\eta > 0$.

It can be seen that, for $i = 1, \ldots, N$, (6.66a) is obtained from the purely decentralized control law (6.49) by replacing $(q_0, \omega_0, \dot{\omega}_0)$ with $(\eta_i, F_0 v_i, F_0 S_0 v_i)$, which is provided by the distributed observer consisting of (6.66b) and (6.66c). Thus, the control law is in the form of (6.51) with $\varphi_0 = (q_0, v_0)$.

We have the following result.

Theorem 6.9 *Given systems (6.48), (6.50), and a switching communication graph $\bar{\mathscr{G}}_{\sigma(t)}$, under Assumptions 4.1 and 6.2, Problem 6.3 is solvable by the control law (6.66).*

Proof Note that, for each $i = 1, \ldots, N$, (6.66a) is obtained from (6.40) by replacing $\psi_0, \zeta_0, \zeta_0^d$ in (6.40) with η_i, ξ_i, ξ_i^d in (6.66) and (6.65), and (6.66b) is in the form (6.33) with $(\psi_0, \zeta_0) = (\eta_i, F_0 v_i)$ and $\Delta = \mu_\eta \sum_{j=0}^N a_{ij}(t)(\eta_j - \eta_i)$ which decays to zero exponentially since $(\eta_j - q_0) - (\eta_i - q_0)$ decays to zero exponentially. To apply Theorem 6.5, let us verify (6.34) with $(\psi_0, \zeta_0, \zeta_0^d) = (\eta_i, \xi_i, \xi_i^d)$. In fact, (6.34a) holds since $\lim_{t\to\infty}(\eta_i(t) - q_0(t)) = 0$. Next, we have

$$\xi_i - \omega_0 = F_0 v_i - F_0 v_0, \ \xi_i^d - \dot{\omega}_0 = F_0 S_0 v_i - F_0 S_0 v_0$$

and

$$\xi_i^d - \dot{\xi}_i = F_0 S_0 v_i - F_0 \left(S_0 v_i + \mu_v \sum_{j=0}^N a_{ij}(t)(v_j - v_i) \right) = -\mu_v F_0 \sum_{j=0}^N a_{ij}(t)(v_j - v_i).$$

Therefore,

$$\lim_{t\to\infty} (\xi_i(t) - \omega_0(t)) = 0, \ \lim_{t\to\infty} (\xi_i^d(t) - \dot{\omega}_0(t)) = 0, \ \lim_{t\to\infty} (\xi_i^d(t) - \dot{\xi}_i(t)) = 0.$$

Thus, (6.34) holds. Invoking Corollary 6.1 and Theorem 6.5 completes the proof. □

Next, to obtain an adaptive distributed observer-based control law, let

$$\xi_i = F_i v_i \tag{6.67a}$$
$$\xi_i^d = F_i S_i v_i \tag{6.67b}$$
$$q_{\epsilon i} = \eta_i^* \odot q_i \tag{6.67c}$$
$$\omega_{\epsilon i} = \omega_i - \mathbf{C}(q_{\epsilon i})\xi_i \tag{6.67d}$$
$$\Omega_{\epsilon i} = \omega_{\epsilon i} + \kappa_i \hat{q}_{\epsilon i} \tag{6.67e}$$

where $\kappa_i, k_i > 0$.

Then, by composing the adaptive distributed observer (6.64) with the purely decentralized control law (6.49) gives the following control law:

$$u_i = \omega_i^\times J_i \omega_i - J_i \left(\omega_{\epsilon i}^\times \mathbf{C}(q_{\epsilon i})\xi_i + \frac{\kappa_i}{2}(\bar{q}_{\epsilon i} I_3 + \hat{q}_{\epsilon i}^\times)\omega_{\epsilon i} - \mathbf{C}(q_{\epsilon i})\xi_i^d \right) - k_i \Omega_{\epsilon i} \tag{6.68a}$$

$$\dot{\eta}_i = \frac{1}{2}\eta_i \odot \mathbf{Q}(\xi_i) + \mu_\eta \sum_{j=0}^N a_{ij}(t)(\eta_j - \eta_i) \tag{6.68b}$$

$$\dot{S}_i = \mu_S \sum_{j=0}^{N} a_{ij}(t)(S_j - S_i) \tag{6.68c}$$

$$\dot{F}_i = \mu_F \sum_{j=0}^{N} a_{ij}(t)(F_j - F_i) \tag{6.68d}$$

$$\dot{v}_i = S_i v_i + \mu_v \sum_{j=0}^{N} a_{ij}(t)(v_j - v_i) \tag{6.68e}$$

where $\eta_0 = q_0$, $\mu_S, \mu_F, \mu_v, \mu_\eta > 0$.

We have the following result.

Theorem 6.10 *Given systems* (6.48), (6.50), *and a switching communication graph* $\bar{\mathcal{G}}_{\sigma(t)}$, *under Assumptions* 4.1 *and* 6.2, *Problem* 6.3 *is solvable by the control law* (6.68).

Proof Note that, for each $i = 1, \ldots, N$, (6.68a) is obtained from (6.40) by replacing $\psi_0, \zeta_0, \zeta_0^d$ in (6.40) with η_i, ξ_i, ξ_i^d in (6.68) and (6.67), and (6.68b) is in the form (6.33) with $(\psi_0, \zeta_0) = (\eta_i, F_i v_i)$ and $\Delta = \mu_\eta \sum_{j=0}^{N} a_{ij}(t)(\eta_j - \eta_i)$ which decays to zero exponentially. We now verify (6.34) with $(\psi_0, \zeta_0, \zeta_0^d) = (\eta_i, \xi_i, \xi_i^d)$. Clearly, (6.34a) holds since $\lim_{t\to\infty}(\eta_i(t) - q_0(t)) = 0$. Also, we have

$$\xi_i - \omega_0 = F_i v_i - F_0 v_0, \; \xi_i^d - \dot{\omega}_0 = F_i S_i v_i - F_0 S_0 v_0 \tag{6.69}$$

and

$$\begin{aligned} \xi_i^d - \dot{\xi}_i &= F_i S_i v_i - \dot{F}_i v_i - F_i \left(S_i v_i + \mu_v \sum_{j=0}^{N} a_{ij}(t)(v_j - v_i) \right) \\ &= -\mu_F \sum_{j=0}^{N} a_{ij}(t)(F_j - F_i) v_i - \mu_v F_i \sum_{j=0}^{N} a_{ij}(t)(v_j - v_i). \end{aligned} \tag{6.70}$$

Therefore,

$$\lim_{t\to\infty} (\xi_i(t) - \omega_0(t)) = 0, \; \lim_{t\to\infty} (\xi_i^d(t) - \dot{\omega}_0(t)) = 0, \; \lim_{t\to\infty} (\xi_i^d(t) - \dot{\xi}_i(t)) = 0. \tag{6.71}$$

That is, (6.34) holds. Invoking Corollary 6.1 and Theorem 6.5 completes the proof. □

Again, the control law (6.68) is in the form of (6.51) with φ_0 consisting of the components of q_0, v_0, S_0, and F_0.

6.4.3.2 Case 2: For $i = 1, \ldots, N$, J_i Is Uncertain and ω_i, q_i Are Available

Again, let us tentatively assume that, for each $i = 1, \ldots, N$, the control u_i of every subsystem of (6.48) can access q_0, ω_0, and $\dot{\omega}_0$. Then, a straightforward extension of control law (6.24) gives the following purely decentralized control law:

$$u_i = -\chi_{ei}\hat{\Theta}_i - k_i \Omega_{ei}, \tag{6.72a}$$

$$\dot{\hat{\Theta}}_i = \Lambda_i^{-1}\chi_{ei}^{\mathsf{T}}\Omega_{ei} \tag{6.72b}$$

where Λ_i is some positive definite gain matrix and $k_i > 0$, and

$$\chi_{ei} = -\omega_i^{\times}L(\omega_i) + L\left(\omega_{ei}^{\times}\mathbf{C}(q_{ei})\omega_0 - \mathbf{C}(q_{ei})\dot{\omega}_0 + \frac{\kappa_i}{2}(\hat{q}_{ei}^{\times} + \bar{q}_{ei}I_3)\omega_{ei}\right)$$

with $\kappa_i > 0$.

Next, we further compose the adaptive distributed observer (6.64) with the purely decentralized control law (6.72) to obtain the following adaptive distributed observer-based control law:

$$u_i = -\chi_{\epsilon i}\hat{\Theta}_i - k_i \Omega_{\epsilon i}, \tag{6.73a}$$

$$\dot{\hat{\Theta}}_i = \Lambda_i^{-1}\chi_{\epsilon i}^{\mathsf{T}}\Omega_{\epsilon i} \tag{6.73b}$$

$$\dot{\eta}_i = \frac{1}{2}\eta_i \odot \mathbf{Q}(\xi_i) + \mu_\eta \sum_{j=0}^{N} a_{ij}(t)(\eta_j - \eta_i) \tag{6.73c}$$

$$\dot{S}_i = \mu_S \sum_{j=0}^{N} a_{ij}(t)(S_j - S_i) \tag{6.73d}$$

$$\dot{F}_i = \mu_F \sum_{j=0}^{N} a_{ij}(t)(F_j - F_i) \tag{6.73e}$$

$$\dot{v}_i = S_i v_i + \mu_v \sum_{j=0}^{N} a_{ij}(t)(v_j - v_i) \tag{6.73f}$$

where $\eta_0 = q_0$, $\mu_S, \mu_F, \mu_v, \mu_\eta > 0$, and Λ_i is positive definite, ξ_i, ξ_i^d, $q_{\epsilon i}$, $\omega_{\epsilon i}$, $\Omega_{\epsilon i}$ are as defined in (6.67), and

$$\chi_{\epsilon i} = -\omega_i^{\times}L(\omega_i) + L\left(\omega_{\epsilon i}^{\times}\mathbf{C}(q_{\epsilon i})\xi_i - \mathbf{C}(q_{\epsilon i})\xi_i^d + \frac{\kappa_i}{2}(\hat{q}_{\epsilon i}^{\times} + \bar{q}_{\epsilon i}I_3)\omega_{\epsilon i}\right).$$

Theorem 6.11 *Given systems (6.48), (6.50), and a switching communication graph $\bar{\mathcal{G}}_{\sigma(t)}$, under Assumptions 4.1 and 6.2, Problem 6.3 is solvable by the control law (6.73).*

Proof Note that, for each $i = 1, \ldots, N$, the first two equations of the control law (6.73) are obtained from (6.42) by replacing $\psi_0, \zeta_0, \zeta_0^d$ in (6.42) with η_i, ξ_i, ξ_i^d in (6.68) and (6.67), and (6.73c) is in the form (6.33) with $(\psi_0, \zeta_0) = (\eta_i, F_i v_i)$ and $\Delta = \mu_\eta \sum_{j=0}^{N} a_{ij}(t)(\eta_j - \eta_i)$ which decays to zero exponentially. We now verify (6.34) with $(\psi_0, \zeta_0, \zeta_0^d) = (\eta_i, \xi_i, \xi_i^d)$. Clearly, (6.34a) holds since $\lim_{t\to\infty}(\eta_i(t) - q_0(t)) = 0$.

Note that (6.69) and (6.70) hold, which gives (6.71). Thus, Invoking Corollary 6.1 and Theorem 6.6 completes the proof. □

Finally, since the satisfaction Assumption 6.2 implies the satisfaction of Assumption 6.1, a straightforward extension of Theorem 6.7 gives the following result regarding the issue of the convergence of $\tilde{\Theta}_i(t)$, $i = 1, \ldots, N$, to the origin.

Theorem 6.12 *Let $\pi_0 = \omega_0^{\mathsf{T}} L(\dot{\omega}_0)$. Under Assumption 6.2, if $\pi_0^{\mathsf{T}}(t)$ is persistently exciting, then, for $i = 1, \ldots, N$, $\lim_{t\to\infty} \tilde{\Theta}_i(t) = 0$.*

6.4.3.3 Case 3: For $i = 1, \ldots, N$, J_i Is Known and Only q_i Is Available

Again, we tentatively assume that, for each $i = 1, \ldots, N$, the control u_i of every subsystem of (6.48) can access q_0, ω_0, and $\dot{\omega}_0$. Then, a straightforward extension of control law (6.29) gives the following purely decentralized attitude feedback control law:

$$u_i = -k_{i1}\hat{q}_{ei} - k_{i2}\hat{\alpha}_i + J_i \mathbf{C}(q_{ei})\dot{\omega}_0 + \left(\mathbf{C}(q_{ei})\omega_0\right)^{\times} J_i \mathbf{C}(q_{ei})\omega_0 \tag{6.74a}$$

$$\dot{\rho}_i = \frac{1}{2}\rho_i \odot \mathbf{Q}(\beta_i) \tag{6.74b}$$

where, for $i = 1, \ldots, N$, $k_{i1}, k_{i2} > 0$, and

$$\alpha_i = \rho_i^* \odot q_{ei} \tag{6.75a}$$

$$\beta_i = \Gamma_i \hat{\alpha}_i \tag{6.75b}$$

with $\Gamma_i \in \mathbb{R}^{3\times 3}$ being some positive definite gain matrix.

Next, we further compose the adaptive distributed observer (6.64) with the purely decentralized control law (6.74) to obtain the following adaptive distributed observer-based attitude feedback control law:

$$u_i = -k_{i1}\hat{q}_{\epsilon i} - k_{i2}\hat{\alpha}_i + J_i\mathbf{C}(q_{\epsilon i})\xi_i^d + \big(\mathbf{C}(q_{\epsilon i})\xi_i\big)^\times J_i\mathbf{C}(q_{\epsilon i})\xi_i \tag{6.76a}$$

$$\dot{\rho}_i = \frac{1}{2}\rho_i \odot \mathbf{Q}(\beta_i) \tag{6.76b}$$

$$\dot{\eta}_i = \frac{1}{2}\eta_i \odot \mathbf{Q}(\xi_i) + \mu_\eta \sum_{j=0}^{N} a_{ij}(t)(\eta_j - \eta_i) \tag{6.76c}$$

$$\dot{S}_i = \mu_S \sum_{j=0}^{N} a_{ij}(t)(S_j - S_i) \tag{6.76d}$$

$$\dot{F}_i = \mu_F \sum_{j=0}^{N} a_{ij}(t)(F_j - F_i) \tag{6.76e}$$

$$\dot{v}_i = S_i v_i + \mu_v \sum_{j=0}^{N} a_{ij}(t)(v_j - v_i) \tag{6.76f}$$

where $\eta_0 = q_0$, $\mu_S, \mu_F, \mu_v, \mu_\eta, \kappa_i, k_{i1}, k_{i2} > 0$, ξ_i, ξ_i^d, $q_{\epsilon i}$, $\omega_{\epsilon i}$ are as defined in (6.67), and

$$\alpha_i = \rho_i^* \odot q_{\epsilon i}$$
$$\beta_i = \Gamma_i \hat{\alpha}_i$$

with $\Gamma_i \in \mathbb{R}^{3\times 3}$ being some positive definite gain matrix.

We have the following result.

Theorem 6.13 *Given systems* (6.48), (6.50), *and a switching communication graph* $\bar{\mathscr{G}}_{\sigma(t)}$, *under Assumptions 4.1 and 6.2, Problem 6.3 is solvable by the control law* (6.76).

Proof Note that, for each $i = 1, \ldots, N$, the first two equations of the control law (6.76) are obtained from (6.45) by replacing ψ_0, ζ_0, ζ_0^d in (6.45) with η_i, ξ_i, ξ_i^d in (6.76) and (6.67), and (6.76c) is in the form (6.33) with $(\psi_0, \zeta_0) = (\eta_i, F_i v_i)$ and $\Delta = \mu_\eta \sum_{j=0}^{N} a_{ij}(t)(\eta_j - \eta_i)$ which decays to zero exponentially.

Note that (6.69) and (6.70) hold, which gives (6.71). Thus, invoking Corollary 6.1 and Theorem 6.8 completes the proof.

6.5 Numerical Examples

In this section, we evaluate the performance of various adaptive distributed observer-based control laws for three cases, respectively, using numerical examples. Consider a group of four rigid spacecraft systems, whose nominal inertial matrices are given as follows:

$$J_1 = J_2 = \begin{bmatrix} 0.729 & & \\ & 0.54675 & \\ & & 0.625 \end{bmatrix} \text{kg} \cdot \text{m}^2$$
$$J_3 = J_4 = \begin{bmatrix} 0.3645 & & \\ & 0.2734 & \\ & & 0.3125 \end{bmatrix} \text{kg} \cdot \text{m}^2. \tag{6.77}$$

The leader system for generating the desired angular velocity is as follows:

$$\dot{v}_0 = \begin{bmatrix} 0 & 1 & & & & \\ -1 & 0 & & & & \\ & & 0 & 2 & & \\ & & -2 & 0 & & \\ & & & & 0 & 3 \\ & & & & -3 & 0 \end{bmatrix} v_0 = S_0 v_0$$
$$\omega_0 = \begin{bmatrix} 1 & 0 & & & & \\ & & 1 & 0 & & \\ & & & & 1 & 0 \end{bmatrix} v_0 = F_0 v_0$$

which satisfies Assumption 6.2. With the initial condition $v_0(0) = \text{col}(0, 1, 0, 1, 0, 1)$, the leader system will generate the following desired angular velocity:

$$\omega_0(t) = \begin{bmatrix} \sin t \\ \sin 2t \\ \sin 3t \end{bmatrix} \text{rad/s}$$

which will be used for simulation.

The communication graph is shown in Fig. 6.3, where the switching signal is defined as follows:

$$\sigma(t) = \begin{cases} 1 \text{ if } 0.4s \leq t < 0.4s + 0.1 \\ 2 \text{ if } 0.4s + 0.1 \leq t < 0.4s + 0.2 \\ 3 \text{ if } 0.4s + 0.2 \leq t < 0.4s + 0.3 \\ 4 \text{ if } 0.4s + 0.3 \leq t < 0.4s + 0.4 \end{cases}$$

where $s = 0, 1, 2, \ldots$. It can be seen that Assumption 4.1 is satisfied.

Thus, by Theorems 6.10, 6.11, and 6.13, Problem 6.3 for the three cases can be solved by the control law (6.68), (6.73), and (6.76), respectively.

Example 6.1 Performance Evaluation for Case 1: In this case, the values of the four inertial matrices are exactly known. The controller is given by (6.68), where the control gains are selected to be $\mu_S = 10$, $\mu_F = 10$, $\mu_v = 100$, $\mu_\eta = 100$, $\kappa_i = 2$, $k_i = 1$. The initial conditions are given by

Fig. 6.3 The switching communication graph $\bar{\mathcal{G}}_{\sigma(t)}$ satisfying Assumption 4.1

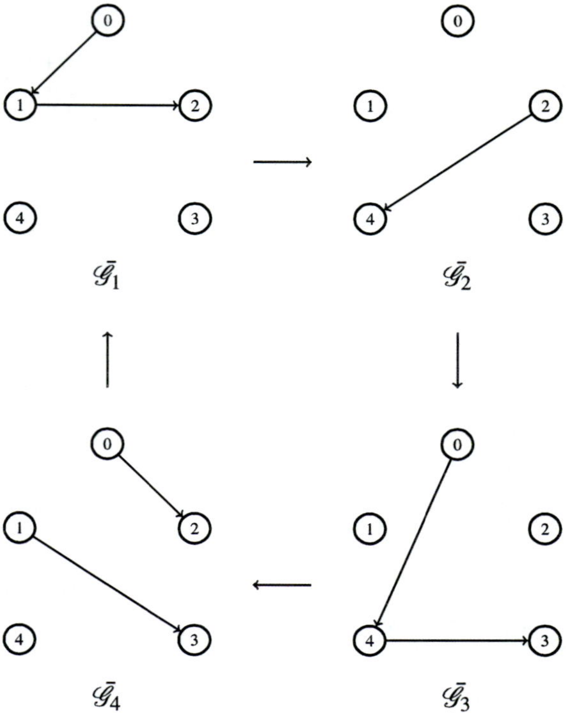

$$
\begin{aligned}
&q_0(0) = \mathrm{col}(0, 0, 0, 1), q_1(0) = \mathrm{col}(\sin\frac{\pi}{8}, 0, 0, \cos\frac{\pi}{8}) \\
&q_2(0) = \mathrm{col}(0, \sin\frac{\pi}{6}, 0, \cos\frac{\pi}{6}), q_3(0) = \mathrm{col}(0, 0, \sin\frac{\pi}{4}, \cos\frac{\pi}{4}) \\
&q_4(0) = \mathrm{col}(1, 0, 0, 0), \omega_i(0) = 0, S_i(0) = 0 \\
&F_i(0) = 0, v_i(0) = 0, \eta_i(0) = 0.
\end{aligned}
\tag{6.78}
$$

The tracking errors of attitude and angular velocity are shown by Fig. 6.4. As expected, all quantities satisfactorily converge to their desired values.

Example 6.2 Performance Evaluation for Case 2: For this case, the four inertial matrices are uncertain with their nominal values given in (6.77). The controller is given by (6.73), where the control gains are selected to be $\mu_S = 10$, $\mu_F = 10$, $\mu_v = 10$, $\mu_\eta = 10$, $\kappa_i = 2$, $k_i = 1$, $\Lambda_i = 0.2I_6$. We assume the same initial conditions as in (6.78) and additionally $\hat{\Theta}_i(0) = 0$. The tracking errors of attitude and angular velocity are shown by Fig. 6.5.

To ascertain the convergence of $\tilde{\Theta}_i(t)$, $i = 1, \ldots, 4$, let us calculate $\pi_0(t)$ as follows:

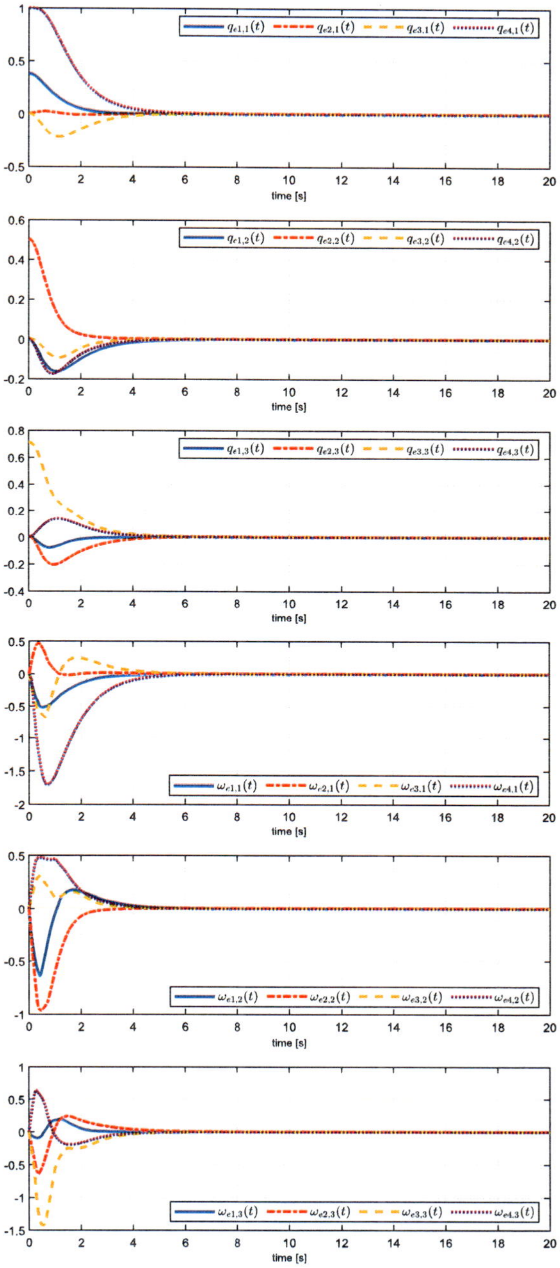

Fig. 6.4 The response of $q_{ei}(t)$ and $\omega_{ei}(t)$ under the control law (6.68)

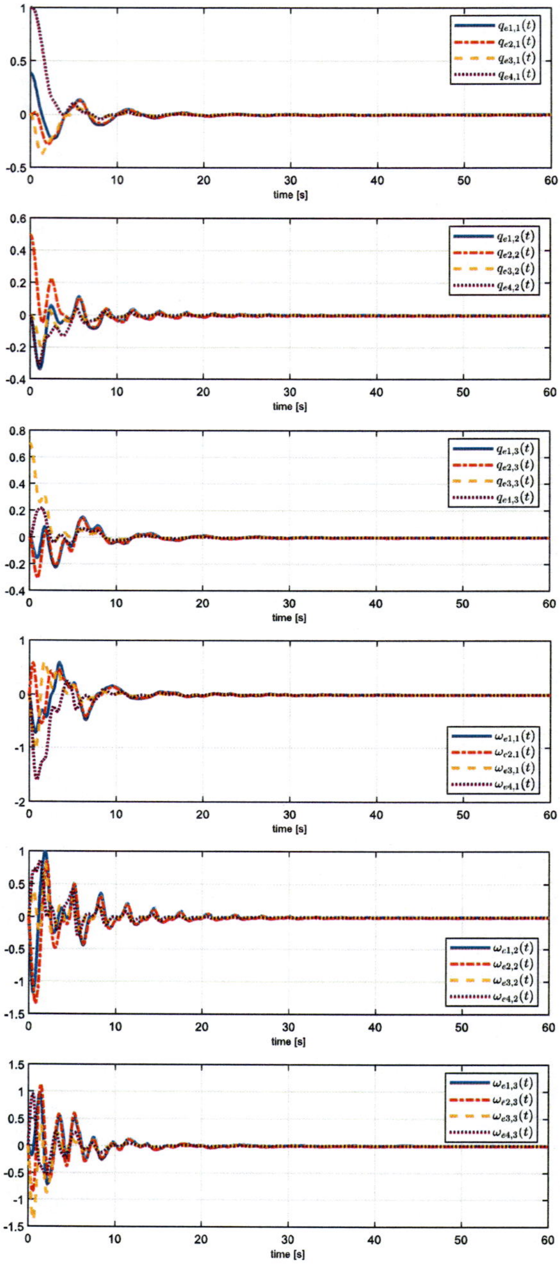

Fig. 6.5 The response of $q_{ei}(t)$ and $\omega_{ei}(t)$ under the control law (6.73)

$$
\begin{aligned}
\pi_0 &= \omega_0^{\mathsf{T}} L(\dot{\omega}_0) \\
&= \begin{pmatrix} \sin t & \sin 2t & \sin 3t \end{pmatrix} \begin{pmatrix} \cos t & 0 & 0 & 0 & 3\cos 3t & 2\cos 2t \\ 0 & 2\cos 2t & 0 & 3\cos 3t & 0 & \cos t \\ 0 & 0 & 3\cos 3t & 2\cos 2t & \cos t & 0 \end{pmatrix} \\
&= \begin{pmatrix} \pi_{01} & \pi_{02} & \pi_{03} & \pi_{04} & \pi_{05} & \pi_{06} \end{pmatrix}
\end{aligned}
$$

where $\pi_{01} = \frac{1}{2}\sin 2t$, $\pi_{02} = \sin 4t$, $\pi_{03} = \frac{3}{2}\sin 6t$, $\pi_{04} = 3\sin 2t\cos 3t + 2\sin 3t\cos 2t$, $\pi_{05} = 3\sin t\cos 3t + \sin 3t\cos t$, and $\pi_{06} = 2\sin t\cos 2t + \sin 2t\cos t$. Clearly, π_0 is periodic and all the entries are linearly independent. Thus, it is persistently exciting. By Theorem 6.12, for $i = 1, \ldots, 4$, $\lim_{t\to\infty} \tilde{\Theta}_i(t) = 0$, which is confirmed in Fig. 6.6.

Example 6.3 Performance Evaluation for Case 3: In this case, we evaluate the attitude feedback controller given by (6.76), where the control gains are selected to be $\mu_S = 10$, $\mu_F = 10$, $\mu_v = 100$, $\mu_\eta = 100$, $k_{i1} = 1$, $k_{i2} = 1$, and $\Gamma_i = I_3$. We assume the same initial conditions as in (6.78) and additionally $\rho_i(0) = \text{col}(1, 1, 1, 1)$. The tracking errors of attitude and angular velocity are shown by Fig. 6.7.

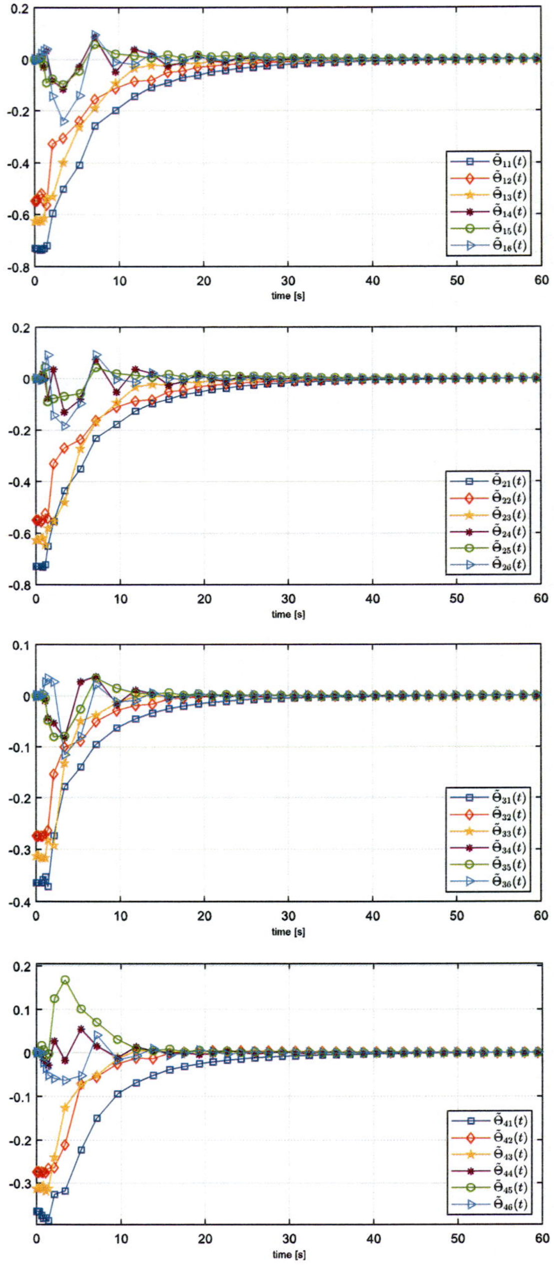

Fig. 6.6 The response of $\tilde{\Theta}_{ij}(t) = \hat{\Theta}_{ij}(t) - \Theta_{ij}$ under the control law (6.73)

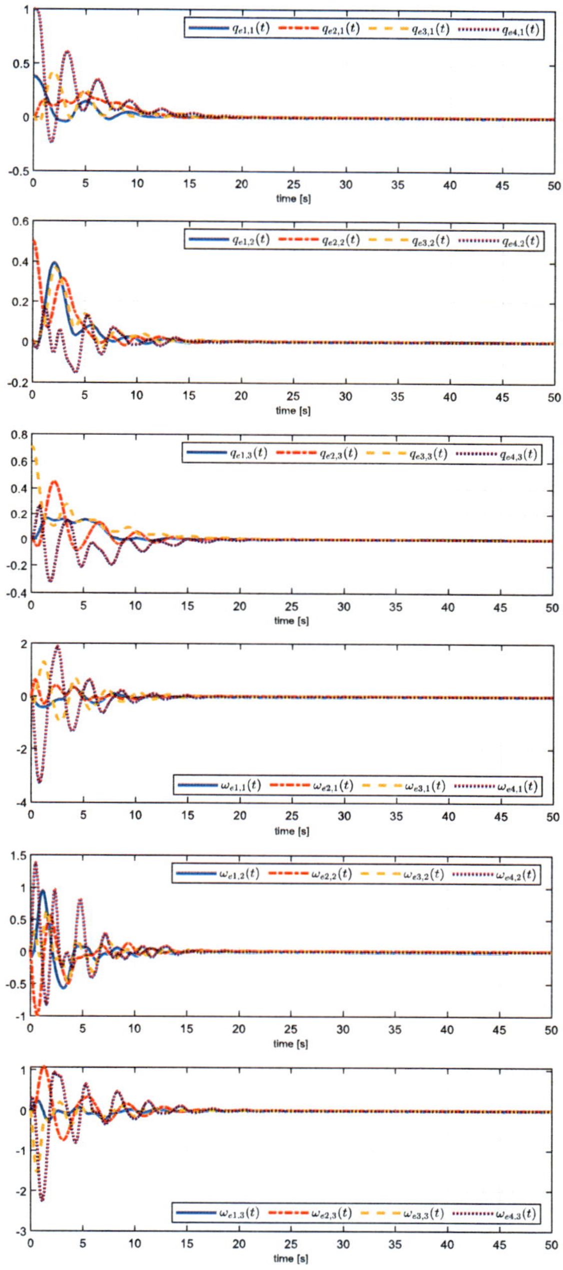

Fig. 6.7 The response of $q_{ei}(t)$ and $\omega_{ei}(t)$ under the control law (6.76)

6.6 Notes and References

Comprehensive expositions on attitude parametrization, kinematics, and dynamics can be found from [1–4]. In particular, the proof of $C(a, \phi) = C_{FI}$ is motivated by the outline in p. 12 of [2]. Lemma 6.2 is adapted from Lemma 3.1 of [5]. Cases 1, 2, and 3 for the standard control of a single rigid-body system can be found from, say, [6–8], respectively. Early works on leaderless attitude synchronization can be found in [9–13]. References [9, 13] also considered the attitude tracking under the assumption that the desired attitude and angular velocity are known to all the rigid-body systems. The leader-following attitude synchronization problem was studied in [14, 15] with various assumptions on the communication networks. Further results along the line of the work of [14, 15] were given in [16, 17]. The leader-following attitude consensus problem over static communication networks for Cases 1, 2, and 3 using the distributed observer approach was solved in [18–20], respectively. These results were extended to the case of switching communication networks in [21–23]. The cooperative control of multiple rigid-body systems with time delay was studied in [24, 25], and the finite time cooperative control of multiple rigid-body systems was considered in [26, 27].

The leader-following consensus problem of multiple rigid spacecraft systems over jointly connected switching networks was studied by the event-triggered control law in [28]. The leader-following consensus with disturbance rejection of multiple uncertain spacecraft systems over static networks was studied in [29]. The inertial parameters of the rigid spacecraft in the example are taken from [30].

The leader-following formation control problem of multiple rigid-body systems was studied in [31].

References

1. Goldstein H, Poole CP, Safko JL (2001) Classical mechanics. Addison-Wesley, New York
2. Hughes P (1986) Spacecraft attitude dynamics. Wiley, New York
3. Sidi MJ (1997) Spacecraft dynamics and control. Cambridge University Press, Cambridge
4. Stuelpnagel J (1964) On the parametrization of the three-dimensional rotation group. SIAM Rev 6(4):422–430
5. Chen Z, Huang J (2009) Attitude tracking and disturbance rejection of rigid spacecraft by adaptive control. IEEE Trans Autom Control 54(3):600–605
6. Yuan JS-C (1988) Closed-loop manipulator control using quaternion feedback. IEEE J Robot Autom 4(4):434–440
7. Ahmed J, Coppola VT, Bernstein D (1998) Adaptive asymptotic tracking of spacecraft attitude motion with inertia matrix identification. J Guid Control Dyn 21(5):684–691
8. Tayebi A (2008) Unit quaternion-based output feedback for the attitude tracking problem. IEEE Trans Autom Control 53(6):1516–1520
9. Abdessameud A, Tayebi A (2009) Attitude synchronization of a group of spacecraft without velocity measurements. IEEE Trans Autom Control 54(11):2642–2648
10. Bai H, Arcak M, Wen JT (2008) Rigid body attitude coordination without inertial frame information. Automatica 44(12):3170–3175

11. Lawton JR, Beard RW (2002) Synchronized multiple spacecraft rotations. Automatica 38(8):1359–1364
12. Sarlette A, Sepulchre R, Leonard NE (2009) Autonomous rigid body attitude synchronization. Automatica 45(2):572–577
13. VanDyke MC, Hall CD (2006) Decentralized coordinated attitude control within a formation of spacecraft. J Guid Control Dyn 29(5):1101–1109
14. Ren W (2007) Distributed attitude alignment in spacecraft formation flying. Int J Adapt Control Signal Process 21(2):95–113
15. Ren W (2010) Distributed cooperative attitude synchronization and tracking for multiple rigid bodies. IEEE Trans Control Syst Technol 18(2):383–392
16. Zou A, Kumar KD, Hou Z (2012) Attitude coordination control for a group of spacecraft without velocity measurements. IEEE Trans Control Syst Technol 20(5):1160–1174
17. Zou A, Kumar KD (2013) Quaternion-based distributed output feedback attitude coordination control for spacecraft formation flying. J Guid Control Dyn 36(2):548–556
18. Cai H, Huang J (2014) The leader-following attitude control of multiple rigid spacecraft systems. Automatica 50:1109–1115
19. Cai H, Huang J (2016) Leader-following adaptive consensus of multiple uncertain rigid spacecraft systems. Sci China Inf Sci 59:1–13
20. Cai H, Huang J (2016) Leader-following attitude consensus of multiple rigid body systems by attitude feedback control. Automatica 69:87–92
21. Liu T, Huang J (2018) Leader-following attitude consensus of multiple rigid body systems subject to jointly connected switching networks. Automatica 92:63–71
22. Wang T, Huang J (2020) Leader-following adaptive consensus of multiple uncertain rigid body systems over jointly connected networks. Unmanned Syst 8(2):85–93
23. Wang T, Huang J (2020) Consensus of multiple spacecraft systems over switching networks by attitude feedback. IEEE Trans Aerosp Electron Syst 56(3):2018–2025
24. Abdessameud A, Tayebi A, Polushin IG (2012) Attitude synchronization of multiple rigid bodies with communication delays. IEEE Trans Autom Control 57(9):2405–2411
25. Li S, Du H, Shi P (2014) Distributed attitude control for multiple spacecraft with communication delays. IEEE Trans Aerosp Electron Syst 50(3):1765–1773
26. Du H, Li S, Qian C (2011) Finite-time attitude tracking control of spacecraft with application to attitude synchronization. IEEE Trans Autom Control 56(11):2711–2717
27. Peng X, Geng Z, Sun J (2020) The specified finite-time distributed observers-based velocity-free attitude synchronization for rigid bodies on SO(3). IEEE Trans Syst, Man, Cybern: Syst 50(4):1610–1621
28. Wang T, Huang J (2021) Leader-following event-triggered adaptive practical consensus of multiple rigid spacecraft systems over jointly connected networks. IEEE Trans Neural Netw Learn Syst 32(12): 5623–5632
29. Cai H, Huang J (2017) Leader-following attitude consensus of multiple uncertain spacecraft systems subject to external disturbance. Int J Robust Nonlinear Control 27(5):742–760
30. Wang PKC, Hadaegh FY, Lau K (1999) Synchronized formation rotation and attitude control of multiple free-flying spacecraft. J Guid Control Dyn 22(1):28–35
31. Wang T, Huang J (2022) Time-varying formation control with attitude synchronization of multiple rigid body systems. Int J Robust Nonlinear Control 32(1): 181–204

Chapter 7
Output Regulation

The output regulation problem aims to design a feedback control law for a plant such that the output of the plant can asymptotically track a class of reference inputs in the presence of a class of disturbances while ensuring the internal stability of the closed-loop system. Here, both the class of the reference inputs and the class of the disturbances are generated by an autonomous system called the exosystem. The output regulation problem is a classical control problem and also the special case of the cooperative output regulation problem of multi-agent systems to be studied in the subsequent chapters. In this chapter, we treat three types of the output regulation problem for linear systems. The first one studies the output regulation problem for exactly known linear systems via the so-called feedforward design method, which makes use of the solution of the regulator equations to synthesize a control law. The second one handles the structurally stable output regulation problem for linear systems with small parameter variations via the so-called p-copy internal model approach. The last one is called the robust output regulation problem which is capable of dealing with arbitrarily large parametric uncertainties for minimum phase linear systems via the so-called canonical internal model approach.

7.1 A Typical Scenario

Figure 7.1 shows a typical feedback control configuration where the given plant is subject to an external disturbance $d(t)$, and a control law making use of the measurement output $y_m(t)$ is designed so that the closed-loop system is internally stable in the sense to be described in the next section, and the performance output of the plant $y(t)$ asymptotically tracks a given reference input $r(t)$, i.e.,

$$\lim_{t \to \infty} e(t) = \lim_{t \to \infty} (y(t) - r(t)) = 0. \tag{7.1}$$

H. Cai et al., *Cooperative Control of Multi-agent Systems*, Advances in Industrial Control,
https://doi.org/10.1007/978-3-030-98377-2_7

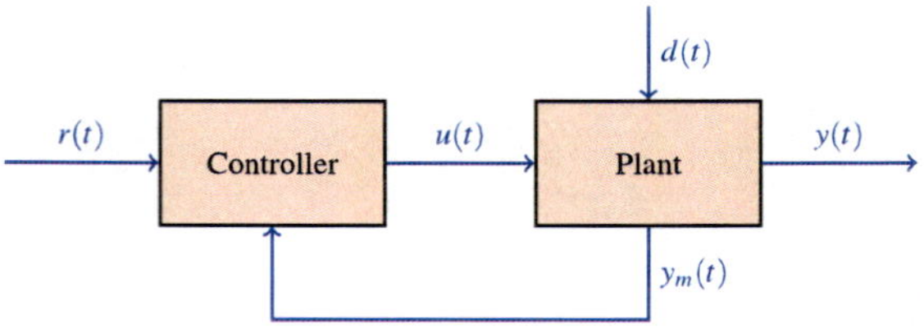

Fig. 7.1 Feedback control configuration

Thus, the problem can also be called the asymptotic tracking and disturbance rejection problem.

For a linear plant, its state equation with u, d as inputs and y, y_m as outputs is given as follows:

$$\dot{x} = Ax + Bu + E_d d \tag{7.2a}$$

$$y = Cx + Du + F_d d \tag{7.2b}$$

$$y_m = C_m x + D_m u + F_{dm} d \tag{7.2c}$$

where $x \in \mathbb{R}^n$ is the plant state, $u \in \mathbb{R}^m$ is the plant input, $y \in \mathbb{R}^p$ is the plant performance output, $y_m \in \mathbb{R}^{p_m}$ is the measurement output, $d \in \mathbb{R}^{q_1}$ is the disturbance, and $r \in \mathbb{R}^{q_2}$ is the reference input.

In practice, the reference input to be tracked and the disturbance to be rejected are not exactly known signals. For example, a disturbance in the form of a sinusoidal function can have any amplitude and initial phase, or even any frequency, and a reference input in the form of a step function can have arbitrary magnitude. For this reason, it is desired that a single controller is able to handle a class of prescribed reference inputs and/or a class of prescribed disturbances. In this chapter, we assume that the reference inputs are generated by the following linear autonomous differential equation:

$$\dot{v}_r = S_r v_r, \ \ v_r(0) = v_{r0} \tag{7.3a}$$

$$r = C_r v_r \tag{7.3b}$$

and the disturbances are generated by the following linear autonomous differential equations:

$$\dot{v}_d = S_d v_d, \ \ v_d(0) = v_{d0} \tag{7.4a}$$

$$d = C_d v_d \tag{7.4b}$$

where v_{r0} and v_{d0} are arbitrary initial states. The above autonomous systems can generate a large class of functions, e.g., a combination of step functions of arbitrary magnitudes, ramp functions of arbitrary slopes, and sinusoidal functions of arbitrary amplitudes and initial phases. Let

$$v = \begin{bmatrix} v_r \\ v_d \end{bmatrix}, \quad S = \begin{bmatrix} S_r & 0 \\ 0 & S_d \end{bmatrix}. \tag{7.5}$$

Then systems (7.3) and (7.4) can be lumped together as follows:

$$\dot{v} = Sv, \quad v(0) = \begin{bmatrix} v_{r0} \\ v_{d0} \end{bmatrix} \triangleq v_0 \tag{7.6}$$

where $v(t) \in \mathbb{R}^q$ is the exogenous signal that can represent both reference inputs and disturbances. Thus, the plant state, the measurement output, and the tracking error can be put into the following form:

$$\dot{x} = Ax + Bu + Ev, \quad x(0) = x_0 \tag{7.7a}$$

$$e = Cx + Du + Fv \tag{7.7b}$$

$$y_m = C_m x + D_m u + F_m v \tag{7.7c}$$

where $A \in \mathbb{R}^{n \times n}$, $B \in \mathbb{R}^{n \times m}$, $C \in \mathbb{R}^{p \times n}$, $D \in \mathbb{R}^{p \times m}$, $C_m \in \mathbb{R}^{p_m \times n}$, $D_m \in \mathbb{R}^{p_m \times m}$, and

$$\begin{bmatrix} E \\ F \\ F_m \end{bmatrix} = \begin{bmatrix} 0 & E_d C_d \\ -C_r & F_d C_d \\ 0 & F_{dm} C_d \end{bmatrix}$$

with $E \in \mathbb{R}^{n \times q}$, $F \in \mathbb{R}^{p \times q}$, and $F_m \in \mathbb{R}^{p_m \times q}$.

In (7.7), we have replaced the performance output $y(t)$ in (7.2) with the tracking error $e(t)$. Thus the problem of asymptotic tracking of $y(t)$ to $r(t)$ of (7.2) is converted to the problem of asymptotic regulation of $e(t)$ to the origin when $e(t)$ is viewed as the regulated output or error output of (7.7). Therefore, it suffices to study the regulation problem described by (7.7) while keeping in mind that the system (7.6), called the *exosystem* in the sequel, can generate either the reference inputs or the disturbances or the both. Thus, the problem of asymptotic tracking and disturbance rejection can be simply called as the *output regulation* problem when the reference inputs and the disturbances are generated by (7.6).

In (7.7), the plant is defined by nine finite-dimensional constant matrices. In the subsequent sections, we will consider three cases. First, all the nine matrices of (7.7) are known exactly. In this case, we say the plant (7.7) is known exactly. Second, every entry of these matrices is allowed to take any value in an open neighborhood of its nominal value. Third, every entry of these matrices is allowed to take any value in an arbitrarily large prescribed compact subset containing its nominal value. The above three cases are called output regulation problem, structurally stable output regulation problem, and robust output regulation problem, respectively. Solvability of these three problems will be dealt with by feedforward control approach, p-copy internal model approach, and canonical internal model approach, respectively.

Remark 7.1 As pointed out in Chap. 1, typically, the performance output y is a function of the measurement output y_m, and the regulated output e is a function of the performance output and the reference input. Thus, typically, they are both measurable. For linear systems, y and e being measurable means that there exist known constant matrices P_1 and P_2 such that $y = P_1 y_m$ and $e = P_2 y_m$.

7.2 Linear Output Regulation: Feedforward Design

In this section, we study the linear output regulation problem for the class of linear time-invariant systems described by (7.7) where all nine matrices in (7.7) are exactly known. For the sake of generality, we will not assume that the exogenous signal $v = \mathrm{col}(v_r, v_d)$ as indicated in (7.5). Rather, we assume it is generated by the following exosystem:

$$\dot{v} = Sv, \quad v(0) = v_0 \tag{7.8}$$

where $v(t) \in \mathbb{R}^q$, and S is a known constant matrix.

As seen in the last section, some components of the exogenous signal v, say, the reference inputs, are measurable, but some other components of the exogenous signal v, say, the external disturbances, may not be measurable. For convenience, we denote $v_u \in \mathbb{R}^{q_u}$ and $v_m \in \mathbb{R}^{q_m}$ the unmeasurable and measurable components of v, where $0 \leq q_u, q_m \leq q$ with $q_u + q_m = q$. Then, without loss of generality, we can assume v_u and v_m are generated, respectively, by the following systems:

$$\dot{v}_u = S_u v_u, \quad \dot{v}_m = S_m v_m \tag{7.9}$$

for some constant matrices S_u and S_m. (7.9) is in the form of (7.8) with $v = \mathrm{col}(v_u, v_m)$ and $S = \mathscr{D}(S_u, S_m)$. To emphasize that v may contain both measurable and unmeasurable components, we can rewrite the plant (7.7) as follows:

$$\dot{x} = Ax + Bu + E_u v_u + E_m v_m, \quad x(0) = x_0 \tag{7.10a}$$

$$e = Cx + Du + F_{eu} v_u + F_{em} v_m \tag{7.10b}$$

$$y_m = C_m x + D_m u + F_{mu} v_u + F_{mm} v_m \tag{7.10c}$$

where $E = [E_u \ E_m]$, $F = [F_{eu} \ F_{em}]$, and $F_m = [F_{mu} \ F_{mm}]$.

The class of feedback control laws in this section take the following general form:

$$u = K_z z + K_y y_m \tag{7.11a}$$

$$\dot{z} = \mathscr{G}_1 z + \mathscr{G}_2 y_m \tag{7.11b}$$

where $z \in \mathbb{R}^{n_z}$ with n_z to be specified later, $(K_z, K_y, \mathcal{G}_1, \mathcal{G}_2)$ are constant matrices of appropriate dimensions. We call (7.11) the dynamic measurement output feedback control law.

The dynamic measurement output feedback control law (7.11) contains the following four types of control laws as special cases.

1. Full Information Static State Feedback when $y_m = \text{col}(x, v)$ and $n_z = 0$:
$$u = K_1 x + K_2 v \tag{7.12}$$
where (K_1, K_2) are constant matrices of appropriate dimensions.
2. Strictly Proper Measurement Output Feedback when $K_y = 0$:
$$u = K_z z \tag{7.13a}$$
$$\dot{z} = \mathcal{G}_1 z + \mathcal{G}_2 y_m. \tag{7.13b}$$
3. Error Output Feedback when $y_m = e$ and $K_y = 0$:
$$u = K_z z \tag{7.14a}$$
$$\dot{z} = \mathcal{G}_1 z + \mathcal{G}_2 e. \tag{7.14b}$$
4. Error Output Feedback plus Feedforward when $y_m = \text{col}(e, v)$:
$$u = K_z z + K_v v \tag{7.15a}$$
$$\dot{z} = \mathcal{G}_1 z + \mathcal{G}_e e + \mathcal{G}_v v \tag{7.15b}$$
where $K_y = [0_{m\times p}, K_v]$, and $\mathcal{G}_2 = [\mathcal{G}_e, \mathcal{G}_v]$.

Needless to say that the control law (7.11) contains cases other than the control laws (7.12)–(7.15).

From the last equation of the plant (7.7) and the first equation of the control law (7.11), the control input u satisfies
$$u = K_z z + K_y(C_m x + D_m u + F_m v) \tag{7.16}$$
or
$$(I_m - K_y D_m)u = K_y C_m x + K_z z + K_y F_m v.$$

Therefore, the control law (7.11) is uniquely defined if and only if $I_m - K_y D_m$ is nonsingular. It can be easily verified that the control laws (7.12)–(7.15) all satisfy $K_y D_m = 0$. Thus, in what follows, we will assume $K_y D_m = 0$ though it suffices to assume $I_m - K_y D_m$ is nonsingular. As a result, (7.16) reduces to the following:
$$u = K_y C_m x + K_z z + K_y F_m v$$

and the control law (7.11) can be put as follows:

$$u = K_y C_m x + K_z z + K_y F_m v$$
$$\dot{z} = \mathcal{G}_2(C_m + D_m K_y C_m)x + (\mathcal{G}_1 + \mathcal{G}_2 D_m K_z)z + \mathcal{G}_2(F_m + D_m K_y F_m)v.$$

Thus, the closed-loop system composed of the plant (7.7) and the control law (7.11) can be written as follows:

$$\dot{x}_c = A_c x_c + B_c v \tag{7.17a}$$
$$e = C_c x_c + D_c v \tag{7.17b}$$

where $x_c = \text{col}(x, z)$ and

$$A_c = \begin{bmatrix} A + BK_y C_m & BK_z \\ \mathcal{G}_2(C_m + D_m K_y C_m) & \mathcal{G}_1 + \mathcal{G}_2 D_m K_z \end{bmatrix}, \quad B_c = \begin{bmatrix} E + BK_y F_m \\ \mathcal{G}_2(F_m + D_m K_y F_m) \end{bmatrix}$$
$$C_c = [C + DK_y C_m \quad DK_z], \quad D_c = F + DK_y F_m.$$

In particular, for the full information control law (7.12), we have $y_m = \text{col}(x, v)$ and $n_z = 0$. Thus, $x_c = x$, $K_y C_m = K_1$, $K_y F_m = K_2$. As a result, we have

$$A_c = A + BK_1, \quad B_c = E + BK_2$$
$$C_c = C + DK_1, \quad D_c = F + DK_2.$$

We now describe the linear output regulation problem as follows.

Problem 7.1 Given the plant (7.10) and the exosystem (7.9), find a control law of the form (7.11) such that the closed-loop system has the following properties:

- **Property 7.1** The matrix A_c is Hurwitz;
- **Property 7.2** For any $x_c(0)$ and $v(0)$, $\lim_{t\to\infty} e(t) = 0$.

At the outset, we list some standard assumptions as follows:

Assumption 7.1 S has no eigenvalues with negative real parts.

Assumption 7.2 The pair (A, B) is stabilizable.

Assumption 7.3 The pair $\left([C_m \quad F_{mu}], \begin{bmatrix} A & E_u \\ 0 & S_u \end{bmatrix}\right)$ is detectable.

Assumption 7.4 The following linear matrix equations:

$$XS = AX + BU + E \tag{7.18a}$$
$$0 = CX + DU + F \tag{7.18b}$$

admit a solution pair (X, U).

Remark 7.2 Assumption 7.2 is made so that Property 7.1, that is, the Hurwitzness of A_c, can be achieved by the state feedback control. Assumption 7.3 together with Assumption 7.2 is to render the Hurwitzness of A_c by the measurement output feedback control. Assumption 7.1 is made only for convenience and loses no generality. In fact, if Assumption 7.1 is violated, then, without loss of generality, we can assume $S = \mathscr{D}(S_1, S_2)$ where S_1 satisfies Assumption 7.1, and all the eigenvalues of S_2 have negative real parts. Thus, if a control law of the form (7.11) solves Problem 7.1 with the exosystem being given by $\dot{v}_1 = S_1 v_1$, then the same control law solves Problem 7.1 with the original exosystem $\dot{v} = Sv$. This is because Property 7.1 is guaranteed by Assumption 7.2 and/or Assumption 7.3, and, as long as the closed-loop system satisfies Property 7.1, Property 7.2 will not be affected by exogenous signals that exponentially decay to zero.

Remark 7.3 Equations (7.18) are known as the regulator equations. It will be shown in Lemma 7.2 that, under Assumption 7.1, Problem 7.1 is solvable by a control law of the form (7.11) only if the regulator equations are solvable. Moreover, if Problem 7.1 is solvable, then, necessarily, the trajectory of the closed-loop system starting from any initial condition is such that

$$\lim_{t\to\infty}(x(t) - Xv(t)) = 0 \text{ and } \lim_{t\to\infty}(u(t) - Uv(t)) = 0.$$

Therefore, Xv and Uv are the steady-state state and the steady-state input of the closed-loop system at which the tracking error $e(t)$ is identically zero. Thus, the steady state behavior of the closed-loop system is characterized by the solution of the regulator equations.

As we will see later, the design of a control law for solving the output regulation problem involves solving the regulator equations. Let us first present a solvability condition for the solvability of the regulator equations.

Theorem 7.1 *For any matrices E and F, the regulator equations* (7.18) *are solvable if and only if the following assumption holds.*

Assumption 7.5

$$\text{rank}\begin{bmatrix} A - \lambda_i(S)I_n & B \\ C & D \end{bmatrix} = n + p, \; i = 1, \ldots, q.$$

Proof The regulator equations (7.18) can be put into the following form:

$$\begin{bmatrix} I_n & 0_{n\times m} \\ 0_{p\times n} & 0_{p\times m} \end{bmatrix}\begin{bmatrix} X \\ U \end{bmatrix} S - \begin{bmatrix} A & B \\ C & D \end{bmatrix}\begin{bmatrix} X \\ U \end{bmatrix} = \begin{bmatrix} E \\ F \end{bmatrix}. \tag{7.19}$$

Using the properties of the Kronecker product as can be found in Sect. A.1, we can transform (7.19) into a standard linear algebraic equation of the following form:

$$Qx = b$$

where

$$Q = S^{\mathsf{T}} \otimes \begin{bmatrix} I_n & 0_{n\times m} \\ 0_{p\times n} & 0_{p\times m} \end{bmatrix} - I_q \otimes \begin{bmatrix} A & B \\ C & D \end{bmatrix}$$

$$x = \text{vec}\left(\begin{bmatrix} X \\ U \end{bmatrix}\right), \quad b = \text{vec}\left(\begin{bmatrix} E \\ F \end{bmatrix}\right).$$

Here the notation vec$(\cdot)$ denotes a vector-valued function of a matrix such that, for any $X \in \mathbb{R}^{n\times m}$,

$$\text{vec}\,(X) = \begin{bmatrix} X_1 \\ \vdots \\ X_m \end{bmatrix}$$

where, for $i = 1, \ldots, m$, X_i is the ith column of X. Thus, Eq. (7.19) is solvable for any matrices E and F if and only if Q has full row rank. To obtain the condition under which Q has full row rank, we assume, without loss of generality, that S is in the following Jordan form:

$$S = \begin{bmatrix} J_1 & 0 & 0 & \cdots & 0 \\ 0 & J_2 & 0 & \cdots & 0 \\ \vdots & \vdots & \vdots & \ddots & \vdots \\ 0 & 0 & 0 & \cdots & J_k \end{bmatrix}$$

where J_i has dimension n_i such that $n_1 + n_2 + \cdots + n_k = q$ and is given by

$$J_i = \begin{bmatrix} \lambda_i(S) & 1 & 0 \cdots & 0 & 0 \\ 0 & \lambda_i(S) & 1 \cdots & 0 & 0 \\ \vdots & \vdots & \vdots \ddots & \vdots & \vdots \\ 0 & 0 & 0 \cdots & \lambda_i(S) & 1 \\ 0 & 0 & 0 \cdots & 0 & \lambda_i(S) \end{bmatrix}.$$

A simple calculation shows that Q is a block lower triangular matrix of k blocks with its ith, $1 \le i \le k$, diagonal block having the following form:

$$\begin{bmatrix} \lambda_i(S)\mathscr{E} - \mathscr{A} & 0 & 0 \cdots & 0 & 0 \\ \mathscr{E} & \lambda_i(S)\mathscr{E} - \mathscr{A} & 0 \cdots & 0 & 0 \\ \vdots & \vdots & \vdots \ddots & \vdots & \vdots \\ 0 & 0 & 0 \cdots & \lambda_i(S)\mathscr{E} - \mathscr{A} & 0 \\ 0 & 0 & 0 \cdots & \mathscr{E} & \lambda_i(S)\mathscr{E} - \mathscr{A} \end{bmatrix}$$

where

$$\mathscr{E} = \begin{bmatrix} I_n & 0_{n\times m} \\ 0_{p\times n} & 0_{p\times m} \end{bmatrix}, \quad \mathscr{A} = \begin{bmatrix} A & B \\ C & D \end{bmatrix}.$$

Clearly, Q has full row rank if and only if Assumption 7.5 holds. □

Remark 7.4 For a particular pair of (E, F), the regulator equations may still have solutions even if Assumption 7.5 fails. This happens when

$$\text{vec}\left(\begin{bmatrix} E \\ F \end{bmatrix}\right) \in \text{Im}(Q)$$

where $\text{Im}(Q)$ denotes the range space of the matrix Q.

To study the solvability of Problem 7.1, let us define the following matrix equations associated with the closed-loop system (7.17):

$$X_c S = A_c X_c + B_c \tag{7.20a}$$

$$0 = C_c X_c + D_c. \tag{7.20b}$$

We call (7.20) the closed-loop regulator equations.

We now present the following lemma on the closed-loop system.

Lemma 7.1 *Suppose, under the control law (7.11), the closed-loop system (7.17) satisfies Property 7.1, i.e., A_c is Hurwitz. Then, the closed-loop system (7.17) also satisfies Property 7.2, that is, $\lim_{t\to\infty} e(t) = 0$, if there exists a matrix X_c that satisfies the closed-loop regulator equations (7.20). Moreover, under the additional Assumption 7.1, the closed-loop system (7.17) also satisfies Property 7.2 only if the closed-loop regulator equations (7.20) admit a unique solution X_c.*

Proof Suppose there exists a matrix X_c that satisfies (7.20). Then the variable $\bar{x}_c = x_c - X_c v$ satisfies

$$\dot{\bar{x}}_c = A_c \bar{x}_c \tag{7.21a}$$

$$e = C_c \bar{x}_c. \tag{7.21b}$$

Since A_c is Hurwitz, $\lim_{t\to\infty} \bar{x}_c(t) = 0$, and hence $\lim_{t\to\infty} e(t) = 0$.

On the other hand, since the closed-loop system satisfies Property 7.1, A_c is Hurwitz. Thus, by Proposition A.3 in Sect. A.1, Assumption 7.1 and the Hurwitzness of A_c guarantee the existence of X_c satisfying the Sylvester equation (7.20a). Let $\bar{x}_c = x_c - X_c v$. Then,

$$\begin{aligned}\dot{\bar{x}}_c &= A_c \bar{x}_c \\ e &= C_c \bar{x}_c + (C_c X_c + D_c) v.\end{aligned}$$

Since A_c is Hurwitz, $\lim_{t\to\infty} \bar{x}_c(t) = 0$. Since the closed-loop system also satisfies Property 7.2, we have

$$\lim_{t\to\infty} e(t) = 0.$$

Thus,

$$\lim_{t\to\infty} (C_c X_c + D_c) v(t) = 0$$

for all $v(t) = e^{St} v(0)$ with any $v(0) \in \mathbb{R}^q$. Due to Assumption 7.1, $v(t)$ does not decay to zero for $v(0) \neq 0$. Therefore, necessarily, $C_c X_c + D_c = 0$. □

Remark 7.5 By Proposition A.3 in Sect. A.1, the solvability of the Sylvester equation (7.20a) is guaranteed as long as the eigenvalues of A_c do not coincide with those of S. Thus, Assumption 7.1 is not necessary for the sufficient part of Lemma 7.1. It suffices to require that the eigenvalues of A_c do not coincide with those of S.

Lemma 7.2 *Under Assumption 7.1, suppose there exists a control law of the form (7.11) such that the closed-loop system satisfies Properties 7.1 and 7.2. Then there exist matrices X and U that satisfy the following regulator equations:*

$$XS = AX + BU + E \tag{7.22a}$$

$$0 = CX + DU + F. \tag{7.22b}$$

Proof Suppose there exists a control law of the form (7.11) such that the closed-loop system satisfies Properties 7.1 and 7.2. Then, by Lemma 7.1, there exists a matrix X_c that satisfies the matrix equations (7.20), i.e., the following equations:

$$\begin{aligned} X_c S &= \begin{bmatrix} A + BK_y C_m & BK_z \\ \mathcal{G}_2(C_m + D_m K_y C_m) & \mathcal{G}_1 + \mathcal{G}_2 D_m K_z \end{bmatrix} X_c + \begin{bmatrix} E + BK_y F_m \\ \mathcal{G}_2(F_m + D_m K_y F_m) \end{bmatrix} \\ 0 &= [C + DK_y C_m \quad DK_z] X_c + F + DK_y F_m. \end{aligned} \tag{7.23}$$

Partition X_c as

$$X_c = \begin{bmatrix} X \\ Z \end{bmatrix}$$

where $X \in \mathbb{R}^{n\times q}$ and $Z \in \mathbb{R}^{n_z \times q}$. Then we can expand (7.23) as follows:

$$
\begin{aligned}
XS &= (A + BK_yC_m)X + BK_zZ + (E + BK_yF_m) \\
ZS &= \mathcal{G}_2(C_m + D_mK_yC_m)X + (\mathcal{G}_1 + \mathcal{G}_2D_mK_z)Z + \mathcal{G}_2(F_m + D_mK_yF_m) \\
0 &= (C + DK_yC_m)X + DK_zZ + (F + DK_yF_m)
\end{aligned}
$$

which is the same as

$$XS = AX + B(K_yC_mX + K_zZ + K_yF_m) + E \tag{7.24a}$$

$$ZS = \mathcal{G}_2(C_m + D_mK_yC_m)X + (\mathcal{G}_1 + \mathcal{G}_2D_mK_z)Z + \mathcal{G}_2(F_m + D_mK_yF_m) \tag{7.24b}$$

$$0 = CX + D(K_yC_mX + K_zZ + K_yF_m) + F. \tag{7.24c}$$

Letting $U = K_yC_mX + K_zZ + K_yF_m$ in (7.24a) and (7.24c) shows that X and U satisfy the regulator equations (7.22). □

Now let us first consider the full information case where the control law (7.12) is defined by two constant matrices K_1 and K_2. The two matrices K_1 and K_2 will be called the feedback gain and the feedforward gain, respectively.

Theorem 7.2 *Under Assumption 7.2, let the feedback gain K_1 be such that $(A + BK_1)$ is Hurwitz. Then, Problem 7.1 is solvable by the full information control law (7.12) if Assumption 7.4 holds and the feedforward gain K_2 is given by*

$$K_2 = U - K_1X. \tag{7.25}$$

Proof Under Assumption 7.2, there exists K_1 such that $A_c = A + BK_1$ is Hurwitz. Thus, under the control law (7.12), Property 7.1 is satisfied. Under Assumption 7.4, let $\bar{x} = x - Xv$ and K_2 be given by (7.25). Then we have

$$
\begin{aligned}
\dot{\bar{x}} &= (A + BK_1)\bar{x} \\
e &= (C + DK_1)\bar{x}.
\end{aligned}
$$

Since $(A + BK_1)$ is Hurwitz, $\bar{x}(t)$ and hence $e(t)$ will approach zero as t tends to infinity. Thus, Property 7.2 is also satisfied. □

Remark 7.6 By Lemma 7.2, Assumption 7.4 is also necessary for the solvability of Problem 7.1 by the full information control law (7.12) if Assumption 7.1 also holds.

We now turn to the construction of the measurement output feedback control law (7.11). Since we have already known how to synthesize the full information control law, which depends on the plant state x and the exosystem state v, naturally, we seek to synthesize a measurement output feedback control law by estimating the state x

and the unmeasurable exogenous signal v_u. To this end, lump the state x and the unmeasured exogenous signals v_u together to obtain the following system:

$$\begin{bmatrix} \dot{x} \\ \dot{v}_u \end{bmatrix} = \begin{bmatrix} A & E_u \\ 0 & S_u \end{bmatrix} \begin{bmatrix} x \\ v_u \end{bmatrix} + \begin{bmatrix} B \\ 0 \end{bmatrix} u + \begin{bmatrix} E_m \\ 0 \end{bmatrix} v_m$$
$$y_m = [C_m \ F_{mu}] \begin{bmatrix} x \\ v_u \end{bmatrix} + D_m u + F_{mm} v_m.$$

Employing the well-known Luenberger observer theory suggests the following observer-based control law:

$$u = [K_1 \quad K_{2u}]z + K_{2m} v_m \tag{7.26a}$$
$$\dot{z} = \begin{bmatrix} A & E_u \\ 0 & S_u \end{bmatrix} z + \begin{bmatrix} B \\ 0 \end{bmatrix} u + \begin{bmatrix} E_m \\ 0 \end{bmatrix} v_m + L(y_m - [C_m \quad F_{mu}]z - D_m u - F_{mm} v_m) \tag{7.26b}$$

where $K_{2u} \in \mathbb{R}^{m \times q_u}$, $K_{2m} \in \mathbb{R}^{m \times q_m}$, and $L \in \mathbb{R}^{(n+q_u) \times p_m}$ is the so-called observer gain matrix. The control law (7.26) can be put in the following form:

$$u = K_z z + K_{2m} v_m \tag{7.27a}$$
$$\dot{z} = \mathscr{G}_1 z + \mathscr{G}_{21} y_m + \mathscr{G}_{22} v_m \tag{7.27b}$$

where

$$K_z = [K_1 \quad K_{2u}]$$
$$\mathscr{G}_1 = \begin{bmatrix} A & E_u \\ 0 & S_u \end{bmatrix} + \begin{bmatrix} B \\ 0 \end{bmatrix} K_z - L(\begin{bmatrix} C_m & F_{mu} \end{bmatrix} + D_m K_z)$$
$$\mathscr{G}_{21} = L, \quad \mathscr{G}_{22} = \begin{bmatrix} B \\ 0 \end{bmatrix} K_{2m} + \begin{bmatrix} E_m \\ 0 \end{bmatrix} - L(F_{mm} + D_m K_{2m}).$$

Since v_m is measurable, there exists a matrix C_v such that $v_m = C_v y_m$. Thus the control law (7.27) can be further put into the standard form (7.11) with $K_y = K_{2m} C_v$ and $\mathscr{G}_2 = \mathscr{G}_{21} + \mathscr{G}_{22} C_v$.

Theorem 7.3 *Under Assumptions 7.2–7.4, Problem 7.1 is solvable by the measurement output feedback control law (7.26) where K_1 and $K_2 = [K_{2u}, K_{2m}]$ are the same as those in Theorem 7.2, and the observer gain $L = col(L_1, L_2)$ is such that*

$$A_L \triangleq \begin{bmatrix} A & E_u \\ 0 & S_u \end{bmatrix} - \begin{bmatrix} L_1 \\ L_2 \end{bmatrix} \begin{bmatrix} C_m & F_{mu} \end{bmatrix} = \begin{bmatrix} A - L_1 C_m & E_u - L_1 F_{mu} \\ -L_2 C_m & S_u - L_2 F_{mu} \end{bmatrix}$$

is Hurwitz.

Proof First note that, by Assumption 7.2, there exists a state feedback gain K_1 such that $(A + BK_1)$ is Hurwitz, and, by Assumption 7.3, there exist matrices L_1 and L_2 such that A_L is Hurwitz. With the coordinate transformations $\bar{x} = x - Xv$, $\bar{u} = u - Uv$, and

$$\bar{z} = \begin{bmatrix} x \\ v_u \end{bmatrix} - z,$$

it can be verified that

$$\begin{aligned}
\bar{u} &= [K_1, K_{2u}]\begin{bmatrix} x \\ v_u \end{bmatrix} - K_z\bar{z} + K_{2m}v_m - Uv \\
&= -K_z\bar{z} + K_1x + K_2v - (K_2 + K_1X)v \\
&= -K_z\bar{z} + K_1\bar{x} \\
\dot{\bar{x}} &= Ax + Bu + Ev - XSv \\
&= A\bar{x} + B\bar{u} \\
&= (A + BK_1)\bar{x} - BK_z\bar{z}
\end{aligned}$$

and

$$\begin{aligned}
\dot{\bar{z}} &= \begin{bmatrix} \dot{x} \\ \dot{v}_u \end{bmatrix} - \dot{z} \\
&= \begin{bmatrix} A & E_u \\ 0 & S_u \end{bmatrix}\begin{bmatrix} x \\ v_u \end{bmatrix} + \begin{bmatrix} B \\ 0 \end{bmatrix}u + \begin{bmatrix} E_m \\ 0 \end{bmatrix}v_m - \begin{bmatrix} A & E_u \\ 0 & S_u \end{bmatrix}z - \begin{bmatrix} B \\ 0 \end{bmatrix}u - \begin{bmatrix} E_m \\ 0 \end{bmatrix}v_m \\
&\quad -L(y_m - [C_m \;\; F_{mu}]z - D_mu - F_{mm}v_m) \\
&= \begin{bmatrix} A & E_u \\ 0 & S_u \end{bmatrix}\bar{z} - L(C_mx + F_{mu}v_u - [C_m \;\; F_{mu}]z) \\
&= \begin{bmatrix} A & E_u \\ 0 & S_u \end{bmatrix}\bar{z} - L[C_m \;\; F_{mu}]\bar{z} \\
&= A_L\bar{z}.
\end{aligned}$$

Then, in terms of $\bar{x}$ and $\bar{z}$, the closed-loop system is governed by

$$\dot{\bar{x}} = (A + BK_1)\bar{x} - BK_z\bar{z} \tag{7.28a}$$

$$\dot{\bar{z}} = A_L\bar{z}. \tag{7.28b}$$

Let A_c be the closed-loop system matrix. Since $\lambda(A_c) = \lambda(A + BK_1) \cup \lambda(A_L)$, Property 7.1 is satisfied.

To show $\lim_{t\to\infty} e(t) = 0$, first note that (7.28) implies that $\lim_{t\to\infty} \bar{x}(t) = 0$ and $\lim_{t\to\infty} \bar{z}(t) = 0$. Then the proof is complete by noting that

$$
\begin{aligned}
e &= Cx + Du + Fv \\
&= C(x - Xv) + D(u - Uv) + (CX + DU + F)v \\
&= C(x - Xv) + D(u - Uv) \\
&= (C + DK_1)\bar{x} - DK_z\bar{z}.
\end{aligned}
$$

□

Remark 7.7 From the proof of Theorem 7.3, Assumption 7.1 is not needed in proving the solvability of Problem 7.1 by a dynamic measurement output feedback control law of the form (7.11).

Specializing (7.27) to the two special cases with $v = v_u$ and $v = v_m$, respectively, gives the following two corollaries of Theorem 7.3.

Corollary 7.1 *Under Assumptions 7.2–7.4 with $v = v_u$, Problem 7.1 is solvable by the following observer-based feedback control law:*

$$u = [K_1 \;\; K_2]z \tag{7.29a}$$

$$\dot{z} = \begin{bmatrix} A & E \\ 0 & S \end{bmatrix} z + \begin{bmatrix} B \\ 0 \end{bmatrix} u + L(y_m - [C_m \;\; F_m]z - D_m u) \tag{7.29b}$$

where $L \in \mathbb{R}^{(n+q)\times p_m}$ is such that $\begin{bmatrix} A & E \\ 0 & S \end{bmatrix} - L[C_m \;\; F_m]$ is Hurwitz.

Corollary 7.2 *Under Assumptions 7.2–7.4 with $v = v_m$, Problem 7.1 is solvable by the following measurement output feedback plus feedforward control law:*

$$u = K_1 z + K_2 v \tag{7.30a}$$

$$\dot{z} = Az + Bu + Ev + L(y_m - C_m z - D_m u - F_m v) \tag{7.30b}$$

where $L \in \mathbb{R}^{n\times p_m}$ is such that $A - LC$ is Hurwitz.

Example 7.1 Consider the disturbance rejection problem for the rotational/translational actuator (RTAC) system[1] whose dynamics are given as follows:

$$\dot{x} = f(x) + g_1(x)u + g_2(x)v$$

with $x = \text{col}(x_1, x_2, x_3, x_4)$,

$$f(x) = \begin{bmatrix} x_2 \\ \frac{-x_1+\varepsilon x_4^2 \sin x_3}{1-\varepsilon^2\cos^2 x_3} \\ x_4 \\ \frac{\varepsilon \cos x_3 (x_1 - \varepsilon x_4^2 \sin x_3)}{1-\varepsilon^2\cos^2 x_3} \end{bmatrix}, \quad g_1(x) = \begin{bmatrix} 0 \\ \frac{-\varepsilon \cos x_3}{1-\varepsilon^2\cos^2 x_3} \\ 0 \\ \frac{1}{1-\varepsilon^2\cos^2 x_3} \end{bmatrix}, \quad g_2(x) = \begin{bmatrix} 0 & 0 \\ \frac{1}{1-\varepsilon^2\cos^2 x_3} & 0 \\ 0 & 0 \\ \frac{-\varepsilon \cos x_3}{1-\varepsilon^2\cos^2 x_3} & 0 \end{bmatrix}$$

[1] The details of RTAC system can be found in, say, [18].

where ε is a positive parameter.

The exogenous signal v is generated by the following exosystem:

$$\dot{v} = \begin{bmatrix} 0 & \omega \\ -\omega & 0 \end{bmatrix} v \triangleq Sv$$

with $v = \mathrm{col}(v_1, v_2)$.

Simple calculations give

$$\frac{\partial f}{\partial x}(0) = \begin{bmatrix} 0 & 1 & 0 & 0 \\ \frac{-1}{1-\varepsilon^2} & 0 & 0 & 0 \\ 0 & 0 & 0 & 1 \\ \frac{\varepsilon}{1-\varepsilon^2} & 0 & 0 & 0 \end{bmatrix}, \quad g_1(0) = \begin{bmatrix} 0 \\ \frac{-\varepsilon}{1-\varepsilon^2} \\ 0 \\ \frac{1}{1-\varepsilon^2} \end{bmatrix}, \quad g_2(0) = \begin{bmatrix} 0 & 0 \\ \frac{1}{1-\varepsilon^2} & 0 \\ 0 & 0 \\ \frac{-\varepsilon}{1-\varepsilon^2} & 0 \end{bmatrix}$$

and thus the linearized model of the RTAC system is given by

$$\begin{aligned} \dot{x} &= \begin{bmatrix} 0 & 1 & 0 & 0 \\ \frac{-1}{1-\varepsilon^2} & 0 & 0 & 0 \\ 0 & 0 & 0 & 1 \\ \frac{\varepsilon}{1-\varepsilon^2} & 0 & 0 & 0 \end{bmatrix} x + \begin{bmatrix} 0 \\ \frac{-\varepsilon}{1-\varepsilon^2} \\ 0 \\ \frac{1}{1-\varepsilon^2} \end{bmatrix} u + \begin{bmatrix} 0 & 0 \\ \frac{1}{1-\varepsilon^2} & 0 \\ 0 & 0 \\ \frac{-\varepsilon}{1-\varepsilon^2} & 0 \end{bmatrix} v \\ &\triangleq Ax + Bu + Ev \\ e &= \begin{bmatrix} 1 & 0 & 0 & 0 \end{bmatrix} x \triangleq Cx \\ y_m &= \begin{bmatrix} 1 & 0 & 0 & 0 \\ 0 & 0 & 1 & 0 \end{bmatrix} x \triangleq C_m x. \end{aligned}$$

It can be verified that

$$\left(\begin{bmatrix} C & 0_{1\times 2} \end{bmatrix}, \begin{bmatrix} A & E \\ 0 & S \end{bmatrix} \right)$$

is not detectable. Thus, the problem cannot be solved by an error output feedback controller. Nevertheless, it can be verified that

$$\left(\begin{bmatrix} C_m & 0_{2\times 2} \end{bmatrix}, \begin{bmatrix} A & E \\ 0 & S \end{bmatrix} \right)$$

is detectable. Thus, the problem can be solved by a dynamic measurement output feedback controller of the form (7.29). To synthesize such a controller, simple calculation gives the solution of the regulator equations as follows:

$$X = \begin{bmatrix} 0 & 0 \\ 0 & 0 \\ \frac{-1}{\varepsilon\omega^2} & 0 \\ 0 & \frac{-1}{\varepsilon\omega} \end{bmatrix}, \quad U = \begin{bmatrix} \frac{1}{\varepsilon} & 0 \end{bmatrix}.$$

To evaluate the performance of the controller by simulation, let $\varepsilon = 0.2$ and $\omega = 3$. Letting the eigenvalues of the matrices $A + BK_1$ and

$$\begin{bmatrix} A & E \\ 0 & S \end{bmatrix} - L \begin{bmatrix} C_m & 0_{2\times 2} \end{bmatrix}$$

be given by

$$\{-1, -2, -3, -4\}$$

and

$$\{-1, -1.2, -1.4, -1.6, -1.8, -2\}$$

respectively, yields the control gain

$$K_1 = [47.8 \; -192 \; -23.04 \; -48]$$

and the observer gain

$$L = \begin{bmatrix} 9.2060 & 16.0713 \\ 18.0013 & 74.1847 \\ -1.2375 & -0.2060 \\ -3.1566 & -12.6343 \\ -46.0868 & -32.4572 \\ -49.9909 & -196.5398 \end{bmatrix}.$$

Finally, let $K_2 = U - K_1 X = [-7.8, -80]$.

With the initial values $v(0) = \text{col}(0, 1), x(0) = \text{col}(0.1, 0, 0, 0), z(0) = 0$, Fig. 7.2 shows the time response of the four state variables of the plant under the dynamic measurement output feedback control law (7.29). As expected, the tracking error $e(t) = x_1(t)$ tends to the origin asymptotically.

In practice, the parameter ε may not be known precisely. Suppose the actual value of ε is 0.21. Then, with the same initial values, the tracking performance of the same control law (7.29) is shown in Fig. 7.3. It can be seen that the tracking error e will not converge to the origin. This is because the feedforward gain K_2 relies on the value of ε. When $\varepsilon = 0.21$, the corresponding K_2 should be equal to $[-7.3765, -75.8651]$. Thus, we can see that, if the plant parameters are not known exactly, the feedforward control approach cannot exactly achieve asymptotic tracking and disturbance rejection.

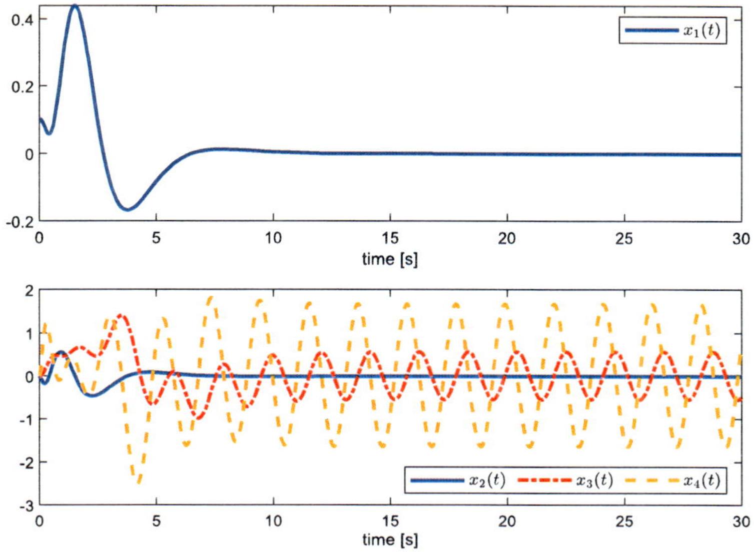

Fig. 7.2 The response of $x_i(t)$ under the control law (7.29) when $\varepsilon = 0.2$

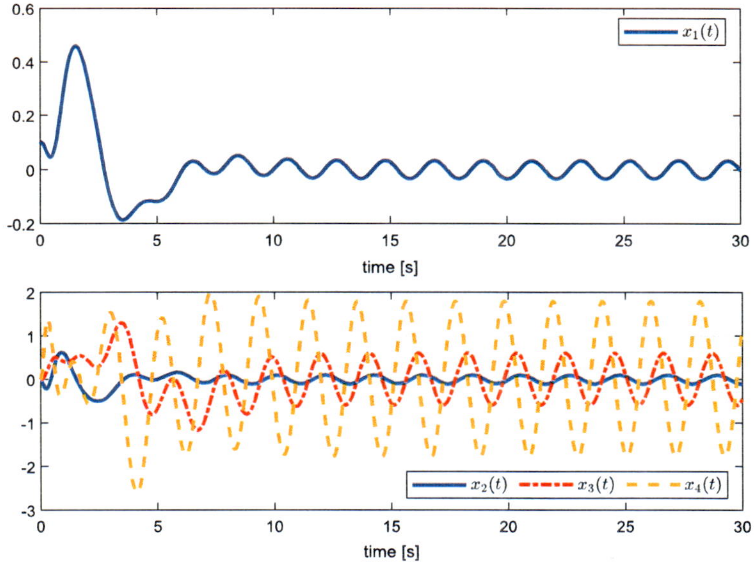

Fig. 7.3 The response of $x_i(t)$ under the control law (7.29) when $\varepsilon = 0.21$

7.3 Linear Structurally Stable Output Regulation: *p*-copy Internal Model

Since the feedforward control approach cannot exactly achieve asymptotic tracking and disturbance rejection for a plant with uncertain parameters, in this section, we will further consider the output regulation problem for uncertain linear systems of the following form:

$$\dot{x} = A(w)x + B(w)u + E(w)v, \ x(0) = x_0 \tag{7.31a}$$

$$e = C(w)x + D(w)u + F(w)v \tag{7.31b}$$

$$y_m = C_m(w)x + D_m(w)u + F_m(w)v \tag{7.31c}$$

where $x \in \mathbb{R}^n$, $u \in \mathbb{R}^m$, $e \in \mathbb{R}^p$, $y_m \in \mathbb{R}^{p_m}$ are the state, the control input, the regulated output, and the measurement output, respectively, $w \in \mathbb{R}^{n_w}$ denotes the uncertain parameter vector, $v \in \mathbb{R}^q$ is also assumed to be generated by the same exosystem (7.8), and all the matrix functions $A(w)$, $B(w)$, $C(w)$, $D(w)$, $E(w)$, and $F(w)$ are continuous in w. Throughout this section, we suppose $w = 0$ is the nominal value of the uncertain parameter w, and we use $\mathbb{W}$ to denote a generic open neighborhood of the origin of $\mathbb{R}^{n_w}$.

For convenience, let

$$A = A(0), \quad B = B(0), \quad E = E(0), \quad C = C(0), \quad D = D(0)$$
$$F = F(0), \quad C_m = C_m(0), \quad D_m = D_m(0), \quad F_m = F_m(0).$$

That is, the matrices A, B, E, C, D, F, C_m, D_m, F_m represent the nominal parts of the plant.

We consider two classes of feedback control laws in this section, namely, the dynamic state feedback control and the dynamic measurement output feedback control, which are also the special cases of the general form (7.11).

1. Dynamic State Feedback:

$$u = K_1 x + K_2 z \tag{7.32a}$$

$$\dot{z} = \mathcal{G}_1 z + \mathcal{G}_2 e \tag{7.32b}$$

where $z \in \mathbb{R}^{n_z}$ with n_z to be specified later, and $(K_1, K_2, \mathcal{G}_1, \mathcal{G}_2)$ are constant matrices of appropriate dimensions.

2. Dynamic Measurement Output Feedback:

$$u = Kz \tag{7.33a}$$

$$\dot{z} = \mathcal{G}_1 z + \mathcal{G}_2 y_m \tag{7.33b}$$

where $z \in \mathbb{R}^{n_z}$ with n_z to be specified later, and $(K, \mathcal{G}_1, \mathcal{G}_2)$ are constant matrices of appropriate dimensions.

If $y_m = e$, then (7.33) reduces to the following dynamic error output feedback control law:

$$u = Kz \tag{7.34a}$$

$$\dot{z} = \mathcal{G}_1 z + \mathcal{G}_2 e. \tag{7.34b}$$

Denote the closed-loop system consisting of the plant (7.31) and the control law (7.32) or (7.33) as follows:

$$\dot{x}_c = A_c(w)x_c + B_c(w)v \tag{7.35a}$$

$$e = C_c(w)x_c + D_c(w)v \tag{7.35b}$$

where, under the dynamic state feedback control law (7.32), $x_c = \text{col}(x, z)$, and

$$A_c(w) = \begin{bmatrix} A(w) + B(w)K_1 & B(w)K_2 \\ \mathcal{G}_2(C(w) + D(w)K_1) & \mathcal{G}_1 + \mathcal{G}_2 D(w)K_2 \end{bmatrix} \tag{7.36a}$$

$$B_c(w) = \begin{bmatrix} E(w) \\ \mathcal{G}_2 F(w) \end{bmatrix} \tag{7.36b}$$

$$C_c(w) = [C(w) + D(w)K_1 \quad D(w)K_2] \tag{7.36c}$$

$$D_c(w) = F(w) \tag{7.36d}$$

and, under the dynamic output feedback control law (7.33), $x_c = \text{col}(x, z)$, and

$$A_c(w) = \begin{bmatrix} A(w) & B(w)K \\ \mathcal{G}_2 C_m(w) & \mathcal{G}_1 + \mathcal{G}_2 D_m(w)K \end{bmatrix} \tag{7.37a}$$

$$B_c(w) = \begin{bmatrix} E(w) \\ \mathcal{G}_2 F_m(w) \end{bmatrix} \tag{7.37b}$$

$$C_c(w) = [C(w) \quad D(w)K] \tag{7.37c}$$

$$D_c(w) = F(w). \tag{7.37d}$$

For convenience, let $A_c = A_c(0)$, $B_c = B_c(0)$, $C_c = C_c(0)$, $D_c = D_c(0)$. We will simply use (A_c, B_c, C_c, D_c) to denote the closed-loop system composed of the nominal plant and the control laws. Then the output regulation problem for the uncertain system (7.31) and the exosystem (7.8) is described as follows.

Problem 7.2 Given the plant (7.31) and the exosystem (7.8), find a control law of the form (7.32) or (7.33) such that the closed-loop system (7.35) has the following properties:

- **Property 7.3** The matrix A_c is Hurwitz;

- **Property 7.4** For any $x_c(0)$ and $v(0)$, there exists an open neighborhood $\mathbb{W}$ of $w = 0$ such that, for all $w \in \mathbb{W}$, the trajectories of (7.35) satisfy $\lim_{t\to\infty} e(t) = 0$.

Remark 7.8 If Property 7.3 is satisfied, then there exists an open neighborhood $\mathbb{W}$ of $w = 0$ such that, for each $w \in \mathbb{W}$, $A_c(w)$ is Hurwitz. Nevertheless, Property 7.4 is not implied by Property 7.2 and it cannot be achieved by the approach in Sect. 7.2 as will be made more clear in Remark 7.11. Since Property 7.2 is known as output regulation property, we call Property 7.4 structurally stable output regulation property, meaning that output regulation property is maintained in spite of small variation of the unknown parameter vector w. As a result, Problem 7.2 is often called robust output regulation problem or, more precisely, structurally stable output regulation problem.

In this section, in addition to Assumptions 7.1, 7.2, and 7.5, we need two more assumptions as follows.

Assumption 7.6 The pair (C_m, A) is detectable.

Assumption 7.7 There exists a matrix $Q \in \mathbb{R}^{p\times p_m}$ such that $e = Qy_m$.

Remark 7.9 Again, as pointed out in Remark 7.2, Assumptions 7.1 is made for convenience and is not necessary. Assumption 7.6 can be viewed as a special case of Assumption 7.3 when $v = v_m$, and it together with Assumption 7.2 is such that the matrix A_c can be made Hurwitz by a dynamic measurement output feedback control. Assumption 7.7 guarantees that

$$\lim_{t\to\infty} e(t) = 0 \text{ if } \lim_{t\to\infty} y_m(t) = 0.$$

Such a condition is known as the readability condition. As pointed out in Remark 7.1, typically, the error output e is measurable. Thus, the readability condition imposes no restriction.

A straightforward extension of Lemma 7.1 is given as follows.

Lemma 7.3 *Suppose, under the control law (7.32) or (7.33), the closed-loop system (7.35) satisfies Property 7.3, i.e., A_c is Hurwitz. Then, the closed-loop system (7.35) also satisfies Property 7.4 if, for each $w \in \mathbb{W}$, there exists a matrix $X_c(w)$ that satisfies the following closed-loop regulator equations:*

$$X_c(w)S = A_c(w)X_c(w) + B_c(w) \tag{7.38a}$$
$$0 = C_c(w)X_c(w) + D_c(w). \tag{7.38b}$$

Moreover, under the additional Assumption 7.1, the closed-loop system (7.35) also satisfies Property 7.4 only if, for each $w \in \mathbb{W}$, there exists a unique matrix $X_c(w)$ that satisfies the closed-loop regulator equations (7.38).

Remark 7.10 As in Lemma 7.1, the role of the solution $X_c(w)$ of (7.38) is such that the variable $\bar{x}_c = x_c - X_c(w)v$ satisfies

$$\dot{\bar{x}}_c = A_c(w)\bar{x}_c \tag{7.39a}$$

$$e = C_c(w)\bar{x}_c. \tag{7.39b}$$

Thus, for each $w \in \mathbb{W}$, the stability of the system (7.39a) implies, for each $w \in \mathbb{W}$, $\lim_{t\to\infty} e(t) = 0$.

Remark 7.11 Assumption 7.5 implies the existence of an open neighborhood $\mathbb{W}$ of $w = 0$ such that, for all $w \in \mathbb{W}$,

$$\text{rank}\begin{bmatrix} A(w) - \lambda_i(S)I_n & B(w) \\ C(w) & D(w) \end{bmatrix} = n + p, \; i = 1, \ldots, q.$$

Thus, by Theorem 7.1, for each $w \in \mathbb{W}$, the following regulator equations:

$$X(w)S = A(w)X(w) + B(w)U(w) + E(w) \tag{7.40a}$$

$$0 = C(w)X(w) + D(w)U(w) + F(w) \tag{7.40b}$$

admit a solution pair $(X(w), U(w))$. Recall that the feedforward approach in Sect. 7.2 relies on the availability of the solution of the regulator equations (7.18), which, in the current case, relies on the unknown parameter w. Thus, the approach in Sect. 7.2 is not feasible for Problem 7.2. In what follows, we will introduce the so-called internal model approach to handle the output regulation problem for uncertain plants.

Definition 7.1 Consider the exosystem (7.8). A pair of matrices (G_1, G_2) is said to be a p-copy internal model of S if the pair takes the following form:

$$G_1 = \mathscr{D}\underbrace{(\beta_1, \ldots, \beta_p)}_{p-tuple}, \quad G_2 = \mathscr{D}\underbrace{(\sigma_1, \ldots, \sigma_p)}_{p-tuple}$$

where, for $i = 1, \ldots, p$, β_i is any constant square matrix such that the minimal polynomial of S divides the characteristic polynomial of β_i, and σ_i is any constant column vector such that (β_i, σ_i) is controllable. The dynamic compensator

$$\dot{\xi} = G_1\xi + G_2 e \tag{7.41}$$

is called a p-copy internal model of the exosystem (7.8).

Remark 7.12 Given any matrix S and any integer $p > 0$, let

$$\alpha_m(\lambda) = \lambda^l + \alpha_l\lambda^{l-1} + \cdots + \alpha_2\lambda + \alpha_1$$

be the minimal polynomial of S,

$$G_1 = I_p \otimes \beta, \quad G_2 = I_p \otimes \sigma \tag{7.42}$$

where $\beta \in \mathbb{R}^{l \times l}$ is any matrix whose characteristic polynomial is $\alpha_m(\lambda)$, and $\sigma \in \mathbb{R}^l$ is any column vector such that the pair (β, σ) is controllable. Then, the pair (G_1, G_2) is a p-copy internal model of S. We call (7.42) the minimal p-copy internal model of S.

Lemma 7.4 *Suppose the matrix pair (G_1, G_2) is a p-copy internal model of the matrix S and $Y \in \mathbb{R}^{p \times q}$ is any constant matrix. If the following matrix equation:*

$$XS - G_1 X = G_2 Y \tag{7.43}$$

has a solution, then $Y = 0$.

Proof Since G_1 and G_2 are block diagonal matrices, we can assume $p = 1$ without loss of generality. That is, $G_1 = \beta_1$ and $G_2 = \sigma_1$. Let the characteristic polynomial of G_1 be

$$\det(\lambda I - G_1) = \lambda^s + a_s \lambda^{s-1} + \cdots + a_2 \lambda + a_1.$$

Since (G_1, G_2) is controllable, they can be put in the controllable canonical form as follows:

$$G_1 = \begin{bmatrix} 0_{(s-1)\times 1} & I_{s-1} \\ -a_1 & A_{s-1} \end{bmatrix}, \quad G_2 = \begin{bmatrix} 0_{(s-1)\times 1} \\ 1 \end{bmatrix}$$

where $A_{s-1} = [-a_2, \ldots, -a_s]$.

Denote $X = \text{col}(X_1, \ldots, X_s)$, where $X_i \in \mathbb{R}^{1 \times q}$. Then, expanding (7.43) gives the following equations:

$$\begin{aligned} X_1 S - X_2 &= 0 \\ X_2 S - X_3 &= 0 \\ &\vdots \\ X_{s-1} S - X_s &= 0 \\ X_s S + a_1 X_1 + \cdots + a_s X_s &= Y. \end{aligned} \tag{7.44}$$

From the first $(s-1)$ equations of (7.44), we have

$$X_i = X_1 S^{i-1}, \; i = 2, \ldots, s. \tag{7.45}$$

Substituting (7.45) into the last equation of (7.44) gives

$$X_1 (S^s + a_s S^{s-1} + \cdots + a_2 S + a_1 I_q) = Y.$$

Since the minimal polynomial of S divides the characteristic polynomial of G_1, we have $Y = 0$. □

We call the system composed of the nominal plant of system (7.31) and the internal model (7.41) the augmented system, which is of the following form:

$$\dot{x} = Ax + Bu + Ev \tag{7.46a}$$
$$\dot{\xi} = G_1\xi + G_2 e \tag{7.46b}$$
$$e = Cx + Du + Fv \tag{7.46c}$$
$$y_m = C_m x + D_m u + F_m v. \tag{7.46d}$$

Then we have the following key lemma.

Lemma 7.5 *The augmented system (7.46) has the following two properties:*

(i) Under Assumption 7.1, suppose that there exists a static state feedback control law

$$u = K_1 x + K_2 \xi \tag{7.47}$$

that stabilizes the augmented plant (7.46) when v is set to 0. Then the control law (7.32) with $z = \xi$ and

$$\mathscr{G}_1 = G_1,\ \mathscr{G}_2 = G_2 \tag{7.48}$$

solves the structurally stable output regulation problem of system (7.31).

(ii) Under Assumptions 7.1 and 7.7, suppose that there exists a dynamic output feedback control law

$$u = K_1\hat{x} + K_2\xi \tag{7.49a}$$
$$\dot{\hat{x}} = S_1\hat{x} + S_2\xi + S_3 y_m \tag{7.49b}$$

with $\hat{x} \in \mathbb{R}^n$ that stabilizes the augmented plant (7.46) when v is set to 0. Then the control law (7.33) with $z = col(\hat{x}, \xi)$ and

$$K = [K_1,\ K_2],\ \mathscr{G}_1 = \begin{bmatrix} S_1 & S_2 \\ 0 & G_1 \end{bmatrix},\ \mathscr{G}_2 = \begin{bmatrix} S_3 \\ G_2 Q \end{bmatrix} \tag{7.50}$$

solves the structurally stable output regulation problem of system (7.31).

Proof Note that the nominal part of the closed-loop system (7.35) can also be viewed as the composition of the augmented system (7.46) and the static state feedback control law (7.47) (respectively, the dynamic output feedback control law (7.49)). Thus, A_c in (7.35) is Hurwitz in both cases, that is, the closed-loop system (7.35) satisfies Property 7.3. It remains to show that Property 7.4 also holds for both cases (i) and (ii). By Lemma 7.3 and Remark 7.10, we only need to show the existence of $X_c(w)$ that satisfies, for any $w \in \mathbb{W}$,

$$X_c(w)S = A_c(w)X_c(w) + B_c(w) \tag{7.51a}$$
$$0 = C_c(w)X_c(w) + D_c(w). \tag{7.51b}$$

Case (i): From (7.36) and noting (7.48), we have

$$A_c(w) = \begin{bmatrix} A(w) + B(w)K_1 & B(w)K_2 \\ G_2(C(w) + D(w)K_1) & G_1 + G_2D(w)K_2 \end{bmatrix}, \quad B_c(w) = \begin{bmatrix} E(w) \\ G_2F(w) \end{bmatrix}$$
$$C_c(w) = [C(w) + D(w)K_1 \quad D(w)K_2], \quad D_c(w) = F(w).$$

Under Assumption 7.1, the spectra of S and $A_c(w)$ are disjoint. Thus, by Proposition A.3 in Sect. A.1, the Sylvester equation (7.51a) admits a unique solution $X_c(w)$. Thus, we only need to show $X_c(w)$ also satisfies (7.51b). For this purpose, let $X_c(w) = \begin{bmatrix} X(w) \\ Z(w) \end{bmatrix}$ with $X(w) \in \mathbb{R}^{n \times q}$ and $Z(w) \in \mathbb{R}^{n_z \times q}$. Then the Sylvester equation (7.51a) can be expanded as follows:

$$X(w)S = A(w)X(w) + B(w)(K_1X(w) + K_2Z(w)) + E(w) \tag{7.52a}$$
$$Z(w)S = G_1Z(w) + G_2Y(w) \tag{7.52b}$$

where

$$Y(w) = (C(w) + D(w)K_1)X(w) + D(w)K_2Z(w) + F(w) = C_c(w)X_c(w) + D_c(w).$$

Since Eq. (7.52b) is in the form of (7.43), by Lemma 7.4, $Y(w) = 0$, which implies (7.51b).

Case (ii): From (7.37), using (7.50), and noting Assumption 7.7 implies $C(w) = QC_m(w)$, $D(w) = QD_m(w)$, $F(w) = QF_m(w)$, we have

$$A_c(w) = \begin{bmatrix} A(w) & B(w)K_1 & B(w)K_2 \\ S_3C_m(w) & S_1 + S_3D_m(w)K_1 & S_2 + S_3D_m(w)K_2 \\ G_2C(w) & G_2D(w)K_1 & G_1 + G_2D(w)K_2 \end{bmatrix}$$
$$B_c(w) = \begin{bmatrix} E(w) \\ S_3F_m(w) \\ \mathcal{G}_2F(w) \end{bmatrix}$$
$$C_c(w) = [C(w) \quad D(w)K_1 \quad D(w)K_2], \quad D_c(w) = F(w).$$

Similar to Case (i), under Assumption 7.1, the spectra of S and $A_c(w)$ are disjoint. Thus, by Proposition A.3 in Sect. A.1, the Sylvester equation (7.51a) admits a unique solution $X_c(w)$. Thus, we only need to show $X_c(w)$ also satisfies (7.51b). For this purpose, let $X_c(w) = \begin{bmatrix} X(w) \\ \hat{Z}(w) \\ \bar{Z}(w) \end{bmatrix}$ with $X(w) \in \mathbb{R}^{n \times q}$, $\hat{Z} \in \mathbb{R}^{n \times q}$, and $\bar{Z}(w) \in \mathbb{R}^{n_z \times q}$. Then the Sylvester equation (7.51a) can be expanded as follows:

$$X(w)S = A(w)X(w) + B_w(K_1\hat{Z}(w) + K_2\bar{Z}(w)) + E(w) \tag{7.53a}$$
$$\hat{Z}(w)S = S_1\hat{Z}(w) + S_2\bar{Z}(w) + S_3Y_m(w) \tag{7.53b}$$
$$\bar{Z}(w)S = G_1\bar{Z}(w) + G_2Y(w) \tag{7.53c}$$

where

$$Y_m(w) = C_m(w)X(w) + D_m(w)(K_1\hat{Z}(w) + K_2\bar{Z}(w)) + F_m(w)$$

and

$$Y(w) = C(w)X(w) + D(w)(K_1\hat{Z}(w) + K_2\bar{Z}(w)) + F(w) = C_c(w)X_c(w) + D_c(w).$$

Since Eq. (7.53c) is in the form of (7.43), by Lemma 7.4, $Y(w) = 0$, which again implies (7.51b). □

Lemma 7.5 has converted Problem 7.2 of the original plant (7.31) to the stabilization problem of the augmented system (7.46). We will further show that if the pair (G_1, G_2) is a p-copy internal model of the matrix S with G_1 satisfying the following condition:

$$\text{rank}\begin{bmatrix} A - \lambda I_n & B \\ C & D \end{bmatrix} = n + p, \ \forall \lambda \in \lambda(G_1), \tag{7.54}$$

then, the augmented system (7.46) indeed can be stabilized by a static state feedback control law (7.47) or by a dynamic output feedback control law (7.49). For this purpose, we establish the following lemma.

Lemma 7.6 *Under Assumption 7.2, let the pair (G_1, G_2) be a p-copy internal model of the matrix S with G_1 satisfying (7.54). Then the pair*

$$\left(\begin{bmatrix} A & 0 \\ G_2C & G_1 \end{bmatrix}, \begin{bmatrix} B \\ G_2D \end{bmatrix}\right) \tag{7.55}$$

is stabilizable.

Proof Let

$$M(\lambda) = \begin{bmatrix} A - \lambda I_n & 0 & B \\ G_2C & G_1 - \lambda I_{n_z} & G_2D \end{bmatrix}.$$

By the well-known PBH (Popov–Belevitch–Hautus) test, the pair (7.55) is stabilizable if and only if

$$\text{rank}\, M(\lambda) = n + n_z, \quad \text{for all } \lambda \in \bar{\mathbb{C}}_+.$$

Since (A, B) is stabilizable, rank $[A - \lambda I_n \;\; B] = n$, for all $\lambda \in \bar{\mathbb{C}}_+$. Also, $\det(G_1 - \lambda I_{n_z}) \neq 0, \forall \lambda \notin \lambda(G_1)$. Thus

$$\text{rank} M(\lambda) = n + n_z, \quad \forall \lambda \notin \lambda(G_1) \text{ and } \forall \lambda \in \bar{\mathbb{C}}_+. \tag{7.56}$$

Write $M(\lambda) = M_1(\lambda) M_2(\lambda)$ where

$$M_1(\lambda) = \begin{bmatrix} I_n & 0 & 0 \\ 0 & G_2 & G_1 - \lambda I_{n_z} \end{bmatrix}, \quad M_2(\lambda) = \begin{bmatrix} A - \lambda I_n & 0 & B \\ C & 0 & D \\ 0 & I_{n_z} & 0 \end{bmatrix}.$$

Since (G_1, G_2) is controllable, for all $\lambda \in \mathbb{C}$, $M_1(\lambda)$ has rank $n + n_z$. Since G_1 satisfies (7.54), $M_2(\lambda)$ has rank $n + n_z + p$ for all $\lambda \in \lambda(G_1)$. Hence, by Sylvester's inequality[2],

$$\begin{aligned} n + n_z \geq \text{rank } M(\lambda) &\geq (n + n_z) + (n + n_z + p) - (n + n_z + p) \\ &= n + n_z, \quad \forall \lambda \in \lambda(G_1). \end{aligned} \tag{7.57}$$

Combining (7.56) and (7.57) gives

$$\text{rank} M(\lambda) = n + n_z, \quad \forall \lambda \in \bar{\mathbb{C}}_+.$$

Thus, the pair (7.55) is stabilizable. □

Remark 7.13 If the pair (G_1, G_2) is the minimal p-copy internal model of the matrix S, then S and G_1 have the same spectrum. In this case, Assumption 7.5 implies that matrix G_1 satisfies (7.54). On the other hand, if condition (7.54) is not satisfied, that is, for some $\lambda \in \lambda(G_1)$, the rank of $M_2(\lambda)$ is less than $n + n_z + p$, then the pair (7.55) may not be stabilizable. Thus, under Assumption 7.1, Assumption 7.5 is necessary for the pair (7.55) to be stabilizable.

Combining Lemmas 7.5 and 7.6 leads to the solvability conditions for Problem 7.2 by a dynamic state feedback control law of the form (7.32) and a dynamic measurement output feedback control law of the form (7.33), respectively.

Theorem 7.4 *(i) Under Assumptions 7.1, 7.2, and 7.5, let the pair (G_1, G_2) be a minimal p-copy internal model of the matrix S and let $K = \begin{bmatrix} K_1 & K_2 \end{bmatrix}$ be such that*

$$\begin{bmatrix} A & 0 \\ G_2 C & G_1 \end{bmatrix} + \begin{bmatrix} B \\ G_2 D \end{bmatrix} K$$

is Hurwitz. Then, Problem 7.3 is solvable by a dynamic state feedback control law of the form (7.32), where

[2] rank A + rank $B - n \leq$ rank $AB \leq \min\{\text{rank } A, \text{rank } B\}$ for any matrices $A \in \mathbb{R}^{m\times n}$ and $B \in \mathbb{R}^{n\times p}$.

$$\mathscr{G}_1 = G_1, \ \mathscr{G}_2 = G_2.$$

(ii) Under additional Assumptions 7.6 and 7.7, let $K = \begin{bmatrix} K_1 & K_2 \end{bmatrix}$ *be such that*

$$\begin{bmatrix} A & 0 \\ G_2C & G_1 \end{bmatrix} + \begin{bmatrix} B \\ G_2D \end{bmatrix} K$$

is Hurwitz, and L be such that $A - LC_m$ *is Hurwitz. Then, Problem 7.3 is solvable by a dynamic output feedback control law of the form (7.33), where*

$$\mathscr{G}_1 = \begin{bmatrix} S_1 & S_2 \\ 0 & G_1 \end{bmatrix}, \ \mathscr{G}_2 = \begin{bmatrix} S_3 \\ G_2Q \end{bmatrix}$$

with $S_1 = A + BK_1 - L(C_m + D_mK_1)$, $S_2 = (B - LD_m)K_2$, $S_3 = L$.

Proof Part (i). Since the pair (G_1, G_2) is the minimal p-copy internal model of S, Assumption 7.5 implies that G_1 satisfies (7.54). By Lemma 7.6, under Assumptions 7.2 and 7.5, the pair (7.55) is stabilizable. Thus, there exists (K_1, K_2) such that

$$A_c = \begin{bmatrix} A + BK_1 & BK_2 \\ G_2(C + DK_1) & G_1 + G_2DK_2 \end{bmatrix} \tag{7.58}$$

is Hurwitz. By Part (i) of Lemma 7.5, under Assumption 7.1, the dynamic state feedback control law (7.32) solves Problem 7.2.

Part (ii). Let (K_1, K_2, G_1, G_2) be the same as those in (7.58). By Assumption 7.6, there exists a constant matrix $L \in \mathbb{R}^{n \times p_m}$ such that $A - LC_m$ is Hurwitz. With $S_1 = A + BK_1 - L(C_m + D_mK_1)$, $S_2 = (B - LD_m)K_2$, $S_3 = L$, the nominal closed-loop system matrix A_c takes the following form:

$$A_c = \begin{bmatrix} A & BK_1 & BK_2 \\ LC_m & A + BK_1 - LC_m & BK_2 \\ G_2C & G_2DK_1 & G_1 + G_2DK_2 \end{bmatrix}. \tag{7.59}$$

In (7.59), subtracting the second row from the first row and adding the first column to the second column gives

$$\begin{bmatrix} A - LC_m & 0 & 0 \\ LC_m & A + BK_1 & BK_2 \\ G_2C & G_2(C + DK_1) & G_1 + G_2DK_2 \end{bmatrix}. \tag{7.60}$$

Thus the spectrum of (7.60) is the union of that of (7.58) and that of $A - LC_m$. That is, A_c as defined by (7.59) is Hurwitz. Thus, by Part (ii) of Lemma 7.5, under Assumptions 7.1 and 7.7, the control law (7.33) solves Problem 7.2. □

Remark 7.14 The specific form of the matrices S_i, $i = 1, 2, 3$, in the second equation of (7.49) is motivated by the Luenberger observer for the following linear system:

$$\dot{x} = Ax + Bu, \;\; y_m = C_m x + D_m u. \tag{7.61}$$

In fact, the Luenberger observer of (7.61) is

$$\dot{\hat{x}} = A\hat{x} + Bu + L(y_m - (C_m\hat{x} + D_m u)). \tag{7.62}$$

Substituting the first equation of (7.49) into (7.62) gives

$$\begin{aligned} \dot{\hat{x}} &= (A + BK_1 - L(C_m + D_m K_1))\hat{x} + (B - LD_m)K_2\xi + Ly_m \\ &= S_1\hat{x} + S_2\xi + S_3 y_m. \end{aligned}$$

Example 7.2 We consider the same example as in Sect. 7.2 with $\omega = 3$. We assume the nominal and actual values of ε are $\varepsilon_0 = 0.2$ and $\varepsilon = 0.21$, respectively. To synthesize a controller of the form (7.32), let

$$G_1 = \begin{bmatrix} 0 & 1 \\ -\omega^2 & 0 \end{bmatrix}, \; G_2 = \begin{bmatrix} 0 \\ 1 \end{bmatrix}.$$

The control gain K is chosen to be $K = \begin{bmatrix} 106.96 & 21.38 & -1.03 & -4.36 & -717.3 & -96.7 \end{bmatrix}$ so that the eigenvalues of $\begin{bmatrix} A & 0 \\ G_2C & G_1 \end{bmatrix} + \begin{bmatrix} B \\ G_2D \end{bmatrix} K$ are located at

$$\{-1, -1.2, -1.4, -1.6, -1.8, -2\}.$$

With the initial values $v(0) = \text{col}(0, 1)$, $x(0) = \text{col}(0.1, 0, 0, 0)$, $z(0) = 0$, the performance of the dynamic state feedback control law (7.32) is shown in Fig. 7.4.

Next, we consider the dynamic measurement output feedback control law (7.33). Let K, G_1, and G_2 be the same as designed above. Moreover, let

$$L = \begin{bmatrix} 2.4 & -0.03 \\ 0.36 & -0.04 \\ -0.04 & 2.79 \\ 0.16 & 1.92 \end{bmatrix}$$

so that the eigenvalues of $A - LC_m$ are located at

$$\{-1, -1.2, -1.4, -1.6\}.$$

Again, choose the initial values as $v(0) = \text{col}(0, 1)$, $x(0) = \text{col}(0.1, 0, 0, 0)$, $z(0) = 0$. The performance of the dynamic measurement output feedback control law (7.33) is shown in Fig. 7.5.

In contrast to the feedforward control law (7.29), it can be seen from Figs. 7.4 and 7.5 that the structurally stable control laws (7.32) and (7.33) can tolerate small parameter uncertainty.

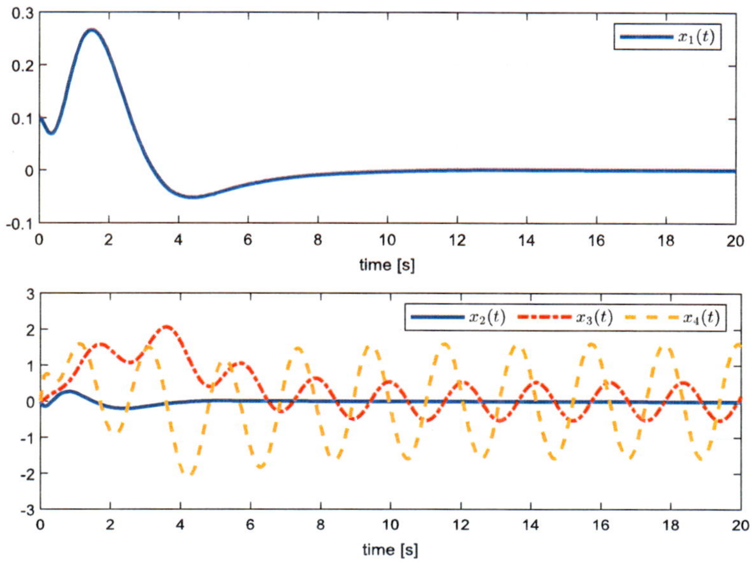

Fig. 7.4 The response of $x_i(t)$ under the control law (7.32) when $\varepsilon = 0.21$

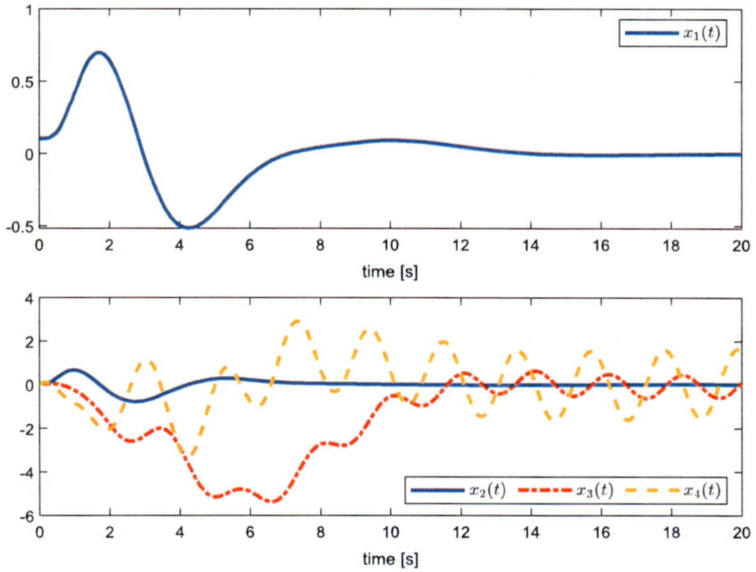

Fig. 7.5 The response of $x_i(t)$ under the control law (7.33) when $\varepsilon = 0.21$

7.4 Linear Robust Output Regulation: Canonical Internal Model

The structurally stable output regulation problem studied in the last section can only handle sufficiently small variations of the unknown parameter vector w. This is because the role of the p-copy internal model is to convert the output regulation problem of the uncertain plant (7.31) to the stabilization problem of the augmented system (7.46). Since the stability of the matrix $A_c(0)$ only guarantees the stability of the matrix $A_c(w)$ for sufficiently small w, the approach in the last section cannot handle large uncertain parameter vector w. In this section, we will further consider the robust output regulation problem of the uncertain plant (7.31) with $y_m = e$ for the case where $w \in \mathbb{W} \subseteq \mathbb{R}^{n_w}$ for some arbitrarily large compact subset $\mathbb{W}$.

To handle such a large uncertain parameter vector w, we need to convert the robust output regulation problem of the uncertain plant (7.31) to the robust stabilization problem of some augmented error system to be defined shortly. For this purpose, we need to employ the canonical internal model as summarized in Sect. A.6.

We focus on the class of dynamic error output feedback control law (7.34) assuming $m = p$. The closed-loop system composed of the plant (7.31) and the control law (7.34) takes the following form:

$$\begin{aligned} \dot{x}_c &= A_c(w)x_c + B_c(w)v \\ e &= C_c(w)x_c + D_c(w)v \end{aligned}$$

where $x_c = \operatorname{col}(x, z)$ and

$$\begin{aligned} A_c(w) &= \begin{bmatrix} A(w) & B(w)K \\ \mathcal{G}_2 C(w) & \mathcal{G}_1 + \mathcal{G}_2 D(w)K \end{bmatrix}, \quad B_c(w) = \begin{bmatrix} E(w) \\ \mathcal{G}_2 F(w) \end{bmatrix} \\ C_c(w) &= [C(w) \quad D(w)K], \ D_c(w) = F(w). \end{aligned}$$

Then the problem is formulated as follows.

Problem 7.3 Given the plant (7.31), the exosystem (7.8), and a compact subset $\mathbb{W} \subseteq \mathbb{R}^{n_w}$ containing the origin $w = 0$, find an error output feedback control law of the form (7.34) such that, for any $w \in \mathbb{W}$, the closed-loop system has the following two properties:

- **Property 7.5** $A_c(w)$ is Hurwitz;
- **Property 7.6** For any $x_c(0)$ and $v(0)$, $\lim_{t\to\infty} e(t) = 0$.

In contrast with Problem 7.2, Problem 7.3 requires both the stability property and the regulation property hold for all w in a prescribed compact subset $\mathbb{W}$ which can be arbitrarily large. We call Problem 7.3 the robust output regulation problem and will apply the canonical internal model approach to handle this problem. For this purpose, we need the following assumption:

Assumption 7.8 For any $w \in \mathbb{W}$, the regulator equations

$$X(w)S = A(w)X(w) + B(w)U(w) + E(w) \quad (7.63a)$$
$$0 = C(w)X(w) + D(w)U(w) + F(w) \quad (7.63b)$$

admit a unique solution pair $(X(w), U(w))$.

Let $\alpha_m(\lambda) = \lambda^l + \alpha_l \lambda^{l-1} + \cdots + \alpha_2 \lambda + \alpha_1$ be the minimal polynomial of S, and

$$\Phi_0 = \begin{bmatrix} 0 & 1 & \cdots & 0 \\ \vdots & \vdots & \ddots & \vdots \\ 0 & 0 & \cdots & 1 \\ -\alpha_1 & -\alpha_2 & \cdots & -\alpha_l \end{bmatrix}, \quad \Psi_0 = \begin{bmatrix} 1 \; 0 \cdots 0 \end{bmatrix}. \quad (7.64)$$

Let $S_i(w) = \text{col}(U_i(w), U_i(w)S, \ldots, U_i(w)S^{l-1})$, where $U_i(w)$ is the ith row of $U(w)$. Then, it can be verified, noting $\alpha_m(S) = 0$, that

$$S_i(w)S = \Phi_0 S_i(w) \quad (7.65a)$$
$$U_i(w) = \Psi_0 S_i(w). \quad (7.65b)$$

Let $\Upsilon(w) = \text{col}(S_1(w), \ldots, S_m(w))$, $\Phi = I_m \otimes \Phi_0$, and $\Psi = I_m \otimes \Psi_0$. Then, it can be further verified that

$$\Upsilon(w)S = \Phi\Upsilon(w) \quad (7.66a)$$
$$U(w) = \Psi\Upsilon(w). \quad (7.66b)$$

Definition 7.2 Under Assumption 7.8, let $M \in \mathbb{R}^{lm \times lm}$ and $Q \in \mathbb{R}^{lm \times m}$ be any controllable pair with M being Hurwitz, both independent of w. If there exist matrices $\Lambda(w) \in \mathbb{R}^{lm \times q}$ and $\Gamma \in \mathbb{R}^{m \times lm}$ such that

$$\Lambda(w)S = M\Lambda(w) + QU(w), \quad U(w) = \Gamma\Lambda(w). \quad (7.67)$$

Then the following dynamic compensator:

$$\dot{\eta} = M\eta + Qu \quad (7.68)$$

is called a canonical internal model of the plant (7.31) and the exosystem (7.8) with the estimated steady-state input $\hat{\mathbf{u}} = \Gamma\eta$.

Remark 7.15 Under Assumptions 7.1 and 7.8, the matrix triplet (M, Q, Γ) in Definition 7.2 always exists. In fact, let $M_0 \in \mathbb{R}^{l \times l}$ and $Q_0 \in \mathbb{R}^l$ be any controllable pair such that M_0 is Hurwitz and the spectra of M_0 and Φ_0 are disjoint. Since (Ψ_0, Φ_0) in (7.64) is observable, by Proposition A.3 in Sect. A.1, the Sylvester equation

$$T_0\Phi_0 = M_0 T_0 + Q_0\Psi_0$$

admits a unique nonsingular solution $T_0 \in \mathbb{R}^{l\times l}$. Let $M = I_m \otimes M_0$, $Q = I_m \otimes Q_0$, $T = I_m \otimes T_0$. Then it can be verified that (M, Q, T) satisfies the following Sylvester equation:

$$T\Phi = MT + Q\Psi. \tag{7.69}$$

Let $\Lambda(w) = T\Upsilon(w)$ and $\Gamma = \Psi T^{-1}$. Then using the equation (7.66b) shows that $\Gamma\Lambda(w) = \Psi T^{-1}T\Upsilon(w) = \Psi\Upsilon(w) = U(w)$, and using (7.69) gives

$$T\Phi\Upsilon(w) = MT\Upsilon(w) + Q\Psi\Upsilon(w). \tag{7.70}$$

Using (7.66) in (7.70) gives

$$T\Upsilon(w)S = MT\Upsilon(w) + QU(w) \tag{7.71}$$

which is the same as the first equation of (7.67).

It is noted that, under Assumption 7.1, none of the eigenvalues of Φ_0 has negative real part. Thus, the spectra of Φ_0 and any Hurwitz matrix M_0 are disjoint.

Attaching the canonical internal model (7.68) to the linear system (7.31) yields the so-called augmented system as follows:

$$\dot{x} = A(w)x + B(w)u + E(w)v \tag{7.72a}$$
$$\dot{\eta} = M\eta + Qu \tag{7.72b}$$
$$e = C(w)x + D(w)u + F(w)v. \tag{7.72c}$$

To understand the role of the internal model, combining the regulator equations (7.63) and (7.71) gives

$$X(w)S = A(w)X(w) + B(w)U(w) + E(w) \tag{7.73a}$$
$$T\Upsilon(w)S = MT\Upsilon(w) + QU(w) \tag{7.73b}$$
$$0 = C(w)X(w) + D(w)U(w) + F(w). \tag{7.73c}$$

Equation (7.73) can be viewed as the regulator equations for the augmented system (7.72), which have a solution pair $(X(w), T\Upsilon(w))$ and $U(w)$.

Performing on the augmented system (7.72) the following coordinate and input transformation:

$$\bar{x} = x - X(w)v \tag{7.74a}$$
$$\bar{\eta} = \eta - T\Upsilon(w)v \tag{7.74b}$$
$$\bar{u} = u - \Psi T^{-1}\eta \tag{7.74c}$$

converts the augmented system (7.72) to the following so-called augmented error system:

$$\dot{\bar{x}} = A(w)\bar{x} + B(w)\bar{u} + B(w)\Psi T^{-1}\bar{\eta} \tag{7.75a}$$

$$\dot{\bar{\eta}} = (M + Q\Psi T^{-1})\bar{\eta} + Q\bar{u} \tag{7.75b}$$

$$e = C(w)\bar{x} + D(w)\bar{u} + D(w)\Psi T^{-1}\bar{\eta}. \tag{7.75c}$$

Lemma 7.7 *Under Assumptions 7.1 and 7.8, if, for all $w \in \mathbb{W}$, the following dynamic error output feedback control law:*

$$\bar{u} = K\zeta \tag{7.76a}$$

$$\dot{\zeta} = F_1\zeta + F_2 e \tag{7.76b}$$

where $\zeta \in \mathbb{R}^{n_\zeta}$ for some integer n_ζ stabilizes the augmented error system (7.75), then the following dynamic output feedback control law:

$$u = K\zeta + \Psi T^{-1}\eta \tag{7.77a}$$

$$\dot{\zeta} = F_1\zeta + F_2 e \tag{7.77b}$$

$$\dot{\eta} = M\eta + Qu \tag{7.77c}$$

solves Problem 7.3 for the system (7.31).

Proof By Remark 7.15, under Assumptions 7.1 and 7.8, the canonical internal model of the form (7.68) exists. Note that (7.77) is in the form of (7.34) with $z = \text{col}(\zeta, \eta)$. It can be verified that, due to the coordinate and input transformation (7.74), the closed-loop system (7.35) composed of the system (7.31) and the control law (7.77) is the same as the composition of the augmented error system (7.75) and the dynamic output feedback control law (7.76). Thus, the robust stabilization of (7.75) via (7.76) means that, for all $w \in \mathbb{W}$, $A_c(w)$ in (7.35) is Hurwitz, that is, the closed-loop system (7.35) satisfies Property 7.5. Thus, for all $w \in \mathbb{W}$, we have $\lim_{t\to\infty} \bar{x}(t) = 0$, $\lim_{t\to\infty} \zeta(t) = 0$, and $\lim_{t\to\infty} \bar{\eta}(t) = 0$, and hence $\lim_{t\to\infty} \bar{u}(t) = 0$. Therefore, from Eq. (7.75c), for all $w \in \mathbb{W}$, $\lim_{t\to\infty} e(t) = 0$, that is, Property 7.6 holds. □

Remark 7.16 Lemma 7.7 has summarized a systematic approach to handling Problem 7.3. This approach actually follows a general framework for the robust output regulation problem for uncertain nonlinear systems described in Sect. A.6. In fact, under Assumption 7.8, Assumption A.2 in Sect. A.6 is satisfied with $\mathbf{x}(v, w) = X(w)v$, $\mathbf{u}(v, w) = U(w)v$, $f(x, u, v, w) = A(w)x + B(w)u + E(w)v$, $h(x, u, v, w) = C(w)x + D(w)u + F(w)v$, and $\mathbf{a}(v, \sigma) = Sv$.

Since $\mathbf{u}(v, w) = U(w)v$ satisfies both Assumptions A.2 and A.5, by Proposition A.6, there exists a canonical internal model of the form (7.68) with the estimated steady-state input $\hat{\mathbf{u}} = \Psi T^{-1}\eta$, which is a special case of the general characterization of the internal model given in Definition A.6. It can also be seen that the augmented system (7.72) and the relation (7.73) are special cases of the augmented system (A.32) and the relation (A.33), respectively, and the transformation (7.74) and the augmented error system (7.75) are the special cases of the transformation (A.34) and the augmented error system (A.35), respectively, and Lemma 7.7 is the special case of Proposition A.5.

Remark 7.17 Even though Lemma 7.7 has converted Problem 7.3 for the system (7.31) into the robust stabilization problem of the augmented error system (7.75) by the dynamic output feedback control law of the form (7.76), the robust stabilization problem of the augmented error system (7.75) is in general not tractable. Thus, in what follows, we will only focus on a special class of linear systems (7.31) with $y_m = e$ satisfying the following assumption.

Assumption 7.9

(i) The system (7.31) is single-input single-output , i.e., $m = p = 1$, and $D(w) = 0$.
(ii) For all $w \in \mathbb{W}$, $C(w)A^k(w)B(w) = 0$ for $k = 0, 1, \ldots, r-2$, and $b(w) \triangleq C(w)A^{r-1}(w)B(w) \neq 0$.
(iii) For $k = 0, 1, \ldots, r-2$, $C(w)A^k(w)E(w) = 0$ for all $w \in \mathbb{W}$.
(iv) For any $w \in \mathbb{W}$, the system (7.31) with v set to zero is minimum phase with u as input and e as output.

Remark 7.18 Parts (i) to (iii) of Assumption 7.9 mean that the system (7.31) has a uniform relative degree r for all $w \in \mathbb{W}$ with input u and output e, which is not greater than the relative degree of the same system with input v and output e. Since $\mathbb{W}$ is compact, by the continuity of $b(w)$, it has the same sign for all $w \in \mathbb{W}$. Without loss of generality, we assume that, for all $w \in \mathbb{W}$, $b(w) > 0$. Otherwise, let $\hat{u} = -u$ and $\hat{b}(w) = -b(w)$. Then $b(w)u = \hat{b}(w)\hat{u}$ with $\hat{b}(w) > 0$ for all $w \in \mathbb{W}$.

As a result of Parts (i) to (iii) of Assumption 7.9, we claim that there exists a matrix $N(w) \in \mathbb{R}^{(n-r)\times n}$ such that, for all $w \in \mathbb{W}$, $N(w)B(w) = 0$ and

$$\mathbf{T}_1(w) = \begin{bmatrix} N(w) \\ C(w) \\ C(w)A(w) \\ \vdots \\ C(w)A^{r-1}(w) \end{bmatrix}$$

is nonsingular.

To show the above claim, first note that if, for any $w \in \mathbb{W}$, there exist constants $a_0(w), \ldots, a_{r-2}(w)$ such that

$$a_0(w)C(w) + a_1(w)C(w)A(w) + \cdots + a_{r-2}(w)C(w)A^{r-2}(w) = 0$$

then post-multiplying the above equality by $A^k(w)B(w)$ for $k = 1, \ldots, r-1$ successively, and noticing that $C(w)A^k(w)B(w) = 0$ for $k = 0, 1, \ldots, r-2$ and $C(w)A^{r-1}(w)\ B(w) \neq 0$ yields $a_0(w) = a_1(w) = \cdots = a_{r-2}(w) = 0$. Thus, for any

$w \in \mathbb{W}$, the row vectors $C(w)$, $C(w)A(w)$, ..., $C(w)A^{r-2}(w)$ are linearly independent.

Second, since $B(w) \in \mathbb{R}^{n\times 1}$, we can always find the other $n - r$ row vectors $N_1(w), \ldots, N_{n-r}(w)$, such that $N_i(w)B(w) = 0, i = 1, \ldots, n - r$, and $N_1(w), \ldots, N_{n-r}(w), C(w), C(w)A(w), \ldots, C(w)A^{r-2}(w)$ are linearly independent, for all $w \in \mathbb{W}$. Let $N(w) = \text{col}(N_1(w), \ldots, N_{n-r}(w))$. Then it remains to show all the rows of $\mathbf{T}_1(w)$ are linearly independent. If it is not, then, for some $w \in \mathbb{W}$, $C(w)A^{r-1}(w)$ can be represented as a linear combination of $N_1(w), \ldots, N_{n-r}(w)$, $C(w)$, $C(w)A(w)$, ..., $C(w)A^{r-2}(w)$. Noticing $N(w)B(w) = 0$ and $C(w)A^k(w)B(w) = 0$ for $k = 0, 1, \ldots, r - 2$ gives $C(w)A^{r-1}(w)B(w) = 0$, thus reaching a contradiction.

Let $\hat{\chi}_0 = N(w)x$, $\chi_k = C(w)A(w)^{k-1}x$, $k = 1, \ldots, r$, and $\hat{\chi} = \text{col}(\hat{\chi}_0, \chi_1, \ldots, \chi_r)$. Then, we have $\hat{\chi} = \mathbf{T}_1(w)x$ and

$$\dot{\hat{\chi}}_0 = N(w)A(w)\mathbf{T}_1^{-1}(w)\hat{\chi} + N(w)E(w)v \tag{7.78a}$$

$$\dot{\chi}_k = \chi_{(k+1)}, \ k = 1, \ldots, r - 1 \tag{7.78b}$$

$$\dot{\chi}_r = C(w)A^r(w)\mathbf{T}_1^{-1}(w)\hat{\chi} + b(w)u + C(w)A^{r-1}(w)E(w)v \tag{7.78c}$$

$$e = \chi_1 + F(w)v. \tag{7.78d}$$

We call (7.78) the normal form of the system (7.31). It is possible to further simplify (7.78). For this purpose, rewrite (7.78a) and (7.78c) as follows:

$$\begin{aligned}\dot{\hat{\chi}}_0 &= A_1(w)\hat{\chi}_0 + D_1(w)\chi_1 + \cdots + D_r(w)\chi_r + E_0(w)v\\ \dot{\chi}_r &= A_3(w)\hat{\chi}_0 + G_1(w)\chi_1 + \cdots + G_r(w)\chi_r + b(w)u + E_r(w)v\end{aligned}$$

where $N(w)A(w)\mathbf{T}_1^{-1}(w) = [A_1(w) \ \ D_1(w) \ \ \cdots \ \ D_r(w)]$, $E_0(w) = N(w)E(w)$, $C(w)A^r(w)\mathbf{T}_1^{-1}(w) = [A_3(w) \ \ G_1(w) \ \ \cdots \ \ G_r(w)]$, and $E_r(w) = C(w)A^{r-1}(w) E(w)$.

Let

$$\chi_0 = \hat{\chi}_0 + B_2(w)\chi_1 + \cdots + B_r(w)\chi_{r-1}$$

where

$$\begin{aligned}B_r(w) &= -D_r(w)\\ B_{r-1}(w) &= A_1(w)B_r(w) - D_{r-1}(w)\\ &\ \vdots\\ B_2(w) &= A_1(w)B_3(w) - D_2(w)\end{aligned}$$

and let $A_2(w) = D_1(w) - A_1(w)B_2(w)$. Then, it can be verified that

$$\begin{aligned}\dot{\chi}_0 &= A_1(w)\chi_0 + A_2(w)\chi_1 + E_0(w)v\\ \dot{\chi}_r &= A_3(w)\chi_0 + \sum_{k=1}^{r} c_k(w)\chi_k + b(w)u + E_r(w)v\end{aligned}$$

where $c_i(w) = G_i(w) - A_3(w)B_{i+1}(w)$, $i = 1, \ldots, r-1$, and $c_r(w) = G_r(w)$. Let

$$\mathbf{T}_2(w) = \begin{bmatrix} I_{n-r} & B_2(w) & \cdots & B_r(w) & 0\\ 0 & 1 & \cdots & 0 & 0\\ \vdots & \vdots & \vdots & \vdots & \vdots\\ 0 & 0 & \cdots & 0 & 1\end{bmatrix}$$

$$\mathbf{T}(w) = \mathbf{T}_2(w)\mathbf{T}_1(w)$$

and $\chi = \text{col}(\chi_0, \chi_1, \cdots, \chi_r)$. Then $\chi = \mathbf{T}(w)x$. Thus, we have the following result.

Lemma 7.8 *Under Parts (i) to (iii) of Assumption 7.9, for any $w \in \mathbb{W}$, with the linear transformation $\chi = \mathbf{T}(w)x$, the system (7.31) is equivalent to*

$$\dot{\chi}_0 = A_1(w)\chi_0 + A_2(w)\chi_1 + E_0(w)v \tag{7.79a}$$

$$\dot{\chi}_k = \chi_{(k+1)},\ k = 1, \ldots, r-1 \tag{7.79b}$$

$$\dot{\chi}_r = A_3(w)\chi_0 + \sum_{k=1}^{r} c_k(w)\chi_k + b(w)u + E_r(w)v \tag{7.79c}$$

$$e = \chi_1 + F(w)v. \tag{7.79d}$$

Moreover, under Part (iv) of Assumption 7.9, $A_1(w)$ is Hurwitz for all $w \in \mathbb{W}$.

Proof In fact, (7.79) is the direct result of the two coordinate transformations $\chi = \mathbf{T}_2(w)\mathbf{T}_1(w)x = \mathbf{T}(w)x$. To show, for all $w \in \mathbb{W}$, $A_1(w)$ is Hurwitz, let

$$\hat{A}(w) = \begin{bmatrix} A_1(w) & A_2(w) & 0 & \cdots & 0\\ 0 & 0 & 1 & \cdots & 0\\ \vdots & \vdots & \vdots & \vdots & \vdots\\ 0 & 0 & \cdots & 0 & 1\\ A_3(w) & c_1(w) & c_2(w) & \cdots & c_r(w)\end{bmatrix},\ \hat{B}(w) = \begin{bmatrix} 0\\ \vdots\\ 0\\ b(w)\end{bmatrix}$$

$$\hat{C}(w) = \begin{bmatrix} 0_{1\times(n-r)} & 1 & 0 & \cdots & 0\end{bmatrix}.$$

Then, the zeros of (7.79) is given by the roots of the following polynomial:

$$\begin{vmatrix} sI - \hat{A}(w) & \hat{B}(w)\\ \hat{C}(w) & 0\end{vmatrix}$$

which, after straightforward calculation, coincide with the eigenvalues of $sI - A_1(w)$. By Part (iv) of Assumption 7.9, $A_1(w)$ is Hurwitz for all $w \in \mathbb{W}$. □

Denote the solution of the regulator equations associated with (7.79) and the exosystem (7.8) by $X(w)$ and $U(w)$. Then, by inspection, we have

$$X(w) = \begin{bmatrix} X_0(w) \\ -F(w) \\ -F(w)S \\ \vdots \\ -F(w)S^{r-1} \end{bmatrix}$$

$$U(w) = \frac{-F(w)S^r - A_3(w)X_0(w) + \sum_{k=1}^{r} c_k(w)F(w)S^{k-1} - E_r(w)}{b(w)}$$

where $X_0(w)$ is the unique solution of the following Sylvester equation:

$$X_0(w)S = A_1(w)X_0(w) - A_2(w)F(w) + E_0(w). \tag{7.80}$$

Since the spectra of S and $A_1(w)$ are disjoint for all $w \in \mathbb{W}$, Proposition A.3 in Sect. A.1 guarantees the solvability of (7.80).

Now attaching the canonical internal model (7.68) to (7.79) gives the augmented system of the form (7.72), and performing on the augmented system the coordinate and input transformation of the form (7.74) gives the augmented error system of the form (7.75) as follows:

$$\dot{\bar{\chi}}_0 = A_1(w)\bar{\chi}_0 + A_2(w)\bar{\chi}_1 \tag{7.81a}$$

$$\dot{\bar{\chi}}_k = \bar{\chi}_{(k+1)}, \ k = 1, \ldots, r-1 \tag{7.81b}$$

$$\dot{\bar{\chi}}_r = A_3(w)\bar{\chi}_0 + \sum_{k=1}^{r} c_k(w)\bar{\chi}_k + b(w)\bar{u} + b(w)\Psi T^{-1}\bar{\eta} \tag{7.81c}$$

$$\dot{\bar{\eta}} = (M + Q\Psi T^{-1})\bar{\eta} + Q\bar{u} \tag{7.81d}$$

$$e = \bar{\chi}_1. \tag{7.81e}$$

By Lemma 7.7, it suffices to robustly stabilize the system (7.81) by a control law relying on e only. We will employ the so-called high-gain feedback control method to do so. For this purpose, we first establish the following technical lemma.

Lemma 7.9 *Consider the linear uncertain system*

$$\begin{bmatrix} \dot{x} \\ \dot{y} \end{bmatrix} = \begin{bmatrix} M_1(w) & N_1(w,\nu) \\ N_2(w,\nu) & N_3(w,\nu) + \nu M_2(w) \end{bmatrix} \begin{bmatrix} x \\ y \end{bmatrix} \tag{7.82}$$

where $\nu \in \mathbb{R}$, $w \in \mathbb{W}$, and, for all $w \in \mathbb{W}$, $M_1(w) \in \mathbb{R}^{n\times n}$ and $M_2(w) \in \mathbb{R}^{m\times m}$ are Hurwitz. Suppose, for some $\varepsilon > 0$, $\nu_0 > 0$, $\|N_i(w,\nu)\| \le \varepsilon$, $i = 1, 2, 3$, for all $\nu \ge$

ν_0, and all $w \in \mathbb{W}$. Let $P_1(w)$ and $P_2(w)$ be such that, for all $w \in \mathbb{W}$,

$$M_1^{\mathsf{T}}(w)P_1(w) + P_1(w)M_1(w) \le -I_n \tag{7.83a}$$

$$M_2^{\mathsf{T}}(w)P_2(w) + P_2(w)M_2(w) \le -I_m. \tag{7.83b}$$

Define

$$\nu^* = 2\varepsilon\sigma + 4\varepsilon^2\sigma^2 \tag{7.84}$$

where $\sigma = \max_{w\in\mathbb{W}}\{\|P_1(w)\|, \|P_2(w)\|\}$. Then, for all $\nu > \nu^*$, the origin of (7.82) is asymptotically stable for every $w \in \mathbb{W}$.

Proof Since $M_1(w)$ and $M_2(w)$ are Hurwitz for any $w \in \mathbb{W}$, there exist positive definite matrices $P_1(w)$ and $P_2(w)$ such that (7.83) holds. Let

$$V(x, y) = x^{\mathsf{T}}P_1(w)x + y^{\mathsf{T}}P_2(w)y.$$

Then $\beta_1\|\text{col}(x, y)\|^2 \le V(x, y) \le \beta_2\|\text{col}(x, y)\|^2$, where

$$\beta_1 = \min_{w\in\mathbb{W}}\{\underline{\lambda}(P_1(w)), \underline{\lambda}(P_2(w))\} > 0$$

$$\beta_2 = \max_{w\in\mathbb{W}}\{\bar{\lambda}(P_1(w)), \bar{\lambda}(P_2(w))\} > 0.$$

Furthermore, the derivative of $V(x, y)$ along system (7.82) satisfies

$$\dot{V}(x, y) \le -\begin{bmatrix} x \\ y \end{bmatrix}^{\mathsf{T}} \begin{bmatrix} \Xi_{11} & \Xi_{12} \\ \Xi_{21} & \Xi_{22} \end{bmatrix} \begin{bmatrix} x \\ y \end{bmatrix}$$

where

$$\begin{aligned}
\Xi_{11} &= I_n \\
\Xi_{12} &= -P_1(w)N_1(w, \nu) - N_2^{\mathsf{T}}(w, \nu)P_2(w) \\
\Xi_{21} &= -P_2(w)N_2(w, \nu) - N_1^{\mathsf{T}}(w, \nu)P_1(w) \\
\Xi_{22} &= \nu I_m - P_2(w)N_3(w, \nu) - N_3^{\mathsf{T}}(w, \nu)P_2(w).
\end{aligned}$$

To guarantee that $\dot{V}(x, y) < -\mu\|\text{col}(x, y)\|^2$ for all $\text{col}(x, y) \in \mathbb{R}^{n+m}$ and some positive real number μ, by Schur complement, it suffices to determine ν so that

$$\begin{aligned}
&\nu I_m - P_2(w)N_3(w, \nu) - N_3^{\mathsf{T}}(w, \nu)P_2(w) \\
&\quad - [P_2(w)N_2(w, \nu) + N_1^{\mathsf{T}}(w, \nu)P_1(w)]^{\mathsf{T}}[P_1(w)N_1(w, \nu) + N_2^{\mathsf{T}}(w, \nu)P_2(w)] > 0
\end{aligned}$$

which holds as long as $\nu > \nu^*$ with ν^* being given by (7.84). By the Lyapunov direct method, the origin of linear uncertain system (7.82) is asymptotically stable for every $w \in \mathbb{W}$. □

We further perform the following coordinate transformation on (7.81):

$$\tilde{\eta} = \bar{\eta} - \frac{1}{b(w)} Q(\xi_1 \bar{\chi}_1 + \xi_2 \bar{\chi}_2 + \cdots + \xi_{r-1} \bar{\chi}_{r-1} + \bar{\chi}_r)$$

where the coefficients ξ_k, $k = 1, 2, \ldots, r-1$ are such that the polynomial $s^{r-1} + \xi_{r-1} s^{r-2} + \cdots + \xi_2 s + \xi_1$ is stable, and obtain the equivalent form of system (7.81) as follows:

$$\dot{\bar{\chi}}_0 = A_1(w)\bar{\chi}_0 + A_2(w)\bar{\chi}_1 \tag{7.85a}$$

$$\dot{\tilde{\eta}} = \hat{A}_3(w)\bar{\chi}_0 + \sum_{k=1}^{r} \hat{c}_k(w)\bar{\chi}_k + M\tilde{\eta} \tag{7.85b}$$

$$\dot{\bar{\chi}}_k = \bar{\chi}_{k+1}, \ k = 1, \ldots, r-1 \tag{7.85c}$$

$$\dot{\bar{\chi}}_r = A_3(w)\bar{\chi}_0 + \sum_{k=1}^{r} \bar{c}_k(w)\bar{\chi}_k + b(w)\bar{u} + b(w)\Psi T^{-1}\tilde{\eta} \tag{7.85d}$$

where, for $k = 1, \ldots, r$, $\xi_r \triangleq 1$ and $\xi_0 \triangleq 0$,

$$\begin{aligned} \bar{c}_k(w) =& c_k(w) + \Psi T^{-1} Q \xi_k \\ \hat{c}_k(w) =& \frac{1}{b(w)}(M Q \xi_k - Q \xi_{k-1} - Q c_k(w)) \\ \hat{A}_3(w) =& -\frac{1}{b(w)} Q A_3(w). \end{aligned}$$

Let $\bar{\chi} = \mathrm{col}(\bar{\chi}_1, \ldots, \bar{\chi}_{r-1})$. Then system (7.85) can be rewritten into the following compact form:

$$\begin{bmatrix} \dot{\bar{\chi}} \\ \dot{\bar{\chi}}_0 \\ \dot{\tilde{\eta}} \\ \dot{\bar{\chi}}_r \end{bmatrix} = \mathscr{A}(w) \begin{bmatrix} \bar{\chi} \\ \bar{\chi}_0 \\ \tilde{\eta} \\ \bar{\chi}_r \end{bmatrix} + \mathscr{B}(w)\bar{u} \tag{7.86}$$

where

$$\mathscr{A}(w) = \begin{bmatrix} \Lambda_0 & 0 & 0 & G \\ A_2(w)D & A_1(w) & 0 & 0 \\ \hat{c}(w) & \hat{A}_3(w) & M & \hat{c}_r(w) \\ \bar{c}(w) & A_3(w) & b(w)\Psi T^{-1} & \bar{c}_r(w) \end{bmatrix}, \ \mathscr{B}(w) = \begin{bmatrix} 0 \\ 0 \\ 0 \\ b(w) \end{bmatrix}$$

with $\bar{c}(w) = [\bar{c}_1(w), \ldots, \bar{c}_{r-1}(w)]$, $\hat{c}(w) = [\hat{c}_1(w), \ldots, \hat{c}_{r-1}(w)]$, and

$$D = [1, 0, \ldots, 0], \; G = [0, \ldots, 0, 1]^\mathsf{T}, \; \Lambda_0 = \begin{bmatrix} 0 & 1 & \cdots & 0 \\ \vdots & \vdots & \ddots & \vdots \\ 0 & 0 & \cdots & 1 \\ 0 & 0 & \cdots & 0 \end{bmatrix}.$$

Lemma 7.10 *Consider the system (7.86) where $A_1(w)$ and M are Hurwitz, and the polynomial $s^{r-1} + \xi_{r-1}s^{r-2} + \cdots + \xi_2 s + \xi_1$ is stable. Then, there exists $K^* > 0$ such that, for any $K > K^*$, the following matrix:*

$$\mathscr{A}(w) - K\mathscr{B}(w)[\Xi, 0, 0, 1] \triangleq A(w, K) \tag{7.87}$$

where $\Xi = [\xi_1, \xi_2, \ldots, \xi_{r-1}]$ is Hurwitz for every $w \in \mathbb{W}$. Thus, the following static state feedback control law:

$$\bar{u} = -K(\xi_1\bar{\chi}_1 + \xi_2\bar{\chi}_2 + \cdots + \xi_{r-1}\bar{\chi}_{r-1} + \bar{\chi}_r) \tag{7.88}$$

stabilizes the augmented error system (7.86) (and hence (7.81)) for every $w \in \mathbb{W}$.

Proof The control law (7.88) can be written as

$$\bar{u} = -K[\Xi, 0, 0, 1]\begin{bmatrix} \bar{\chi} \\ \bar{\chi}_0 \\ \bar{\eta} \\ \bar{\chi}_r \end{bmatrix}.$$

Therefore, the system matrix of the closed-loop system composed of system (7.86) and control law (7.88) is given by (7.87). Let

$$T_A = \begin{bmatrix} I & 0 & 0 & 0 \\ 0 & I & 0 & 0 \\ 0 & 0 & I & 0 \\ \Xi & 0 & 0 & 1 \end{bmatrix}.$$

Then,

$$T_A^{-1} = \begin{bmatrix} I & 0 & 0 & 0 \\ 0 & I & 0 & 0 \\ 0 & 0 & I & 0 \\ -\Xi & 0 & 0 & 1 \end{bmatrix}$$

and

$$\begin{aligned}\bar{A}(w, K) &\triangleq T_A A(w, K) T_A^{-1} = T_A(\mathscr{A}(w) - K\mathscr{B}(w)[\Xi, 0, 0, 1])T_A^{-1} \\ &= T_A \mathscr{A}(w) T_A^{-1} - K\mathscr{B}(w)[0, 0, 0, 1].\end{aligned}$$

Simple calculation gives

$$T_A \mathscr{A}(w) T_A^{-1} = \begin{bmatrix} \Lambda_0 - G\Xi & 0 & 0 & G \\ A_2(w)D & A_1(w) & 0 & 0 \\ \hat{c}(w) - \hat{c}_r(w)\Xi & \hat{A}_3(w) & M & \hat{c}_r(w) \\ \bar{c}(w) + \Xi\Lambda_0 - \bar{c}_r(w)\Xi - \Xi G\Xi & A_3(w) & b(w)\Psi T^{-1} & \bar{c}_r(w) + \Xi G \end{bmatrix}$$

where

$$\Lambda_0 - G\Xi = \begin{bmatrix} 0 & 1 & \cdots & 0 \\ \vdots & \vdots & \ddots & \vdots \\ 0 & 0 & \cdots & 1 \\ -\xi_1 & -\xi_2 & \cdots & -\xi_{r-1} \end{bmatrix}$$

is Hurwitz. Notice that $\bar{A}(w, K)$ can be put in the following form:

$$\bar{A}(w, K) = \begin{bmatrix} M_1(w) & N_1(w) \\ N_2(w) & N_3(w) + KM_2(w) \end{bmatrix}$$

where

$$\begin{aligned} M_1(w) &= \begin{bmatrix} \Lambda_0 - G\Xi & 0 & 0 \\ A_2(w)D & A_1(w) & 0 \\ \hat{c}(w) - \hat{c}_r(w)\Xi & \hat{A}_3(w) & M \end{bmatrix}, \quad N_1(w) = \begin{bmatrix} G \\ 0 \\ \hat{c}_r(w) \end{bmatrix} \\ N_2(w) &= \begin{bmatrix} \bar{c}(w) + \Xi\Lambda_0 - \bar{c}_r(w)\Xi - \Xi G\Xi & A_3(w) & b(w)\Psi T^{-1} \end{bmatrix} \\ N_3(w) &= \bar{c}_r(w) + \Xi G, \quad M_2(w) = -b(w). \end{aligned} \tag{7.89}$$

Under the conditions of the lemma, it is ready to see that $M_1(w)$ and $M_2(w)$ are Hurwitz for all $w \in \mathbb{W}$. Since $N_i(w)$, $i = 1, 2, 3$ are independent of K, all the conditions in Lemma 7.9 are satisfied. Let $\varepsilon_1 = \max_{w\in\mathbb{W}, i=1,2,3}\{\|N_i(w)\|\}$ with $N_i(w)$, $i = 1, 2, 3$, being given by (7.89), $P_1(w)$ and $P_2(w)$ be the positive definite matrices such that, for all $w \in \mathbb{W}$

$$\begin{aligned} M_1(w)^{\mathsf{T}} P_1(w) + P_1(w) M_1(w) &\leq -I_{n+l-1} \\ 2M_2(w)P_2(w) &\leq -1. \end{aligned}$$

$\sigma_1 = \max_{w\in\mathbb{W}} \{\|P_1(w)\|, |P_2(w)|\}$, and

$$K^* = \max\{1, 2\varepsilon_1\sigma_1 + 4\varepsilon_1^2\sigma_1^2\}. \tag{7.90}$$

Then, by Lemma 7.9, for any $K > K^*$, $\bar{A}(w, K)$ and hence $A(w, K)$ are Hurwitz for all $w \in \mathbb{W}$, i.e., the origin of the closed-loop system is asymptotically stable for all $w \in \mathbb{W}$. □

Since the control law (7.88) depends on the states $\bar{\chi}_k$, $k = 1, \ldots, r$, and hence v and w, it is not feasible. Next, we will further consider synthesizing an output feedback control law which relies on e only. For this purpose, let h be some positive number, and let

$$A_o(h) = \begin{bmatrix} -h\delta_r & 1\ 0 \cdots 0 \\ -h^2\delta_{r-1} & 0\ 1 \cdots 0 \\ \vdots & \vdots\ \vdots\ \vdots\ \vdots \\ -h^{r-1}\delta_2 & 0\ 0 \cdots 1 \\ -h^r\delta_1 & 0\ 0 \cdots 0 \end{bmatrix}, \quad B_o(h) = \begin{bmatrix} h\delta_r \\ h^2\delta_{r-1} \\ \vdots \\ h^{r-1}\delta_2 \\ h^r\delta_1 \end{bmatrix}$$

where δ_j, $j = 1, 2, \ldots, r$, are such that the polynomial $s^r + \delta_r s^{r-1} + \cdots + \delta_2 s + \delta_1$ is stable. We call the following dynamic compensator:

$$\dot{\zeta} = A_o(h)\zeta + B_o(h)e$$

a high-gain observer of the state $\text{col}(\bar{\chi}, \bar{\chi}_r)$.

Then we have the following result.

Lemma 7.11 *Consider the system (7.86) where $A_1(w)$ and M are Hurwitz, and the polynomials $s^{r-1} + \xi_{r-1}s^{r-2} + \cdots + \xi_2 s + \xi_1$ and $s^r + \delta_r s^{r-1} + \cdots + \delta_2 s + \delta_1$ are both stable. There exist positive numbers K^* and h^* such that, for any $K > K^*$ and $h > h^*$, the following dynamic output feedback control law:*

$$\bar{u} = -K(\xi_1\zeta_1 + \xi_2\zeta_2 + \cdots + \xi_{r-1}\zeta_{r-1} + \zeta_r) \tag{7.91a}$$
$$\dot{\zeta} = A_o(h)\zeta + B_o(h)e \tag{7.91b}$$

where $\zeta = col(\zeta_1, \ldots, \zeta_r) \in \mathbb{R}^r$ stabilizes the augmented system (7.86) (and hence (7.81)) for every $w \in \mathbb{W}$.

Proof Note that, from (7.81), e and its kth derivatives are

$$e = \bar{\chi}_1, \quad e^{(k)} = \bar{\chi}_{k+1}, \quad k = 1, \ldots, r-1. \tag{7.92}$$

Let $\vartheta_k = h^{r-k}(e^{(k-1)} - \zeta_k)$, $k = 1, \ldots, r$, $\vartheta = \text{col}(\vartheta_1, \ldots, \vartheta_r)$, and $\phi = \text{col}(\bar{\chi}, \bar{\chi}_0, \bar{\eta}, \bar{\chi}_r)$. Then the control law (7.91a) can be rewritten as follows:

$$\begin{aligned} \bar{u} =& - K(\xi_1\bar{\chi}_1 + \xi_2\bar{\chi}_2 + \cdots + \xi_{r-1}\bar{\chi}_{r-1} + \bar{\chi}_r) \\ &+ K(\xi_1 h^{1-r}\vartheta_1 + \xi_2 h^{2-r}\vartheta_2 + \cdots + \xi_{r-1}h^{-1}\vartheta_{r-1} + \vartheta_r) \\ =& - K[\Xi, 0, 0, 1]\phi + K[\Xi, 1]D_h^{-1}\vartheta \end{aligned}$$

where $D_h = \mathscr{D}(h^{r-1}, h^{r-2}, \ldots, 1)$. As a result, from (7.86),

$$\dot{\phi} = A(w, K)\phi + Z_1(K, h, w)\vartheta$$

where $Z_1(K, h, w) = K\mathscr{B}(w)[\Xi, 1]D_h^{-1}$. On the other hand,

$$\begin{aligned}
\dot{\vartheta}_k =& h^{r-k}(e^{(k)} - \dot{\zeta}_k)\\
=& h^{r-k}(e^{(k)} - (-h^k\delta_{r+1-k}\zeta_1 + \zeta_{k+1} + h^k\delta_{r+1-k}e))\\
=& h^{r-k}((e^{(k)} - \zeta_{k+1}) - h^k\delta_{r+1-k}(e - \zeta_1))\\
=& h(\vartheta_{k+1} - \delta_{r+1-k}\vartheta_1),\ k = 1, \ldots, r-1\\
\dot{\vartheta}_r =& e^{(r)} - \dot{\zeta}_r\\
=& -h^r\delta_1(e - \zeta_1) + e^{(r)}\\
=& -h\delta_1\vartheta_1 + e^{(r)}\\
=& -h\delta_1\vartheta_1 + \dot{\bar{\chi}}_r.
\end{aligned}$$

Putting them together gives

$$\begin{aligned}
\dot{\vartheta} &= hA_o(1)\vartheta + E\dot{\phi}\\
&= (hA_o(1) + Z_3(w, K, h))\vartheta + Z_2(w, K)\phi
\end{aligned}$$

where

$$E = \begin{bmatrix} 0_{(r-1)\times(n+l-1)} & 0_{(r-1)\times 1} \\ 0_{1\times(n+l-1)} & 1 \end{bmatrix}$$

and $Z_2(w, K) = EA(w, K)$, $Z_3(w, K, h) = EZ_1(K, h, w)$.

Then the closed-loop system composed of the system (7.86) and the dynamic output feedback control law (7.91) can be put into the following compact form:

$$\begin{bmatrix} \dot{\phi} \\ \dot{\vartheta} \end{bmatrix} = \begin{bmatrix} A(w, K) & Z_1(w, K, h) \\ Z_2(w, K) & hA_o(1) + Z_3(w, K, h) \end{bmatrix} \begin{bmatrix} \phi \\ \vartheta \end{bmatrix}. \tag{7.93}$$

Let the positive number K^* be given by (7.90). From Lemma 7.10, $A(w, K)$ is Hurwitz for any $K > K^*$ and any $w \in \mathbb{W}$. Since the polynomial $s^r + \delta_r s^{r-1} + \cdots + \delta_2 s + \delta_1$ is stable, $A_o(1)$ is Hurwitz. Further, note that (7.93) takes the form of (7.82) with $\nu = h$, and $M_1(w) = A(w, K)$, $N_1(w, \nu) = Z_1(w, K, h)$, $N_2(w, \nu) = Z_2(w, K)$, $N_3(w, \nu) = Z_3(w, K, h)$, $M_2(w) = A_o(1)$, and $\|D_h^{-1}\| \leq \|D_1^{-1}\|$ for any $h \geq 1$. Therefore, for any $h \geq 1$, there exists $\varepsilon_2 > 0$, independent of h, such that $\|N_i(w, \nu)\|$
$\leq \varepsilon_2$, $i = 1, 2, 3$. Let $\bar{P}_1(w)$ and $\bar{P}_2$ be the positive definite matrices such that, for all $w \in \mathbb{W}$

$$A(w, K)^{\mathsf{T}} \bar{P}_1(w) + \bar{P}_1(w) A(w, K) \leq -I_{n+l}$$
$$A_o(1)^{\mathsf{T}} \bar{P}_2 + \bar{P}_2 A_o(1) \leq -I_r$$

$\sigma_2 = \max_{w \in \mathbb{W}} \{\|\bar{P}_1(w)\|, \|\bar{P}_2\|\}$, and

$$h^* = \max\{1, 2\varepsilon_2\sigma_2 + 4\varepsilon_2^2\sigma_2^2\}.$$

Then, by Lemma 7.9, for any $h > h^*$, the origin of the closed-loop system (7.93) is asymptotically stable for every $w \in \mathbb{W}$. □

Finally, the solvability of Problem 7.3 is summarized in the following theorem.

Theorem 7.5 *Given the class of linear systems* (7.31)*, the exosystem* (7.8)*, and any compact subset* $\mathbb{W}$*, under Assumptions 7.1, 7.8, and 7.9, Problem 7.3 is solvable by a dynamic error output feedback control law of the following form:*

$$u = \Psi T^{-1}\eta - K(\xi_1\zeta_1 + \xi_2\zeta_2 + \cdots + \xi_{r-1}\zeta_{r-1} + \zeta_r) \tag{7.94a}$$
$$\dot{\eta} = M\eta + Qu \tag{7.94b}$$
$$\dot{\zeta} = A_o(h)\zeta + B_o(h)e. \tag{7.94c}$$

Example 7.3 Consider a linear system of the form (7.31) with $y_m = e$. The system matrices are as follows:

$$A(w) = \begin{bmatrix} c_1(w) & 1 & 0 \\ 0 & 0 & 1 \\ 1 & c_2(w) & c_3(w) \end{bmatrix}, \quad B(w) = \begin{bmatrix} 0 \\ 0 \\ c_4(w) \end{bmatrix}, \quad E(w) = \begin{bmatrix} 0 & 0 \\ 0 & 0 \\ 0 & 0 \end{bmatrix},$$
$$C(w) = \begin{bmatrix} 0 & 1 & 0 \end{bmatrix}, \quad D(w) = 0, \quad F(w) = \begin{bmatrix} -1 & 0 \end{bmatrix}$$

where $w \in \mathbb{R}^4$. The exosystem system is given by (7.8) with $S = \begin{bmatrix} 0 & 1 \\ -1 & 0 \end{bmatrix}$.

Express the parameter vector $c(w) = \text{col}(c_1(w), c_2(w), c_3(w), c_4(w))$ as $c(w) = c(0) + w$, where $c(0) = \text{col}(c_1(0), c_2(0), c_3(0), c_4(0))$ is the nominal part of $c(w)$ and $w = \text{col}(w_1, w_2, w_3, w_4)$ is the uncertainty of $c(w)$. Let $c_1(0) = c_2(0) = c_3(0) = -2$, $c_4(0) = 2$, and $w_1 \in [-1, 1]$, $w_2 \in [-1, 1.5]$, $w_3 \in [-2, 1]$, $w_4 \in [-1, 2]$.

It can be verified that Assumptions 7.1, 7.8, and 7.9 are all satisfied. Then, by Theorem 7.5, it is possible to design a dynamic output feedback controller of the form (7.94) to solve the robust output regulation problem. More specifically, it is noted that the linear system has the uniform relative degree 2 with input u and output $e = C(w)x$, so $A_o(h)$ and $B_o(h)$ can be chosen as $A_o(h) = \begin{bmatrix} -h\delta_2 & 1 \\ -h^2\delta_1 & 0 \end{bmatrix}$ and $B_o(h) = \begin{bmatrix} h\delta_2 \\ h^2\delta_1 \end{bmatrix}$. Since $S = \begin{bmatrix} 0 & 1 \\ -1 & 0 \end{bmatrix}$, from (7.64) to (7.66), we have $\Phi = \begin{bmatrix} 0 & 1 \\ -1 & 0 \end{bmatrix}$, $\Psi = \begin{bmatrix} 1 \\ 0 \end{bmatrix}^{\mathsf{T}}$. Consider the controllable pair (M, Q) with $M = \begin{bmatrix} 0 & 1 \\ -\varrho_1 & -\varrho_2 \end{bmatrix}$ and $Q = \begin{bmatrix} 0 \\ 1 \end{bmatrix}$, where $\varrho_1, \varrho_2 > 0$. Solving the Sylvester equa-

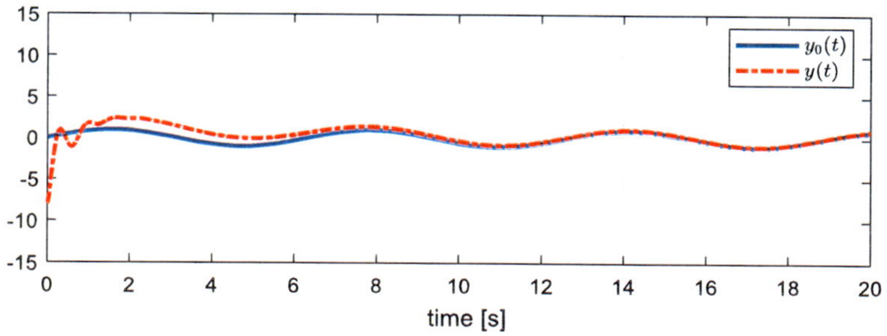

Fig. 7.6 The response of $y_0(t)$ and the response of $y(t)$ under the control law (7.94)

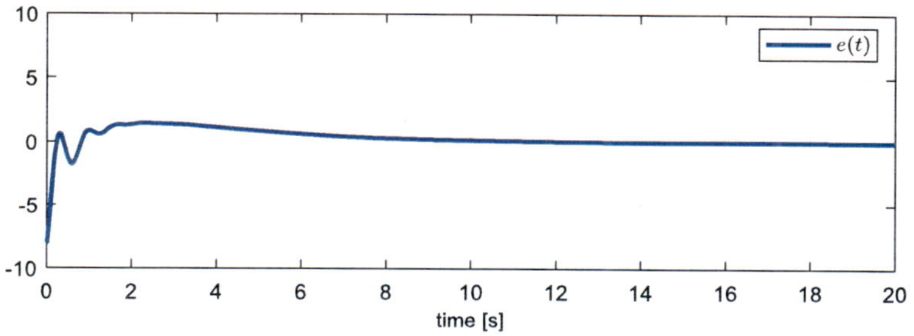

Fig. 7.7 The response of $e(t)$ under the control law (7.94)

tion (7.69) gives $T^{-1} = \begin{bmatrix} \varrho_1 - 1 & \varrho_2 \\ -\varrho_2 & \varrho_1 - 1 \end{bmatrix}$. Therefore, $\Psi T^{-1} = \begin{bmatrix} \varrho_1 - 1 & \varrho_2 \end{bmatrix}$. Then we can obtain an output feedback controller of the form (7.94) with $\xi_1 = \xi_2 = 1$, $\varrho_1 = 6$, $\varrho_2 = 18$, $K = 2$, $\delta_1 = 30$, $\delta_2 = 10$, and $h = 50$. The simulation results are shown in Figs. 7.6 and 7.7, where the actual values of the uncertain parameters are given by $w = [0.5, 0.7, 0.6, 0.5]^{\mathsf{T}}$. It is verified that the controller has successfully achieved the objective of robust output regulation.

7.5 Notes and References

The classic output regulation problem for linear systems was thoroughly studied during the 1970s. Early results can be found in [1, 2] which focused only on the special case where both the reference input and disturbance are step functions. The study on the general case can be found in [3–11], to name just a few. In particular, it was shown in [7] that the readability condition, that is, Assumption 7.7 is a necessary condition for the solvability of the structurally stable output regulation problem by a dynamic measurement output feedback control law.

The most salient outcome on the structurally stable output regulation is the so-called internal model principle which enables the conversion of the output regulation problem into an eigenvalue placement problem for an augmented linear system [5, 8], which is a generalization of the well-known PID controller. The canonical internal model was first proposed in [12], and later it was extensively used to deal with the robust output regulation problem for nonlinear systems [13–15].

Section 7.2 is an improvement of Sects. 7.2 and 7.3 of the book chapter [16], and Sect. 7.4 is an expansion of [14, Sect. 7.5].

The details on the high-gain observer can be found in [17].

References

1. Johnson CD (1971) Accommodation of external disturbances in linear regulator and servomechanism problem. IEEE Trans Autom Control 16(6):535–644
2. Smith HW, Davison EJ (1972) Design of industrial regulators: integral feedback and feedforward control. Proc IEEE 199(8):1210–1216
3. Cheng L, Pearson JB (1978) Frequency-domain synthesis of multivariable linear regulators. IEEE Trans Autom Control 23(1):3–15
4. Davison EJ (1972) The output control of linear time-invariant multivariable systems with unmeasurable arbitrary disturbance. IEEE Trans Autom Control 17(5):621–630
5. Davison EJ (1975) A generalization of the output control of linear time-invariant multivariable systms with unmeasureable arbitrary disturbance. IEEE Trans Autom Control 20(6):788–792
6. Davison EJ (1976) The robust control of a servomechanism problem for linear time-invariant multivariable systems. IEEE Trans Autom Control 21(1):25–34
7. Francis BA, Wonham WM (1976) The internal model principle of control theory. Automatica 12(5):457–465
8. Francis BA (1977) The linear multivariable regulator problem. SIAM J Control Optim 15(3):486–505
9. Wonham WM, Pearson JB (1974) Regulation and internal stabilization in linear multivariable systems. SIAM J Control Optim 12(1):5–18
10. Wonham WM (1985) Linear multivariable control: a geometric approach, 3rd edn. Springer, New York
11. Knoblich HW, Isidori A, Flokerzi D (1993) Topics in control theory. Birkhauser, Boston
12. Nikiforov VO (1998) Adaptive non-linear tracking with complete compensation of unknown disturbances. Eur J Control 4(2):132–139
13. Byrnes CI, Delli Priscoli F, Isidori A (1997) Output regulation of uncertain nonlinear systems. Birkhauser, Boston
14. Chen Z, Huang J (2015) Stabilization and regulation of nonlinear systems: a robust and adaptive approach. Springer International Publishing, Switzerland
15. Huang J (2004) Nonlinear output regulation problem: theory and applications. SIAM, Philadelphia
16. Huang J (2016) Certainty equivalence, separation principle, and cooperative output regulation of multi-agent systems by the distributed observer approach. Control Compl Syst: Theory and Appl 14:421–449
17. Khalil H (2002) Nonlinear systems. Prentice Hall, New Jersey
18. Bupp RT, Bernstein DS, Coppola VT (1998) A benchmark problem for nonlinear control design. Int J Robust Nonlinear Control 8(4–5):307–310

Chapter 8
Cooperative Output Regulation of Linear Multi-agent Systems by Distributed Observer Approach

In this chapter, we turn to the cooperative output regulation problem of linear multi-agent systems by distributed observer approach. In Sect. 8.1, the cooperative output regulation problem of linear multi-agent systems is formulated and a fundamental lemma is established. Then, the cooperative output regulation problem of linear multi-agent systems is solved by the distributed observer approach and the adaptive distributed observer approach in Sects. 8.2 and 8.3, respectively.

8.1 Linear Cooperative Output Regulation

In this section, we consider the cooperative output regulation problem for a group of linear systems as follows:

$$\dot{x}_i = A_i x_i + B_i u_i + E_i v_0 \tag{8.1a}$$
$$e_i = C_i x_i + D_i u_i + F_i v_0 \tag{8.1b}$$
$$y_{mi} = C_{mi} x_i + D_{mi} u_i + F_{mi} v_0, \ i = 1, \ldots, N \tag{8.1c}$$

where $x_i \in \mathbb{R}^{n_i}$, $e_i \in \mathbb{R}^{p_i}$, $y_{mi} \in \mathbb{R}^{p_{mi}}$, and $u_i \in \mathbb{R}^{m_i}$ are the state, error output, measurement output, and input of the ith subsystem, and $v_0 \in \mathbb{R}^q$ is the exogenous signal generated by the same exosystem as (4.6), and is repeated below for convenience.

$$\dot{v}_0 = S_0 v_0 \tag{8.2a}$$
$$y_{m0} = W_0 v_0 \tag{8.2b}$$

where $y_{m0} \in \mathbb{R}^{p_0}$ is the measurement output of the exosystem.

H. Cai et al., *Cooperative Control of Multi-agent Systems*, Advances in Industrial Control,
https://doi.org/10.1007/978-3-030-98377-2_8

It can be seen that when $N = 1$, (8.1) reduces to system (7.7) in Chap. 7 with $v = v_0$.

Various assumptions are as follows.

Assumption 8.1 S_0 has no eigenvalues with negative real parts.

Assumption 8.2 For $i = 1, \ldots, N$, the pairs (A_i, B_i) are stabilizable.

Assumption 8.3 For $i = 1, \ldots, N$, the pairs (C_{mi}, A_i) are detectable.

Assumption 8.4 The linear matrix equations

$$X_i S_0 = A_i X_i + B_i U_i + E_i \tag{8.3a}$$
$$0 = C_i X_i + D_i U_i + F_i, \ i = 1, \ldots, N \tag{8.3b}$$

have solution pairs (X_i, U_i).

Remark 8.1 As pointed out in Remark 7.2, Assumption 8.1 is made for convenience and will not lose generality. Assumption 8.3 can be viewed as a special case of Assumption 7.3 when v_0 is measurable and we are only interested in seeking the measurement output feedback plus feedforward control law of the form (7.30). The system (8.1) is still in the form of (7.7) with $x = \text{col}(x_1, \ldots, x_N)$, $u = \text{col}(u_1, \ldots, u_N)$, $y_m = \text{col}(y_{m1}, \ldots, y_{mN})$, $e = \text{col}(e_1, \ldots, e_N)$. Thus, if the state v_0 of the exosystem can be used by the control u_i of each follower, then, by Theorem 7.2, under Assumptions 8.2 and 8.4, the output regulation problem of the system (8.1) and the exosystem (8.2) can be solved by the following purely decentralized full information control law:

$$u_i = K_{1i} x_i + K_{2i} v_0, \ i = 1, \ldots, N \tag{8.4}$$

where K_{1i} are such that $A_i + B_i K_{1i}$ are Hurwitz, and $K_{2i} = U_i - K_{1i} X_i$. Similarly, under the additional Assumption 8.3, there exist $\ell_i \in \mathbb{R}^{n_i \times p_{mi}}$ such that $A_i - \ell_i C_{mi}$ are Hurwitz. Then, by Corollary 7.2, under Assumptions 8.2–8.4, the output regulation problem of the system (8.1) and the exosystem (8.2) can be solved by the following purely decentralized measurement output feedback plus feedforward control law:

$$u_i = K_{1i} z_i + K_{2i} v_0, \ i = 1, \ldots, N \tag{8.5a}$$
$$\dot{z}_i = A_i z_i + B_i u_i + E_i v_0 + \ell_i (y_{mi} - C_{mi} z_i - D_{mi} u_i - F_{mi} v_0). \tag{8.5b}$$

The control law (8.4) and the control law (8.5) can be unified into the following form:

$$u_i = K_{zi} z_i + K_{yi} y_{mi} + K_{vi} v_0 \tag{8.6a}$$
$$\dot{z}_i = \mathcal{G}_{1i} z_i + \mathcal{G}_{2i} y_{mi} + \mathcal{G}_{3i} v_0, \ i = 1, \ldots, N \tag{8.6b}$$

where $z_i \in \mathbb{R}^{n_{zi}}$, K_{zi}, K_{yi}, K_{vi}, $\mathcal{G}_{1i}$, $\mathcal{G}_{2i}$, $\mathcal{G}_{3i}$ are some constant matrices. The control law (8.6) clearly contains (8.5) as a special case and it also contains (8.4) as a special case if the dimension of z_i is zero and $y_{mi} = x_i$. Noting, as in Chap. 7, for $i = 1, \ldots, N$, $K_{yi} D_{mi} = 0$, the closed-loop system composed of (8.1) and (8.6) can be put in the following form:

$$\dot{x}_{ci} = A_{ci} x_{ci} + B_{ci} v_0 \tag{8.7a}$$

$$e_i = C_{ci} x_{ci} + D_{ci} v_0 \tag{8.7b}$$

where $x_{ci} = \text{col}(x_i, z_i)$,

$$\begin{aligned} A_{ci} &= \begin{bmatrix} A_i + B_i K_{yi} C_{mi} & B_i K_{zi} \\ \mathcal{G}_{2i}(C_{mi} + D_{mi} K_{yi} C_{mi}) & \mathcal{G}_{1i} + \mathcal{G}_{2i} D_{mi} K_{zi} \end{bmatrix} \\ B_{ci} &= \begin{bmatrix} E_i + B_i K_{yi} F_{mi} + B_i K_{vi} \\ \mathcal{G}_{2i}(F_{mi} + D_{mi}(K_{yi} F_{mi} + K_{vi})) + \mathcal{G}_{3i} \end{bmatrix} \\ C_{ci} &= [C_i + D_i K_{yi} C_{mi} \quad D_i K_{zi}], \quad D_{ci} = F_i + D_i K_{yi} F_{mi} + D_i K_{vi}. \end{aligned} \tag{8.8}$$

If, for each $i = 1, \ldots, N$, the control law (8.6) solves the output regulation problem of the ith subsystem of (8.1), then A_{ci} is Hurwitz, and, by Lemma 7.1, there exists a unique matrix X_{ci} that satisfies

$$X_{ci} S_0 = A_{ci} X_{ci} + B_{ci} \tag{8.9a}$$

$$0 = C_{ci} X_{ci} + D_{ci}. \tag{8.9b}$$

Let $\bar{x}_{ci}(t) = x_{ci}(t) - X_{ci} v_0(t)$. Then the closed-loop system composed of (8.1) and (8.6) satisfies

$$\begin{aligned} \dot{\bar{x}}_{ci} &= A_{ci} \bar{x}_{ci} \\ e_i &= C_{ci} \bar{x}_{ci}. \end{aligned}$$

Nevertheless, as the leader-following consensus problem studied in Chap. 3, in practice, the communication among different subsystems of (8.1) is subject to some constraints due to, say, the physical distance among these subsystems. Thus, the measurement output y_{m0} of the exosystem may not be available for the control u_i of all the followers. As a result, the tracking error e_i may not be available for the control u_i of all the followers. To describe the communication constraints among various subsystems, as in Chap. 3, we view the system (8.1) and the system (8.2) together as a multi-agent system with (8.2) as the leader and the N subsystems of (8.1) as the followers, respectively. Let $\bar{\mathcal{G}}_{\sigma(t)} = (\bar{\mathcal{V}}, \bar{\mathcal{E}}_{\sigma(t)})$ with $\bar{\mathcal{V}} = \{0, 1, \ldots, N\}$ and $\bar{\mathcal{E}}_{\sigma(t)} \subseteq \bar{\mathcal{V}} \times \bar{\mathcal{V}}$ for all $t \geq 0$ be a switching communication graph, where the node 0 is associated with the leader system (8.2) and the node i, $i = 1, \ldots, N$, is associated with the ith subsystem of the system (8.1). Here $\sigma(t) : [0, \infty) \mapsto \mathcal{P}$ where $\mathcal{P} = \{1, \ldots, \rho\}$ for some positive integer ρ is a piecewise constant switching signal with dwell time τ. For $i = 1, \ldots, N$, $j = 0, 1, \ldots, N$, $i \neq j$, $(j, i) \in \bar{\mathscr{E}}_{\sigma(t)}$ if and only if u_i can use y_{mj} for control at time instant t. Let $\bar{\mathscr{N}}_i(t) = \{j : (j, i) \in \bar{\mathscr{E}}_{\sigma(t)}\}$ denote the neighbor set of the agent i at time instant t.

The case where the network topology is static can be viewed as a special case of switching network topology when the switching index set contains only one element. We will use the simplified notation $\bar{\mathscr{G}}$ to denote a static communication graph.

Given the purely decentralized control law (8.6), we can construct the following class of distributed feedback control laws:

$$u_i = K_{zi} z_i + K_{yi} y_{mi} + K_{vi} \xi_i \tag{8.10a}$$

$$\dot{z}_i = \mathscr{G}_{1i} z_i + \mathscr{G}_{2i} y_{mi} + \mathscr{G}_{3i} \xi_i \tag{8.10b}$$

$$\dot{\xi}_i = g_i(\xi_i, y_{mi}, y_{mj}, j \in \bar{\mathscr{N}}_i(t)) \tag{8.10c}$$

where K_{zi}, K_{yi}, K_{vi}, $\mathscr{G}_{1i}$, $\mathscr{G}_{2i}$, and $\mathscr{G}_{3i}$ are gain matrices, g_i is a linear function in its arguments and will be designed using some type of distributed observers of the leader system studied in Chap. 4. In particular, g_i is time-varying if the graph $\bar{\mathscr{G}}_{\sigma(t)}$ is. It can be seen that, at each time $t \geq 0$, for any $i = 1, \ldots, N$, u_i can make use of y_{m0} if and only if the leader is a neighbor of the subsystem i. Thus, (8.10) is a distributed control law.

We now describe our problem as follows.

Problem 8.1 Given the systems (8.1), (8.2), and a switching communication graph $\bar{\mathscr{G}}_{\sigma(t)}$, find a distributed control law of the form (8.10) such that the closed-loop system has the following two properties:

- **Property 8.1** The origin of the closed-loop system with v_0 set to zero is asymptotically stable;
- **Property 8.2** For any initial condition $x_i(0)$, $z_i(0)$, $\xi_i(0)$, $i = 1, \ldots, N$, and $v_0(0)$, the solution of the closed-loop system satisfies

$$\lim_{t \to \infty} e_i(t) = 0, \; i = 1, \ldots, N.$$

Clearly, the solvability of the above problem not only depends on the dynamics of the systems (8.1) and (8.2), but also the property of the switching communication graph $\bar{\mathscr{G}}_{\sigma(t)}$. As in Chaps. 3 and 4, we focus on the case where the graph $\bar{\mathscr{G}}_{\sigma(t)}$ satisfies Assumption 4.1. Also, like in Chaps. 3 and 4, let $\mathscr{G}_{\sigma(t)} = (\mathscr{V}, \mathscr{E}_{\sigma(t)})$ denote the subgraph of $\bar{\mathscr{G}}_{\sigma(t)}$ where $\mathscr{V} = \{1, \ldots, N\}$, and $\mathscr{E}_{\sigma(t)} \subseteq \mathscr{V} \times \mathscr{V}$ is obtained from $\bar{\mathscr{E}}_{\sigma(t)}$ by removing all edges between the node 0 and the nodes in $\mathscr{V}$. In some cases, we also need Assumption 4.3. We treat the static graph as a special case of the switching graph, and treat Assumption 4.2 as the special case of Assumption 4.1.

To show that a control law of the form (8.10) is able to solve the cooperative output regulation problem of (8.1), we will first establish the following lemma.

Lemma 8.1 *Under Assumptions 8.1–8.4, suppose the purely decentralized control law (8.6) solves the output regulation problem of the system (8.1). Let* $\delta_{ui} : [0,\infty) \mapsto \mathbb{R}^{m_i}$ *and* $\delta_{zi} : [0,\infty) \mapsto \mathbb{R}^{n_{zi}}$ *be two piecewise continuous functions such that* $\lim_{t\to\infty} \delta_{ui}(t) = 0$, *and* $\lim_{t\to\infty} \delta_{zi}(t) = 0$. *Then the following control law:*

$$u_i = K_{zi} z_i + K_{yi} y_{mi} + K_{vi} v_0 + \delta_{ui} \tag{8.11a}$$
$$\dot{z}_i = \mathcal{G}_{1i} z_i + \mathcal{G}_{2i} y_{mi} + \mathcal{G}_{3i} v_0 + \delta_{zi},\ i = 1, \ldots, N \tag{8.11b}$$

is such that, for any initial condition of the closed-loop system, the solution is uniformly bounded over $[0,\infty)$ *for uniformly bounded* $v_0(t)$, *and* $\lim_{t\to\infty} e_i(t) = 0$.

Proof Since we have assumed $K_{yi} D_{mi} = 0$, the closed-loop system composed of (8.1) and (8.11) can be put in the following form:

$$\begin{aligned}
\dot{x}_{ci} &= A_{ci} x_{ci} + B_{ci} v_0 + B_{ui}\delta_{ui} + B_{zi}\delta_{zi}\\
e_i &= C_{ci} x_{ci} + D_{ci} v_0 + D_i \delta_{ui}
\end{aligned}$$

where $x_{ci} = \mathrm{col}(x_i, z_i)$, A_{ci}, B_{ci}, C_{ci}, D_{ci} are as defined in (8.8), and $B_{ui} = \mathrm{col}(B_i, \mathcal{G}_{2i} D_{mi})$ and $B_{zi} = \mathrm{col}(0_{n_i \times n_{z_i}}, I_{n_{z_i}})$. Since the purely decentralized control law (8.6) solves the output regulation problem of the system (8.1), by Lemma 7.1, there exists a unique matrix X_{ci} that satisfies (8.9). Let $\bar{x}_{ci} = x_{ci} - X_{ci} v_0$. Then the closed-loop system composed of (8.1) and (8.11) satisfies

$$\begin{aligned}
\dot{\bar{x}}_{ci} &= A_{ci}\bar{x}_{ci} + B_{ui}\delta_{ui} + B_{zi}\delta_{zi}\\
e_i &= C_{ci}\bar{x}_{ci} + D_i \delta_{ui}.
\end{aligned}$$

By Lemma 2.5, we have $\lim_{t\to\infty} \bar{x}_{ci}(t) = 0$, and hence $\lim_{t\to\infty} e_i(t) = 0$. Moreover, for uniformly bounded $v_0(t)$ over $[0,\infty)$, $x_{ci}(t) = \bar{x}_{ci}(t) + X_{ci} v_0(t)$ is also uniformly bounded over $[0,\infty)$. □

From the proof of Lemma 8.1, we can immediately obtain the following result.

Corollary 8.1 *Under Assumptions 8.1–8.4, suppose a control law of the form (8.6) solves Problem 8.1. Let* $\mathcal{S}(t) \in \mathbb{R}^{n_\xi \times n_\xi}$ *be piecewise continuous and uniformly bounded over* $[0,\infty)$ *and is such that the origin of* $\dot{\xi} = \mathcal{S}(t)\xi$ *is asymptotically stable, and* K_{ui}, K_{zi} *be any constant matrices with conformable dimensions. Then, under the following control law:*

$$\begin{aligned}
u_i &= K_{zi} z_i + K_{yi} y_{mi} + K_{vi} v_0 + K_{ui}\xi\\
\dot{z}_i &= \mathcal{G}_{1i} z_i + \mathcal{G}_{2i} y_{mi} + \mathcal{G}_{3i} v_0 + K_{zi}\xi\\
\dot{\xi} &= \mathcal{S}(t)\xi
\end{aligned}$$

the closed-loop system also satisfies the two properties in Problem 8.1.

8.2 Distributed Observer-Based Approach

Let us again recall the distributed observer (4.11) for the leader system (8.2) over the switching communication graph $\bar{\mathscr{G}}_{\sigma(t)}$ as follows:

$$\dot{v}_i = S_0 v_i + \mu L_0 \left(\sum_{j=0}^{N} a_{ij}(t) W_0 (v_j - v_i) \right), \; i = 1, \dots, N \tag{8.12}$$

where $\mu > 0$, $L_0 \in \mathbb{R}^{q \times p_0}$, and $\bar{\mathscr{A}}_{\sigma(t)} = [a_{ij}(t)]_{i,j=0}^{N}$ is the weighted adjacency matrix of $\bar{\mathscr{G}}_{\sigma(t)}$.

Corresponding to the two purely decentralized control laws (8.4) and (8.5), we can synthesize two types of distributed control laws as follows:

1. Distributed full information control law:

$$u_i = K_{1i} x_i + K_{2i} v_i \tag{8.13a}$$

$$\dot{v}_i = S_0 v_i + \mu L_0 \left(\sum_{j=0}^{N} a_{ij}(t) W_0 (v_j - v_i) \right), \; i = 1, \dots, N \tag{8.13b}$$

   where $K_{1i} \in \mathbb{R}^{m_i \times n_i}$ is such that $A_i + B_i K_{1i}$ is Hurwitz, $K_{2i} = U_i - K_{1i} X_i$, and μ is some positive constant.
2. Distributed measurement output feedback plus feedforward control law :

$$u_i = K_{1i} z_i + K_{2i} v_i \tag{8.14a}$$

$$\dot{z}_i = A_i z_i + B_i u_i + E_i v_i + \ell_i (y_{mi} - C_{mi} z_i - D_{mi} u_i - F_{mi} v_i) \tag{8.14b}$$

$$\dot{v}_i = S_0 v_i + \mu L_0 \left(\sum_{j=0}^{N} a_{ij}(t) W_0 (v_j - v_i) \right), \; i = 1, \dots, N \tag{8.14c}$$

   where $\ell_i \in \mathbb{R}^{n_i \times p_{mi}}$ is such that $A_{Li} = A_i - \ell_i C_{mi}$ is Hurwitz.

Remark 8.2 Both of the control laws (8.13) and (8.14) are synthesized based on the certainty equivalence principle in the sense that they are obtained from the purely decentralized control laws (8.4) and (8.5) by replacing v_0 in (8.4) and (8.5) with v_i, respectively, where v_i is generated by a distributed observer of the form (8.12).

Let $\tilde{v}_i = v_i - v_0$, $i = 1, \dots, N$, and $\tilde{v} = \mathrm{col}(\tilde{v}_1, \dots, \tilde{v}_N)$. Then, as shown in Chap. 4, $\tilde{v}$ is governed by the following equation:

$$\dot{\tilde{v}} = \left((I_N \otimes S_0) - \mu (H_{\sigma(t)} \otimes L_0 W_0) \right) \tilde{v}. \tag{8.15}$$

We now consider the solvability of Problem 8.1.

Lemma 8.2 *Suppose the origin of the error system (8.15) is asymptotically stable. Then,*

(i) Under Assumptions 8.1, 8.2, and 8.4, Problem 8.1 is solvable by the distributed full information control law (8.13);

(ii) Under the additional Assumption 8.3, Problem 8.1 is solvable by the distributed measurement output feedback plus feedforward control law (8.14).

Proof Part (i): The control law (8.13) can be put in the following form:

$$\begin{aligned} u_i &= K_{1i}x_i + K_{2i}v_0 + K_{2i}\tilde{v}_i, \ i = 1, \ldots, N \\ \dot{\tilde{v}} &= \left((I_N \otimes S_0) - \mu(H_{\sigma(t)} \otimes L_0 W_0)\right)\tilde{v} \end{aligned}$$

where $\lim_{t\to\infty} \tilde{v}(t) = 0$. Since the purely decentralized full information control law (8.4) solves Problem 8.1, the proof is complete upon using Corollary 8.1 with $\xi = \tilde{v}$.

Part (ii): Denote the control law (8.5) by

$$\begin{aligned} u_i &= k_i(z_i, v_0), \ i = 1, \ldots, N \\ \dot{z}_i &= g_i(z_i, k_i(z_i, v_0), y_{mi}, v_0) \end{aligned}$$

and the control law (8.14) by

$$\begin{aligned} u_i &= k_i(z_i, v_i), \quad i = 1, \ldots, N, \\ \dot{z}_i &= g_i(z_i, k_i(z_i, v_i), y_{mi}, v_i), \\ \dot{\tilde{v}} &= \left((I_N \otimes S_0) - \mu(H_{\sigma(t)} \otimes L_0 W_0)\right)\tilde{v}. \end{aligned}$$

Then it is ready to verify that

$$k_i(z_i, v_i) = k_i(z_i, \tilde{v}_i + v_0) = k_i(z_i, v_0) + K_{2i}\tilde{v}_i \tag{8.16}$$

and

$$\begin{aligned} g_i(z_i, k_i(z_i, v_i), y_{mi}, v_i) &= g_i(z_i, k_i(z_i, v_0) + K_{2i}\tilde{v}_i, y_{mi}, \tilde{v}_i + v_0) \\ &= g_i(z_i, k_i(z_i, v_0), y_{mi}, v_0) + \Gamma_i\tilde{v}_i \end{aligned} \tag{8.17}$$

where $\Gamma_i = B_i K_{2i} + E_i - \ell_i(F_{mi} + D_{mi}K_{2i})$.

Since, under the additional Assumption 8.3, the control law (8.5) solves Problem 8.1, and $\lim_{t\to\infty} \tilde{v}_i(t) = 0$ for $i = 1, \ldots, N$, the proof follows from Corollary 8.1 with $\xi = \tilde{v}$. □

By Theorem 4.1, under Assumptions 4.1, 4.3, 4.4, and 4.6, there exists a constant matrix $L_0 \in \mathbb{R}^{q \times p_{m0}}$ such that the origin of the error system (8.15) is exponentially stable. Combining Lemma 8.2 with Theorem 4.1 leads to the following result regarding the switching network case.

Theorem 8.1 *Under Assumptions 4.1, 4.3, 4.4, 4.6, 8.1, 8.2 and 8.4, Problem 8.1 is solvable by a distributed full information control law of the form* (8.13), *and, under the additional Assumption 8.3, Problem 8.1 is solvable by a distributed measurement output feedback plus feedforward control law of the form* (8.14).

Remark 8.3 The observer gain L_0 can be calculated as described in Remark 4.3. In particular, if S_0 is neutrally stable, then the observer gain $L_0 = \mu_v P_0 W_0^{\mathrm{T}}$ where $P_0 > 0$ satisfies $P_0 S_0^{\mathrm{T}} + S_0 P_0 = 0$ and $\mu_v > 0$. If $y_{m0} = v_0$, then, by Theorem 4.2, Assumption 4.3 is not needed, Assumption 4.4 is satisfied automatically, and Assumption 4.6 can be relaxed to Assumption 4.5. In this case, the observer gain $L_0 = \mu I_q$ with any $\mu > 0$.

Combining Lemma 8.2 with Theorems 4.3 and 4.4 leads to the following result regarding the static network case.

Theorem 8.2 *Under Assumptions 4.2, 4.4, 8.1, 8.2, and 8.4, Problem 8.1 is solvable by a distributed full information control law of the form (8.13), and, under the additional Assumption 8.3, Problem 8.1 is solvable by a distributed measurement output feedback plus feedforward control law of the form (8.14).*

Remark 8.4 From Theorems 4.3 and 4.4, the observer gain $L_0 = \mu P_0 W_0^{\mathrm{T}}$ where P_0 is the unique positive definite solution of the equation $P_0 S_0^{\mathrm{T}} + S_0 P_0 - P_0 W_0^{\mathrm{T}} W_0 P_0 + I_q = 0$, and $\mu \geq \frac{1}{2}\lambda_H^{-1}$. If $y_{m0} = v_0$, then Assumption 4.4 is satisfied automatically. In this case, the observer gain $L_0 = \mu I_q$ with $\mu > \bar{\lambda}_{S_0} \underline{\lambda}_H^{-1}$. If none of the eigenvalues of S_0 has positive real part, then μ can be any positive real number.

Example 8.1 In this example, we further study the leader-following formation problem described in Example 1.1. Let us start with a group of four mobile robots whose motion equations are as follows:

$$
\begin{aligned}
\dot{x}_i &= \nu_i \cos(\theta_i) && (8.18a)\\
\dot{y}_i &= \nu_i \sin(\theta_i) && (8.18b)\\
\dot{\theta}_i &= \omega_i && (8.18c)\\
m_i \dot{\nu}_i &= f_i && (8.18d)\\
J_i \dot{\omega}_i &= \tau_i, \quad i = 1, \ldots, 4 && (8.18e)
\end{aligned}
$$

where (x_i, y_i, θ_i) are the Cartesian position and orientation of the robot center, respectively; (ν_i, ω_i) are the linear and angular velocities, respectively; m_i is the mass; J_i is the mass moment of inertia; $u_i = \mathrm{col}(f_i, \tau_i)$ is the control input, where f_i is the force and τ_i is the torque applied to the robot. Let us define the performance output of the system (8.18) as

$$
h_i = \begin{bmatrix} x_{hi} \\ y_{hi} \end{bmatrix} = \begin{bmatrix} x_i \\ y_i \end{bmatrix} + d_i \begin{bmatrix} \cos(\theta_i) \\ \sin(\theta_i) \end{bmatrix}
$$

which represents the robot hand position off the wheel axis of the ith mobile robot by a constant distance d_i.

With the following standard input transformation:

$$u_i = \begin{bmatrix} \frac{1}{m_i}\cos(\theta_i) & -\frac{d_i}{J_i}\sin(\theta_i) \\ \frac{1}{m_i}\sin(\theta_i) & \frac{d_i}{J_i}\cos(\theta_i) \end{bmatrix}^{-1} \begin{bmatrix} \bar{u}_{xi} + \nu_i\omega_i\sin(\theta_i) + d_i\omega_i^2\cos(\theta_i) \\ \bar{u}_{yi} - \nu_i\omega_i\cos(\theta_i) + d_i\omega_i^2\sin(\theta_i) \end{bmatrix} \quad (8.19)$$

where $\bar{u}_i = \text{col}(\bar{u}_{xi}, \bar{u}_{yi})$ is the new input, system (8.18) is converted into the double integrator form (1.28) (or equivalently (1.29)).

As described in Example 1.1, the solvability of the cooperative output regulation problem of system (1.28) with leader system (1.30) and tracking error (1.31) implies the solvability of the leader-following formation problem of the robot system (8.18) with the leader's trajectory generated by (1.30), which satisfies Assumption 4.5.

To make our problem more interesting, we consider the switching network dictated by the following switching signal:

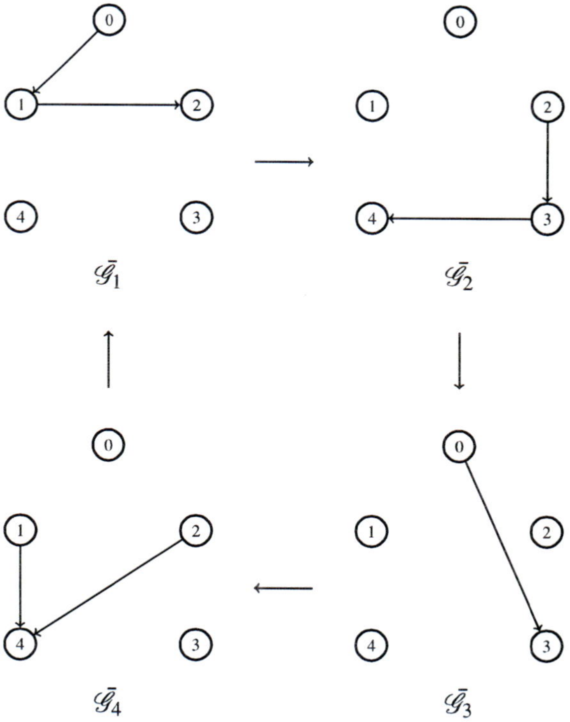

Fig. 8.1 The switching communication graph $\bar{\mathcal{G}}_{\sigma(t)}$

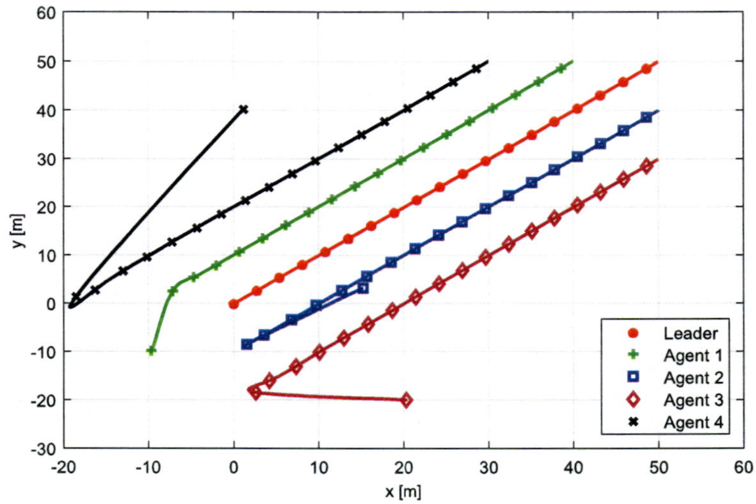

Fig. 8.2 Trajectories of the four robots and the leader

$$\sigma(t) = \begin{cases} 1 & \text{if } sT \leq t < (s + \frac{1}{4})T \\ 2 & \text{if } (s + \frac{1}{4})T \leq t < (s + \frac{1}{2})T \\ 3 & \text{if } (s + \frac{1}{2})T \leq t < (s + \frac{3}{4})T \\ 4 & \text{if } (s + \frac{3}{4})T \leq t < (s + 1)T \end{cases}$$

where $s = 0, 1, 2, \ldots$. The four graphs $\bar{\mathcal{G}}_i$, $i = 1, 2, 3, 4$ are illustrated in Fig. 8.1, where the node 0 is associated with the leader, and other nodes are associated with the followers. It can be verified that this switching graph satisfies Assumption 4.1.

It is ready to see that Assumptions 8.1 and 8.2 are satisfied. Moreover, Assumption 8.4 is also satisfied by verifying that $X_i = I_4$ and $U_i = [0, 0] \otimes I_2$ satisfy the regulator equations (8.3). Let $K_{1i} = [-1.2, -1.4] \otimes I_2$ and $\mu = 1$. Then $K_{2i} = U_i - K_{1i}X_i = [1.2, 1.4] \otimes I_2 = -K_{1i}$. By Theorem 8.1 with $y_{m0} = v_0$, we obtain the following distributed full information feedback controller:

$$\bar{u}_i = K_{1i}x_i + K_{2i}v_i, \ i = 1, \ldots, N \tag{8.20a}$$

$$\dot{v}_i = S_0 v_i + \mu \left(\sum_{j=0}^{N} a_{ij}(t)(v_j - v_i) \right) \tag{8.20b}$$

where $x_i = \text{col}(x_{hi} - x_{di}, y_{hi} - y_{di}, p_{xi}, p_{yi})$. Finally, the controller composed of (8.19) and (8.20) solves the leader-following formation problem of the original system (8.18).

Assume that the leader signal is $h_0(t) = \text{col}(t, t)$, and the desired relative displacements are given by $h_{d1} = \text{col}(-10, 0)$, $h_{d2} = \text{col}(0, -10)$, $h_{d3} = \text{col}(0, -20)$,

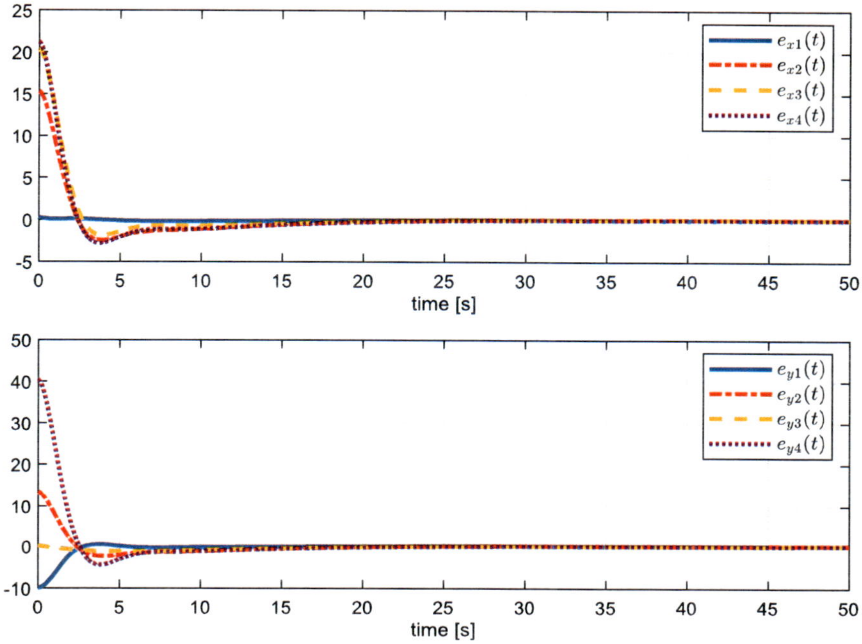

Fig. 8.3 The response of $e_{xi}(t)$ and $e_{yi}(t)$ under the control law (8.19) and (8.20)

$h_{d4} = \text{col}(-20, 0)$. The simulation is conducted for the following values of various parameters: $m_i = 1\,\text{kg}$, $J_i = 0.15\,\text{kg}\cdot\text{m}^2$, $d_i = 0.3\,\text{m}$, $T = 2\text{s}$. The initial positions of the four robots are $(x_1(0), y_1(0)) = (-10, -10)$, $(x_2(0), y_2(0)) = (15, 3)$, $(x_3(0), y_3(0)) = (20, -20)$, $(x_4(0), y_4(0)) = (1, 40)$, and all other initial values are randomly chosen over the interval $[0, 1]$. Figure 8.2 shows the trajectories of the leader and the four robots. Figure 8.3 shows the goal-seeking errors of the four robots all approach zero. Figures 8.4 and 8.5 show the orientations, and the linear and angular velocities of the four robots.

8.3 Adaptive Distributed Observer-Based Approach

The control laws (8.13) and (8.14) make use of the distributed observer, which assume that the control of every follower knows the system matrix S_0 of the leader. To remove this assumption, we will further consider, in this section, employing the adaptive distributed observer (4.26) for the leader system (8.2) to solve the cooperative output regulation problem. Since the control law to be designed in this section will result in

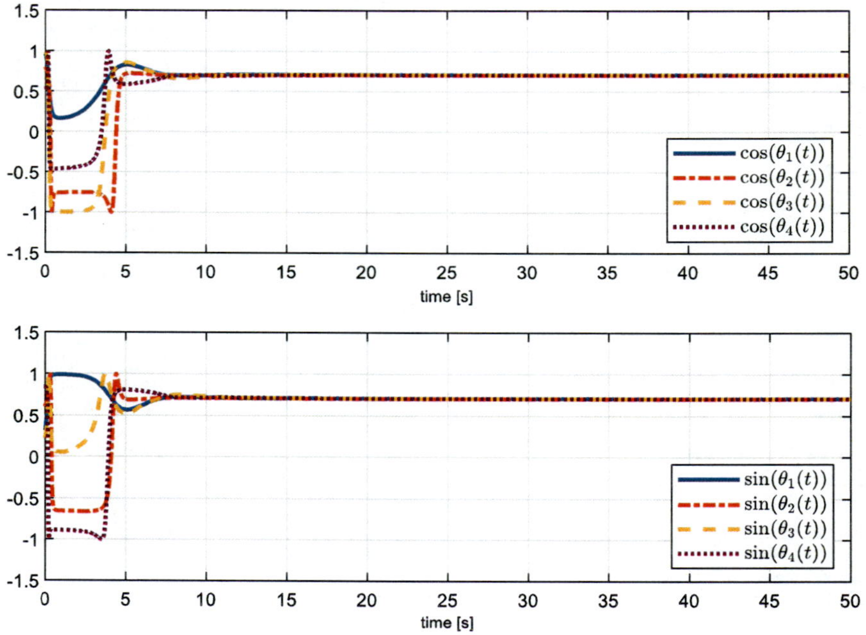

Fig. 8.4 The response of $\cos(\theta_i(t))$ and $\sin(\theta_i(t))$ under the control law (8.19) and (8.20)

a closed-loop system whose origin is not an equilibrium point when v_0 is set to zero, we need to modify Problem 8.1 to the following:

Problem 8.2 Given the systems (8.1), (8.2), and a switching communication graph $\bar{\mathcal{G}}_{\sigma(t)}$, find a distributed control law of the form (8.10) such that, for any initial

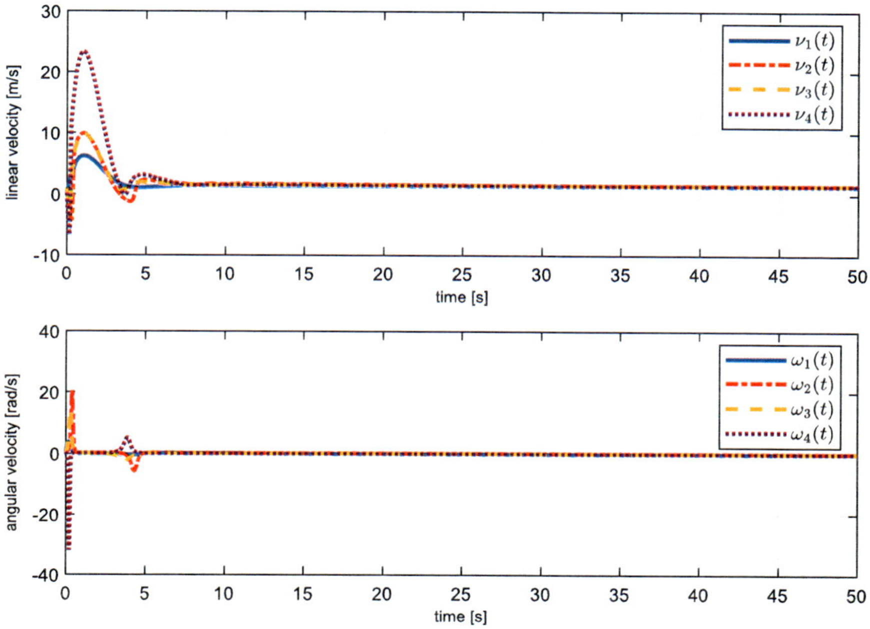

Fig. 8.5 The response of $\nu_i(t)$ and $\omega_i(t)$ under the control law (8.19) and (8.20)

condition $x_i(0)$, $z_i(0)$, $\xi_i(0)$, $i = 1, \ldots, N$, and $v_0(0)$, the closed-loop system has the following two properties:

- **Property 8.3** The solution of the closed-loop system exists for all $t \geq 0$ and is uniformly bounded over $[0, \infty)$ for uniformly bounded $v_0(t)$;
- **Property 8.4** The solution of the closed-loop system is such that

$$\lim_{t\to\infty} e_i(t) = 0, \ i = 1, \ldots, N.$$

Clearly, if the closed-loop system satisfies Property 8.1, it also satisfies Property 8.3.

To save the notation, let us adopt the simpler adaptive distributed observer where $y_{m0} = v_0$. Also, we assume F_0 is known by every follower. In this case, the adaptive distributed observer takes the form (4.34) and, for convenience, is repeated as follows:

$$\dot{S}_i = \mu_S \sum_{j=0}^{N} a_{ij}(t)(S_j - S_i) \tag{8.21a}$$

$$\dot{v}_i = S_i v_i + \mu_v \left(\sum_{j=0}^{N} a_{ij}(t)(v_j - v_i) \right), \quad i = 1, \ldots, N \tag{8.21b}$$

where μ_S, μ_v are some positive constants, $S_i \in \mathbb{R}^{q\times q}$, and $v_i \in \mathbb{R}^q$.

Let $\tilde{S}_i = S_i - S_0$, $\tilde{v}_i = v_i - v_0$, $\tilde{S} = \text{col}(\tilde{S}_1, \ldots, \tilde{S}_N)$, $\tilde{v} = \text{col}(\tilde{v}_1, \ldots, \tilde{v}_N)$, $\tilde{S}_d = \mathscr{D}(\tilde{S}_1, \ldots, \tilde{S}_N)$, Then, we have the following error system:

$$\dot{\tilde{S}} = -\mu_S \left(H_{\sigma(t)} \otimes I_q\right) \tilde{S} \tag{8.22a}$$

$$\begin{aligned} \dot{\tilde{v}} &= ((I_N \otimes S_0) - \mu_v(H_{\sigma(t)} \otimes I_q))\tilde{v} + \tilde{S}_d\tilde{v} + \tilde{S}_d(1_N \otimes v_0) \\ &\triangleq A_{\sigma(t)}\tilde{v} + \tilde{S}_d\tilde{v} + \tilde{S}_d(1_N \otimes v_0) \end{aligned} \tag{8.22b}$$

where $A_{\sigma(t)} = (I_N \otimes S_0) - \mu_v(H_{\sigma(t)} \otimes I_q)$.

Nevertheless, since the solution of the regulator equations (8.3) associated with every follower system relies on the exact information of the system matrix S_0 of the leader system, we cannot obtain the solution of (8.3) without knowing the system matrix S_0. To overcome this difficulty, we need to develop an algorithm to asymptotically obtain the solution of (8.3) based on a sequence $S_i(t)$, $i = 1, 2, \ldots, N$, which approaches S_0 asymptotically. For this purpose, recall from the proof of Theorem 7.1 that, for $i = 1, \ldots, N$, the regulator equations (8.3) can be put into the following form:

$$\begin{bmatrix} I_{n_i} & 0 \\ 0 & 0 \end{bmatrix} \begin{bmatrix} X_i \\ U_i \end{bmatrix} S_0 - \begin{bmatrix} A_i & B_i \\ C_i & D_i \end{bmatrix} \begin{bmatrix} X_i \\ U_i \end{bmatrix} = \begin{bmatrix} E_i \\ F_i \end{bmatrix}, \tag{8.23}$$

which can be transformed into the following standard form:

$$Q_i \chi_i = b_i \tag{8.24}$$

where

$$Q_i = S_0^{\mathsf{T}} \otimes \begin{bmatrix} I_{n_i} & 0 \\ 0 & 0 \end{bmatrix} - I_q \otimes \begin{bmatrix} A_i & B_i \\ C_i & D_i \end{bmatrix} \tag{8.25}$$

and

$$\chi_i = \text{vec}\left(\begin{bmatrix} X_i \\ U_i \end{bmatrix} \right), \quad b_i = \text{vec}\left(\begin{bmatrix} E_i \\ F_i \end{bmatrix} \right).$$

On the other hand, consider the following equation:

$$\begin{bmatrix} I_{n_i} & 0 \\ 0 & 0 \end{bmatrix} \begin{bmatrix} X_i(t) \\ U_i(t) \end{bmatrix} S_i(t) - \begin{bmatrix} A_i & B_i \\ C_i & D_i \end{bmatrix} \begin{bmatrix} X_i(t) \\ U_i(t) \end{bmatrix} = \begin{bmatrix} E_i \\ F_i \end{bmatrix}, \tag{8.26}$$

which can be transformed into the following standard form:

$$\mathscr{Q}_i(t)\chi_i(t) = b_i \tag{8.27}$$

where

$$\mathscr{Q}_i(t) = S_i^{\mathsf{T}}(t) \otimes \begin{bmatrix} I_{n_i} & 0 \\ 0 & 0 \end{bmatrix} - I_q \otimes \begin{bmatrix} A_i & B_i \\ C_i & D_i \end{bmatrix} \tag{8.28}$$

and

$$\chi_i(t) = \text{vec}\left(\begin{bmatrix} X_i(t) \\ U_i(t) \end{bmatrix}\right), \ b_i = \text{vec}\left(\begin{bmatrix} E_i \\ F_i \end{bmatrix}\right).$$

Then, we will show that, for $i = 1, \ldots, N$, the solution $\chi_i(t)$ of the equation (8.27) will tend to χ_i exponentially as $t \to \infty$.

Before going on, we recall that, given a function $F(t) \in \mathbb{R}^{n\times m}$, if there exist $\alpha, \beta > 0$ such that $\|F(t)\| \leq \beta e^{-\alpha t}$, then we say $F(t) \to 0$ as $t \to \infty$ exponentially at the rate of at least α. Now, we are ready to present the following result.

Lemma 8.3 *Let $A \in \mathbb{R}^{m\times n}$, $b \in \mathbb{R}^m$, and rank$(A) =$ rank$([A, b]) = k$ for some positive integers $m, n, k \geq 1$. Let $\mathscr{A}(t) \in \mathbb{R}^{m\times n}$ be any piecewise continuous time-varying matrix and uniformly bounded over $[t_0, \infty)$ such that $\tilde{A}(t) \triangleq (\mathscr{A}(t) - A) \to 0$ as $t \to \infty$ exponentially at the rate of at least α. Then, for any $x(t_0) \in \mathbb{R}^n$ and $\varepsilon > 0$, the solution $x(t)$ of the following system:*

$$\dot{x} = -\varepsilon\mathscr{A}(t)^{\mathsf{T}}(\mathscr{A}(t)x - b) \tag{8.29}$$

is uniformly bounded over $t \geq t_0$ and is such that, for some $x^* \in \mathbb{R}^n$ satisfying $Ax^* = b$, $\lim_{t\to\infty}(x(t) - x^*) = 0$ exponentially. In particular, there exists a known $\varepsilon^* > 0$ such that, if $\varepsilon > \varepsilon^*$, then $\lim_{t\to\infty}(x(t) - x^*) = 0$ exponentially at the rate of at least α.

Proof Let $V_x = \frac{1}{2}x^{\mathsf{T}}x$. Since $\mathscr{A}(t)$ is uniformly bounded over $[t_0, \infty)$, along (8.29), we have

$$\begin{aligned}
\dot{V}_x &= x^{\mathsf{T}}\left(-\varepsilon\mathscr{A}(t)^{\mathsf{T}}(\mathscr{A}(t)x - b)\right) \\
&= x^{\mathsf{T}}(-\varepsilon\mathscr{A}(t)^{\mathsf{T}}\mathscr{A}(t)x + \varepsilon\mathscr{A}(t)^{\mathsf{T}}b) \\
&= -\varepsilon x^{\mathsf{T}}\mathscr{A}(t)^{\mathsf{T}}\mathscr{A}(t)x + \varepsilon x^{\mathsf{T}}\mathscr{A}(t)^{\mathsf{T}}b \\
&\leq \varepsilon x^{\mathsf{T}}\mathscr{A}(t)^{\mathsf{T}}b \leq \varepsilon\|x\| \cdot \|\mathscr{A}(t)^{\mathsf{T}}b\| \leq a_1\sqrt{V_x}
\end{aligned}$$

for some $a_1 > 0$. Therefore, by the comparison lemma, i.e., Theorem A.3, for any $V_x(0) > 0$, $V_x(t) \leq (a_1t + a_2)^2/4$ for some $a_2 \in \mathbb{R}$, which in turn implies $\|x(t)\| \leq |a_1t + a_2|/\sqrt{2}$. By the singular value decomposition method, there exists an orthogonal matrix $P \in \mathbb{R}^{n\times n}$ such that $AP = \begin{bmatrix}\bar{A} & 0_{m\times(n-k)}\end{bmatrix}$ where $\bar{A} \in \mathbb{R}^{m\times k}$. Then we have

$$P^{\mathsf{T}}A^{\mathsf{T}}AP = \begin{bmatrix} \bar{A}^{\mathsf{T}}\bar{A} & 0_{k\times(n-k)} \\ 0_{(n-k)\times k} & 0_{(n-k)\times(n-k)} \end{bmatrix} \tag{8.30}$$

and

$$P^\mathsf{T} A^\mathsf{T} b = \begin{bmatrix} \bar{A}^\mathsf{T} b \\ 0_{(n-k)\times 1} \end{bmatrix}. \tag{8.31}$$

Since $\bar{A}$ has full column rank, there exists a $\bar{x}_1^* \in \mathbb{R}^k$ such that $\bar{A}\bar{x}_1^* = b$. Let $\bar{x}_2^*$ be any column vector of dimension $(n-k)$, and $x^* = P \begin{bmatrix} \bar{x}_1^* \\ \bar{x}_2^* \end{bmatrix}$. Then

$$Ax^* = APP^\mathsf{T} x^* = \begin{bmatrix} \bar{A} & 0 \end{bmatrix} \begin{bmatrix} \bar{x}_1^* \\ \bar{x}_2^* \end{bmatrix} = b. \tag{8.32}$$

Next, let $\bar{x} = P^\mathsf{T} x$. Then,

$$\begin{aligned}
\dot{\bar{x}} &= P^\mathsf{T}(-\varepsilon \mathscr{A}(t)^\mathsf{T} \mathscr{A}(t)x + \varepsilon \mathscr{A}(t)^\mathsf{T} b) \\
&= -\varepsilon P^\mathsf{T} \mathscr{A}(t)^\mathsf{T} \mathscr{A}(t)x + \varepsilon P^\mathsf{T} \mathscr{A}(t)^\mathsf{T} b \\
&= -\varepsilon P^\mathsf{T} A^\mathsf{T} Ax + \varepsilon P^\mathsf{T} A^\mathsf{T} Ax - \varepsilon P^\mathsf{T} \mathscr{A}(t)^\mathsf{T} \mathscr{A}(t)x \\
&\quad + \varepsilon P^\mathsf{T} \mathscr{A}(t)^\mathsf{T} b - \varepsilon P^\mathsf{T} A^\mathsf{T} b + \varepsilon P^\mathsf{T} A^\mathsf{T} b \\
&= -\varepsilon P^\mathsf{T} A^\mathsf{T} AP\bar{x} + \varepsilon P^\mathsf{T}(A^\mathsf{T} A - \mathscr{A}(t)^\mathsf{T} \mathscr{A}(t))x \\
&\quad + \varepsilon P^\mathsf{T} \tilde{A}(t)^\mathsf{T} b + \varepsilon P^\mathsf{T} A^\mathsf{T} b \\
&= -\varepsilon P^\mathsf{T} A^\mathsf{T} AP\bar{x} + \varepsilon P^\mathsf{T} A^\mathsf{T} b + d(t)
\end{aligned} \tag{8.33}$$

where $d(t) = \varepsilon P^\mathsf{T}(A^\mathsf{T} A - \mathscr{A}(t)^\mathsf{T} \mathscr{A}(t))x + \varepsilon P^\mathsf{T} \tilde{A}(t)^\mathsf{T} b$. Clearly, $\lim_{t\to\infty} d(t) = 0$ exponentially at the rate of at least α. Next, let $\bar{x} = \mathrm{col}(\bar{x}_1, \bar{x}_2)$ and $d(t) = \mathrm{col}(d_1(t), d_2(t))$ with $\bar{x}_1, d_1(t) \in \mathbb{R}^k$ and $\bar{x}_2, d_2(t) \in \mathbb{R}^{n-k}$. Then by (8.30) and (8.31), (8.33) can be rewritten as

$$\dot{\bar{x}}_1 = -\varepsilon \bar{A}^\mathsf{T} \bar{A}\bar{x}_1 + \varepsilon \bar{A}^\mathsf{T} b + d_1(t) \tag{8.34a}$$

$$\dot{\bar{x}}_2 = d_2(t). \tag{8.34b}$$

Clearly, for any $\bar{x}_2(t_0)$, there exists $\bar{x}_2^* \in \mathbb{R}^{n-k}$ such that $\lim_{t\to\infty}(\bar{x}_2(t) - \bar{x}_2^*) = 0$ exponentially at the rate of at least α. Let $\tilde{x}_1 = \bar{x}_1 - \bar{x}_1^*$. Then we have

$$\begin{aligned}
\dot{\tilde{x}}_1 &= -\varepsilon \bar{A}^\mathsf{T} \bar{A}\bar{x}_1 + \varepsilon \bar{A}^\mathsf{T} b + d_1(t) \\
&= -\varepsilon \bar{A}^\mathsf{T} \bar{A}\tilde{x}_1 - \varepsilon \bar{A}^\mathsf{T} \bar{A}\bar{x}_1^* + \varepsilon \bar{A}^\mathsf{T} b + d_1(t) \\
&= -\varepsilon \bar{A}^\mathsf{T} \bar{A}\tilde{x}_1 - \varepsilon \bar{A}^\mathsf{T}(\bar{A}\bar{x}_1^* - b) + d_1(t) \\
&= -\varepsilon \bar{A}^\mathsf{T} \bar{A}\tilde{x}_1 + d_1(t).
\end{aligned}$$

Since $-\bar{A}^\mathsf{T}\bar{A}$ is Hurwitz, and $\lim_{t\to\infty} d_1(t) = 0$ exponentially at the rate of at least α, by Lemma 2.7, for any $\varepsilon > 0$, $\lim_{t\to\infty} \tilde{x}_1(t) = 0$ exponentially. In particular, let $\varepsilon^* =$

$\alpha\underline{\lambda}^{-1}_{(\bar{A}^{\mathsf{T}}\bar{A})}$. Then, by Lemma 2.7 again, with $\varepsilon > \varepsilon^*$, $\lim_{t\to\infty}\tilde{x}_1(t) = 0$ exponentially at the rate of at least α.

Let $x^* = P\begin{bmatrix}\bar{x}_1^* \\ \bar{x}_2^*\end{bmatrix}$. Then $Ax^* = b$ from (8.32) and

$$\begin{aligned}\lim_{t\to\infty}(x(t) - x^*) &= \lim_{t\to\infty}\left(P\bar{x}(t) - P\begin{bmatrix}\bar{x}_1^* \\ \bar{x}_2^*\end{bmatrix}\right) \\ &= \lim_{t\to\infty} P\begin{bmatrix}\bar{x}_1(t) - \bar{x}_1^* \\ \bar{x}_2(t) - \bar{x}_2^*\end{bmatrix} = 0\end{aligned}$$

exponentially at the rate of at least α. □

To introduce the next result, we further introduce some notation. For any column vector $X \in \mathbb{R}^{nq}$ for some positive integers n and q, $M_n^q(X) = [X_1, \ldots, X_q]$, where, for $i = 1, \ldots, q$, $X_i \in \mathbb{R}^n$, and are such that $X = \mathrm{col}(X_1, \ldots, X_q)$.

Then, we have the following result.

Lemma 8.4 *Suppose* $\lim_{t\to\infty}\tilde{S}_i(t) = 0$ *exponentially at the rate of at least* α. *Under Assumption 8.4, for any* $\mu_\zeta > 0$, *for* $i = 1, \ldots, N$, *and any initial condition* $\zeta_i(t_0)$, *the solution* ζ_i *of each of the following equations:*

$$\dot{\zeta}_i = -\mu_\zeta \mathscr{Q}_i^{\mathsf{T}}(t)(\mathscr{Q}_i(t)\zeta_i - b_i) \tag{8.35}$$

is uniformly bounded over $[t_0, \infty)$, *and is such that, for some solution* $\begin{bmatrix}X_i \\ U_i\end{bmatrix}$ *of the regulator equations (8.3),*

$$\lim_{t\to\infty}\left(\Xi_i(t) - \begin{bmatrix}X_i \\ U_i\end{bmatrix}\right) = 0$$

where $\Xi_i = M^q_{(n_i+m_i)}(\zeta_i)$, *exponentially. Moreover, there exists a known* $\mu_\zeta^* > 0$ *such that, if* $\mu_\zeta > \mu_\zeta^*$, *then,*

$$\lim_{t\to\infty}\left(\Xi_i(t) - \begin{bmatrix}X_i \\ U_i\end{bmatrix}\right) = 0$$

exponentially at the rate of at least α.

Proof For $i = 1, \ldots, N$, system (8.35) is in the form (8.29) with $\zeta_i = x$, $\mathscr{Q}_i(t) = \mathscr{A}(t)$, $b_i = b$, $\mu_\zeta = \varepsilon$. Under Assumption 8.4, for $i = 1, \ldots, N$, the linear algebraic equations (8.24) are solvable. Since $\lim_{t\to\infty}\tilde{S}_i(t) = 0$ exponentially at the rate of at least α, by (8.25) and (8.28), $\lim_{t\to\infty}(\mathscr{Q}_i(t) - Q_i) = 0$ exponentially at the rate of at least α. Noting

$$\Xi_i(t) - \begin{bmatrix}X_i \\ U_i\end{bmatrix} = M^q_{(n_i+m_i)}(\zeta_i(t) - \chi_i)$$

and applying Lemma 8.3 to (8.35) for $i = 1, \ldots, N$ completes the proof. □

Similar to (8.13) and (8.14), we can synthesize two types of distributed control laws as follows:

1. Distributed full information control law:

$$u_i = K_{1i}x_i + K_{2i}(t)v_i \tag{8.36a}$$

$$\dot{S}_i = \mu_S \sum_{j=0}^{N} a_{ij}(t)(S_j - S_i) \tag{8.36b}$$

$$\mathcal{Q}_i = S_i^{\mathsf{T}} \otimes \begin{bmatrix} I_{n_i} & 0 \\ 0 & 0 \end{bmatrix} - I_q \otimes \begin{bmatrix} A_i & B_i \\ C_i & D_i \end{bmatrix} \tag{8.36c}$$

$$\dot{\zeta}_i = -\mu_\zeta \mathcal{Q}_i^{\mathsf{T}}(\mathcal{Q}_i\zeta_i - b_i) \tag{8.36d}$$

$$\dot{v}_i = S_i v_i + \mu_v \left(\sum_{j=0}^{N} a_{ij}(t)(v_j - v_i) \right), \quad i = 1, \ldots, N, \tag{8.36e}$$

where $K_{1i} \in \mathbb{R}^{m_i \times n_i}$ is such that $A_i + B_i K_{1i}$ is Hurwitz, $K_{2i}(t) = [-K_{1i} \;\; I_{m_i}]\Xi_i(t)$ with $\Xi_i(t) = M^q_{(n_i+m_i)}(\zeta_i(t))$, μ_S, μ_ζ, μ_v are some positive constants, and $S_i \in \mathbb{R}^{q\times q}$.

2. Distributed measurement output feedback plus feedforward control law:

$$u_i = K_{1i}z_i + K_{2i}(t)v_i \tag{8.37a}$$

$$\dot{z}_i = A_i z_i + B_i u_i + E_i v_i + \ell_i(y_{mi} - C_{mi}z_i - D_{mi}u_i - F_{mi}v_i) \tag{8.37b}$$

$$\dot{S}_i = \mu_S \sum_{j=0}^{N} a_{ij}(t)(S_j - S_i) \tag{8.37c}$$

$$\mathcal{Q}_i(t) = S_i^{\mathsf{T}}(t) \otimes \begin{bmatrix} I_{n_i} & 0 \\ 0 & 0 \end{bmatrix} - I_q \otimes \begin{bmatrix} A_i & B_i \\ C_i & D_i \end{bmatrix} \tag{8.37d}$$

$$\dot{\zeta}_i = -\mu_\zeta \mathcal{Q}_i^{\mathsf{T}}(t)(\mathcal{Q}_i(t)\zeta_i - b_i) \tag{8.37e}$$

$$\dot{v}_i = S_i v_i + \mu_v \left(\sum_{j=0}^{N} a_{ij}(t)(v_j - v_i) \right), \quad i = 1, \ldots, N, \tag{8.37f}$$

where $\ell_i \in \mathbb{R}^{n_i \times p_{mi}}$ is such that $(A_i - \ell_i C_{mi})$ is Hurwitz.

Lemma 8.5 *Consider system (8.22). Suppose* $\lim_{t\to\infty} \tilde{v}(t) = 0$, *and* $\lim_{t\to\infty} \tilde{S}(t) = 0$ *exponentially at the rate of at least* α *with* $\alpha > \bar{\lambda}_{S_0}$. *Then,*

(i) Under Assumptions 8.1, 8.2, and 8.4, there exists some $\mu_\zeta^* > 0$ *such that, for any* $\mu_\zeta > \mu_\zeta^*$, *Problem 8.1 is solvable by the distributed full information control law (8.36);*

(ii) Under the additional Assumption 8.3, for the same μ_ζ as in Part (i), Problem 8.2 is solvable by the distributed measurement output feedback plus feedforward control law (8.37).

Proof We need to show that the closed-loop system under (8.36) or (8.37) satisfies Properies 8.3 and 8.4. By Lemma 8.4, under Assumption 8.4, for any $\mu_\zeta > 0$, the solution of (8.35) is such that, for some solution $\begin{bmatrix} X_i \\ U_i \end{bmatrix}$ of the regulator equations, $\lim_{t\to\infty}\left(\Xi_i(t) - \begin{bmatrix} X_i \\ U_i \end{bmatrix}\right) = 0$ exponentially.

Part (i): Under Assumptions 8.1, 8.2, and 8.4, let $K_{1i} \in \mathbb{R}^{m_i \times n_i}$ be such that $A_i + B_i K_{1i}$ is Hurwitz, and $K_{2i} = U_i - K_{1i} X_i, i = 1, \ldots, N$. By Remark 8.1, the purely decentralized full information control law (8.4) solves Problem 8.1.

By Lemma 8.4 again, there exists some $\mu_\zeta^* > 0$ such that, for any $\mu_\zeta > \mu_\zeta^*$, $\lim_{t\to\infty}\left(\Xi_i(t) - \begin{bmatrix} X_i \\ U_I \end{bmatrix}\right) = 0$ exponentially at the rate of at least α, which implies that $(K_{2i}(t) - K_{2i})$ decays to zero exponentially at the rate of at least α. Since $\alpha > \bar{\lambda}_{S_0}$, $(K_{2i}(t) - K_{2i})v_i(t)$ decays to zero exponentially. Let $k_i(x_i, v_0) = K_{1i}x_i + K_{2i}v_0$. Then, the control law (8.36a) can be put in the following form:

$$\begin{aligned} u_i &= K_{1i}x_i + K_{2i}(t)v_i \\ &= K_{1i}x_i + K_{2i}v_0 + K_{2i}(t)v_i - K_{2i}v_i + K_{2i}v_i - K_{2i}v_0 \\ &= K_{1i}x_i + K_{2i}v_0 + (K_{2i}(t) - K_{2i})v_i + K_{2i}\tilde{v}_i \\ &= k_i(x_i, v_0) + \delta_{ui}(t) \end{aligned} \tag{8.38}$$

where $\delta_{ui}(t) = (K_{2i}(t) - K_{2i})v_i + K_{2i}\tilde{v}_i$. Since $\lim_{t\to\infty}\delta_{ui}(t) = 0$, by Lemma 8.1, Property 8.4 is satisfied, and $x_i(t)$ is uniformly bounded over $[0, \infty)$ when $v_0(t)$ is uniformly bounded. Noting that, by the assumptions of this lemma, for any initial condition, the solution of (8.36b) to (8.36e) is uniformly bounded over $[0, \infty)$ for uniformly bounded $v_0(t)$ completes the proof.

Part (ii): Under the additional Assumption 8.3, the purely decentralized measurement output feedback plus feedforward control law (8.5) solves Problem 8.1. For $i = 1, \ldots, N$, denote the right-hand side of the control law (8.5a) and (8.5b) by $k_i(z_i, v_0)$ and $g_i(z_i, k_i(z_i, v_0), y_{mi}, v_0)$, respectively. Then, the right-hand sides of equations (8.37a) and (8.37b) are

$$k_i(z_i, v_i) = k_i(z_i, \tilde{v}_i + v_0) = k_i(z_i, v_0) + \delta_{ui}(t) \tag{8.39}$$

and, respectively,

$$\begin{aligned} g_i(z_i, k_i(z_i, v_i), y_{mi}, v_i) &= g_i(z_i, k_i(z_i, v_0) + \delta_{ui}(t), y_{mi}, \tilde{v}_i + v_0) \\ &= g_i(z_i, k_i(z_i, v_0), y_{mi}, v_0) + \delta_{zi}(t) \end{aligned} \tag{8.40}$$

where $\delta_{zi}(t) = B_i\delta_{ui}(t) + E_i\tilde{v}_i - \ell_i(F_{mi}\tilde{v}_i + D_{mi}\delta_{ui}(t))$. Like in the proof of Part (i), for any $\mu_\zeta > \mu_\zeta^*$, $\lim_{t\to\infty}\delta_{zi}(t) = 0$. Thus, by Lemma 8.1, Property 8.4 is satisfied, and $x_i(t)$ and $z_i(t)$ are uniformly bounded over $[0, \infty)$ when $v_0(t)$ is uniformly bounded. Noting that, by the assumptions of this lemma, for any initial condition, the solution of (8.37c) to (8.37f) is uniformly bounded over $[0, \infty)$ for uniformly bounded $v_0(t)$ completes the proof. □

Combining Lemma 8.5 with Theorem 4.8 leads to the following result regarding the switching network case.

Theorem 8.3 *Under Assumptions 4.1, 4.5, 8.1, 8.2, and 8.4, Problem 8.2 is solvable by a distributed full information control law of the form (8.36) with any $\mu_S, \mu_v > 0$, and sufficiently large μ_ζ, and, under the additional Assumption 8.3, Problem 8.2 is solvable by a distributed measurement output feedback plus feedforward control law of the form (8.37) with any $\mu_S, \mu_v > 0$, and sufficiently large μ_ζ.*

Combining Lemma 8.5 with Theorem 4.6 leads to the following result regarding the static network case.

Theorem 8.4 *Under Assumptions 4.2, 8.1, 8.2, and 8.4, Problem 8.2 is solvable by a distributed full information control law of the form (8.36) with any $\mu_S, \mu_v > \bar{\lambda}_{S_0}\underline{\lambda}_H^{-1}$, and sufficiently large μ_ζ, and, under the additional Assumption 8.3, Problem 8.1 is solvable by a distributed measurement output feedback plus feedforward control law of the form (8.37) with any $\mu_S, \mu_v > \bar{\lambda}_{S_0}\underline{\lambda}_H^{-1}$, and sufficiently large μ_ζ.*

Remark 8.5 From the proof of Lemma 8.3, μ_ζ can be calculated as follows. Let $P_i \in \mathbb{R}^{q(n_i+m_i)\times q(n_i+m_i)}$ be an orthogonal matrix such that

$$Q_i P_i = \left[\bar{Q}_i \; 0_{q(n_i+p_i)\times(q(n_i+m_i)-k_i)}\right]$$

where $\bar{Q}_i \in \mathbb{R}^{q(n_i+p_i)\times k_i}$ has full column rank. Then $\mu_\zeta > \alpha\underline{\lambda}_{\bar{Q}_i^T\bar{Q}_i}^{-1}$ for $i = 1, \ldots, N$.

8.4 An Application to the Power Tracking Problem of a Grid-Connected Spatially Concentrated Microgrid

Consider the power tracking problem of a grid-connected spatially concentrated microgrid as shown in Fig. 8.6. The microgrid consists of non-dispatchable distributed generators (NDDGs), such as solar and wind power, loads, and a cluster of dispatchable distributed generators (DDGs), such as batteries and flying wheels. To smooth the power injected into the main grid, the DDG cluster receives command signal for power output from higher level control to deal with the intermittent power generation of the NDDGs by absorbing or releasing electrical energy. Let ω and $V_m = \mathrm{col}(V_{md}, V_{mq})$ denote the angular frequency and the bus voltage of the main grid, respectively, which are assumed to be known. Suppose the DDG cluster is made up of N DDGs. In the direct-quadrature frame, the dynamics of the ith DDG can be written as follows:

$$\begin{aligned} \dot{x}_i &= A_{\psi i} x_i + B_{\psi i} u_{i,dq} \\ y_i &= C_i x_i \end{aligned}$$

where

$$A_{\psi i} = \begin{bmatrix} -R_{\psi i}/L_{\psi i} & -\omega & 0 & 0 \\ \omega & -R_{\psi i}/L_{\psi i} & 0 & 0 \\ \omega_c V_{md} & \omega_c V_{mq} & -\omega_c & 0 \\ -\omega_c V_{mq} & \omega_c V_{md} & 0 & -\omega_c \end{bmatrix}, \; B_{\psi i} = \begin{bmatrix} 1/L_{\psi i} & 0 \\ 0 & 1/L_{\psi i} \\ 0 & 0 \\ 0 & 0 \end{bmatrix}$$

$$C_i = \begin{bmatrix} 0 & 0 & 1 & 0 \\ 0 & 0 & 0 & 1 \end{bmatrix}.$$

Here, ω_c denotes the cutoff frequency of the low-pass filter. For the ith DDG, $R_{\psi i}$ and $L_{\psi i}$ denote the resistance and inductance of the L filter as shown in Fig. 8.6, and $y_i = \text{col}(P_i, Q_i)$ where P_i and Q_i denote the active and reactive power outputs, respectively.

The objective is to design a distributed control law to enable the two components of the performance output y_i of each subsystem to asymptotically track the desired active power $P_0(t)$ and the desired reactive power $Q_0(t)$, respectively. To be specific, we assume that the desired active power $P_0(t)$ is a ramp function as follows:

$$P_0(t) = at + b \tag{8.41}$$

where a, b are two constants, and the desired reactive power $Q_0(t)$ is set to be identically equal to zero.

This problem can be formulated as a cooperative output regulation problem. Let the exosystem be as follows:

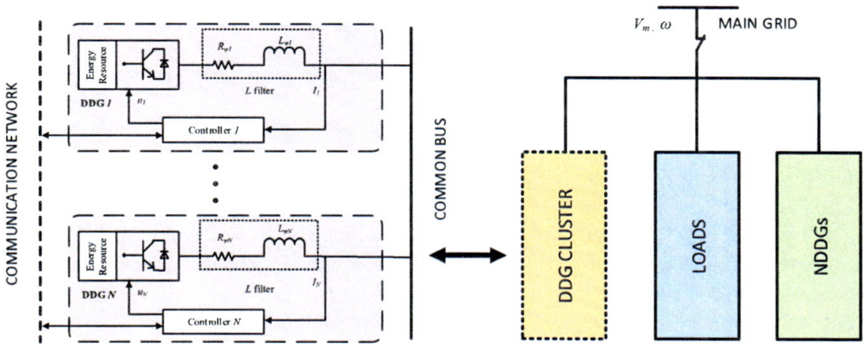

Fig. 8.6 Grid-connected spatially concentrated microgrid

$$
\begin{aligned}
\dot{v}_0 &= \begin{bmatrix} 0 & a \\ 0 & 0 \end{bmatrix} v_0, \;\; v_0(0) = \begin{bmatrix} v_{01}(0) \\ v_{02}(0) \end{bmatrix} \\
y_{m0} &= v_0 \\
y_0 &= F_0 v_0
\end{aligned} \tag{8.42}
$$

where

$$
F_0 = \begin{bmatrix} 1 & 0 \\ 0 & 0 \end{bmatrix}.
$$

Then, simple calculation gives

$$
v_0(t) = \begin{bmatrix} a v_{02}(0)t + v_{01}(0) \\ v_{02}(0) \end{bmatrix}.
$$

Thus, letting

$$
v_0(0) = \begin{bmatrix} b \\ 1 \end{bmatrix}
$$

gives

$$
v_0(t) = \begin{bmatrix} at + b \\ 1 \end{bmatrix}
$$

and hence

$$
y_0(t) = F_0 v_0 = \begin{bmatrix} 1 & 0 \\ 0 & 0 \end{bmatrix} \begin{bmatrix} at + b \\ 1 \end{bmatrix} = \begin{bmatrix} P_0(t) \\ Q_0(t) \end{bmatrix}.
$$

Thus, if we let $e_i = C_i x_i + F_i v_0$ where $F_i = -F_0$. Then, the power tracking control problem described above is formulated as the cooperative output regulation problem of the following multi-agent system:

$$
\begin{aligned}
\dot{x}_i &= A_{\psi i} x_i + B_{\psi i} u_{i,dq} \\
e_i &= C_i x_i + F_i v_0 \\
y_{mi} &= x_i
\end{aligned}
$$

and the leader system (8.42). It can be verified that Assumptions 4.5, 8.1, and 8.2 are satisfied.

Example 8.2 To design a specific controller, let $N = 4$ and other system parameters be given by $\omega = 314\,\text{rad/s}$, $\omega_c = 31.4\,\text{rad/s}$, $V_{md} = 460\,\text{V}$, $V_{mq} = 0\,\text{V}$, $R_i = 0.1\Omega$, $L_i = 1.35\,\text{mH}$. The communication network is given by Fig. 8.7, where the node 0 is associated with the exosystem, and other nodes are associated with the DDGs. The switching communication graph $\bar{\mathcal{G}}_{\sigma(t)}$ switches among $\bar{\mathcal{G}}_i$, $i = 1, 2, 3, 4$ every 0.1 s. Thus, the graph satisfies Assumption 4.1. Also, it can be verified that Assumptions 8.1 and 8.4 are also satisfied with

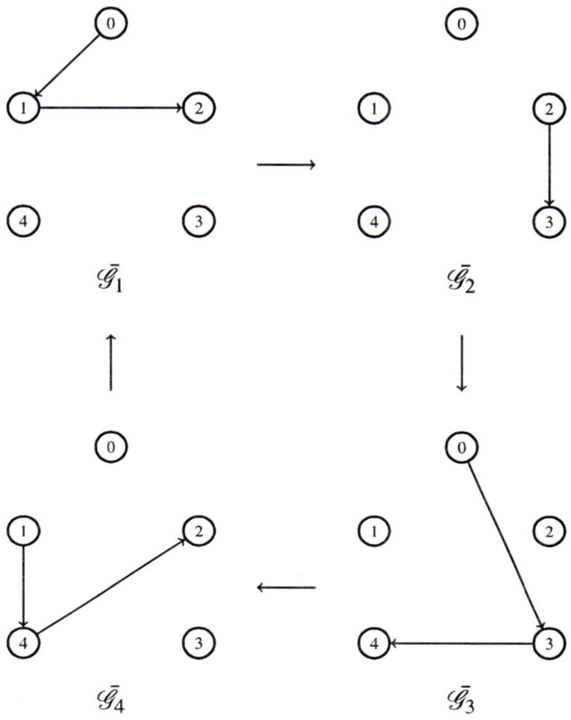

Fig. 8.7 The switching communication graph $\bar{\mathcal{G}}_{\sigma(t)}$

$$X_i = \begin{bmatrix} \frac{V_{md}}{V_{md}^2+V_{mq}^2} & \frac{aV_{md}}{\omega_c(V_{md}^2+V_{mq}^2)} \\ \frac{V_{mq}}{V_{md}^2+V_{mq}^2} & \frac{aV_{mq}}{\omega_c(V_{md}^2+V_{mq}^2)} \\ 1 & 0 \\ 0 & 0 \end{bmatrix} = \begin{bmatrix} \frac{1}{460} & \frac{a}{14444} \\ 0 & 0 \\ 1 & 0 \\ 0 & 0 \end{bmatrix}$$

$$U_i = \begin{bmatrix} \frac{R_{\psi i}V_{md}+L_{\psi i}\omega V_{mq}}{V_{md}^2+V_{mq}^2} & \frac{a(R_{\psi i}+\omega_c L_{\psi i})V_{md}+a\omega L_{\psi i}V_{mq}}{\omega_c(V_{md}^2+V_{mq}^2)} \\ \frac{R_{\psi i}V_{mq}-L_{\psi i}\omega V_{md}}{V_{md}^2+V_{mq}^2} & \frac{a(R_{\psi i}+\omega_c L_{\psi i})V_{mq}-a\omega L_{\psi i}V_{md}}{\omega_c(V_{md}^2+V_{mq}^2)} \end{bmatrix} = \begin{bmatrix} \frac{0.1}{460} & \frac{0.14239a}{14444} \\ -\frac{0.4239}{460} & -\frac{0.4239a}{14444} \end{bmatrix}.$$

To make the simulation more interesting, we allow a takes different values as follows:

$$a = \begin{cases} 50\text{W/s}, & t \in [0\text{s}, 10\text{s}]; \\ 0\text{W/s}, & t \in [10\text{s}, 20\text{s}]; \\ -50\text{W/s}, & t \in [20\text{s}, 30\text{s}]. \end{cases}$$

The control gains for the control law (8.36) are selected to be $\mu_s = 10$, $\mu_v = 10$ and $\mu_\zeta = 20$. Moreover, let

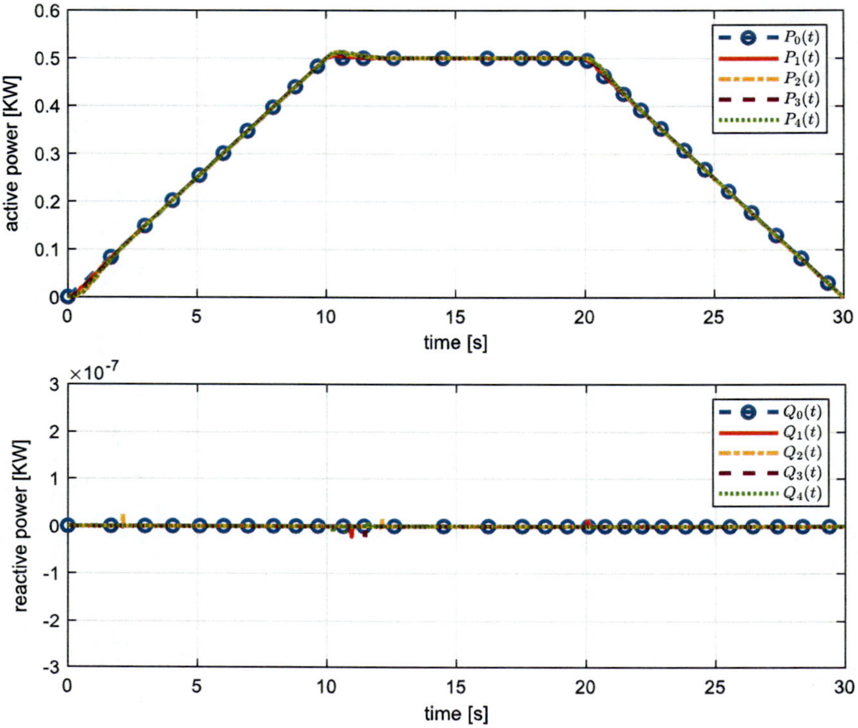

Fig. 8.8 The response of $(P_0(t), Q_0(t))$ and the response of $(P_i(t), Q_i(t))$ under the control law (8.36)

$$K_{1i} = \begin{bmatrix} -0.8026 & 0.4239 & -0.0093 & 0 \\ -0.4239 & -0.2626 & 0 & -0.0011 \end{bmatrix}$$

so that $\lambda(A_i + B_i K_{1i}) = \{-100, -200, -300, -400\}$. Then $K_{2i}(t) = [-K_{1i}\ I_{m_i}]\Xi_i(t)$ with $\Xi_i(t) = M^q_{(n_i+m_i)}(\zeta_i(t))$ and

$$\begin{aligned} K_{2i} &= U_i - K_{1i}X_i \\ &= \begin{bmatrix} \frac{0.1}{460} & \frac{0.14239a}{14444} \\ -\frac{0.4239}{460} & -\frac{0.4239a}{14444} \end{bmatrix} - \begin{bmatrix} -0.8026 & 0.4239 & -0.0093 & 0 \\ -0.4239 & -0.2626 & 0 & -0.0011 \end{bmatrix} \begin{bmatrix} \frac{1}{460} & \frac{a}{14444} \\ 0 & 0 \\ 1 & 0 \\ 0 & 0 \end{bmatrix} \\ &= \begin{bmatrix} 0.01126 & \frac{0.94499a}{14444} \\ 0 & 0 \end{bmatrix}. \end{aligned}$$

The time profiles of the active and reactive power outputs of all the DDGs are shown in Fig. 8.8. The time profiles of the active power tracking errors of all the

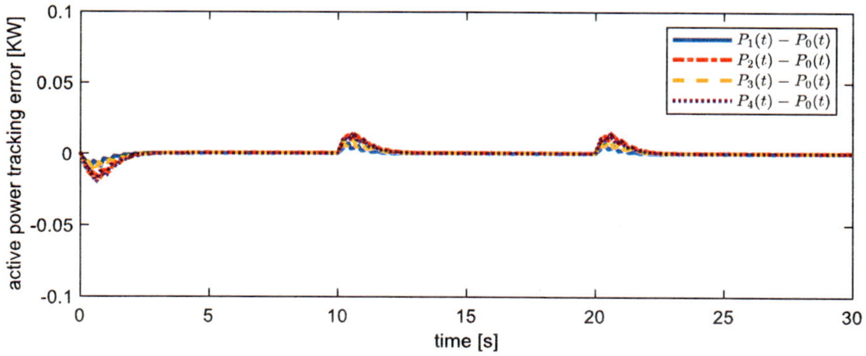

Fig. 8.9 The response of $P_i(t) - P_0(t)$ under the control law (8.36)

DDGs are shown in Fig. 8.9. Since $Q_0(t) = 0$, the time profiles of the reactive power tracking errors of all the DDGs are the same as those of the second figure of Fig. 8.8. It can be seen that successful power tracking has been achieved.

8.5 Notes and References

The linear cooperative output regulation problem was first formally formulated in [1] and solved over static networks using the distributed observer approach, and then solved over jointly connected switching networks in [2]. Reference [3] further studied the same problem as [1] via the dynamic measurement output feedback design by combining the distributed observer and the Luenberger observer. The linear cooperative output regulation problem was studied over static networks using the adaptive distributed observer approach in [4]. A systematic distributed observer design framework for cooperative output regulation problem was presented in [5]. Sections 8.1 and 8.2 are expansions of the results in [1, 2, 5]. Section 8.3 is based on the results in [4].

Extension of the cooperative output regulation problem to other types of linear multi-agent systems can be found, say, in [6–8] for discrete-time linear multi-agent systems, in [9–11] for time-delay multi-agent systems, and in [12, 13] for singular multi-agent systems. The cooperative output regulation problem with uncertain leader systems and switched leader systems were considered in [14, 15], respectively. A variety of implementation issues of the distributed observer were reported, such as reinforcement learning gain design [16], event-triggered cooperative output regulation [17], sampled-data cooperative output regulation [18], non-overshooting design [19], transient synchronization design [20], input saturation design [21], fault-tolerant control design [22].

Reference [23] solved the leader-following formation problem for a group of four mobile robots for the case where the leader signal is a constant vector. The model of Example 8.2 is a simplified version of the model of the DDG from [24] by assuming that the bus voltage is known, and thus can be pre-compensated through the control input. Reference [25] considered the reference power input in polynomial form. The approach proposed in this book can also deal with polynomial reference signals.

It is noted that the leaderless output synchronization problem was studied using the technique of output regulation theory in [26, 27].

References

1. Su Y, Huang J (2012) Cooperative output regulation of linear multi-agent systems. IEEE Trans Autom Control 57(4):1062–1066
2. Su Y, Huang J (2012) Cooperative output regulation with application to multi-agent consensus under switching network. IEEE Trans Syst Man Cybern B Cybern 42(3):864–875
3. Su Y, Huang J (2012) Cooperative output regulation of linear multi-agent systems by output feedback. Syst Control Lett 61(12):1248–1253
4. Cai H, Lewis FL, Hu G, Huang J (2017) The adaptive distributed observer approach to the cooperative output regulation of linear multi-agent systems. Automatica 75:299–305
5. Huang J (2016) Certainty equivalence, separation principle, and cooperative output regulation of multi-agent systems by the distributed observer approach. Control Complex Syst: Theory and Appl 14:421–449
6. Huang J (2017) The cooperative output regulation problem of discrete-time linear multi-agent systems by the adaptive distributed observer. IEEE Trans Autom Control 62(4):1979–1984
7. Liu T, Huang J (2018) Adaptive cooperative output regulation of discrete-time linear multi-agent systems by a distributed feedback control law. IEEE Trans Autom Control 63(12):4383–4390
8. Liu T, Huang J (2021) Discrete-time distributed observers over jointly connected switching networks and an application. IEEE Trans Autom Control 66(4):1918–1924
9. Lu M, Huang J (2016) Cooperative output regulation problem for linear time-delay multi-agent systems under switching network. Neurocomputing 190:132–139
10. Yan Y, Huang J (2016) Cooperative output regulation of discrete-time linear time-delay multi-agent systems. IET Control Theory Appl 10(16):2019–2026
11. Yan Y, Huang J (2017) Cooperative output regulation of discrete-time linear time-delay multi-agent systems under switching network. Neurocomputing 241:108–114
12. Ma Q, Xu S, Lewis FL (2016) Cooperative output regulation of singular heterogeneous multi-agent systems. IEEE Trans Cybernet 46(6):1471–1475
13. Wang S, Huang J (2018) Cooperative output regulation of singular multi-agent systems under switching network by standard reduction. IEEE Trans Circuits Syst I - Regul Papers 65(4):1377–1385
14. Wang S, Huang J (2021) Cooperative output regulation of linear multi-agent systems subject to an uncertain leader system. Int J Control 94(4):952–960
15. Yuan C, Wu F (2018) Cooperative output regulation of multiagent systems with switched leader dynamics. Int J Syst Sci 49(7):1463–1477
16. Yang Y, Modares H, Wunsch DC II, Yin Y (2018) Leader-follower output synchronization of linear heterogeneous systems with active leader using reinforcement learning. IEEE Trans Neural Netw Learn Syst 29(6):2139–2153
17. Hu W, Liu L, Feng G (2019) Event-triggered cooperative output regulation of linear multi-agent systems under jointly connected topologies. IEEE Trans Autom Control 64(3):1317–1322

18. Liu W, Huang J (2021) Sampled-data cooperative output regulation of linear multi-agent systems. Int J Robust Nonlinear Control 31(10):4805–4822
19. Schmid R, Aghbolagh HD (2019) Nonovershooting cooperative output regulation of linear multiagent systems by dynamic output feedback. IEEE Trans Control of Netw Syst 6(2):526–536
20. Seyboth GS, Ren W, Allgöwer F (2016) Cooperative control of linear multi-agent systems via distributed output regulation and transient synchronization. Automatica 68:132–139
21. Shi L, Li Y, Lin Z (2018) Semi-global leader-following output consensus of heterogeneous multi-agent systems with input saturation. Int J Robust Nonlinear Control 28:4916–4930
22. Deng C, Yang G (2019) Distributed adaptive fault-tolerant control approach to cooperative output regulation for linear multi-agent systems. Automatica 103:62–68
23. Ren W, Atkins E (2007) Distributed multi-vehicle coordinated control via local information exchange. Int J Robust Nonlinear Control 17(10–11):1002–1033
24. Cai H, Hu G (2019) Distributed robust hierarchical power sharing control of grid-connected spatially concentrated AC microgrid. IEEE Trans Control Syst Technol 27(3):1012–1022
25. Wang G, Ciobotaru M, Agelidis VG (2014) Power smoothing of large solar PV plant using hybrid energy storage. IEEE Trans Sustain Energy 5(3):834–842
26. Scardovi L, Sepulchre R (2009) Synchronization in networks of identical linear systems. Automatica 45(11):2557–2562
27. Wieland P, Sepulchre R, Allgöwer F (2011) An internal model principle is necessary and sufficient for linear output synchronization. Automatica 47(5):1068–1074

Chapter 9
Cooperative Robust Output Regulation of Linear Multi-agent Systems

The cooperative output regulation problem studied in Chap. 8 is an extension of the output regulation problem for linear systems (7.7) to multi-agent systems (8.1). Various distributed observer-based control laws, which rely on the solution of the regulator equations, can be viewed as extensions of the feedforward control law studied in Sect. 7.2. Thus, like the feedforward control law, the distributed observer-based control laws alone cannot handle the model uncertainty. For this reason, in this chapter, we will further study the cooperative robust output regulation problem for linear uncertain multi-agent systems by distributed internal model approach. This chapter is organized as follows. Section 9.1 formulates the cooperative structurally stable output regulation problem. Section 9.2 presents the solution of the cooperative structurally stable output regulation problem for linear uncertain multi-agent systems by distributed p-copy internal model. Section 9.3 further studies the cooperative robust output regulation problem for a class of linear uncertain minimum phase multi-agent systems by distributed canonical internal model. The materials in Sects. 9.2 and 9.3 are extensions of those in Sects. 7.3 and 7.4, respectively, from a single plant to a multi-agent system.

9.1 Cooperative Structurally Stable Output Regulation

Consider a group of linear systems

$$\dot{x}_i = A_i(w)x_i + B_i(w)u_i + E_i(w)v, \ x_i(0) = x_{i0} \tag{9.1a}$$

$$y_i = C_i(w)x_i \tag{9.1b}$$

$$y_{mi} = y_i \tag{9.1c}$$

H. Cai et al., *Cooperative Control of Multi-agent Systems*, Advances in Industrial Control,
https://doi.org/10.1007/978-3-030-98377-2_9

where, for $i = 1, \ldots, N$, $x_i \in \mathbb{R}^{n_i}$, $u_i \in \mathbb{R}^m$, $y_i \in \mathbb{R}^p$, $y_{mi} \in \mathbb{R}^p$ are the state, the control input, the performance output, and the measurement output, respectively. $v \in \mathbb{R}^q$ is the exogenous signal representing the disturbance and/or the reference input, and $w \in \mathbb{R}^{n_w}$ is the plant uncertain parameter vector. The exogenous signal v is generated by the following exosystem:

$$\dot{v} = Sv, \quad y_0 = -F(w)v \tag{9.2}$$

where $y_0 \in \mathbb{R}^p$ is its reference output. The tracking error for each subsystem of (9.1) is defined by

$$e_i = y_i - y_0 = C_i(w)x_i + F(w)v. \tag{9.3}$$

Here $A_i(w) \in \mathbb{R}^{n_i \times n_i}$, $B_i(w) \in \mathbb{R}^{n_i \times m}$, $C_i(w) \in \mathbb{R}^{p \times n_i}$, $E_i(w) \in \mathbb{R}^{n_i \times q}$, and $F(w) \in \mathbb{R}^{p \times q}$ are all continuous in w. Like in Chap. 7, we assume in this chapter that $w \in \mathbb{W} \subseteq \mathbb{R}^{n_w}$ for some nonempty subset $\mathbb{W}$ containing the origin $w = 0$ which is the nominal value of w.

It can be seen that, for each $i = 1, \ldots, N$, system (9.1) and system (9.3) together are in a special form of (7.31) in that the measurement output y_{mi} is independent of the input u_i and the state v of the exosystem, and is equal to the performance output y_i.

For convenience, let

$$A_i = A_i(0), \quad B_i = B_i(0), \quad C_i = C_i(0), \quad E_i = E_i(0), \quad F = F(0).$$

That is, the matrices A_i, B_i, C_i, E_i, and F represent the nominal parts of the plant and the exosystem.

Then we need the following assumptions:

Assumption 9.1 There exist a positive integer n and matrices A, B, C such that $n_i = n$, $A_i = A$, $B_i = B$, and $C_i = C$.

Assumption 9.2 S has no eigenvalues with negative real parts.

Assumption 9.3 The pair (A, B) is stabilizable.

Assumption 9.4 The pair (C, A) is detectable.

Assumption 9.5

$$\text{rank} \begin{bmatrix} A - \lambda_i(S)I_n & B \\ C & 0 \end{bmatrix} = n + p, \ i = 1, \ldots, q.$$

Remark 9.1 Assumption 9.1 means that all the followers have the same nominal dynamics. Assumptions 9.2–9.5 are standard ones for guaranteeing the solvability of the linear robust output regulation problem for each subsystem of system (9.1) and the exosystem (9.2) by the p-copy internal model control scheme studied in Chap. 7. More specifically, by a straightforward extension of the results in Sect. 7.3, under Assumptions 9.2, 9.3, and 9.5, the cooperative structurally stable output regulation problem of the multi-agent system (9.1) and the exosystem (9.2) can be solved by a dynamic state feedback control law of the following form:

$$\begin{aligned} u_i &= K_{1i}x_i + K_{2i}z_i \\ \dot{z}_i &= \mathcal{G}_{1i}z_i + \mathcal{G}_{2i}e_i, \ i = 1, \dots, N \end{aligned}$$

where $K_{1i}, K_{2i}, \mathcal{G}_{1i}, \mathcal{G}_{2i}$ are constant matrices of appropriate dimensions. Under the additional Assumption 9.4, the cooperative structurally stable output regulation problem of the multi-agent system (9.1) and the exosystem (9.2) can be solved by a dynamic output feedback control law of the following form:

$$\begin{aligned} u_i &= K_i z_i \\ \dot{z}_i &= \mathcal{G}_{1i}z_i + \mathcal{G}_{2i}e_i, \ i = 1, \dots, N. \end{aligned}$$

It is also noted that, as pointed out in Remark 7.2, Assumption 9.2 is made for convenience and will not lose generality.

Nevertheless, as in Chap. 8, the communication among different subsystems of (9.1) is subject to some constraints due to, say, the physical distance among these subsystems. Thus, the reference output y_0 of the exosystem (9.2) may not be available for the control u_i of every subsystem of system (9.1). As a result, the tracking error e_i may not be available for the control u_i of every subsystem of system (9.1). To describe the communication constraints among various subsystems, like in Chap. 8, we view systems (9.1) and (9.2) together as a multi-agent system of $N + 1$ agents with the exosystem (9.2) as the leader and all subsystems of system (9.1) as the followers. The communication topology of this multi-agent system is described in the same way as in Chap. 8. However, we assume in this chapter that the communication graph is static and denoted by $\bar{\mathcal{G}}$. Also, like in Chap. 8, we make the following standard assumption on $\bar{\mathcal{G}}$.

Assumption 9.6 $\bar{\mathcal{G}}$ contains a spanning tree with the node 0 as its root.

To describe our control law, we first introduce the virtual tracking error defined as follows:

$$e_{vi} = \sum_{j=0}^{N} a_{ij}(e_i - e_j) = \sum_{j=0}^{N} a_{ij}(y_i - y_j)$$

where $[a_{ij}]_{i,j=0}^{N}$ is a weighted adjacency matrix of $\bar{\mathcal{G}}$.

In terms of the virtual tracking error, we define two types of control laws as follows:

1. Distributed state feedback control law:

$$u_i = K_{1i} x_{mi} + K_{2i} z_i \tag{9.4a}$$
$$\dot{z}_i = G_{1i} z_i + G_{2i} e_{vi} \tag{9.4b}$$

where $z_i \in \mathbb{R}^{n_z}$ with n_z to be specified later and K_{1i}, K_{2i}, G_{1i}, G_{2i} are some constant matrices to be designed, and $x_{mi} = \sum_{j=1}^{N} a_{ij}(x_i - x_j) + a_{i0} x_i$.
2. Distributed output feedback control law:

$$u_i = K_{1i} \zeta_{mi} + K_{2i} z_i \tag{9.5a}$$
$$\dot{z}_i = G_{1i} z_i + G_{2i} e_{vi} \tag{9.5b}$$
$$\dot{\zeta}_i = F_{1i} e_{vi} + F_{2i} \zeta_i + F'_{2i} \zeta_{mi} + F_{3i} z_i \tag{9.5c}$$

where $z_i \in \mathbb{R}^{n_z}$ with n_z to be specified later and $K_{1i}, K_{2i}, G_{1i}, G_{2i}, F_{1i}, F_{2i}, F'_{2i}, F_{3i}$ are some constant matrices to be designed, and $\zeta_{mi} = \sum_{j=1}^{N} a_{ij}(\zeta_i - \zeta_j) + a_{i0} \zeta_i$.

We are now ready to formulate the linear cooperative structurally stable output regulation problem.

Problem 9.1 (*Linear Cooperative Structurally Stable Output Regulation*) Given the multi-agent system (9.1), the exosystem (9.2) and the corresponding communication graph $\mathcal{G}$, find a distributed feedback control law of the form (9.4) (or (9.5)) such that the closed-loop system has the following properties.

- **Property 9.1** The origin of the closed-loop system is asymptotically stable at $w = 0$ when v is set to zero;
- **Property 9.2** For any initial state of the closed-loop system, and any $v(0)$, there is some open neighborhood $\mathbb{W}$ of the origin of $\mathbb{R}^{n_w}$ such that, for any $w \in \mathbb{W}$, the trajectory of the closed-loop system satisfies $\lim_{t \to \infty} e_i(t) = 0$, $i = 1, \ldots, N$.

As in Sect. 7.3, under Property 9.1, for some open neighborhood $\mathbb{W}$ of the origin of $\mathbb{R}^{n_w}$, the origin of the closed-loop system is asymptotically stable for all $w \in \mathbb{W}$ when v is set to zero. For convenience, in the next section, we use $\mathbb{W}$ to denote a generic open neighborhood of the origin of $\mathbb{R}^{n_w}$ such that the origin of the closed-loop system is asymptotically stable for all $w \in \mathbb{W}$ when v is set to zero.

9.2 Solvability of Cooperative Structurally Stable Output Regulation

We will use a so-called distributed p-copy internal model approach to handle our problem. Let (G_1, G_2) be the minimal p-copy internal model of S as defined in Remark 7.12, and $e_v = \text{col}(e_{v1}, \ldots, e_{vN})$. Also, let $\bar{G}_1 = I_N \otimes G_1$ and $\bar{G}_2 = I_N \otimes G_2$. Then, the pair $(\bar{G}_1, \bar{G}_2)$ is the minimal pN-copy internal model of the matrix S, and the following dynamic compensator:

$$\dot{\zeta} = \bar{G}_1\zeta + \bar{G}_2 e_v$$

is called a distributed minimal pN-copy internal model of the exosystem.

Lemma 9.1 *Under Assumption 9.6, for* $i = 1, \ldots, N$, $\lim_{t\to\infty} e_i(t) = 0$ *if and only if* $\lim_{t\to\infty} e_{vi}(t) = 0$.

Proof Let $e = \mathrm{col}(e_1, \ldots, e_N)$. Then it can be verified that

$$e_v = (H \otimes I_p)e \tag{9.6}$$

where H is as defined in (2.5). By Lemma 2.1, under Assumption 9.6, H is nonsingular. Therefore, $\lim_{t\to\infty} e_i(t) = 0$ if and only if $\lim_{t\to\infty} e_{vi}(t) = 0$. □

Let $x = \mathrm{col}(x_1, \ldots, x_N)$. Then $e = \mathscr{D}(C_1(w), \ldots, C_N(w))x + 1_N \otimes F(w)$. From (2.5) and noting $\mathscr{L}1_N = 0$, we have

$$H1_N = \Delta 1_N. \tag{9.7}$$

Thus, we have

$$\begin{aligned} e_v &= (H \otimes I_p)e \\ &= (H \otimes I_p)(\mathscr{D}(C_1(w), \ldots, C_N(w)x + (1_N \otimes F(w))v) \\ &= (H \otimes I_p)(\mathscr{D}(C_1(w), \ldots, C_N(w))x + ((H1_N) \otimes F(w))v \\ &= (H \otimes I_p)(\mathscr{D}(C_1(w), \ldots, C_N(w))x + ((\Delta 1_N) \otimes F(w))v \\ &= \bar{C}(w)x + \bar{F}(w)v \end{aligned} \tag{9.8}$$

where $\bar{C}(w) = (H \otimes I_p)\mathscr{D}(C_1(w), \ldots, C_N(w))$ and $\bar{F}(w) = (\Delta 1_N) \otimes F(w)$.

Let $u = \mathrm{col}(u_1, \ldots, u_N)$, and

$$\begin{aligned} \bar{A}(w) &= \mathscr{D}(A_1(w), \ldots, A_N(w)), \ \bar{B}(w) = \mathscr{D}(B_1(w), \ldots, B_N(w)) \\ \bar{E}(w) &= \mathrm{col}(E_1(w), \ldots, E_N(w)). \end{aligned}$$

Further, let

$$\bar{A} = \bar{A}(0), \quad \bar{B} = \bar{B}(0), \quad \bar{C} = \bar{C}(0), \quad \bar{E} = \bar{E}(0), \quad \bar{F} = \bar{F}(0).$$

Then we can define an auxiliary nominal augmented system as follows:

$$\dot{x} = \bar{A}x + \bar{B}u + \bar{E}v \tag{9.9a}$$

$$\dot{z} = \bar{G}_1 z + \bar{G}_2 e_v \tag{9.9b}$$

$$e_v = \bar{C}x + \bar{F}v. \tag{9.9c}$$

The auxiliary nominal augmented system (9.9) has the following property:

Lemma 9.2 *Under Assumptions 9.1, 9.2, 9.3, and 9.5, let the pair (G_1, G_2) be the minimum p-copy internal model of the matrix S. Then the pair*

$$\left(\begin{bmatrix} \bar{A} & 0 \\ \bar{G}_2\bar{C} & \bar{G}_1 \end{bmatrix}, \begin{bmatrix} \bar{B} \\ 0 \end{bmatrix}\right) \tag{9.10}$$

is stabilizable if and only if Assumption 9.6 holds.

Proof Since $\bar{A} = I_N \otimes A$ and $\bar{B} = I_N \otimes B$, Assumption 9.3 implies that the pair $(\bar{A}, \bar{B})$ is stabilizable. It can also be seen that the pair $(\bar{G}_1, \bar{G}_2)$ is the minimal pN-copy internal model of the matrix S. Thus Assumption 9.5 implies that

$$\text{rank} \begin{bmatrix} I_N \otimes A - \lambda I_{nN} & I_N \otimes B \\ I_N \otimes C & 0 \end{bmatrix} = (n+p)N, \quad \text{for all } \lambda \in \lambda(\bar{G}_1) = \lambda(G_1). \tag{9.11}$$

Since

$$\begin{bmatrix} \bar{A} - \lambda I_{nN} & \bar{B} \\ \bar{C} & 0 \end{bmatrix} = \begin{bmatrix} I_{nN} & 0 \\ 0 & H \otimes I_p \end{bmatrix} \cdot \begin{bmatrix} I_N \otimes A - \lambda I_{nN} & I_N \otimes B \\ I_N \otimes C & 0 \end{bmatrix},$$

it follows from (9.11) that

$$\text{rank} \begin{bmatrix} \bar{A} - \lambda I_{nN} & \bar{B} \\ \bar{C} & 0 \end{bmatrix} = (n+p)N, \quad \text{for all } \lambda \in \lambda(\bar{G}_1) \tag{9.12}$$

if and only if H is nonsingular, or if and only if the communication graph $\bar{\mathcal{G}}$ satisfies Assumption 9.6. By Lemma 7.6, the pair (9.10) is stabilizable if Assumption 9.6 holds. On the other hand, if H is singular, (9.12) does not hold. By Lemma 7.6 and Remark 7.13, under Assumption 9.2, the pair (9.10) is not stabilizable if Assumption 9.6 is not satisfied. □

According to Lemma 9.2, under Assumptions 9.1–9.5, the auxiliary augmented nominal system (9.9) is stabilizable and detectable with the state col(x, z), input u, and virtual tracking error e_v. From Lemma 7.5, if any static state feedback control law of the following form:

$$u = \bar{K}_1 x + \bar{K}_2 z \tag{9.13}$$

or dynamic output feedback control law of the following form:

$$u = \bar{K}_1 \zeta + \bar{K}_2 z \tag{9.14a}$$
$$\dot{\zeta} = \bar{F}_1 e_v + \bar{F}_2 \zeta + \bar{F}_3 z \tag{9.14b}$$

stabilizes the auxiliary augmented system (9.9), then the dynamic state feedback control law of the following form:

$$
\begin{aligned}
u &= \bar{K}_1 x + \bar{K}_2 z \\
\dot{z} &= \bar{G}_1 z + \bar{G}_2 e_v
\end{aligned}
$$

or the dynamic output feedback control law of the following form:

$$
\begin{aligned}
u &= \bar{K}_1 \zeta + \bar{K}_2 z \\
\dot{\zeta} &= \bar{F}_1 e_v + \bar{F}_2 \zeta + \bar{F}_3 z \\
\dot{z} &= \bar{G}_1 z + \bar{G}_2 e_v
\end{aligned}
$$

solves the structurally stable output regulation problem of the following composite plant:

$$\dot{x} = \bar{A}(w)x + \bar{B}(w)u + \bar{E}(w)v \tag{9.15a}$$

$$\dot{v} = Sv \tag{9.15b}$$

$$e_v = \bar{C}(w)x + \bar{F}(w)v \tag{9.15c}$$

viewing e_v as the regulated output of (9.15).

The following lemma shows that the following distributed state feedback control law:

$$u_i = K_1 \left(\sum_{j=1}^{N} a_{ij}(x_i - x_j) + a_{i0}x_i \right) + K_2 z_i$$

does stabilize system (9.9). It is noted that the compact form of the above control law is in the form of (9.13) with $\bar{K}_1 = H \otimes K_1$ and $\bar{K}_2 = I_N \otimes K_2$.

Lemma 9.3 *Let (G_1, G_2) with $G_1 \in \mathbb{R}^{pl \times pl}$ and $G_2 \in \mathbb{R}^{pl}$ be the minimal p-copy internal model of S. Then, under Assumptions 9.1–9.3, and 9.5, there exist matrices K_1 and K_2 such that*

$$A_c \triangleq \begin{bmatrix} I_N \otimes A + H \otimes (BK_1) & I_N \otimes (BK_2) \\ H \otimes (G_2 C) & I_N \otimes G_1 \end{bmatrix} \tag{9.16}$$

is Hurwitz if and only if Assumption 9.6 is satisfied.

Proof (If part:) Denote the eigenvalues of H by λ_i, $i = 1, \ldots, N$. Let T_1 be the nonsingular matrix such that $J_H = T_1 H T_1^{-1}$ is the Jordan form of H, and is an upper triangular matrix. Let $T_2 = \begin{bmatrix} T_1 \otimes I_n & 0 \\ 0 & T_1 \otimes I_{pl} \end{bmatrix}$. Then

$$\bar{A}_c \triangleq T_2 A_c T_2^{-1} = \begin{bmatrix} I_N \otimes A + J_H \otimes (BK_1) & I_N \otimes (BK_2) \\ J_H \otimes (G_2 C) & I_N \otimes G_1 \end{bmatrix}.$$

Partition $I_{N(n+pl)}$ as $I_{N(n+pl)} = \mathrm{col}(R_1, \ldots, R_N, W_1, \ldots, W_N)$ where $R_i \in \mathbb{R}^{n \times (N(n+pl))}$ and $W_i \in \mathbb{R}^{(pl) \times (N(n+pl))}$ for $i = 1, \ldots, N$. Let $T_3 = \mathrm{col}(R_1, W_1, \ldots, R_N, W_N)$, which is nonsingular. Then $\hat{A}_c \triangleq T_3 \bar{A}_c T_3^{-1}$ is an upper block triangular matrix whose diagonal blocks are

$$\hat{A}_{ci} \triangleq \begin{bmatrix} A + \lambda_i BK_1 & BK_2 \\ \lambda_i G_2 C & G_1 \end{bmatrix}, \quad i = 1, \ldots, N. \tag{9.17}$$

Under Assumption 9.6, by Lemma 2.1, for all $i = 1, \ldots, N$, λ_i has positive real part. Let $T_{4i} = \begin{bmatrix} I_n & 0 \\ 0 & \lambda_i^{-1} I_{pl} \end{bmatrix}$ and $\tilde{A}_{ci} \triangleq \begin{bmatrix} A + \lambda_i BK_1 & \lambda_i BK_2 \\ G_2 C & G_1 \end{bmatrix}$, $i = 1, \ldots, N$. It can be verified that $T_{4i} \hat{A}_{ci} T_{4i}^{-1} = \tilde{A}_{ci}$. Therefore, A_c is Hurwitz if and only if, for all $i = 1, \ldots, N$, $\hat{A}_{ci}$ and hence $\tilde{A}_{ci}$ are.

Let $Y \triangleq \begin{bmatrix} A & 0 \\ G_2 C & G_1 \end{bmatrix}$, $J \triangleq \begin{bmatrix} B \\ 0 \end{bmatrix}$. Under Assumptions 9.1, 9.3, and 9.5, by Lemma 7.6, (Y, J) is stabilizable. Therefore, by Lemma 2.11, the Riccati equation

$$Y^{\mathsf{T}} P + PY - PJJ^{\mathsf{T}} P + I = 0 \tag{9.18}$$

admits a positive definite solution P. Let

$$[K_1, K_2] \triangleq K = -\nu_1 J^{\mathsf{T}} P \tag{9.19}$$

where $\nu_1 \in \mathbb{R}$ satisfies

$$\nu_1 \mathrm{Re}(\lambda_i) \geq \frac{1}{2}, \quad i = 1, \ldots, N. \tag{9.20}$$

Therefore, $\tilde{A}_{ci} = Y + \lambda_i JK = Y - \lambda_i \nu_1 JJ^{\mathsf{T}} P$. Since $\mathrm{Re}(\lambda_i \nu_1) = \nu_1 \mathrm{Re}(\lambda_i) \geq \frac{1}{2}$, it follows from Lemma 2.12 that, for $i = 1, \ldots, N$, $\tilde{A}_{ci}$ is Hurwitz. Thus, A_c is Hurwitz.

(Only if part:) Suppose the communication graph $\mathscr{G}$ does not satisfy Assumption 9.6. Then by Lemma 2.1, H has at least one eigenvalue at the origin. Without loss of generality, we assume that $\lambda_1 = 0$. Then by (9.17), $\hat{A}_{c1} = \begin{bmatrix} A & BK_2 \\ 0 & G_1 \end{bmatrix}$. Since the spectrum of G_1 is the same as the spectrum of S, under Assumption 9.2, $\hat{A}_{c1}$ and hence A_c cannot be Hurwitz regardless of the choice of $[K_1, K_2]$. The proof is thus completed. □

Notice that, under the state feedback control law (9.4) with $K_{1i} = K_1$ and $K_{2i} = K_2$ given in (9.19), the nominal closed-loop system matrix A_c is given by (9.16). Hence, from Lemmas 9.1 and 9.3, the following main theorem holds.

Theorem 9.1 *Under Assumptions 9.1–9.3, and 9.5, there exist matrices $K_{1i} = K_1$, $K_{2i} = K_2$, $G_{1i} = G_1$, and $G_{2i} = G_2$ such that the cooperative structurally stable output regulation problem is solvable by the distributed state feedback control law (9.4) with (G_1, G_2) being the minimal p-copy internal model of S and (K_1, K_2) given in (9.19) if and only if Assumption 9.6 is satisfied.*

Next, we show that a special form of the distributed output feedback control law (9.14) with $\bar{K}_1 = H \otimes K_1$, $\bar{K}_2 = I_N \otimes K_2$, $\bar{F}_1 = I_N \otimes L$, $\bar{F}_2 = I_N \otimes A + H \otimes (BK_1 - LC)$, and $\bar{F}_3 = I_N \otimes BK_2$, i.e.,

$$u_i = K_1 \left(\sum_{j=1}^{N} a_{ij}(\zeta_i - \zeta_j) + a_{i0}\zeta_i \right) + K_2 z_i \tag{9.21a}$$

$$\dot{\zeta}_i = A\zeta_i + Bu_i - LC \left(\sum_{j=1}^{N} a_{ij}(\zeta_i - \zeta_j) + a_{i0}\zeta_i \right) + Le_{vi} \tag{9.21b}$$

also stabilizes system (9.9).

Lemma 9.4 *Let (G_1, G_2) be the minimal p-copy internal model of S. Then, under Assumptions 9.1–9.5, there exist matrices (K_1, K_2, L) such that*

$$A_c = \begin{bmatrix} I_N \otimes A & H \otimes (BK_1) & I_N \otimes (BK_2) \\ H \otimes (LC) & I_N \otimes A + H \otimes (BK_1 - LC) & I_N \otimes (BK_2) \\ H \otimes (G_2 C) & 0 & I_N \otimes G_1 \end{bmatrix} \tag{9.22}$$

is Hurwitz if and only if Assumption 9.6 is satisfied.

Proof Let

$$T = \begin{bmatrix} I_{Nn} & 0 & 0 \\ 0 & 0 & I_{Npl} \\ -I_{Nn} & I_{Nn} & 0 \end{bmatrix}.$$

Then

$$\begin{aligned} \bar{A}_c &\triangleq TA_cT^{-1} \\ &= \left[\begin{array}{cc|c} I_N \otimes A + H \otimes (BK_1) & I_N \otimes (BK_2) & H \otimes (BK_1) \\ H \otimes (G_2 C) & I_N \otimes G_1 & 0 \\ \hline 0 & 0 & I_N \otimes A - H \otimes (LC) \end{array} \right]. \end{aligned}$$

Let (K_1, K_2) be defined in (9.19). By Lemma 9.3, the first diagonal block matrix $\bar{A}_c$ can be made Hurwitz by this choice of $[K_1, K_2]$ if and only if the communication graph $\bar{\mathcal{G}}$ satisfies Assumption 9.6. Thus, the only if part has been proved.

To complete the proof of if part, it suffices to show that there exists L such that $I_N \otimes A - H \otimes (LC)$ or equivalently $I_N \otimes A^\top - H^\top \otimes (C^\top L^\top)$ is Hurwitz. Since (C, A) is detectable, $(A^\top, C^\top)$ is stabilizable. Therefore, by Theorem 3.2 with $B = C^\top$ and

$$L^\top = \nu_2 C \tilde{P} \tag{9.23}$$

where $\tilde{P}$ is the positive definite matrix satisfying the following Riccati equation:

$$A\tilde{P} + \tilde{P}A^\top - \tilde{P}C^\top C\tilde{P} + I_n = 0 \tag{9.24}$$

and $\nu_2 \in \mathbb{R}$ satisfies

$$\nu_2 \geq \frac{1}{2}\lambda_H^{-1} \tag{9.25}$$

the matrix $I_N \otimes A^\top - H^\top \otimes (C^\top L^\top)$ and hence the matrix $I_N \otimes A - H \otimes (LC)$ are both Hurwitz. The proof is thus completed. □

Let (K_1, K_2) and L be given in (9.19) and (9.23), respectively. Then, under the output feedback control law (9.5) with $K_{1i} = K_1$, $K_{2i} = K_2$, $F_{1i} = L$, $F_{2i} = A$, $F'_{2i} = BK_1 - LC$, $F_{3i} = BK_2$, $G_{1i} = G_1$, and $G_{2i} = G_2$, the nominal closed-loop system matrix A_c is given by (9.22). Hence, from Lemmas 9.1 and 9.4, the following main theorem holds.

Theorem 9.2 *Under Assumptions 9.1–9.5, there exist matrices K_1, K_2, L such that the cooperative structurally stable output regulation problem is solvable by the distributed output feedback control law (9.5) with (G_1, G_2) being the minimal p-copy internal model of S and (K_1, K_2) and L being given in (9.19) and (9.23), respectively, if and only if Assumption 9.6 is satisfied.*

Remark 9.2 The second equation in (9.21) can be viewed as the observer of the state x_i using the relative output information e_{vi}. It is possible to simplify this observer to the following form:

$$\dot{\zeta}_i = A\zeta_i + Bu_i + L(y_i - Cx_i)$$

which relies on the measurement output y_i of the ith subsystem itself. The proof can be obtained similarly to that of Lemma 9.4. Nevertheless, in practice, it is easier to obtain the relative output information e_{vi} than the output y_i.

Example 9.1 Consider a group of four robots modeled as double integrators with parameter uncertainties and external disturbance:

$$\dot{x}_{1i} = x_{2i} \tag{9.26a}$$
$$\dot{x}_{2i} = w_{1i}x_{1i} + w_{2i}x_{2i} + u_i + (1 + w_{3i})v_3 \tag{9.26b}$$
$$y_i = x_{1i} \tag{9.26c}$$
$$e_i = x_{1i} - (v_1 + v_2),\ i = 1, 2, 3, 4, \tag{9.26d}$$

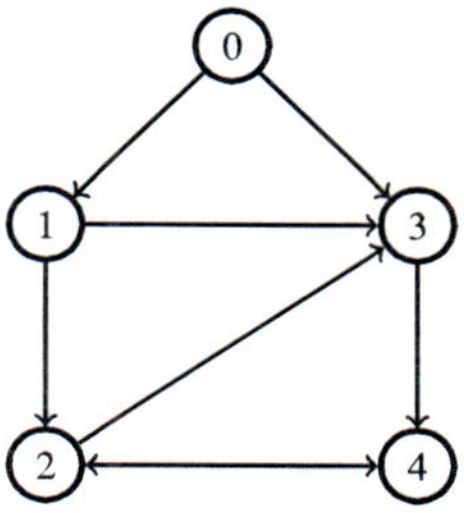

Fig. 9.1 The static communication graph $\bar{\mathscr{G}}$

where the exogenous signal $v = \text{col}(v_1, v_2, v_3)$ is generated by the exosystem of the form (9.2) with $S = \begin{bmatrix} 0 & 1 & 0 \\ -1 & 0 & 0 \\ 0 & 0 & 0 \end{bmatrix}$. System (9.26) is in the form of (9.1) with $A = \begin{bmatrix} 0 & 1 \\ 0 & 0 \end{bmatrix}$, $B = \begin{bmatrix} 0 \\ 1 \end{bmatrix}$, $E_i = \begin{bmatrix} 0 & 0 & 0 \\ 0 & 0 & 1 \end{bmatrix}$, $C = \begin{bmatrix} 1 & 0 \end{bmatrix}$, $F = \begin{bmatrix} -1 & -1 & 0 \end{bmatrix}$. The information exchange among all robots is described by the static communication graph $\bar{\mathscr{G}}$ as shown in Fig. 9.1. By inspection,

$$\mathscr{A} = \begin{bmatrix} 0 & 0 & 0 & 0 \\ 1 & 0 & 0 & 1 \\ 1 & 1 & 0 & 0 \\ 0 & 1 & 1 & 0 \end{bmatrix}, \quad \Delta = \begin{bmatrix} 1 & 0 & 0 & 0 \\ 0 & 0 & 0 & 0 \\ 0 & 0 & 1 & 0 \\ 0 & 0 & 0 & 0 \end{bmatrix}.$$

Then

$$H = \begin{bmatrix} 1 & 0 & 0 & 0 \\ -1 & 2 & 0 & -1 \\ -1 & -1 & 3 & 0 \\ 0 & -1 & -1 & 2 \end{bmatrix}$$

whose eigenvalues are $\{0.7944, 1.0000, 3.1028 + 0.6655\imath, 3.1028 - 0.6655\imath\}$.

It can be verified that Assumptions 9.1–9.5 are all satisfied. Thus, it is possible to solve the problem using our distributed control laws. The design process is given as follows. Let

$$G_1 = \begin{bmatrix} 0 & 1 & 0 \\ 0 & 0 & 1 \\ 0 & -1 & 0 \end{bmatrix}, \quad G_2 = \begin{bmatrix} 0 \\ 0 \\ 1 \end{bmatrix}.$$

By solving the Riccati equation (9.18), we obtain

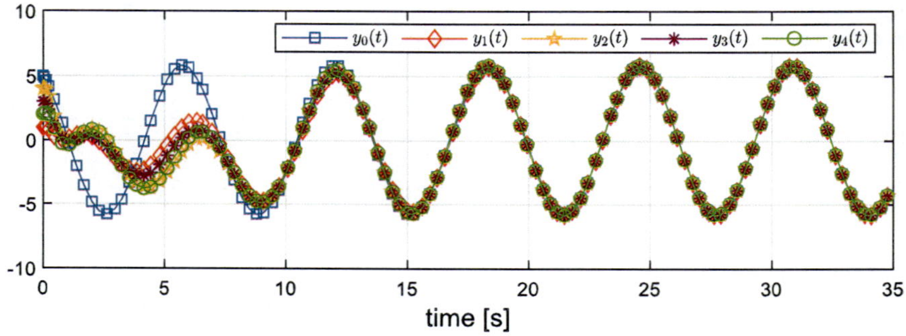

Fig. 9.2 The response of $y_0(t)$ and the response of $y_i(t)$ under the control law (9.4)

$$P = \begin{bmatrix} 10.6869 & 4.2948 & 3.0967 & -0.3431 & 8.7226 \\ 4.2948 & 3.0967 & 1.0000 & -0.6316 & 2.6128 \\ 3.0967 & 1.0000 & 3.6632 & 2.6128 & 4.2948 \\ -0.3431 & -0.6316 & 2.6128 & 5.6194 & 2.9133 \\ 8.7226 & 2.6128 & 4.2948 & 2.9133 & 11.5645 \end{bmatrix}.$$

Let $\nu_1 = 3$ which satisfies (9.20). Then by (9.19), we have

$$[K_1, K_2] = -3 \times \begin{bmatrix} 4.2948 & 3.0967 & 1.0000 & -0.6316 & 2.6128 \end{bmatrix}.$$

By solving the Riccati equation (9.24), we obtain $\tilde{P} = \begin{bmatrix} 1.7321 & 1.0000 \\ 1.0000 & 1.7321 \end{bmatrix}$. Let $\nu_2 = 4$ which satisfies (9.25). Then by (9.23), we have $L = 4 \times \begin{bmatrix} 1.7321 \\ 1.0000 \end{bmatrix}$. The performance of the controllers (9.4) and (9.5) is evaluated by computer simulation with the initial conditions being $x_{11}(0) = 1$, $x_{21}(0) = 4$, $x_{12}(0) = 4$, $x_{22}(0) = 3$, $x_{13}(0) = 3$, $x_{23}(0) = 2$, $x_{14}(0) = 2$, $x_{24}(0) = 1$, $z_1(0) = \text{col}(0, 2, 4)$, $z_2(0) = \text{col}(1, 4, 3)$, $z_3(0) = \text{col}(2, 1, 4)$, $z_4(0) = \text{col}(3, 2, 1)$, $v(0) = \text{col}(4, 1, 1)$, and $\zeta_1(0) = \text{col}(1, 4)$, $\zeta_2(0) = \text{col}(4, 3)$, $\zeta_3(0) = \text{col}(3, 2)$, $\zeta_4(0) = \text{col}(1, 2)$, and the actual values of the uncertain parameters being $w_{ji} = 0.1 \times j \times i$, $j = 1, 2$, $i = 1, 2, 3, 4$, and $w_{3i} = i - 1$, $i = 1, 2, 3, 4$. Figures 9.2, 9.3, 9.4, and 9.5 show the tracking performance. It can be seen that all the regulated outputs of the subsystems converge to the origin asymptotically.

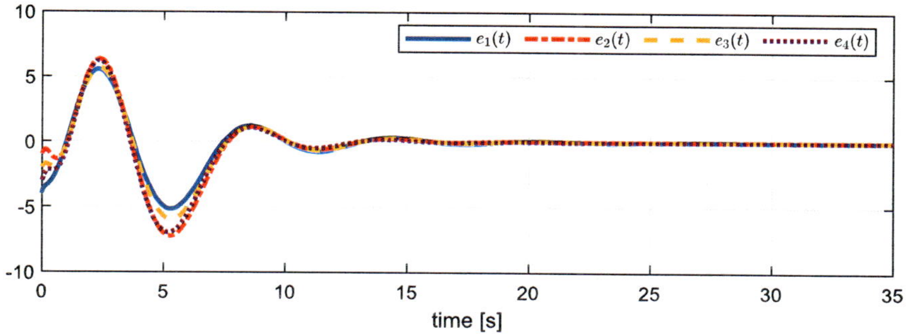

Fig. 9.3 The response of $e_i(t)$ under the control law (9.4)

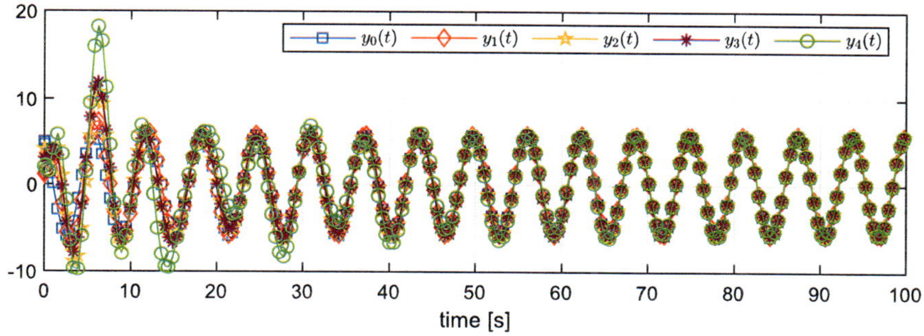

Fig. 9.4 The response of $y_0(t)$ and the response of $y_i(t)$ under the control law (9.5)

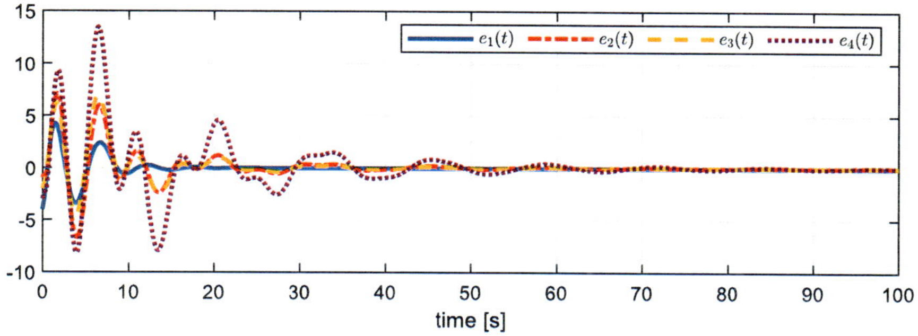

Fig. 9.5 The response of $e_i(t)$ under the control law (9.5)

9.3 Cooperative Robust Output Regulation by Distributed Canonical Internal Model Approach

As shown in the last section, the cooperative structurally stable output regulation problem can only handle sufficiently small variations of the unknown parameter vector w. In this section, we will further consider the cooperative robust output regulation problem of the uncertain multi-agent systems (9.1) for the case where $w \in \mathbb{W} \subseteq \mathbb{R}^{n_w}$ for some compact subset $\mathbb{W}$. For this purpose, we need to extend the canonical internal model approach introduced in Sect. 7.4 to its distributed version.

We consider the following distributed output feedback control law:

$$u_i = K_{1i}\zeta_i + K_{2i}z_i \tag{9.27a}$$

$$\dot{\eta}_i = G_{1i}\eta_i + G_{2i}e_{vi} \tag{9.27b}$$

$$\dot{\zeta}_i = F_{1i}e_{vi} + F_{2i}\zeta_i + F_{3i}\eta_i. \tag{9.27c}$$

Our problem formulation is as follows:

Problem 9.2 (*Linear Cooperative Robust Output Regulation*) Given the multi-agent system (9.1), the exosystem (9.2), a static communication graph $\bar{\mathcal{G}}$, and a compact subset $\mathbb{W} \subseteq \mathbb{R}^{n_w}$, find the distributed feedback control law of the form (9.27) such that, for any $w \in \mathbb{W}$, the closed-loop system has the following two properties:

- **Property 9.3** The origin of the closed-loop system is asymptotically stable when v is set to zero;
- **Property 9.4** For $i = 1, \ldots, N$, $\lim_{t\to\infty} e_i(t) = 0$.

As mentioned in Sect. 7.4, when $\mathbb{W}$ is an arbitrarily prescribed compact subset of $\mathbb{R}^{n_w}$, even for a single system of the general form (9.1), the solvability of the robust output regulation problem is still intractable. For this reason, in what follows, we will only focus on a special class of minimum phase linear systems satisfying the following assumption.

Assumption 9.7 For $i = 1, \ldots, N$,

(i) each subsystem of the system (9.1) is single-input single-output;
(ii) for all $w \in \mathbb{W}$, $C_i(w)A_i(w)^k B_i(w) = 0$, $k = 0, 1, \ldots, r-2$, and $b_i(w) \triangleq C_i(w)A_i(w)^{r-1}B_i(w) \neq 0$;
(iii) for $k = 0, 1, \ldots, r-2$, $C_i(w)A_i(w)^k E_i(w) = 0$ for all $w \in \mathbb{W}$;
(iv) for any $w \in \mathbb{W}$, each subsystem of (9.1) is minimum phase with u_i as input and y_i as output.

Remark 9.3 As pointed out in Remark 7.18, Parts (i) to (iii) of Assumption 9.7 mean that, for all $w \in \mathbb{W}$, the relative degree of the ith subsystem of system (9.1) with input u_i and output y_i is equal to r, which is not greater than the relative degree of the same subsystem with input v and output y_i. Since $\mathbb{W}$ is compact, by the continuity of $b_i(w)$, it has the same sign for all $w \in \mathbb{W}$. Without loss of generality, we assume that, for $i = 1, \ldots, N$, for all $w \in \mathbb{W}$, $b_i(w) > 0$.

From Lemma 7.8, under Parts (i)–(iii) of Assumption 9.7, there exists a matrix $T_i(w)$ such that under the linear transformation

$$T_i(w)x_i \triangleq \mathrm{col}(\chi_{0i}, \chi_{1i}, \ldots, \chi_{ri})$$

where $z_i \in \mathbb{R}^{n_i-r}$, $\chi_{ki} \in \mathbb{R}$, $k = 1, \ldots, r$, system (9.1) together with (9.3) is equivalent to

$$\dot{\chi}_{0i} = A_{1i}(w)\chi_{0i} + A_{2i}(w)\chi_{1i} + E_{0i}(w)v \tag{9.28a}$$
$$\dot{\chi}_{ki} = \chi_{(k+1)i}, \ k = 1, \ldots, r-1 \tag{9.28b}$$
$$\dot{\chi}_{ri} = A_{3i}(w)z_i + \sum_{k=1}^{r} c_{ki}(w)\chi_{ki} + E_{ri}(w)v + b_i(w)u_i \tag{9.28c}$$
$$y_i = \chi_{1i} \tag{9.28d}$$
$$e_i = \chi_{1i} + F(w)v, \ i = 1, \ldots, N. \tag{9.28e}$$

It is noted that, under Part (iv) of Assumption 9.7, $A_{1i}(w)$ is Hurwitz for all $w \in \mathbb{W}$.

System (9.28) itself is an N input and N output linear system. Under Assumptions 9.2 and 9.7, it is possible to find a canonical internal model for each subsystem of (9.28). These internal models together form the internal model for the system composed of (9.28) and (9.3), and hence for the system composed of (9.1) and (9.3).

To find such an internal model for (9.28), as in Sect. 7.4, we note that, since, for $i = 1, \ldots, N$, $A_{1i}(w)$ is Hurwitz for all $w \in \mathbb{W}$, by Proposition A.3 in Sect. A.1, under Assumption 9.2, the Sylvester equation

$$\Pi_i(w)S = A_{1i}(w)\Pi_i(w) + A_{2i}(w)F(w) + E_{0i}(w)$$

admits a unique solution $\Pi_i(w)$. Let

$$X_i(w) = \begin{bmatrix} \Pi_i(w) \\ -F(w) \\ -F(w)S \\ \vdots \\ -F(w)S^{r-1} \end{bmatrix}$$
$$U_i(w) = \frac{-F(w)S^r - A_{3i}(w)\Pi_i(w) + \sum_{k=1}^{r} c_{ki}(w)F(w)S^{k-1} - E_{ri}(w)}{b(w)}.$$

Then, it can be verified that $X_i(w)$ and $U_i(w)$ are the solution of the regulator equations associated with the transformed plant (9.28) and the exosystem (9.2).

Let $\alpha_m(s) = s^l + \alpha_l s^{l-1} + \cdots + \alpha_2 x + \alpha_1$ be the minimal polynomial of S, and

$$\Phi = \begin{bmatrix} 0 & 1 & \cdots & 0 \\ \vdots & \vdots & \ddots & \vdots \\ 0 & 0 & \cdots & 1 \\ -\alpha_1 & -\alpha_2 & \cdots & -\alpha_l \end{bmatrix}, \quad \Psi = \begin{bmatrix} 1 \; 0 \cdots 0 \end{bmatrix}. \tag{9.29}$$

Denote $\Upsilon_i(w) = \text{col}(U_i(w), U_i(w)S, \ldots, U_i(w)S^{l-1})$. Then it can be verified that

$$\Upsilon_i(w)S = \Phi\Upsilon_i(w)$$
$$U_i(w) = \Psi\Upsilon_i(w).$$

Let $M \in \mathbb{R}^{l\times l}$ and $Q \in \mathbb{R}^{l\times 1}$ be any controllable pair with M being Hurwitz. Since, under Assumption 9.2, none of the eigenvalues of Φ has negative real part, the spectra of Φ and any Hurwitz matrix M are disjoint. Therefore, by Proposition A.3, there exists a nonsingular matrix T that satisfies the following Sylvester equation:

$$T\Phi = MT + Q\Psi. \tag{9.30}$$

Let $\Lambda_i(w) = T\Upsilon_i(w)$ and $\Gamma = \Psi T^{-1}$. Then, like in Sect. 7.4, it can be verified that $U_i(w) = \Gamma\Lambda_i(w)$, and $\Lambda_i(w)S = M\Lambda_i(w) + QU_i(w)$. That is, under Assumptions 9.2 and 9.7, from Definition 7.2, for each subsystem i, $i = 1, \ldots, N$, we can define a canonical internal model

$$\dot{\eta}_i = M\eta_i + Qu_i \tag{9.31}$$

with the estimated steady-state input $\hat{\mathbf{u}}_i = \Psi T^{-1}\eta_i$.

Remark 9.4 It is not difficult to see that the N dynamic compensators in (9.31) together constitute an internal model for system (9.28). In fact, let $\eta = \text{col}(\eta_1, \ldots, \eta_N)$, $u = \text{col}(u_1, \ldots, u_N)$, $\Lambda(w) = \text{col}(\Lambda_1(w), \ldots, \Lambda_N(w))$, and $U(w) = \text{col}(U_1(w), \ldots, U_N(w))$. Then $\Lambda(w)S = (I_N \otimes \Phi)\Lambda(w)$ and $U(w) = (I_N \otimes \Psi)\Lambda(w)$. Then the compensator $\dot{\eta} = (I_N \otimes M)\eta + (I_N \otimes Q)u$ is an internal model of system (9.28) with the estimated steady-state input $\hat{\mathbf{u}} = (I_N \otimes \Psi T^{-1})\eta$.

Now attaching the canonical internal model (9.31) to (9.28) gives the augmented system, and performing the following coordinate and input transformation

$$\text{col}(\bar{\chi}_{0i}, \bar{\chi}_{1i}, \ldots, \bar{\chi}_{ri}) = \text{col}(\chi_{0i}, \chi_{1i}, \ldots, \chi_{ri}) - X_i(w)v$$
$$\bar{\eta}_i = \eta_i - T\Upsilon_i(w)v$$
$$\bar{u}_i = u_i - \Psi T^{-1}\eta_i$$

converts the augmented system to the following augmented error system:

$$\dot{\bar{\chi}}_{0i} = A_{1i}(w)\bar{\chi}_{0i} + A_{2i}(w)\bar{\chi}_{1i} \tag{9.32a}$$
$$\dot{\bar{\chi}}_{ki} = \bar{\chi}_{(k+1)i},\ k = 1, \ldots, r-1 \tag{9.32b}$$
$$\dot{\bar{\chi}}_{ri} = A_{3i}(w)\bar{\chi}_{0i} + \sum_{k=1}^{r} c_{ki}(w)\bar{\chi}_{ki} + b_i(w)\bar{u}_i + b_i(w)\Psi T^{-1}\bar{\eta}_i \tag{9.32c}$$
$$\dot{\bar{\eta}}_i = (M + Q\Psi T^{-1})\bar{\eta}_i + Q\bar{u}_i \tag{9.32d}$$
$$e_i = \bar{\chi}_{1i}. \tag{9.32e}$$

The augmented error system (9.32) is still a multi-agent system. Similarly to Lemma 7.7, the following lemma converts Problem 9.2 for the multi-agent system (9.1) into the distributed robust stabilization problem of the augmented error system (9.32) by a dynamic output feedback control law. By Lemma 9.1, the following result is a straightforward extension of Lemma 7.7.

Lemma 9.5 *Under Assumptions 9.2, 9.6, and 9.7, if, for all $w \in \mathbb{W}$, the following dynamic output feedback control law:*

$$\bar{u}_i = K\zeta_i$$
$$\dot{\zeta}_i = F_1\zeta_i + F_2 e_{vi}$$

where $\zeta_i \in \mathbb{R}^{n_\zeta}$ for some integer n_ζ stabilizes the augmented error system (9.32), then, for all $w \in \mathbb{W}$, the following dynamic output feedback control law:

$$u_i = K\zeta_i + \Psi T^{-1}\eta_i \tag{9.33a}$$
$$\dot{\zeta}_i = F_1\zeta_i + F_2 e_{vi} \tag{9.33b}$$
$$\dot{\eta}_i = M\eta_i + Qu_i \tag{9.33c}$$

solves Problem 9.2 for the multi-agent system (9.1).

Remark 9.5 The control law (9.33) takes the form (9.27).

By Lemma 9.5, it suffices to robustly stabilize the system (9.32) by the distributed control law (9.33). We will employ the distributed high-gain feedback control method to do so.

Remark 9.6 It is known that, under Assumption 9.6, H is an $\mathcal{M}$-matrix.[1] Let $H(w) = \mathcal{D}(b_1(w), \ldots, b_N(w))H$. Then, by Lemma A.2 in Sect. A.3, $\underline{\lambda}_{H(w)} \geq \underline{\lambda}_H \min_{1\leq i\leq n}\{b_i(w)\} > 0$. Therefore, we conclude that all the eigenvalues of $H(w)$ have positive real parts if and only if Assumption 9.6 is satisfied.

Note that e_{vi} and its kth derivatives are

[1] The definition of $\mathcal{M}$-matrix is given in Sect. A.2.

$$\begin{aligned} e_{vi} &= \sum_{j=1}^{N} a_{ij}(\chi_{1i} - \chi_{1j}) + a_{i0}(\chi_{1i} - y_0) \\ &= \sum_{j=1}^{N} a_{ij}(\bar{\chi}_{1i} - \bar{\chi}_{1j}) + a_{i0}\bar{\chi}_{1i} \end{aligned} \tag{9.34a}$$

$$\begin{aligned} e_{vi}^{(k)} &= \sum_{j=1}^{N} a_{ij}(\chi_{(k+1)i} - \chi_{(k+1)j}) + a_{i0}(\chi_{(k+1)i} - y_0^{(k)}) \\ &= \sum_{j=1}^{N} a_{ij}(\bar{\chi}_{(k+1)i} - \bar{\chi}_{(k+1)j}) + a_{i0}\bar{\chi}_{(k+1)i}, \ k = 1, \dots, r-1. \end{aligned} \tag{9.34b}$$

We first synthesize a static state feedback control law using e_{vi} and its derivatives to robustly stabilize the system (9.32).

For this purpose, like in Sect. 7.4, we perform the following coordinate transformation:

$$\tilde{\eta}_i = \bar{\eta}_i - \frac{1}{b_i(w)} Q(\xi_1\bar{\chi}_{1i} + \xi_2\bar{\chi}_{2i} + \cdots + \xi_{r-1}\bar{\chi}_{(r-1)i} + \bar{\chi}_{ri})$$

where the coefficients ξ_k, $k = 1, 2, \dots, r-1$, are such that the polynomial $s^{r-1} + \xi_{r-1}s^{r-2} + \cdots + \xi_2 s + \xi_1$ is stable, and obtain the equivalent form of system (9.32) as follows:

$$\dot{\bar{\chi}}_{0i} = A_{1i}(w)\bar{\chi}_{0i} + A_{2i}(w)\bar{\chi}_{1i} \tag{9.35a}$$

$$\dot{\tilde{\eta}}_i = \hat{A}_{3i}(w)\bar{\chi}_{0i} + \sum_{k=1}^{r} \hat{c}_{ki}(w)\bar{\chi}_{ki} + M\tilde{\eta}_i \tag{9.35b}$$

$$\dot{\bar{\chi}}_{ki} = \bar{\chi}_{(k+1)i}, \ k = 1, \dots, r-1 \tag{9.35c}$$

$$\dot{\bar{\chi}}_{ri} = A_{3i}(w)\bar{\chi}_{0i} + \sum_{k=1}^{r} \bar{c}_{ki}(w)\bar{\chi}_{ki} + b_i(w)\bar{u}_i + b_i(w)\Psi T^{-1}\tilde{\eta}_i \tag{9.35d}$$

where, for $k = 1, \dots, r$, $\xi_r \triangleq 1$ and $\xi_0 \triangleq 0$,

$$\begin{aligned} \bar{c}_{ki}(w) =& c_{ki}(w) + \Psi T^{-1} Q\xi_k \\ \hat{c}_{ki}(w) =& \frac{1}{b_i(w)}(MQ\xi_k - Q\xi_{k-1} - Qc_{ki}(w)) \\ \hat{A}_{3i}(w) =& -\frac{1}{b_i(w)} QA_{3i}(w). \end{aligned}$$

Let $\bar{\chi}_i = \mathrm{col}(\bar{\chi}_{1i}, \dots, \bar{\chi}_{(r-1)i})$. Then system (9.35) can be rewritten in the following form:

$$\begin{bmatrix} \dot{\bar{\chi}}_i \\ \dot{\bar{\chi}}_{0i} \\ \dot{\tilde{\eta}}_i \\ \dot{\bar{\chi}}_{ri} \end{bmatrix} = \mathscr{A}_i(w) \begin{bmatrix} \bar{\chi}_i \\ \bar{\chi}_{0i} \\ \tilde{\eta}_i \\ \bar{\chi}_{ri} \end{bmatrix} + \mathscr{B}_i(w)\bar{u}_i \tag{9.36}$$

where

$$\mathscr{A}_i(w) = \begin{bmatrix} \Lambda_0 & 0 & 0 & G \\ A_{2i}(w)D & A_{1i}(w) & 0 & 0 \\ \hat{c}_i(w) & \hat{A}_{3i}(w) & M & \hat{c}_{ri}(w) \\ \bar{c}_i(w) & A_{3i}(w) & b_i(w)\Psi T^{-1} & \bar{c}_{ri}(w) \end{bmatrix}, \quad \mathscr{B}_i(w) = \begin{bmatrix} 0 \\ 0 \\ 0 \\ b_i(w) \end{bmatrix}$$

with $\bar{c}_i(w) = [\bar{c}_{1i}(w), \ldots, \bar{c}_{(r-1)i}(w)]$, $\hat{c}_i(w) = [\hat{c}_{1i}(w), \ldots, \hat{c}_{(r-1)i}(w)]$, and

$$D = [1, 0, \ldots, 0], \; G = [0, \ldots, 0, 1]^{\mathsf{T}}, \; \Lambda_0 = \begin{bmatrix} 0 & 1 & \cdots & 0 \\ \vdots & \vdots & \ddots & \vdots \\ 0 & 0 & \cdots & 1 \\ 0 & 0 & \cdots & 0 \end{bmatrix}.$$

Furthermore, let $\bar{\chi} = \mathrm{col}(\bar{\chi}_1, \ldots, \bar{\chi}_N)$, $\bar{\chi}_0 = \mathrm{col}(\bar{\chi}_{01}, \ldots, \bar{\chi}_{0N})$, $\tilde{\eta} = \mathrm{col}(\tilde{\eta}_1, \ldots, \tilde{\eta}_N)$, $\bar{\chi}_r = \mathrm{col}(\bar{\chi}_{r1}, \ldots, \bar{\chi}_{rN})$, $\phi = \mathrm{col}(\bar{\chi}, \bar{\chi}_0, \tilde{\eta}, \bar{\chi}_r)$, and $\bar{u} = \mathrm{col}(\bar{u}_1, \ldots, \bar{u}_N)$. Then system (9.36) can be put into the following compact form:

$$\dot{\phi} = \mathscr{A}(w)\phi + \mathscr{B}(w)\bar{u} \tag{9.37}$$

where

$$\mathscr{A}(w) = \begin{bmatrix} I_N \otimes \Lambda_0 & 0 & 0 & I_N \otimes G \\ C_{21}(w) & C_{22}(w) & 0 & 0 \\ C_{31}(w) & C_{32}(w) & I_N \otimes M & C_{34}(w) \\ C_{41}(w) & C_{42}(w) & C_{43}(w) & C_{44}(w) \end{bmatrix}, \quad \mathscr{B}(w) = \begin{bmatrix} 0 \\ 0 \\ 0 \\ B_0(w) \end{bmatrix}$$

and

$$\begin{aligned}
C_{21}(w) &= \mathscr{D}(A_{21}(w)D, \ldots, A_{2N}(w)D) \\
C_{22}(w) &= \mathscr{D}(A_{11}(w), \ldots, A_{1N}(w)) \\
C_{31}(w) &= \mathscr{D}(\hat{c}_1(w), \ldots, \hat{c}_N(w)) \\
C_{32}(w) &= \mathscr{D}(\hat{A}_{31}(w), \ldots, \hat{A}_{3N}(w)) \\
C_{34}(w) &= \mathscr{D}(\hat{c}_{r1}(w), \ldots, \hat{c}_{rN}(w)) \\
C_{41}(w) &= \mathscr{D}(\bar{c}_1(w), \ldots, \bar{c}_N(w)) \\
C_{42}(w) &= \mathscr{D}(A_{31}(w), \ldots, A_{3N}(w)) \\
C_{43}(w) &= \mathscr{D}(b_1(w)\Psi T^{-1}, \ldots, b_N(w)\Psi T^{-1})
\end{aligned}$$

$$C_{44}(w) = \mathscr{D}(\bar{c}_{r1}(w), \ldots, \bar{c}_{rN}(w))$$
$$B_0(w) = \mathscr{D}(b_1(w), \ldots, b_N(w)).$$

Lemma 9.6 *Consider the system (9.37) where $A_{1i}(w)$ and M are Hurwitz, and the polynomial $s^{r-1} + \xi_{r-1}s^{r-2} + \cdots + \xi_2 s + \xi_1$ is stable. Then, there exists $K^* > 0$ such that, for any $K > K^*$, the following matrix:*

$$\mathscr{A}(w) - K\mathscr{B}(w)[H \otimes \Xi, 0, 0, H] \triangleq A(w, K) \tag{9.38}$$

where $\Xi = [\xi_1, \xi_2, \ldots, \xi_{r-1}]$ is Hurwitz for every $w \in \mathbb{W}$. Thus, the following distributed static state feedback control law:

$$\bar{u}_i = -K\left(\xi_1 e_{vi} + \xi_2 e_{vi}^{(1)} + \cdots + \xi_{r-1} e_{vi}^{(r-2)} + e_{vi}^{(r-1)}\right) \tag{9.39}$$

stabilizes the augmented system (9.37) (and hence (9.32)) for every $w \in \mathbb{W}$.

Remark 9.7 The centralized stabilizer (7.88) presented in Chap. 7 can be viewed as a special case of (9.39) with $N = 1$ and $a_{10} = 1$.

Proof Noting (9.34), the control law (9.39) can be put into the following compact form:

$$\bar{u} = -K[H \otimes \Xi, 0, 0, H]\phi. \tag{9.40}$$

Therefore, the system matrix of the closed-loop system composed of system (9.37) and control law (9.40) is given by (9.38). Let $s = \sum_{i=1}^{N} n_i - Nr$, and

$$T_A = \begin{bmatrix} I_{(r-1)N} & 0 & 0 & 0 \\ 0 & I_s & 0 & 0 \\ 0 & 0 & I_{lN} & 0 \\ I_N \otimes \Xi & 0 & 0 & I_N \end{bmatrix}.$$

Then,

$$T_A^{-1} = \begin{bmatrix} I_{(r-1)N} & 0 & 0 & 0 \\ 0 & I_s & 0 & 0 \\ 0 & 0 & I_{lN} & 0 \\ -I_N \otimes \Xi & 0 & 0 & I_N \end{bmatrix}$$

and

$$\begin{aligned} \bar{A}(w, K) &\triangleq T_A A(w, K) T_A^{-1} = T_A(\mathscr{A}(w) - K\mathscr{B}(w)[H \otimes \Xi, 0, 0, H])T_A^{-1} \\ &= T_A \mathscr{A}(w) T_A^{-1} - K\mathscr{B}(w)[0, 0, 0, H]. \end{aligned}$$

Simple calculation gives

$$T_A \mathscr{A}(w) T_A^{-1} =$$

$$\begin{bmatrix} I_N \otimes (\Lambda_0 - G\Xi) & 0 & 0 & I_N \otimes G \\ C_{21}(w) & C_{22}(w) & 0 & 0 \\ C_{31}(w) - C_{34}(w)(I_N \otimes \Xi) & C_{32}(w) & I_N \otimes M & C_{34}(w) \\ C_{41}(w) + I_N \otimes (\Xi \Lambda_0) - C_{44}(w)(I_N \otimes \Xi) - I_N \otimes (\Xi G \Xi) & C_{42}(w) & C_{43}(w) & C_{44}(w) + I_N \otimes (\Xi G) \end{bmatrix}$$

where

$$\Lambda_0 - G\Xi = \begin{bmatrix} 0 & 1 & \cdots & 0 \\ \vdots & \vdots & \ddots & \vdots \\ 0 & 0 & \cdots & 1 \\ -\xi_1 & -\xi_2 & \cdots & -\xi_{r-1} \end{bmatrix}$$

is Hurwitz. Thus, $\bar{A}(w, K)$ can be put in the following form:

$$\bar{A}(w, K) = \begin{bmatrix} M_1(w) & N_1(w) \\ N_2(w) & N_3(w) + KM_2(w) \end{bmatrix},$$

where

$$\begin{aligned} M_1(w) &= \begin{bmatrix} I_N \otimes (\Lambda_0 - G\Xi) & 0 & 0 \\ C_{21}(w) & C_{22}(w) & 0 \\ C_{31}(w) - C_{34}(w)(I_N \otimes \Xi) & C_{32}(w) & I_N \otimes M \end{bmatrix}, \quad N_1(w) = \begin{bmatrix} I_N \otimes G \\ 0 \\ C_{34}(w) \end{bmatrix} \\ N_2(w) &= \begin{bmatrix} C_{41}(w) + I_N \otimes (\Xi \Lambda_0) - C_{44}(w)(I_N \otimes \Xi) - I_N \otimes (\Xi G \Xi) & C_{42}(w) & C_{43}(w) \end{bmatrix} \\ N_3(w) &= C_{44}(w) + I_N \otimes (\Xi G), \quad M_2(w) = -H(w) \end{aligned} \tag{9.41}$$

where $H(w) \triangleq B_0(w)H$ is defined in Remark 9.6.

From Remark 9.6, $M_2(w) = -H(w)$ is Hurwitz for all $w \in \mathbb{W}$. On the other hand, by Assumption 9.7 and the choice of Λ_0 and M, $I_N \otimes (\Lambda_0 - G\Xi)$, $C_{22}(w)$ and $I_N \otimes M$ are Hurwitz for all $w \in \mathbb{W}$, so is the block lower triangular matrix $M_1(w)$. Since $N_i(w)$, $i = 1, 2, 3$, are independent of K, all the conditions in Lemma 7.9 are satisfied. Let $\varepsilon_1 = \max_{w \in \mathbb{W}, i=1,2,3}\{\|N_i(w)\|\}$ with $N_i(w)$, $i = 1, 2, 3$, being given by (9.41), $P_1(w)$ and $P_2(w)$ be two positive definite matrices satisfying, for all $w \in \mathbb{W}$

$$\begin{aligned} M_1(w)^\top P_1(w) + P_1(w) M_1(w) &\leq -I_s \\ M_2(w)^\top P_2(w) + P_2(w) M_2(w) &\leq -I_N \end{aligned}$$

$\sigma_1 = \max_{w \in \mathbb{W}}\{\|P_1(w)\|, \|P_2(w)\|\}$, and

$$K^* = \max\{1, 2\varepsilon_1\sigma_1 + 4\varepsilon_1^2\sigma_1^2\}. \tag{9.42}$$

Then, by Lemma 7.9, for any $K > K^*$, $\bar{A}(w, K)$ and hence $A(w, K)$ are Hurwitz for all $w \in \mathbb{W}$, i.e., the origin of the closed-loop system is asymptotically stable for all $w \in \mathbb{W}$. □

The control law (9.39) depends on the derivatives of e_{vi} which in turn depends on the states $\bar{\chi}_{ki}$, $k = 1, \ldots, r$, $i = 1, \ldots, N$, and hence the solution of the regulator equations. Thus, the control law is not feasible. For this reason, we need to further synthesize an output feedback control law that only depends on e_{vi}, $i = 1, \ldots, N$. For this purpose, like in Sect. 7.4, let h be some positive number, and let

$$A_o(h) = \begin{bmatrix} -h\delta_r & 1\ 0 \cdots 0 \\ -h^2\delta_{r-1} & 0\ 1 \cdots 0 \\ \vdots & \vdots\ \vdots\ \vdots\ \vdots \\ -h^{r-1}\delta_2 & 0\ 0 \cdots 1 \\ -h^r\delta_1 & 0\ 0 \cdots 0 \end{bmatrix}, \quad B_o(h) = \begin{bmatrix} h\delta_r \\ h^2\delta_{r-1} \\ \vdots \\ h^{r-1}\delta_2 \\ h^r\delta_1 \end{bmatrix}$$

where δ_j, $j = 1, 2, \ldots, r$, are such that the polynomial $s^r + \delta_r s^{r-1} + \cdots + \delta_2 s + \delta_1$ is stable. Then, we define the following so-called distributed high-gain observer:

$$\dot{\zeta}_i = A_o(h)\zeta_i + B_o(h)e_{vi}$$

which asymptotically estimates the state $\mathrm{col}(e_{vi}, e_{vi}^{(1)}, \ldots, e_{vi}^{(r-1)})$ and is obviously an extension of the standard high-gain observer used in Sect. 7.4. Then it will be shown that the augmented error system (9.37) is robustly stabilizable by a distributed dynamic output feedback control law as summarized below.

Lemma 9.7 *Consider the system* (9.36) *where* $A_{1i}(w)$ *and* M *are Hurwitz, and the polynomials* $s^{r-1} + \xi_{r-1}s^{r-2} + \cdots + \xi_2 s + \xi_1$ *and* $s^r + \delta_r s^{r-1} + \cdots + \delta_2 s + \delta_1$ *are both stable. There exist positive numbers* K^* *and* h^* *such that, for any* $K > K^*$ *and* $h > h^*$, *the following dynamic output feedback control law:*

$$\bar{u}_i = -K(\xi_1\zeta_{1i} + \xi_2\zeta_{2i} + \cdots + \xi_{r-1}\zeta_{(r-1)i} + \zeta_{ri}) \tag{9.43a}$$

$$\dot{\zeta}_i = A_o(h)\zeta_i + B_o(h)e_{vi} \tag{9.43b}$$

where $\zeta_i = \mathrm{col}(\zeta_{1i}, \ldots, \zeta_{ri}) \in \mathbb{R}^r$ *stabilizes the augmented system* (9.36) *(and hence* (9.32)*) for every* $w \in \mathbb{W}$.

Proof Let $\vartheta_{ki} = h^{r-k}(e_{vi}^{(k-1)} - \zeta_{ki})$, $k = 1, \ldots, r$, $\vartheta_i = \mathrm{col}(\vartheta_{1i}, \ldots, \vartheta_{ri})$, and $\vartheta = \mathrm{col}(\vartheta_1, \ldots, \vartheta_N)$. Then the control law (9.43) can be rewritten as follows:

$$\begin{aligned} \bar{u}_i &= -K\left(\xi_1 e_{vi} + \xi_2 e_{vi}^{(1)} + \cdots + \xi_{r-1}e_{vi}^{(r-2)} + e_{vi}^{(r-1)}\right) \\ &\quad + K\left(\xi_1 h^{1-r}\vartheta_{1i} + \xi_2 h^{2-r}\vartheta_{2i} + \cdots + \xi_{r-1}h^{-1}\vartheta_{(r-1)i} + \vartheta_{ri}\right) \\ &= -K\left(\xi_1 e_{vi} + \xi_2 e_{vi}^{(1)} + \cdots + \xi_{r-1}e_{vi}^{(r-2)} + e_{vi}^{(r-1)}\right) + K[\Xi, 1]D_h^{-1}\vartheta_i \end{aligned}$$

whose concatenated form is

$$\bar{u} = -K[H \otimes \Xi, 0, 0, H]\phi + K(I_N \otimes [\Xi, 1]D_h^{-1})\vartheta$$

where $D_h = \mathscr{D}(h^{r-1}, h^{r-2}, \ldots, 1)$. As a result, from (9.37), we have

$$\dot{\phi} = A(w, K)\phi + Z_1(K, h, w)\vartheta \tag{9.44}$$

where $Z_1(K, h, w) = K\bar{B}(w)\left(I_N \otimes [\Xi, 1]D_h^{-1}\right)$. On the other hand,

$$\begin{aligned}
\dot{\vartheta}_{ki} &= h^{r-k}(e_{vi}^{(k)} - \dot{\zeta}_{ki}) \\
&= h^{r-k}(e_{vi}^{(k)} - (-h^k\delta_{r+1-k}\zeta_{1i} + \zeta_{(k+1)i} + h^k\delta_{r+1-k}e_{vi})) \\
&= h^{r-k}((e_{vi}^{(k)} - \zeta_{(k+1)i}) - h^k\delta_{r+1-k}(e_{vi} - \zeta_{1i})) \\
&= h(\vartheta_{(k+1)i} - \delta_{r+1-k}\vartheta_{1i}), \ k = 1, \ldots, r-1 \\
\dot{\vartheta}_{ri} &= e_{vi}^{(r)} - \dot{\zeta}_{ri} = -h^r\delta_1(e_{vi} - \zeta_{1i}) + e_{vi}^{(r)} \\
&= -h\delta_1\vartheta_{1i} + e_{vi}^{(r)} \\
&= -h\delta_1\vartheta_{1i} + \sum_{j=1}^{N} a_{ij}(\dot{\tilde{\chi}}_{ri} - \dot{\tilde{\chi}}_{rj}) + a_{i0}\dot{\tilde{\chi}}_{ri}.
\end{aligned}$$

Putting them together gives

$$\dot{\vartheta}_i = hA_o(1)\vartheta_i + \left(\sum_{j=1}^{N} a_{ij}(E_i\dot{\phi} - E_j\dot{\phi}) + a_{i0}E_i\dot{\phi}\right)$$

where

$$E_i = \begin{bmatrix} 0_{(r-1)\times(\sum_{i=1}^{N} n_i + Nl - N)} & 0_{(r-1)\times N} \\ 0_{1\times(\sum_{i=1}^{N} n_i + Nl - N)} & E_i^0 \end{bmatrix}$$

with E_i^0 being the ith row of the matrix I_N. As a result, some calculations give

$$\dot{\vartheta} = (h(I_N \otimes A_o(1)) + Z_3(w, K, h))\vartheta + Z_2(w, K)\phi \tag{9.45}$$

where $Z_2(w, K) = (H \otimes I_r)EA(w, K)$, $Z_3(w, K, h) = (H \otimes I_r)EZ_1(K, h, w)$ with $E = \mathrm{col}(E_1, \ldots, E_N)$.

Putting the systems (9.44) and (9.45) together gives the following compact form:

$$\begin{bmatrix} \dot{\phi} \\ \dot{\vartheta} \end{bmatrix} = \begin{bmatrix} A(w, K) & Z_1(w, K, h) \\ Z_2(w, K) & Z_3(w, K, h) + h(I_N \otimes A_o(1)) \end{bmatrix} \begin{bmatrix} \phi \\ \vartheta \end{bmatrix}. \tag{9.46}$$

Let the positive number K^* be given by (9.42). From Lemma 9.6, $A(w, K)$ is Hurwitz for any $K > K^*$ and any $w \in \mathbb{W}$. Since the polynomial $s^r + \delta_r s^{r-1} + \cdots + \delta_2 s + \delta_1$ is stable, $A_o(1)$ and hence $I_N \otimes A_o(1)$ are Hurwitz. Further, (9.46) takes the form of (7.82) with $\nu = h$, and $M_1(w) = A(w, K)$, $N_1(w, \nu) = Z_1(w, K, h)$, $N_2(w, \nu) = Z_2(w, K)$, $N_3(w, \nu) = Z_3(w, K, h)$, $M_2(w) = I_N \otimes A_o(1)$, and $\|D_h^{-1}\|$

$\leq \|D_1^{-1}\|$ for any $h \geq 1$. Therefore, for any $h \geq 1$, there exists $\varepsilon_2 > 0$, independent of h, such that $\|N_i(w, v)\| \leq \varepsilon_2, i = 1, 2, 3$. Let $s = \sum_{i=1}^{N} n_i$, $\bar{P}_1(w)$ and $\bar{P}_2$ be two positive definite matrices satisfying, for all $w \in \mathbb{W}$,

$$A(w, K)^{\mathsf{T}} \bar{P}_1(w) + \bar{P}_1(w) A(w, K) \leq -I_{s+Nl}$$
$$(I_N \otimes A_o(1))^{\mathsf{T}} \bar{P}_2 + \bar{P}_2(I_N \otimes A_o(1)) \leq -I_{Nr},$$

$\sigma_2 = \max_{w \in \mathbb{W}}\{\|\bar{P}_1(w)\|, \|\bar{P}_2\|\}$, and

$$h^* = \max\{1, 2\varepsilon_2\sigma_2 + 4\varepsilon_2^2\sigma_2^2\}.$$

Then, by Lemma 7.9, for any $h > h^*$, the origin of the closed-loop system (9.46) is asymptotically stable for every $w \in \mathbb{W}$. The proof is thus completed. □

Remark 9.8 Lemma 9.7 has explicitly given the lower bounds of K^* and h^*. Specifically, K^* is determined by the bound of the compact subset $\mathbb{W}$ only, and h^* is determined by both the value of K and the bound of the compact subset $\mathbb{W}$.

Finally, the solvability of Problem 9.2 is summarized in the following theorem.

Theorem 9.3 *Given the class of linear multi-agent systems (9.1), the exosystem (9.2), and any compact subset $\mathbb{W}$, under Assumptions 9.2 and 9.7, Problem 9.2 is solvable by a distributed dynamic output feedback control law of the following form:*

$$u_i = \Psi T^{-1}\eta_i - K(\xi_1\zeta_{1i} + \xi_2\zeta_{2i} + \cdots + \xi_{r-1}\zeta_{(r-1)i} + \zeta_{ri}) \quad (9.47a)$$
$$\dot{\eta}_i = M\eta_i + Qu_i \quad (9.47b)$$
$$\dot{\zeta}_i = A_o(h)\zeta_i + B_o(h)e_{vi} \quad (9.47c)$$

where $\zeta_i \triangleq \operatorname{col}(\zeta_{1i}, \ldots, \zeta_{ri}) \in \mathbb{R}^r$, Ψ, M, Q, T, K, and h are chosen from (9.30), $A_o(h)$, $B_o(h)$ and the coefficients ξ_i, $i = 1, \ldots, r-1$, and δ_j, $j = 1, \ldots, r$, are the same as those in Lemma 9.7, if and only if the static communication graph $\bar{\mathcal{G}}$ satisfies Assumption 9.6.

Proof The "if part" can be directly obtained from Lemma 9.7 by choosing the same positive numbers K^* and h^*. Here we only give the proof of "only if part". When v is set to zero, the closed-loop system composed of the plant (9.1) and the control law (9.47) is given by

$$\dot{x}_i = A_i(w)x_i + B_i(w)(\Psi T^{-1}\eta_i - K[\Xi, 1]\zeta_i) \quad (9.48a)$$
$$\dot{\eta}_i = (M + Q\Psi T^{-1})\eta_i - KQ[\Xi, 1]\zeta_i \quad (9.48b)$$
$$\dot{\zeta}_i = A_o(h)\zeta_i + B_o(h)\left(\sum_{j=1}^{N} a_{ij}(C_i(w)x_i - C_j(w)x_j) + a_{i0}C_i(w)x_i\right), \quad i = 1, \ldots, N. \quad (9.48c)$$

The closed-loop system (9.48) can be put into the following compact form:

$$\dot{x} = \tilde{A}(w)x + \tilde{B}(w)(I_N \otimes (\Psi T^{-1}))\eta - \tilde{B}(w)K(I_N \otimes [\Xi, 1])\zeta \tag{9.49a}$$
$$\dot{\eta} = (I_N \otimes (M + Q\Psi T^{-1}))\eta - K(I_N \otimes (Q[\Xi, 1]))\zeta \tag{9.49b}$$
$$\dot{\zeta} = (I_N \otimes A_o(h))\zeta + (H \otimes B_o(h))\tilde{C}(w)x \tag{9.49c}$$

where $x = \mathrm{col}(x_1, \ldots, x_N)$, $\eta = \mathrm{col}(\eta_1, \ldots, \eta_N)$, $\zeta = \mathrm{col}(\zeta_1, \ldots, \zeta_N)$, and

$$\tilde{A}(w) = \mathscr{D}(A_1(w), \ldots, A_N(w))$$
$$\tilde{B}(w) = \mathscr{D}(B_1(w), \ldots, B_N(w))$$
$$\tilde{C}(w) = \mathscr{D}(C_1(w), \ldots, C_N(w)).$$

Let T_1 be a nonsingular matrix such that $T_1 H T_1^{-1} \triangleq J$ is the Jordan form of H. Let $\hat{\eta} = (T_1 \otimes I_l)\eta$, and $\hat{\zeta} = (T_1 \otimes I_r)\zeta$. Then system (9.49) is equivalent to

$$\dot{x} = \tilde{A}(w)x + \tilde{B}(w)(T_1^{-1} \otimes \Psi T^{-1})\hat{\eta} - \tilde{B}(w)K(T_1^{-1} \otimes [\Xi, 1])\hat{\zeta} \tag{9.50a}$$
$$\dot{\hat{\eta}} = (I_N \otimes (M + Q\Psi T^{-1}))\hat{\eta} - K(I_N \otimes Q[\Xi, 1])\hat{\zeta} \tag{9.50b}$$
$$\dot{\hat{\zeta}} = (I_N \otimes A_o(h))\hat{\zeta} + (J \otimes B_o(h))(T_1 \otimes I_r)\tilde{C}(w)x. \tag{9.50c}$$

Suppose that the static communication graph $\bar{\mathscr{G}}$ does not satisfy Assumption 9.6. Then H has at least one zero eigenvalue. Then, there exists an integer $1 \leq s \leq N$ such that the entries of the sth row of J are all 0. Thus, system (9.50) contains a subsystem as follows:

$$\dot{\hat{\eta}}_s = (M + Q\Psi T^{-1})\hat{\eta}_s - KQ[\Xi, 1]\hat{\zeta}_s \tag{9.51a}$$
$$\dot{\hat{\zeta}}_s = A_o(h)\hat{\zeta}_s \tag{9.51b}$$

where $\hat{\eta}_s$ and $\hat{\zeta}_s$ are the sth component of $\hat{\eta}$ and $\hat{\zeta}$, respectively. It is noted that system (9.51) has an upper triangular structure. Since the eigenvalues of $M + Q\Psi T^{-1} = T\Phi T^{-1}$ coincide with those of S, by Assumption 9.2, system (9.51) and hence system (9.50) are not asymptotically stable, which is a contradiction. Thus, Assumption 9.6 must be satisfied. □

Example 9.2 Consider the leader-following tracking problem of a group of linear heterogeneous systems of the form (9.1) with $N = 4$. The system matrices are given as, for $i = 1, 2$,

$$A_i(w) = \begin{bmatrix} c_{1i}(w) & 1 & 0 \\ 0 & 0 & 1 \\ 1 & c_{2i}(w) & c_{3i}(w) \end{bmatrix}, \quad B_i(w) = \begin{bmatrix} 0 \\ 0 \\ c_{4i}(w) \end{bmatrix}, \quad C_i(w) = \begin{bmatrix} 0 & 1 & 0 \end{bmatrix},$$

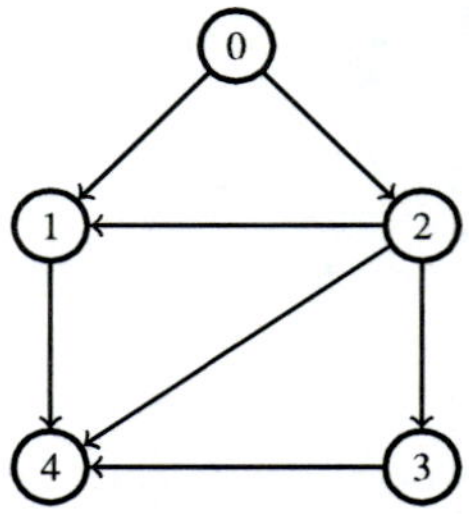

Fig. 9.6 The static communication graph $\bar{\mathscr{G}}$

for $i = 3, 4$,

$$A_i(w) = \begin{bmatrix} c_{1i}(w) & 0 & 0 & 1 \\ 0 & -1 & 0 & 1 \\ 0 & 0 & 0 & 1 \\ 1 & 1 & c_{2i}(w) & c_{3i}(w) \end{bmatrix}, \; B_i(w) = \begin{bmatrix} 0 \\ 0 \\ 0 \\ c_{4i}(w) \end{bmatrix}, \; C_i(w) = \begin{bmatrix} 0 & 0 & 1 & 0 \end{bmatrix}$$

and $E_i(w)$ are zero matrices with conformable dimensions. The reference signal is generated by the following linear exosystem system:

$$\dot{v}_1 = v_2, \; \dot{v}_2 = -v_1, \; y_0 = v_1.$$

The communication graph $\bar{\mathscr{G}}$ is shown by Fig. 9.6, and its Laplacian is

$$\bar{\mathscr{L}} = \begin{bmatrix} 0 & 0 & 0 & 0 & 0 \\ -1 & 2 & -1 & 0 & 0 \\ -1 & 0 & 1 & 0 & 0 \\ 0 & 0 & -1 & 1 & 0 \\ 0 & -1 & -1 & -1 & 3 \end{bmatrix}.$$

Here the node 0 is associated with the leader and the node i, $i = 1, 2, 3, 4$, is associated with the ith follower, respectively. The objective is to design a distributed controller such that the closed-loop system is asymptotically stable for all $w \in \mathbb{W}$ for any prescribed compact subset $\mathbb{W} \subseteq \mathbb{R}^4$, and the tracking errors satisfy $\lim_{t\to\infty} e_i(t) = \lim_{t\to\infty}(y_i(t) - y_0(t)) = 0$.

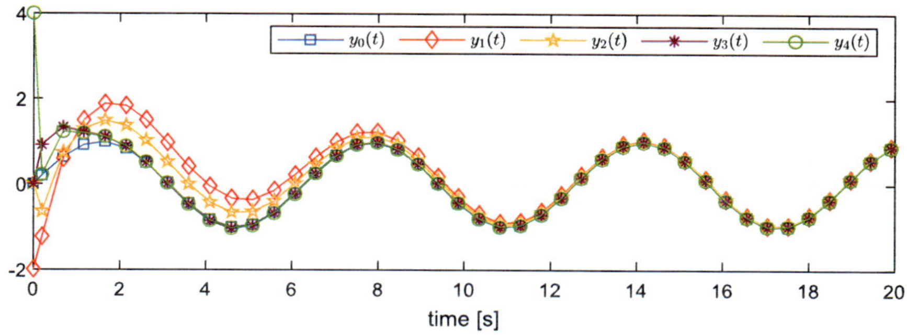

Fig. 9.7 The response of $y_0(t)$ and the response of $y_i(t)$ under the control law (9.47)

Express the parameter vector $c_i(w) = \text{col}(c_{1i}(w), c_{2i}(w), c_{3i}(w), c_{4i}(w))$ as $c_i(w) = c_i(0) + w_i$, where $c_i(0) = \text{col}(c_{1i}(0), c_{2i}(0), c_{3i}(0), c_{4i}(0))$ is the nominal part of $c_i(w)$ and $w_i = \text{col}(w_{1i}, w_{2i}, w_{3i}, w_{4i})$ is the uncertainty of $c_i(w)$. Let $c_{1i}(0) = c_{2i}(0) = c_{3i}(0) = -2$, $c_{4i}(0) = 2$, and $w_{1i} \in [-1, 1]$, $w_{2i} \in [-1, 1.5]$, $w_{3i} \in [-2, 1]$, $w_{4i} \in [-1, 2]$. It can be easily verified that Assumptions 9.2, 9.6, and 9.7 are all satisfied. Then, by Theorem 9.3, it is possible to design a distributed controller of the form (9.47) to solve this problem. More specifically, since each subsystem has the uniform relative degree 2 with input u_i and output y_i, $A_o(h)$ and $B_o(h)$ can be chosen as $A_o(h) = \begin{bmatrix} -h\delta_2 & 1 \\ -h^2\delta_1 & 0 \end{bmatrix}$ and $B_o(h) = \begin{bmatrix} h\delta_2 \\ h^2\delta_1 \end{bmatrix}$. Since $S = \begin{bmatrix} 0 & 1 \\ -1 & 0 \end{bmatrix}$, from (9.29), we have $\Phi = \begin{bmatrix} 0 & 1 \\ -1 & 0 \end{bmatrix}$, $\Psi = \begin{bmatrix} 1 \\ 0 \end{bmatrix}^{\text{T}}$. Consider the controllable pair (M, Q) with $M = \begin{bmatrix} 0 & 1 \\ -\varrho_1 & -\varrho_2 \end{bmatrix}$ and $Q = \begin{bmatrix} 0 \\ 1 \end{bmatrix}$, where $\varrho_1, \varrho_2 > 0$. Solving the Sylvester equation (9.30) gives $T^{-1} = \begin{bmatrix} \varrho_1 - 1 & \varrho_2 \\ -\varrho_2 & \varrho_1 - 1 \end{bmatrix}$, and hence $\Psi T^{-1} = \begin{bmatrix} \varrho_1 - 1 & \varrho_2 \end{bmatrix}$. Then we can obtain a distributed output feedback controller of the form (9.47) with $\xi_1 = \xi_2 = 1$, $\varrho_1 = 6$, $\varrho_2 = 18$, $K = 2$, $\delta_1 = 30$, $\delta_2 = 10$, and $h = 50$.

The simulation result is shown in Figs. 9.7 and 9.8, where the actual values of the uncertain parameters are $w_1 = \text{col}(-0.5, 0.7, -0.6, 0.5)$, $w_2 = \text{col}(0.8, 0.6, 0, 1.2)$, $w_3 = \text{col}(-0.5, 0, -1.1, 1.1)$, $w_4 = \text{col}(0.3, 0.7, -1.9, 0.8)$, and the initial conditions of the closed-loop system are $x_1(0) = \text{col}(-1, 2, -7)$, $x_2(0) = \text{col}(2, 0, -3)$, $x_3(0) = \text{col}(-2, 1, 0, -4)$, $x_4(0) = \text{col}(0, 1, 4, -5)$, $\eta_1(0) = \text{col}(-6, -5)$, $\eta_2(0) = \text{col}(-2, -1)$, $\eta_3(0) = \text{col}(2, 3)$, $\eta_4(0) = \text{col}(6, 7)$ and $\zeta_1(0) = \zeta_2(0) = \zeta_3(0) = \zeta_4(0) = \text{col}(0, 0)$. Both figures verify that the controller has successfully achieved the objective of the robust output regulation.

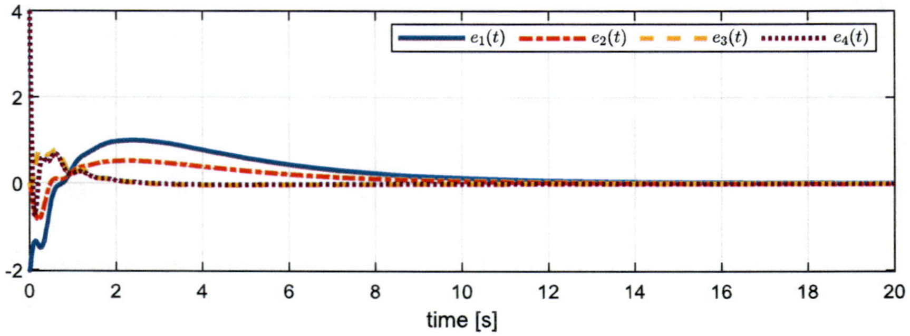

Fig. 9.8 The response of $e_i(t)$ under the control law (9.47)

9.4 Notes and References

The distributed p-copy internal model design was first studied in [1] for static communication graphs $\bar{\mathcal{G}}$ containing no cycle. This assumption was removed in [2] by adopting the LQR-based gain assignment [3]. Some other attempts on this topic can be found in [4]. The same problem for the discrete-time plant and the time-delay plant was treated in [5, 6], respectively.

The robust cooperative output regulation problem for SISO linear minimum phase systems with identical relative degree was first treated in [7], and was extended to SISO linear minimum phase systems with non-identical relative degree in [8, 9], and to MIMO linear systems in [10, 11]. Combining with the potential function approach, this design method is able to handle the rendezvous of uncertain robots [12]. The implementation of the control law in [7] by event-triggered control law was given in [13].

The leaderless output consensus of heterogeneous uncertain linear multi-agent systems was studied by the internal model approach in [14].

References

1. Wang X, Hong Y, Huang J, Jiang Z (2010) A distributed control approach to a robust output regulation problem for multi-agent linear systems. IEEE Trans Autom Control 55(12):2891–2895
2. Su Y, Hong Y, Huang J (2013) A general result on the robust cooperative output regulation for linear uncertain multi-agent systems. IEEE Trans Autom Control 58(5):1275–1279
3. Tuna SE (2008) LQR-based coupling gain for synchronization of linear systems. http://arxiv.org/abs/0801.3390
4. Hong Y, Wang X, Jiang Z (2013) Distributed output regulation of leader-follower multi-agent systems. Int J Robust Nonlinear Control 23(1):48–66
5. Yan Y, Huang J (2018) Cooperative robust output regulation problem for discrete-time linear time-delay multi-agent systems. Int J Robust Nonlinear Control 28(3):1035–1048

6. Lu M, Huang J (2018) Internal model approach to cooperative robust output regulation for linear uncertain time-delay multi-agent systems. Int J Robust Nonlinear Control 28(6):2528–2542
7. Su Y, Huang J (2014) Cooperative robust output regulation of a class of heterogeneous linear uncertain multi-agent systems. Int J Robust Nonlinear Control 24(17):2819–2839
8. Su Y (2016) Output feedback cooperative control for linear uncertain multi-agent systems with nonidentical relative degrees. IEEE Trans Autom Control 61(12):4027–4033
9. Zhang Y, Su Y, Wang X (2021) Distributed adaptive output feedback control for multi-agent systems with unknown dynamics. IEEE Trans Autom Control 66(3):1367–1374
10. Wang X, Su Y, Xu D (2018) Cooperative robust output regulation of linear uncertain multiple multivariable systems with performance constraints. Automatica 95:137–145
11. Zhang Y, Su Y (2018) Cooperative output regulation for linear uncertain MIMO multi-agent systems by output feedback. Sci China Inf Sci 61(9):092206
12. Su Y (2015) Leader-following rendezvous with connectivity preservation and disturbance rejection via internal model approach. Automatica 57:203–212
13. Liu W, Huang J (2017) Event-triggered cooperative robust practical output regulation for a class of linear multi-agent systems. Automatica 85:158–164
14. Kim H, Shim H, Seo JH (2011) Output consensus of heterogeneous uncertain linear multi-agent systems. IEEE Trans Autom Control 56(1):200–206

Chapter 10
Cooperative Robust Output Regulation of Nonlinear Multi-agent Systems Over Static Networks

Having studied the cooperative robust output regulation problem for linear multi-agent systems, we will further study the cooperative robust output regulation problem for nonlinear multi-agent systems. Section A.6 has summarized a framework for handling the robust output regulation problem for general nonlinear systems by internal model approach. Like linear systems, the role of the internal model is to define the so-called augmented error system such that the robust output regulation problem of the original plant can be converted to the robust stabilization problem of the augmented error system. For a single system, one can seek any type of control laws to solve the robust stabilization problem of the augmented error system. However, for multi-agent systems, the augmented error system is also a multi-agent system. Due to the communication constraints, one has to design a distributed control law to stabilize the augmented error system. That is what makes the problem of this chapter more interesting than the robust output regulation problem of a single plant as summarized in Sect. A.6.

It is worth noting that, due to the complexity of nonlinear systems, there is no generic way to handle the robust output regulation problem even for a single nonlinear system. Thus, one has to first classify nonlinear systems, and then, for each class of nonlinear systems, one develops a specific approach to handling this class of nonlinear systems. For this reason, in this chapter, we will focus on the so-called nonlinear multi-agent systems in normal form with unity relative degree over static communication graphs. In the next chapter, we will further study the same problem for the same class of nonlinear multi-agent systems over jointly connected switching communication graphs.

This chapter is organized as follows. Section 10.1 formulates the problem. Section 10.2 solves the problem assuming that the exosystem is known exactly and the initial condition of the exosystem and the unknown parameter vector of the plant belong to some known compact subsets, respectively. Section 10.3 studies the problem assuming that the exosystem is known exactly but the initial condition of

H. Cai et al., *Cooperative Control of Multi-agent Systems*, Advances in Industrial Control,
https://doi.org/10.1007/978-3-030-98377-2_10

the exosystem and the unknown parameter vector of the plant can take any value. Section 10.4 further studies the problem assuming that the exosystem is uncertain, and the initial condition of the exosystem and the unknown parameters in both the plant and the exosystem can take any value. Section 10.5 illustrates these design methods with examples.

10.1 Problem Formulation and Preliminaries

We consider nonlinear multi-agent systems of the following form:

$$\dot{x}_i = F_i(x_i, u_i, v, w) \tag{10.1a}$$

$$y_i = h_i(x_i, v, w) \tag{10.1b}$$

where, for $i = 1, \ldots, N$, $x_i \in \mathbb{R}^{n_i}$, $u_i \in \mathbb{R}^m$, $y_i \in \mathbb{R}^p$ are the state, the control input, and the plant performance output, respectively. $v \in \mathbb{R}^q$ is the exogenous signal representing the disturbance and/or the reference input, and $w \in \mathbb{R}^{n_w}$ is the plant uncertain parameter vector. A general nonlinear exosystem takes the following form:

$$\dot{v} = f_0(v, \sigma) \tag{10.2a}$$

$$y_0 = h_0(v) \tag{10.2b}$$

which governs the motion of the exogenous signal v, and produces the measurable reference output $y_0 \in \mathbb{R}^p$. Here $\sigma \in \mathbb{R}^{n_\sigma}$ is the uncertain parameter vector. Like in Sect. 9.1, for $i = 1, \ldots, N$, the tracking error for the ith subsystem of (10.1) is defined by

$$e_i = y_i - y_0.$$

Also, like linear systems studied in Sect. 9.1, systems (10.1) and (10.2) together are considered as a multi-agent system of $N + 1$ agents with the exosystem (10.2) as the leader and all subsystems of system (10.1) as the followers. Thus, the communication graph $\bar{\mathscr{G}}$ of the systems (10.1) and (10.2) is described in the same way as in Sect. 9.1.

To describe our distributed control law, we recall the virtual tracking error introduced in Sect. 9.1 for studying the cooperative robust output regulation problem for linear multi-agent systems as follows:

$$e_{vi} = \sum_{j=0}^{N} a_{ij}(y_i - y_j).$$

In terms of the virtual tracking error, we describe our distributed control law as follows:

$$u_i = \kappa_i(z_i, e_{vi}) \tag{10.3a}$$

$$\dot{z}_i = \psi_i(z_i, e_{vi}) \tag{10.3b}$$

where $z_i \in \mathbb{R}^{n_{z_i}}$ with n_{z_i} to be specified later.

For convenience, we assume all functions in (10.1), (10.2), and (10.3) are globally defined and sufficiently smooth.

To formulate our problem, let $\mathbb{V}_0 \subseteq \mathbb{R}^q$, $\mathbb{W} \subseteq \mathbb{R}^{n_w}$, and $\mathbb{S} \subseteq \mathbb{R}^{n_\sigma}$ be any subsets with $\mathbb{V}_0$ and $\mathbb{W}$ containing the origin of their respective Euclidian spaces. We allow $\mathbb{S}$ to consist of a single known vector to accommodate the case where the leader system (10.2) is known exactly. It is noted that the three subsets can be almost arbitrary. In particular, we allow $\mathbb{V}_0 = \mathbb{R}^q$ and $\mathbb{W} = \mathbb{R}^{n_w}$.

Then we describe our problem as follows:

Problem 10.1 (*Nonlinear Global Cooperative Robust Output Regulation*) Given the follower system (10.1), the leader system (10.2), the corresponding communication graph $\bar{\mathcal{G}}$, and any compact subset $\Sigma_0 \subseteq \mathbb{V}_0 \times \mathbb{W} \times \mathbb{S}$, find a distributed feedback control law of the form (10.3) such that, for any $\text{col}(v(0), w, \sigma) \in \Sigma_0$ and any initial state, the closed-loop system has the following properties:

- **Property 10.1** The solution of the closed-loop system exists for all $t \geq 0$, and is uniformly bounded over $[0, \infty)$ for uniformly bounded $v_0(t)$;
- **Property 10.2** For $i = 1, \ldots, N$, $\lim_{t\to\infty} e_i(t) = 0$.

As mentioned above, there is no generic way to handle the cooperative robust output regulation problem for the general nonlinear multi-agent system (10.1) and general exosystem (10.2). Thus, in what follows, we focus on the following class of nonlinear multi-agent systems:

$$\dot{z}_i = f_i(z_i, y_i, v, w) \tag{10.4a}$$

$$\dot{y}_i = g_i(z_i, y_i, v, w) + b_i(w)u_i, \ i = 1, \ldots, N \tag{10.4b}$$

where $z_i \in \mathbb{R}^{n_{z_i}}$, $y_i, u_i \in \mathbb{R}$, and $w \in \mathbb{R}^{n_w}$. We assume, for $i = 1, \ldots, N$, $b_i(w) \neq 0$ for all $w \in \mathbb{R}^{n_w}$. System (10.4) is of the form (10.1) with $x_i = \text{col}(z_i, y_i)$, $F_i(x_i, u_i, v, w) = \text{col}(f_i(z_i, y_i, v, w), g_i(z_i, y_i, v, w) + b_i(w)u_i)$, and $h_i(x_i, v, w) = y_i$.

Each subsystem of the system (10.4) is called a nonlinear system in normal form with unity relative degree. The linear system (9.28) studied in Sect. 9.3 is also in this form when $r = 1$. Even though systems of the form (10.4) are quite special, it can describe many practical nonlinear systems such as the Lorenz system shown in Sect. 10.5.

Remark 10.1 Since Σ_0 is a compact subset of $\mathbb{V}_0 \times \mathbb{W} \times \mathbb{S}$, it can be represented as $\Sigma_0 = \bar{\mathbb{V}}_0 \times \mathbb{W}_0 \times \mathbb{S}_0$ for some compact subsets $\bar{\mathbb{V}}_0 \subseteq \mathbb{V}_0$, $\mathbb{W}_0 \subseteq \mathbb{W}$, and $\mathbb{S}_0 \subseteq \mathbb{S}$. By the continuity of $b_i(w)$, it has the same sign for all $w \in \mathbb{W}_0$. Without loss of generality, we assume that, for $i = 1, \ldots, N$, for all $w \in \mathbb{W}_0$, $b_i(w) > 0$. Thus, there exist positive numbers $b_{\min}$ and $b_{\max}$ such that $b_{\min} \leq b_i(w) \leq b_{\max}$ for all $w \in \mathbb{W}_0$, and all $i = 1, \ldots, N$.

Also, we focus on the following exosystem with a linear state equation:

$$\dot{v} = S(\sigma)v \tag{10.5a}$$

$$y_0 = h_0(v) \tag{10.5b}$$

which is a special form of (10.2) with $f_0(v, \sigma) = S(\sigma)v$.

It is assumed that the exosystem (10.5) satisfies the following assumption.

Assumption 10.1 The system (10.5a) is neutrally stable, that is, all the eigenvalues of $S(\sigma)$ are simple with zero real parts for all $\sigma \in \mathbb{S}$.

Assumption 10.1 guarantees that, for any $\sigma \in \mathbb{S}$, and any initial condition $v(0)$, the exogenous signal $v(t)$ is uniformly bounded over $[0, \infty)$. Thus, given any $\mathbb{V}_0 \subseteq \mathbb{R}^q$, there exists a subset $\mathbb{V} \subseteq \mathbb{R}^q$ such that $v(0) \in \mathbb{V}_0$ implies $v(t) \in \mathbb{V}$ for all $t \geq 0$. As a result, given a compact subset $\Sigma_0 \subseteq \mathbb{V}_0 \times \mathbb{W} \times \mathbb{S}$, there exists a compact subset $\Sigma \subseteq \mathbb{V} \times \mathbb{W} \times \mathbb{S}$ such that $(v(0), w, \sigma) \in \Sigma_0$ implies $(v(t), w, \sigma) \in \Sigma$ for all $t \geq 0$.

Remark 10.2 Three cases of the above problem will be studied. The first case is where the exosystem is known exactly and the compact subset $\Sigma_0 \subseteq \mathbb{V}_0 \times \mathbb{W} \times \mathbb{S}$ is known. In this case, $\mathbb{S}$ and hence $\mathbb{S}_0$ consists of a single known vector, and the initial condition $v(0)$ and the unknown parameter w belong to some known compact subsets $\bar{\mathbb{V}}_0$ and $\mathbb{W}_0$, respectively, and, under Assumption 10.1, there exists a known compact subset $\Sigma \subseteq \mathbb{V} \times \mathbb{W} \times \mathbb{S}$ such that $(v(0), w, \sigma) \in \Sigma_0$ implies $(v(t), w, \sigma) \in \Sigma$ for all $t \geq 0$. For this case, the problem can be handled by some robust control technique to be described in Sect. 10.2. The second case is where the exosystem is known exactly again but the compact subset Σ_0 is unknown. In this case, $\mathbb{S}$ and hence $\mathbb{S}_0$ still consists of a single known vector, but $v(0)$ can take any value in $\mathbb{V}_0$ and w can take any value in $\mathbb{W}$ since, for any $(v(0), w, \sigma) \in \mathbb{V}_0 \times \mathbb{W} \times \mathbb{S}$, there exists a compact subset Σ_0 such that $(v(0), w, \sigma) \in \Sigma_0$. The third case is where the exosystem is uncertain, and the compact subset $\Sigma_0 \subseteq \mathbb{V}_0 \times \mathbb{W} \times \mathbb{S}$ is unknown. In this case, the exosystem contains an uncertain parameter vector which can take any value in $\mathbb{S}$, and the initial condition $v(0)$ and the unknown parameter w can also take any values in $\mathbb{V}_0$ and $\mathbb{W}$, respectively. For both the second and third cases, under Assumption 10.1, there exists an unknown compact subset $\Sigma \subseteq \mathbb{V} \times \mathbb{W} \times \mathbb{S}$ such that $(v(0), w, \sigma) \in \Sigma_0$ implies $(v(t), w, \sigma) \in \Sigma$ for all $t \geq 0$. To deal with the second case, we need to further introduce the universal adaptive control technique, and, to deal with the third case, on top of the control law for dealing with the second case, we have to further introduce some parameter identification method to estimate the unknown parameter vector σ.

For all three cases, we need to guarantee the composite system composed of (10.4) and (10.5) admits a canonical internal model similar to what we have used in Chap. 9. For this purpose, we need the following assumption:

Assumption 10.2 There exist sufficiently smooth functions $\mathbf{z}_i(v, w, \sigma)$ with $\mathbf{z}_i(0, 0, \sigma) = 0$ such that, for any $v(t)$ with $v(0) \in \mathbb{V}_0$ and any $\mathrm{col}(w, \sigma) \in \mathbb{W} \times \mathbb{S}$,

$$\frac{\partial \mathbf{z}_i(v, w, \sigma)}{\partial v} S(\sigma)v = f_i(\mathbf{z}_i(v, w, \sigma), h_0(v), v, w), \ i = 1, \ldots, N.$$

Under Assumptions 10.2, for $i = 1, \ldots, N$, let

$$\begin{aligned}
\mathbf{y}_i(v, w, \sigma) &= h_0(v) \\
\mathbf{x}_i(v, w, \sigma) &= \begin{bmatrix} \mathbf{z}_i(v, w, \sigma) \\ \mathbf{y}_i(v, w, \sigma) \end{bmatrix} \\
\mathbf{u}_i(v, w, \sigma) &= b_i^{-1}(w) \left(\frac{\partial h_0(v)}{\partial v} S(\sigma)v - g_i(\mathbf{z}_i(v, w, \sigma), h_0(v), v, w) \right).
\end{aligned}$$

Then it can be verified that $\mathbf{x}_i(v, w, \sigma)$ and $\mathbf{u}_i(v, w, \sigma)$ are the solutions of the following regulator equations associated with systems (10.4) and (10.5):

$$\begin{aligned}
\frac{\partial \mathbf{x}_i(v, w, \sigma)}{\partial v} S(\sigma)v &= \begin{bmatrix} f_i(\mathbf{x}_i(v, w, \sigma), h_0(v), v, w) \\ g_i(\mathbf{z}_i(v, w, \sigma), h_0(v), v, w) + b_i(w)\mathbf{u}_i(v, w, \sigma) \end{bmatrix} \\
0 &= \mathbf{y}_i(v, w, \sigma) - h_0(v).
\end{aligned}$$

Assumption 10.3 The functions $\mathbf{u}_i(v, w, \sigma)$, $i = 1, \ldots, N$, are polynomials in v with coefficients possibly depending on w and σ.

Under Assumption 10.3, for $i = 1, \ldots, N$, $\mathbf{u}_i(v, w, \sigma)$ takes the form (A.41) in Sect. A.6. Also, let $L_{\mathbf{a}}\mathbf{u}_i$ denote the Lie derivative of $\mathbf{u}_i$ along the vector field $\mathbf{a}(v, \sigma) = S(\sigma)v$, i.e., $L_{\mathbf{a}}\mathbf{u}_i = \frac{\partial \mathbf{u}_i(v,w,\sigma)}{\partial v} S(\sigma)v$, and, for $l \geq 2$, let $L_{\mathbf{a}}^l \mathbf{u}_i = L_{\mathbf{a}}(L_{\mathbf{a}}^{l-1}\mathbf{u}_i)$. Then, as shown in Proposition A.6 in Sect. A.6, there exist integers s_i, $i = 1, \ldots, N$, such that $\mathbf{u}_i(v, w, \sigma)$ satisfy, for any $v(t)$ generated by the exosystem, and any $\mathrm{col}(w, \sigma) \in \mathbb{W} \times \mathbb{S}$,

$$L_{\mathbf{a}}^{s_i}\mathbf{u}_i = -a_{1i}(\sigma)\mathbf{u}_i - a_{2i}(\sigma)L_{\mathbf{a}}\mathbf{u}_i - \cdots - a_{s_i i}(\sigma)L_{\mathbf{a}}^{s_i - 1}\mathbf{u}_i \tag{10.6}$$

where $a_{1i}(\sigma), \ldots, a_{s_i i}(\sigma)$, $i = 1, \ldots, N$, are scalars such that the roots of polynomials $P_i(\lambda, \sigma) = \lambda^{s_i} - a_{1i}(\sigma) - a_{2i}(\sigma)\lambda - \cdots - a_{s_i i}(\sigma)\lambda^{s_i - 1}$ are distinct with zero real parts for all $\sigma \in \mathbb{S}$.

Let

$$\tau_i(v, w, \sigma) = \mathrm{col}\left(\mathbf{u}_i, L_{\mathbf{a}}\mathbf{u}_i, \ldots, L_{\mathbf{a}}^{s_i - 1}\mathbf{u}_i\right) \tag{10.7}$$

and

$$\Phi_i(\sigma) = \begin{bmatrix} 0 & 1 & \cdots & 0 \\ \vdots & \vdots & \ddots & \vdots \\ 0 & 0 & \cdots & 1 \\ -a_{1i}(\sigma) & -a_{2i}(\sigma) & \cdots & -a_{s_i i}(\sigma) \end{bmatrix}, \ \Psi_i = \begin{bmatrix} 1 \\ 0 \\ \vdots \\ 0 \end{bmatrix}^{\mathsf{T}}_{s_i \times 1}. \tag{10.8}$$

Let (M_i, Q_i), $i = 1, \ldots, N$, be any controllable pair with $M_i \in \mathbb{R}^{s_i \times s_i}$ a Hurwitz matrix, and $Q_i \in \mathbb{R}^{s_i \times 1}$ a column vector. Let $T_i(\sigma)$ be a nonsingular matrix satisfying, for all $\sigma \in \mathbb{S}$,

$$T_i(\sigma)\Phi_i(\sigma)T_i^{-1}(\sigma) = M_i + Q_i\Gamma_i(\sigma) \tag{10.9}$$

and $\Gamma_i(\sigma) = \Psi_i T_i^{-1}(\sigma)$. Then, as in Sect. 9.3, for each subsystem of (10.4), we can construct a canonical internal model

$$\dot{\eta}_i = M_i\eta_i + Q_i u_i, \ i = 1, \ldots, N \tag{10.10}$$

with the estimated steady-state input $\hat{\mathbf{u}}_i = \Gamma_i(\sigma)\eta_i$.

Moreover, it can be verified that all these internal models (10.10) together constitute the internal model of the overall multi-agent system composed of (10.4) and (10.5).

Attaching the internal model (10.10) to the original follower system (10.4) yields the following augmented system:

$$\dot{z}_i = f_i(z_i, y_i, v, w) \tag{10.11a}$$

$$\dot{\eta}_i = M_i\eta_i + Q_i u_i \tag{10.11b}$$

$$\dot{y}_i = g_i(z_i, y_i, v, w) + b_i(w)u_i, \ i = 1, \ldots, N. \tag{10.11c}$$

Let $\theta_i(v, w, \sigma) = T_i(\sigma)\tau_i(v, w, \sigma)$. Then performing on the augmented system (10.11) the following coordinate transformation:

$$\bar{z}_i = z_i - \mathbf{z}_i(v, w, \sigma) \tag{10.12a}$$

$$\bar{\eta}_i = \eta_i - \theta_i(v, w, \sigma) \tag{10.12b}$$

$$\bar{u}_i = u_i - \Gamma_i(\sigma)\eta_i, \ i = 1, \ldots, N \tag{10.12c}$$

gives the following augmented error system:

$$\dot{\bar{z}}_i = \bar{f}_i(\bar{z}_i, e_i, \mu) \tag{10.13a}$$

$$\dot{\bar{\eta}}_i = (M_i + Q_i\Gamma_i(\sigma))\bar{\eta}_i + Q_i\bar{u}_i \tag{10.13b}$$

$$\dot{e}_i = \bar{g}_i(\bar{z}_i, e_i, \mu) + b_i(w)\Gamma_i(\sigma)\bar{\eta}_i + b_i(w)\bar{u}_i, \ i = 1, \ldots, N \tag{10.13c}$$

where $\mu(t) = \text{col}(v(t), w, \sigma)$, and

$$\begin{aligned}\bar{f}_i(\bar{z}_i, e_i, \mu) &= f_i(\bar{z}_i + \mathbf{z}_i(v, w, \sigma), e_i + h_0(v), v, w) \\ &\quad - f_i(\mathbf{z}_i(v, w, \sigma), h_0(v), v, w) \\ \bar{g}_i(\bar{z}_i, e_i, \mu) &= g_i(\bar{z}_i + \mathbf{z}_i(v, w, \sigma), e_i + h_0(v), v, w) \\ &\quad - g_i(\mathbf{z}_i(v, w, \sigma), h_0(v), v, w).\end{aligned}$$

An important property of the augmented error system (10.13) is that, for all $\mu(t)$ with $\mu(0) \in \mathbb{V}_0 \times \mathbb{W} \times \mathbb{S}$, $\bar{f}_i(0, 0, \mu) = 0$ and $\bar{g}_i(0, 0, \mu) = 0$. Thus, from Proposition A.5 in Sect. A.6, we can directly obtain the following result.

Lemma 10.1 *Under Assumptions 10.1–10.3, for any compact subset $\Sigma_0 \subseteq \mathbb{V}_0 \times \mathbb{W} \times \mathbb{S}$, if a distributed virtual error output feedback control law of the following form:*

$$\bar{u}_i = \bar{\kappa}_i(e_{vi}, \xi_i), \quad i = 1, \ldots, N \tag{10.14a}$$

$$\dot{\xi}_i = g_i(e_{vi}) \tag{10.14b}$$

where $\xi_i \in \mathbb{R}^{n_\xi}$ with n_ξ to be specified later, and $\bar{\kappa}_i(\cdot, \cdot)$ and $g_i(\cdot)$ are smooth functions vanishing at the origin, globally stabilizes the zero solution of the augmented error system (10.13) for any $\text{col}(v(0), w, \sigma) \in \Sigma_0$, then the following control law:

$$u_i = \bar{\kappa}_i(e_{vi}, \xi_i) + \Gamma_i(\sigma)\eta_i, \quad i = 1, \ldots, N \tag{10.15a}$$

$$\dot{\xi}_i = g_i(e_{vi}) \tag{10.15b}$$

$$\dot{\eta}_i = M_i\eta_i + Q_i u_i \tag{10.15c}$$

solves the cooperative global output regulation problem for the composite system composed of (10.4) and (10.5).

Remark 10.3 For the case where $\mathbb{S}$ consists of a single known vector, $\Gamma_i(\sigma)$ is a known constant vector, and thus the control law (10.15) is directly implementable. If $\Gamma_i(\sigma)$ is unknown, then the control law (10.15) is not implementable. We need to further introduce some adaptive control technique in Sect. 10.4 to estimate $\Gamma_i(\sigma)$. Also, we allow the dimension of ξ_i to be equal to 0 so that the control law (10.14) can also represent a distributed static control law $\bar{u}_i = \bar{\kappa}_i(e_{vi})$.

To find a control law of the form (10.14) to stabilize the augmented error system (10.13), let us perform on the augmented error system (10.13) the following coordinate transformation:

$$\tilde{\eta}_i = \bar{\eta}_i - b_i^{-1}(w)Q_i e_i, \quad i = 1, \ldots, N. \tag{10.16}$$

Then the augmented error system (10.13) is transformed to the following form:

$$\dot{\bar{z}}_i = \bar{f}_i(\bar{z}_i, e_i, \mu) \tag{10.17a}$$
$$\dot{\tilde{\eta}}_i = M_i\tilde{\eta}_i + b_i^{-1}(w)M_i Q_i e_i - b_i^{-1}(w)Q_i\bar{g}_i(\bar{z}_i, e_i, \mu) \tag{10.17b}$$
$$\dot{e}_i = \bar{g}_i(\bar{z}_i, e_i, \mu) + b_i(w)\Gamma_i(\sigma)\tilde{\eta}_i + \Gamma_i(\sigma)Q_i e_i + b_i(w)\bar{u}_i,\ i = 1, \ldots, N. \tag{10.17c}$$

What makes the system (10.17) special is that the subsystems (10.17a) and (10.17b) are independent of the control input $\bar{u}_i$ and can thus be viewed as the dynamic uncertainty of (10.17) so that (10.17) can be stabilized by a control law depending on e_i only provided that the dynamic uncertainty has some good property. This good property is guaranteed by the following assumption:

Assumption 10.4 For any compact subset $\Sigma \subseteq \mathbb{V} \times \mathbb{W} \times \mathbb{S}$, and, for $i = 1, \ldots, N$, there exist continuously differentiable functions $V_{0i}(\cdot)$ such that

$$\underline{\alpha}_{0i}(\|\bar{z}_i\|) \leq V_{0i}(\bar{z}_i) \leq \overline{\alpha}_{0i}(\|\bar{z}_i\|)$$

for some class $\mathcal{K}_\infty$ functions $\underline{\alpha}_{0i}(\cdot)$ and $\overline{\alpha}_{0i}(\cdot)$, and, for all $\mu \in \Sigma$, along the trajectory of the $\bar{z}_i$ subsystem (10.17a),

$$\dot{V}_{0i} \leq -\alpha_{0i}(\|\bar{z}_i\|) + \delta_i\gamma_i(\|e_i\|)$$

where δ_i are some positive constants depending on Σ, $\alpha_{0i}(\cdot)$ are some known class $\mathcal{K}_\infty$ functions satisfying $\limsup_{s\to 0^+}(s^2/\alpha_{0i}(s)) < \infty$, and $\gamma_i(\cdot)$ are some known class $\mathcal{K}$ functions satisfying $\limsup_{s\to 0^+}(\gamma_i(s)/s^2) < \infty$.

Like in Chap. 9, we also need the following assumption to guarantee the stabilizability of (10.17) by a distributed control law.

Assumption 10.5 The communication graph $\bar{\mathcal{G}}$ contains a spanning tree with the node 0 as the root.

Also, let $\mathcal{G}$ be a subgraph of $\bar{\mathcal{G}}$ obtained from $\bar{\mathcal{G}}$ by removing the node 0 and all edges incident on the node 0. We need one more assumption on $\mathcal{G}$.

Assumption 10.6 The subgraph $\mathcal{G}$ is undirected.

Remark 10.4 It is known from Theorem A.6 in Sect. A.7 that the Assumption 10.4 guarantees that each subsystem $\dot{\bar{z}}_i = \bar{f}_i(\bar{z}_i, e_i, \mu)$ is input-to-state stable with the state $\bar{z}_i$ and the input e_i for all $\mu \in \Sigma$, and thus it also implies that the system $\dot{\bar{z}}_i = \bar{f}_i(\bar{z}_i, 0, \mu)$ is globally asymptotically stable for all $\mu \in \Sigma$. Assumption 10.5 is standard for handling the cooperative control problem over static communication graphs throughout this book. Assumption 10.6 is made to guarantee the symmetry of the matrix H. Under Assumptions 10.5 and 10.6, the matrix H is positive definite.

Let

$$Z_i = \mathrm{col}(\bar{z}_i, \tilde{\eta}_i).$$

Then the system (10.17) is of the form

$$\dot{Z}_i = \tilde{F}_i(Z_i, e_i, \mu) \tag{10.18a}$$
$$\dot{e}_i = \tilde{g}_i(Z_i, e_i, \mu) + b_i(w)\bar{u}_i, \ i = 1, \ldots, N \tag{10.18b}$$

where

$$\tilde{F}_i(Z_i, e_i, \mu) = \begin{bmatrix} \bar{f}_i(\bar{z}_i, e_i, \mu) \\ M_i\tilde{\eta}_i + b_i^{-1}(w)M_i Q_i e_i - b_i^{-1}(w)Q_i\bar{g}_i(\bar{z}_i, e_i, \mu) \end{bmatrix}$$
$$\tilde{g}_i(Z_i, e_i, \mu) = \bar{g}_i(\bar{z}_i, e_i, \mu) + b_i(w)\Gamma_i(\sigma)\tilde{\eta}_i + \Gamma_i(\sigma)Q_i e_i.$$

Lemma 10.2 *Under Assumptions 10.1–10.4, for $i = 1, \ldots, N$, there exist continuously differentiable functions $U_{1i}(Z_i)$ such that*

$$\underline{\beta}_{1i}(\|Z_i\|) \leq U_{1i}(Z_i) \leq \overline{\beta}_{1i}(\|Z_i\|) \tag{10.19}$$

for some class $\mathcal{K}_\infty$ functions $\underline{\beta}_{1i}(\cdot)$ and $\overline{\beta}_{1i}(\cdot)$, and, for all $\mu \in \Sigma$, along the trajectory of the Z_i-subsystem (10.18a),

$$\dot{U}_{1i}(Z_i) \leq -\|Z_i\|^2 + \tilde{\delta}_i\tilde{\gamma}_i(e_i) \tag{10.20}$$

for some positive constants $\tilde{\delta}_i$ whose values depend on Σ and some known smooth positive definite functions $\tilde{\gamma}_i(\cdot)$, $i = 1, \ldots, N$.

Proof Since Assumption 10.4 holds, for $i = 1, \ldots, N$, by Lemma A.12 in Sect. A.7 with $z_1 = \bar{z}_i$, $z_2 = \tilde{\eta}_i$, and $\mu = \mathrm{col}(v, w, \sigma)$, there exist continuously differentiable functions $U_{1i}(Z_i)$ satisfying (10.19) for some class $\mathcal{K}_\infty$ functions $\underline{\beta}_{1i}(\cdot)$ and $\overline{\beta}_{1i}(\cdot)$, and, for all $\mu \in \Sigma$, (10.20) for some positive constants $\tilde{\delta}_i$ whose values depend on Σ and some known smooth positive definite functions $\tilde{\gamma}_i(\cdot)$. The proof is thus completed. □

Next, let $Z = \mathrm{col}(Z_1, \ldots, Z_N)$, $e = \mathrm{col}(e_1, \ldots, e_N)$, and $\bar{u} = \mathrm{col}(\bar{u}_1, \ldots, \bar{u}_N)$. Then the system (10.18) can be put into the following compact form:

$$\dot{Z} = F(Z, e, \mu) \tag{10.21a}$$
$$\dot{e} = G(Z, e, \mu) + B(w)\bar{u} \tag{10.21b}$$

where $B(w) = \mathscr{D}(b_1(w), \ldots, b_N(w))$, and

$$F(Z, e, \mu) = \begin{bmatrix} \tilde{F}_1(Z_1, e_1, \mu) \\ \vdots \\ \tilde{F}_N(Z_N, e_N, \mu) \end{bmatrix}, \ G(Z, e, \mu) = \begin{bmatrix} \tilde{g}_1(Z_1, e_1, \mu) \\ \vdots \\ \tilde{g}_N(Z_N, e_N, \mu) \end{bmatrix}.$$

We call (10.21a) the inverse dynamics of (10.21), which has the following property.

Lemma 10.3 *Under Assumptions 10.1–10.4, given any smooth functions $\Theta_i(Z_i) > 0$, there exists a continuously differentiable function $V_1(Z)$ such that*

$$\underline{\alpha}_1(\|Z\|) \le V_1(Z) \le \overline{\alpha}_1(\|Z\|) \tag{10.22}$$

for some class $\mathcal{K}_\infty$ functions $\underline{\alpha}_1(\cdot)$ and $\overline{\alpha}_1(\cdot)$, and, for all $\mu \in \Sigma$, along the trajectory of the subsystem (10.21a),

$$\dot{V}_1(Z) \le -\sum_{i=1}^{N} \Theta_i(Z_i)\|Z_i\|^2 + \hat{\delta}\sum_{i=1}^{N} \hat{\gamma}_i(e_i)e_i^2 \tag{10.23}$$

where $\hat{\delta}$ is some positive constant depending on Σ, and $\hat{\gamma}_i(\cdot)$, $i = 1, \ldots, N$, are some known smooth positive functions.

Proof By Lemma 10.2, for $i = 1, \ldots, N$, there exist continuously differentiable functions $U_{1i}(Z_i)$ satisfying (10.19) and, for all $\mu \in \Sigma$, (10.20).

From (10.20), by the changing supply functions technique (Corollary A.2 in Sect. A.7), given any smooth function $\Theta_i(Z_i) > 0$, there exists a continuously differentiable function $V_{1i}(Z_i)$ satisfying $\underline{\alpha}_{1i}(\|Z_i\|) \le V_{1i}(Z_i) \le \overline{\alpha}_{1i}(\|Z_i\|)$ for some class $\mathcal{K}_\infty$ functions $\underline{\alpha}_{1i}(\cdot)$ and $\overline{\alpha}_{1i}(\cdot)$ such that, for all $\mu \in \Sigma$, along the trajectory of the Z_i-subsystem,

$$\dot{V}_{1i}(Z_i) \le -\Theta_i(Z_i)\|Z_i\|^2 + \hat{\delta}_i\hat{\gamma}_i(e_i)e_i^2$$

for some positive constants $\hat{\delta}_i$ depending on Σ and some known smooth functions $\hat{\gamma}_i(\cdot) \ge 1$, $i = 1, \ldots, N$. Let $V_1(Z) = \sum_{i=1}^{N} V_{1i}(Z_i)$. Then, by Lemma A.9 in Sect. A.7, there exist some class $\mathcal{K}_\infty$ functions $\underline{\alpha}_1(\cdot)$ and $\overline{\alpha}_1(\cdot)$ such that (10.22) holds, and, for all $\mu \in \Sigma$, along the trajectory of the subsystem (10.21a), (10.23) holds with $\hat{\delta} = \max_{i=1,\ldots,N}\{\hat{\delta}_i\}$. The proof is thus completed. □

Remark 10.5 If the compact subset Σ is known, then the positive constants δ_i and $\hat{\delta}_i$ are also known. This case will be considered in Sect. 10.2 by employing the robust control approach. In contrast, if Σ is unknown, δ_i and $\hat{\delta}_i$ are unknown. Such a case will be studied in Sect. 10.3 where the matrix $S(\sigma)$ is known and in Sect. 10.4 where the matrix $S(\sigma)$ is uncertain.

10.2 Solvability of Case 1 of the Problem

As mentioned in the last section, in this section, we will handle the first case where the exosystem is known exactly and the compact subset $\Sigma_0 = \bar{\mathbb{V}}_0 \times \mathbb{W}_0 \times \mathbb{S}_0$ is also known. Since the leader system (10.5) contains no uncertain parameters, $\mathbb{S}$ and hence $\mathbb{S}_0$ consist of a known singleton. Since $\bar{\mathbb{V}}_0$ is known, there exists a known compact subset $\bar{\mathbb{V}} \subseteq \mathbb{V}$ such that, for all $v(0) \in \bar{\mathbb{V}}_0$, $v(t) \in \bar{\mathbb{V}}$ for all $t \geq 0$. For simplicity, we can let $\Sigma_0 = \bar{\mathbb{V}}_0 \times \mathbb{W}_0$ which is just a subset of $\mathbb{V}_0 \times \mathbb{W}$, and correspondingly, $\Sigma = \bar{\mathbb{V}} \times \mathbb{W}_0$ is just a known compact subset of $\mathbb{V} \times \mathbb{W}$, which is such that, for all $(v(0), w) \in \Sigma_0$, $(v(t), w) \in \Sigma$ for all $t \geq 0$. Also, for simplicity, let us denote $\Gamma_i(\sigma)$ by Γ_i. Then the function $\tilde{g}_i(Z_i, e_i, \mu)$ in the system (10.18) is simplified as follows:

$$\tilde{g}_i(Z_i, e_i, \mu) = \bar{g}_i(\bar{z}_i, e_i, \mu) + b_i(w)\Gamma_i \tilde{\eta}_i + \Gamma_i Q_i e_i$$

where $\mu(t) = \mathrm{col}(v(t), w) \in \Sigma \subseteq \mathbb{V} \times \mathbb{W}$.

Recall from (10.21) that

$$G(Z, e, \mu) = \mathrm{col}(\tilde{g}_1(Z_1, e_1, \mu), \ldots, \tilde{g}_N(Z_N, e_N, \mu)). \tag{10.24}$$

Since $G(0, 0, \mu) = 0$, for all $\mu \in \Sigma$, by Part (iii) of Lemma A.8 in Sect. A.7, for all $\mu \in \Sigma$,

$$\|G(Z, e, \mu)\|^2 \leq c_0 \left(\sum_{i=1}^{N} \pi_i(Z_i)\|Z_i\|^2 + \sum_{i=1}^{N} \phi_i(e_i)e_i^2 \right) \tag{10.25}$$

for some positive constant c_0 depending on Σ and some known smooth functions $\pi_i(\cdot) \geq 1$ and $\phi_i(\cdot) \geq 1$, $i = 1, \ldots, N$. Let $e_v = \mathrm{col}(e_{v1}, \ldots, e_{vN})$. Recall that $e_v = He$. Then we have the following result regarding the stabilization of the augmented error system (10.21).

Lemma 10.4 *Under Assumptions 10.1–10.6, for some smooth positive functions $\rho_i(\cdot) \geq 1$ and some positive numbers k_i, $i = 1, \ldots, N$, the closed-loop system composed of the system (10.21) and the following distributed static output feedback control law:*

$$\bar{u}_i = -k_i \rho_i(e_{vi}) e_{vi}, \; i = 1, \ldots, N \tag{10.26}$$

has the property that there exists a continuously differentiable function $V_a(Z, e)$ such that

$$\underline{\alpha}_a(\|(Z, e)\|) \leq V_a(Z, e) \leq \overline{\alpha}_a(\|(Z, e)\|) \tag{10.27}$$

for some class $\mathcal{K}_\infty$ functions $\underline{\alpha}_a(\cdot)$ and $\overline{\alpha}_a(\cdot)$, and, for all $\mu \in \Sigma$,

$$\dot{V}_a(Z, e) \leq -\|Z\|^2 - \|e\|^2. \tag{10.28}$$

That is to say, for all $\mu \in \Sigma$, the origin of the closed-loop system is globally asymptotically stable.

Proof Let $V_2(e) = \frac{1}{2}e^{\mathsf{T}}He$. Then V_2 is proper and globally positive definite, and, under the control law (10.26), the derivative of V_2 along the trajectory of the subsystem (10.21b) satisfies

$$\begin{aligned}
\dot{V}_2(e) &= e_v^{\mathsf{T}}\dot{e} \\
&= e_v^{\mathsf{T}}G(Z, e, \mu) - \sum_{i=1}^{N} b_i(w)k_i\rho_i(e_{vi})e_{vi}^2 \\
&\leq \frac{c_0}{4}\|e_v\|^2 + \frac{1}{c_0}\|G(Z, e, \mu)\|^2 - \sum_{i=1}^{N} b_i(w)k_i\rho_i(e_{vi})e_{vi}^2 \\
&\leq \sum_{i=1}^{N}\pi_i(Z_i)\|Z_i\|^2 + \sum_{i=1}^{N}\phi_i(e_i)e_i^2 - \sum_{i=1}^{N}\left(b_i(w)k_i\rho_i(e_{vi}) - \frac{c_0}{4}\right)e_{vi}^2.
\end{aligned} \tag{10.29}$$

Now, let

$$V_a(Z, e) = V_1(Z) + V_2(e) \tag{10.30}$$

where $V_1(Z)$ satisfies (10.23). Then, by Lemma A.9 in Sect. A.7,

$$\underline{\alpha}_a(\|(Z, e)\|) \leq V_a(Z, e) \leq \overline{\alpha}_a(\|(Z, e)\|)$$

for some class $\mathcal{K}_\infty$ functions $\underline{\alpha}_a(\cdot)$ and $\overline{\alpha}_a(\cdot)$, and, by (10.23) and (10.29), for all $\mu \in \Sigma$, along the trajectory of the closed-loop system, we have

$$\begin{aligned}
\dot{V}_a(Z, e) \leq &- \sum_{i=1}^{N}\Theta_i(Z_i)\|Z_i\|^2 + \sum_{i=1}^{N}\delta\hat{\gamma}_i(e_i)e_i^2 \\
&+ \sum_{i=1}^{N}\pi_i(Z_i)\|Z_i\|^2 + \sum_{i=1}^{N}\phi_i(e_i)e_i^2 - \sum_{i=1}^{N}\left(k_ib_i(w)\rho_i(e_{vi}) - \frac{c_0}{4}\right)e_{vi}^2 \\
\leq &- \sum_{i=1}^{N}(\Theta_i(Z_i) - \pi_i(Z_i))\|Z_i\|^2 - \sum_{i=1}^{N}\left(k_ib_i(w)\rho_i(e_{vi}) - \frac{c_0}{4}\right)e_{vi}^2 + \delta\gamma(e)
\end{aligned} \tag{10.31}$$

where $\delta = 1 + \hat{\delta}$, and $\gamma(e) = \sum_{i=1}^{N} \bar{\phi}_i(e_i)e_i^2$ with $\bar{\phi}_i(\cdot)$ being some known smooth functions such that

$$\bar{\phi}_i(e_i) \geq \max\{\hat{\gamma}_i(e_i), \phi_i(e_i)\}. \tag{10.32}$$

By Part (ii) of Lemma A.8 in Sect. A.7 and noting that every term of $\gamma(e)$ contains a factor e_i^2, we have

$$\gamma(e) = \gamma(H^{-1}e_v) \leq \sum_{i=1}^{N} \tilde{\phi}_i(e_{vi})e_{vi}^2 \tag{10.33}$$

for some known smooth functions $\tilde{\phi}_i(\cdot) \geq 1$. Substituting (10.33) into (10.31) gives

$$\dot{V}_a(Z, e) \leq -\sum_{i=1}^{N}(\Theta_i(Z_i) - \pi_i(Z_i))\|Z_i\|^2 - \sum_{i=1}^{N}\left(k_i b_i(w)\rho_i(e_{vi}) - \delta\tilde{\phi}_i(e_{vi}) - \frac{c_0}{4}\right)e_{vi}^2. \tag{10.34}$$

In (10.34), letting $\Theta_i(\cdot)$, k_i, and $\rho_i(\cdot)$, $i = 1, \ldots, N$, be such that

$$\Theta_i(Z_i) \geq \pi_i(Z_i) + 1 \tag{10.35}$$

$$k_i\rho_i(e_{vi}) \geq \frac{1}{b_{\min}}\left(\delta\tilde{\phi}_i(e_{vi}) + \frac{c_0}{4} + \frac{1}{\lambda_{\min}(H^2)}\right) \tag{10.36}$$

gives

$$\dot{V}_a(Z, e) \leq -\sum_{i=1}^{N}\|Z_i\|^2 - \frac{1}{\lambda_{\min}(H^2)}\sum_{i=1}^{N}|e_{vi}|^2 \leq -\|Z\|^2 - \|e\|^2.$$

By the standard Lyapunov argument, for all $\mu \in \Sigma$, the origin of the closed-loop system is globally asymptotically stable. Finally, we note that a simple but conservative way to make (10.36) be satisfied is to let

$$k_i \geq \frac{1}{b_{\min}}\left(\delta + \frac{c_0}{4} + \frac{1}{\lambda_{\min}(H^2)}\right) \tag{10.37}$$

$$\rho_i(e_{vi}) \geq \tilde{\phi}_i(e_{vi}). \tag{10.38}$$

□

Let $\bar{\kappa}_i(e_{vi}) = -k_i\rho_i(e_{vi})e_{vi}$. Then, (10.26) is a special case of (10.14) with $n_\xi = 0$. Combining Lemmas 10.1 and 10.4 gives the following main result in this section.

Theorem 10.1 *Suppose the exosystem and the compact subset Σ_0 are known. Under Assumptions 10.1–10.6, let $\rho_i(\cdot) \geq 1$ and $k_i > 0$ be the same as those in Lemma 10.4,* *and M_i, Q_i, $\Gamma_i(\sigma) = \Gamma_i$ the same as those in (10.9). Then the following distributed dynamic output feedback control law:*

$$u_i = -k_i\rho_i(e_{vi})e_{vi} + \Gamma_i\eta_i \tag{10.39a}$$
$$\dot{\eta}_i = M_i\eta_i + Q_iu_i,\ i = 1, \ldots, N \tag{10.39b}$$

solves the global cooperative robust output regulation problem for the original nonlinear plant (10.4).

Remark 10.6 The control law (10.26) is defined by a positive constant k_i and a known function $\rho_i(\cdot)$. The design process entails two applications of the changing supply functions technique (one is to obtain $U_{1i}(Z_i)$ satisfying (10.20) and the other is to obtain $V_1(Z)$ satisfying (10.23)). However, this systematic approach may result in a large constant k_i and a large function $\rho_i(\cdot)$. In some situations, for example, the system (10.4) is a polynomial nonlinear system, it is possible to reduce this conservatism by adopting a more pragmatic approach.

Indeed, all one needs to do is to choose k_i and $\rho_i(e_{vi})$ to satisfy (10.36). From (10.32) and (10.33), $\tilde{\phi}_i(e_{vi})$ is a polynomial of e_{vi} if $\phi_i(e_i)$ and $\hat{\gamma}_i(e_i)$ are both polynomials of e_i. On one hand, from (10.25), the function $\phi_i(e_i)$ is a polynomial in e_i if $g_i(z_i, y_i, v, w)$ is a polynomial in (z_i, y_i). On the other hand, for a polynomial nonlinear system of the form (10.4), if Assumption 10.4 is satisfied by a positive definite polynomial function $V_{0i}(Z_i)$, then, for any polynomial function $\Theta_i(Z_i)$ satisfying $\Theta_i(Z_i) \geq \pi_i(Z_i) + 1$, one can find a positive definite polynomial function $V_1(Z)$ such that $\dot{V}_1(Z)$ satisfies (10.23) with $\hat{\gamma}_i(e_i)$ a polynomial. Once a polynomial $\tilde{\phi}_i(e_{vi})$ with degree m_0 is obtained, one can find a sufficiently large number k_i and a polynomial function $\rho_i(s) = (1 + s^{2m})$, with m satisfying $2m \geq m_0$, such that (10.36) is satisfied.

It is noted that $V_1(Z)$ does not have to be found by the changing supply functions technique. It can be found by anyway as long as $\dot{V}_1(Z)$ satisfies (10.23) such that $\Theta_i(Z_i) \geq \pi_i(Z_i) + 1$ for all Z_i.

In practice, it is more efficient to work on the components of Z_i. More specifically, denote the dimension of Z_i by n_{Z_i} and partition Z_i as $Z_i = \text{col}(\zeta_{1i}, \ldots, \zeta_{s_i i})$ with $\zeta_{li} \in \mathbb{R}^{n_{z_{li}}}, l = 1, \ldots, s_i$, such that $\sum_{l=1}^{s_i} n_{z_{li}} = n_{Z_i}$. Then, instead of (10.23), we aim to achieve the following:

$$\dot{V}_1(Z) \leq -\sum_{i=1}^{N}\sum_{l=1}^{s_i}\Theta_{li}(\zeta_{li})\|\zeta_{li}\|^2 + \hat{\delta}\sum_{i=1}^{N}\hat{\gamma}_i(e_i)e_i^2. \tag{10.40}$$

Correspondingly, instead of (10.25), by making use of the following inequality:

$$\|G(Z, e, \mu)\|^2 \leq c_0\left(\sum_{i=1}^{N}\sum_{l=1}^{s_i}\pi_{li}(\zeta_{li})\|\zeta_{li}\|^2 + \sum_{i=1}^{N}\phi_i(e_i)e_i^2\right) \tag{10.41}$$

we obtain, instead of (10.34), the following:

$$\dot{V}_a(Z,e) \leq -\sum_{i=1}^{N}\sum_{l=1}^{s_i}(\Theta_{li}(\zeta_{li}) - \pi_{li}(\zeta_{li}))\|\zeta_{li}\|^2 + \sum_{i=1}^{N}\left(k_i b_i(w)\rho_i(e_{vi}) - \delta\tilde{\phi}_i(e_{vi}) - \frac{c_0}{4}\right)e_{vi}^2. \quad (10.42)$$

Thus, as long as $\Theta_{li}(\zeta_{li}) \geq \pi_{li}(\zeta_{li}) + 1$ for all ζ_i, we can still use (10.36) to obtain the number k_i and the function $\rho_i(\cdot)$.

10.3 Solvability of Case 2 of the Problem

In this section, we study the second case where the exosystem is known and the compact subset $\Sigma_0 = \bar{\mathbb{V}}_0 \times \mathbb{W}_0 \subseteq \mathbb{V}_0 \times \mathbb{W}$ is unknown. Like in Sect. 10.2, there exists an unknown compact subset $\bar{\mathbb{V}} \subseteq \mathbb{V}$ such that for all $v(0) \in \bar{\mathbb{V}}_0$, $v(t) \in \bar{\mathbb{V}}$ for all $t \geq 0$. Let $\Sigma = \bar{\mathbb{V}} \times \mathbb{W}_0$. Then, $\Sigma \subseteq \mathbb{V} \times \mathbb{W}$. Thus, for all $(v(0), w) \in \Sigma_0$, $(v(t), w) \in \Sigma$ for all $t \geq 0$. Also, for simplicity, let us denote $\Gamma_i(\sigma)$ by Γ_i.

We have the following result regarding the stabilization of the augmented error system (10.21).

Lemma 10.5 *Under Assumptions 10.1–10.6, for some smooth positive functions $\rho_i(\cdot) \geq 1$, $i = 1, \ldots, N$, the closed-loop system composed of (10.21) and the following distributed dynamic output feedback control law:*

$$\bar{u}_i = -k_i\rho_i(e_{vi})e_{vi}, \quad (10.43a)$$

$$\dot{k}_i = \rho_i(e_{vi})e_{vi}^2, \; i = 1, \ldots, N \quad (10.43b)$$

has the property that there exists a continuously differentiable function $V_b(Z, e, \tilde{k})$ such that

$$\underline{\alpha}_b(\|(Z, e, \tilde{k})\|) \leq V_b(Z, e, \tilde{k}) \leq \overline{\alpha}_b(\|(Z, e, \tilde{k})\|) \quad (10.44)$$

for some class $\mathcal{K}_\infty$ functions $\underline{\alpha}_b(\cdot)$ and $\overline{\alpha}_b(\cdot)$, and, for all $\mu \in \Sigma$,

$$\dot{V}_b(Z, e, \tilde{k}) \leq -\|Z\|^2 - \|e\|^2 \quad (10.45)$$

where $\tilde{k} = \mathrm{col}(\tilde{k}_1, \ldots, \tilde{k}_N)$ with $\tilde{k}_i(t) = k_i(t) - \bar{k}$, $i = 1, \ldots, N$, and with $\bar{k}$ being some positive constant. As a result, for any initial condition, the solution of the closed-loop system composed of (10.21) and (10.43) is uniformly bounded over $[0, \infty)$, and

$$\lim_{t\to\infty}(\|Z(t)\| + \|e(t)\|) = 0. \quad (10.46)$$

Proof Let

$$V_b(Z, e, \tilde{k}) = V_a(Z, e) + \frac{1}{2}\sum_{i=1}^{N} b_i(w)\tilde{k}_i^2 \tag{10.47}$$

where $V_a(Z, e)$ is as defined in (10.30). Then, by Lemma A.9 in Sect. A.7,

$$\underline{\alpha}_b(\|(Z, e, \tilde{k})\|) \leq V_b(Z, e, \tilde{k}) \leq \overline{\alpha}_b(\|(Z, e, \tilde{k})\|)$$

for some class $\mathcal{K}_\infty$ functions $\underline{\alpha}_b(\cdot)$ and $\overline{\alpha}_b(\cdot)$, and, by (10.31), for all $\mu \in \Sigma$, along the trajectory of the closed-loop system, we have

$$\begin{aligned}
\dot{V}_b(Z, e, \tilde{k}) \leq & -\sum_{i=1}^{N} \Theta_i(Z_i)\|Z_i\|^2 + \sum_{i=1}^{N} \hat{\delta}\hat{\gamma}_i(e_i)e_i^2 + \frac{c_0}{4}\sum_{i=1}^{N} e_{vi}^2 \\
& + \sum_{i=1}^{N} \pi_i(Z_i)\|Z_i\|^2 + \sum_{i=1}^{N} \phi_i(e_i)e_i^2 - \sum_{i=1}^{N} b_i(w)k_i\rho_i(e_{vi})e_{vi}^2 + \sum_{i=1}^{N} b_i(w)\tilde{k}_i\dot{k}_i \\
\leq & -\sum_{i=1}^{N}(\Theta_i(Z_i) - \pi_i(Z_i))\|Z_i\|^2 - \sum_{i=1}^{N}\left(\bar{k}b_i(w)\rho_i(e_{vi}) - \frac{c_0}{4}\right)e_{vi}^2 \\
& + \delta\gamma(e) + \sum_{i=1}^{N} b_i(w)\tilde{k}_i(\dot{k}_i - \rho_i(e_{vi})e_{vi}^2)
\end{aligned} \tag{10.48}$$

where $\delta = 1 + \hat{\delta}$, and $\gamma(e) = \sum_{i=1}^{N} \bar{\phi}_i(e_i)e_i^2$ with $\bar{\phi}_i(\cdot)$ being some known smooth functions such that $\bar{\phi}_i(e_i) \geq \max\{\hat{\gamma}_i(e_i), \phi_i(e_i)\}$. Using (10.32) and (10.43b) gives

$$\dot{V}_b(Z, e, \tilde{k}) \leq -\sum_{i=1}^{N}(\Theta_i(Z_i) - \pi_i(Z_i))\|Z_i\|^2 - \sum_{i=1}^{N}\left(\bar{k}b_i(w)\rho_i(e_{vi}) - \delta\tilde{\phi}_i(e_{vi}) - \frac{c_0}{4}\right)e_{vi}^2. \tag{10.49}$$

In (10.49), letting $\Theta_i(\cdot)$ be given by (10.35), and $\bar{k}$ and $\rho_i(\cdot)$, $i = 1, \ldots, N$, satisfy the following:

$$\bar{k}\rho_i(e_{vi}) \geq \frac{1}{b_{\min}}\left(\delta\tilde{\phi}_i(e_{vi}) + \frac{c_0}{4} + \frac{1}{\lambda_{\min}(H^2)}\right) \tag{10.50}$$

gives

$$\dot{V}_b(Z, e, \tilde{k}) \leq -\sum_{i=1}^{N}\|Z_i\|^2 - \frac{1}{\lambda_{\min}(H^2)}\sum_{i=1}^{N}|e_{vi}|^2 \leq -\|Z\|^2 - \|e\|^2.$$

By LaSalle–Yoshizawa Theorem (Theorem A.4 in Sect. A.3), the functions $Z(t)$, $e(t)$, and $\tilde{k}(t)$ are all uniformly bounded over $[0, \infty)$, and (10.46) holds. Again, a simple but conservative way to make (10.50) be satisfied is to let

$$\bar{k} \geq \frac{1}{b_{\min}}\left(\delta + \frac{c_0}{4} + \frac{1}{\lambda_{\min}(H^2)}\right) \tag{10.51}$$

$$\rho_i(e_{vi}) \geq \tilde{\phi}_i(e_{vi}). \tag{10.52}$$

□

Combining Lemmas 10.1 and 10.5 with $k_i = \xi_i$ gives the following result.

Theorem 10.2 *Suppose the exosystem is known and the compact subset Σ_0 is unknown. Under Assumptions 10.1–10.6, let $\rho_i(\cdot) \geq 1$ be the same as those in Lemma 10.5, and M_i, Q_i, $\Gamma_i(\sigma) = \Gamma_i$ the same as those in (10.9). Then the distributed dynamic output feedback control law:*

$$u_i = -k_i \rho_i(e_{vi})e_{vi} + \Gamma_i \eta_i \tag{10.53a}$$

$$\dot{\eta}_i = M_i \eta_i + Q_i u_i \tag{10.53b}$$

$$\dot{k}_i = \rho_i(e_{vi})e_{vi}^2, \ i = 1, \ldots, N \tag{10.53c}$$

solves the global cooperative robust output regulation problem for the original nonlinear plant (10.4).

Remark 10.7 The quantity k_i in the control law (10.53) is called the dynamic gain. It will approach a sufficiently large unknown constant number.

Remark 10.8 The approach described in Remark 10.6 still applies to this case. However, since $\bar{k}$ is unknown, there is no need to calculate it. Thus, one only need to determine the functions $\rho_i(\cdot)$ according to (10.50) assuming $\bar{k}$ can be arbitrarily large. In particular, if $\tilde{\phi}_i(e_{vi})$ is a polynomial of degree $2m$ for some integer $m > 0$, one can always choose $\tilde{\phi}_i(e_{vi}) = k(1 + e_{vi}^{2m})$ for any $k > 0$. It is noted that the value of k will affect the transient response of the closed-loop system.

10.4 Solvability of Case 3 of the Problem

In this section, we study the third case where the exosystem contains an uncertain parameter vector $\sigma \in \mathbb{S}$, and the compact subset $\Sigma_0 = \bar{\mathbb{V}}_0 \times \mathbb{W}_0 \times \mathbb{S}_0 \subseteq \mathbb{V}_0 \times \mathbb{W} \times \mathbb{S}$ is unknown. In this case, there exists an unknown compact subset $\Sigma = \bar{\mathbb{V}} \times \mathbb{W}_0 \times \mathbb{S}_0 \subseteq \mathbb{V} \times \mathbb{W} \times \mathbb{S}$ such that, for all $(v(0), w, \sigma) \in \Sigma_0$, $\mu(t) = \mathrm{col}(v(t), w, \sigma) \in \Sigma$ for all $t \geq 0$. As pointed out in Sect. 10.1, when σ is unknown, the control law (10.53) is not implementable. We need to further introduce the adaptive control technique to estimate $\Gamma_i(\sigma)$. For this purpose, substituting $\bar{u}_i = u_i - \Gamma_i(\sigma)\eta_i$ into (10.18) gives the following system:

$$\dot{Z}_i = \tilde{F}_i(Z_i, e_i, \mu) \tag{10.54a}$$

$$\dot{e}_i = \tilde{g}_i(Z_i, e_i, \mu) + b_i(w)(u_i - \Gamma_i(\sigma)\eta_i). \tag{10.54b}$$

We have the following result.

Theorem 10.3 *Under Assumptions 10.1–10.6, let $\rho_i(\cdot)$ be the same as those in Lemma 10.5, and M_i, Q_i, $\Gamma_i(\sigma)$ the same as those in (10.9). Then, the following distributed dynamic output feedback control law:*

$$u_i = -k_i\rho_i(e_{vi})e_{vi} + \hat{\Gamma}_i\eta_i \tag{10.55a}$$
$$\dot{\eta}_i = M_i\eta_i + Q_iu_i \tag{10.55b}$$
$$\dot{\hat{\Gamma}}_i = -\eta_i^{\mathsf{T}}e_{vi} \tag{10.55c}$$
$$\dot{k}_i = \rho_i(e_{vi})e_{vi}^2,\ i = 1, \ldots, N \tag{10.55d}$$

solves the global cooperative output regulation problem for the original nonlinear plant (10.4).

Proof Let $\tilde{\Gamma}_i(t) = \hat{\Gamma}_i(t) - \Gamma_i(\sigma)$, $i = 1, \ldots, N$. First, consider the following distributed dynamic output feedback control law:

$$\begin{aligned} u_i &= -k_i\rho_i(e_{vi})e_{vi} + \hat{\Gamma}_i\eta_i \\ &= -k_i\rho_i(e_{vi})e_{vi} + \Gamma_i(\sigma)\eta_i + \tilde{\Gamma}_i\eta_i \end{aligned} \tag{10.56a}$$
$$\dot{\hat{\Gamma}}_i = -\eta_i^{\mathsf{T}}e_{vi} \tag{10.56b}$$
$$\dot{k}_i = \rho_i(e_{vi})e_{vi}^2,\ i = 1, \ldots, N. \tag{10.56c}$$

Let

$$V_c(Z, e, \tilde{k}, \tilde{\Gamma}) = V_b(Z, e, \tilde{k}) + \frac{1}{2}\sum_{i=1}^{N} b_i(w)\tilde{\Gamma}_i\tilde{\Gamma}_i^{\mathsf{T}},$$

where $V_b(Z, e, \tilde{k})$ is as defined in (10.47), and $\tilde{\Gamma} = [\tilde{\Gamma}_1, \ldots, \tilde{\Gamma}_N]$. Then, by Lemma A.9 in Sect. A.7, V_c is continuously differentiable satisfying

$$\underline{\alpha}_c(\|(Z, e, \tilde{k}, \tilde{\Gamma})\|) \le V_c(Z, e, \tilde{k}, \tilde{\Gamma}) \le \overline{\alpha}_c(\|(Z, e, \tilde{k}, \tilde{\Gamma})\|)$$

for some class $\mathcal{K}_\infty$ functions $\underline{\alpha}_c(\cdot)$ and $\overline{\alpha}_c(\cdot)$. Moreover, it can be verified that, for all $\mu \in \Sigma$, the derivative of V_c along the trajectory of the system (10.54) under the control law (10.56) satisfies

$$\begin{aligned}
&\dot{V}_c(Z, e, \tilde{k}, \tilde{\Gamma}) \mid_{(10.54)+(10.56)} \\
=&\dot{V}_b(Z, e, \tilde{k}) \mid_{(10.54)+(10.56)} + \sum_{i=1}^{N} b_i(w)\dot{\hat{\Gamma}}_i \tilde{\Gamma}_i^{\mathsf{T}} \\
=&\dot{V}_b(Z, e, \tilde{k}) \mid_{(10.18)+(10.43)} + \sum_{i=1}^{N} b_i(w)\frac{\partial V}{\partial e_i}\tilde{\Gamma}_i \eta_i + \sum_{i=1}^{N} b_i(w)\dot{\hat{\Gamma}}_i \tilde{\Gamma}_i^{\mathsf{T}} \\
\leq& - \|Z\|^2 - \|e\|^2 + \sum_{i=1}^{N} b_i(w)(\eta_i^{\mathsf{T}} e_{vi} + \dot{\hat{\Gamma}}_i)\tilde{\Gamma}_i^{\mathsf{T}} \\
=& - \|Z\|^2 - \|e\|^2.
\end{aligned}$$

By LaSalle–Yoshizawa Theorem again (Theorem A.4 in Sect. A.3), the functions $Z(t)$, $e(t)$, $\tilde{\Gamma}(t)$, and $\tilde{k}(t)$ are all uniformly bounded over $[0, \infty)$, and

$$\lim_{t\to\infty} (\|Z(t)\| + \|e(t)\|) = 0.$$

Thus, the control law (10.55) solves the cooperative global robust output regulation problem for the original plant (10.4). □

It is interesting to ascertain whether or not the estimated parameter vectors $\hat{\Gamma}_i(t)$, $i = 1, \ldots, N$, will converge to the actual parameter vectors $\Gamma_i(\sigma)$. For this purpose, let us give the following lemma.

Lemma 10.6 *Under Assumptions 10.1–10.6, for any initial condition, the solution of the closed-loop system composed of the plant (10.4) and the control law (10.55) satisfies, for $i = 1, \ldots, N$,*

$$\lim_{t\to\infty} \dot{\hat{\Gamma}}_i(t) = 0 \tag{10.57}$$

$$\lim_{t\to\infty} (\hat{\Gamma}_i(t) - \Gamma_i(\sigma))T_i(\sigma)\tau_i(v(t), w, \sigma) = 0. \tag{10.58}$$

Moreover, if the function $\tau_i(v(t), w, \sigma)$ is persistently exciting (PE), where τ_i is given in (10.7), then

$$\lim_{t\to\infty} (\hat{\Gamma}_i(t) - \Gamma_i(\sigma)) = 0. \tag{10.59}$$

Proof Under the assumptions of this lemma, by Theorem 10.3, $\lim_{t\to\infty} e_v(t) = \lim_{t\to\infty} He(t) = 0$ and, for $i = 1, \ldots, N$, $\eta_i(t)$ is uniformly bounded over $[0, \infty)$. By (10.55c), (10.57) holds.

Note that $\lim_{t\to\infty} e_{vi}(t) = 0$ implies $\lim_{t\to\infty} \rho(e_{vi}(t))e_{vi}(t) = 0$. Since $k_i(t)$ is uniformly bounded over $[0, \infty)$, by (10.55a), we have

$$\lim_{t\to\infty} (u_i(t) - \hat{\Gamma}_i(t)\eta_i(t)) = 0. \tag{10.60}$$

On the other hand, since $\lim_{t\to\infty} \bar{g}_i(\bar{z}_i(t), \tilde{\eta}_i(t), e_i(t), \mu(t)) = 0, i = 1, \ldots, N$, and $e_i(t)$, $e_{vi}(t)$, $\ddot{e}_i(t)$ and $\ddot{e}_{vi}(t)$ are all uniformly bounded over $[0, \infty)$, by (10.54b), using Barbalat's lemma gives $\lim_{t\to\infty} \dot{e}_i(t) = 0$, which in turn implies

$$\lim_{t\to\infty} (u_i(t) - \Gamma_i(\sigma)\eta_i(t)) = 0. \tag{10.61}$$

Moreover, (10.60) and (10.61) together imply $\lim_{t\to\infty}(\hat{\Gamma}_i(t) - \Gamma_i(\sigma))\eta_i(t) = 0$. Since

$$\begin{aligned}
&\lim_{t\to\infty} (\eta_i(t) - T_i(\sigma)\tau_i(v(t), w, \sigma)) \\
&= \lim_{t\to\infty} (\eta_i(t) - \theta_i(v(t), w, \sigma)) \\
&= \lim_{t\to\infty} (\tilde{\eta}_i(t) + b_i^{-1}(w)Q_i e_i(t)) = 0
\end{aligned}$$

and $(\hat{\Gamma}_i(t) - \Gamma_i(\sigma))$ is uniformly bounded over $[0, \infty)$, we have (10.58).

Since $(\hat{\Gamma}_i(t) - \Gamma_i(\sigma))T_i(\sigma)$ is continuously differentiable, and $\tau_i(v(t), w, \sigma)$ is uniformly bounded and piecewise continuous over $[0, \infty)$, if $\tau_i(v(t), w, \sigma)$ is PE, by Lemma A.4 in Sect. A.3 with $g^{\mathsf{T}}(t) = (\hat{\Gamma}_i(t) - \Gamma_i(\sigma))T_i(\sigma)$ and $f(t) = \tau_i(v(t), w, \sigma)$, (10.59) holds. □

By Lemma 10.6, if the function $\tau_i(v(t), w, \sigma)$ is persistently exciting (PE), then $\hat{\Gamma}_i(t)$ will converge to $\Gamma_i(\sigma)$ asymptotically. In order to determine whether or not $\tau_i(v(t), w, \sigma)$ is PE, we will further introduce the concept of minimal internal model. A monic polynomial $P_i(\lambda, \sigma)$ is called a global zeroing polynomial of $\mathbf{u}_i(v(t), w, \sigma)$ on $\mathbb{S}$, if, along all the trajectories $v(t)$ and for all $\text{col}(w, \sigma) \in \mathbb{W} \times \mathbb{S}$, $\mathbf{u}_i(v(t), w, \sigma)$ satisfies a differential equation of the form (10.6). A zeroing polynomial $P_i(\lambda, \sigma)$ of $\mathbf{u}_i(v(t), w, \sigma)$ is further called a minimal zeroing polynomial of $\mathbf{u}_i(v(t), w, \sigma)$ on $\mathbb{S}$, if $P_i(\lambda, \sigma)$ is of least degree. An internal model whose dimension is equal to the degree of the minimal zeroing polynomial of $\mathbf{u}_i(v(t), w, \sigma)$ is called the minimal internal model.

Denote the degree of the minimal zeroing polynomial of $\mathbf{u}_i(v(t), w, \sigma)$ by s_i for some positive integer s_i. By Remark A.5 in Sect. A.6, under Assumptions 10.1 and 10.3, the roots of the minimal zeroing polynomial of $\mathbf{u}_i(v(t), w, \sigma)$ are all simple with zero real parts. Denote the roots of the minimal zeroing polynomial of $\mathbf{u}_i(v(t), w, \sigma)$ by $\imath\hat{\omega}_{l_i i}$, $l_i = 1, \ldots, s_i$. Then, we have

$$\mathbf{u}_i(v(t), w, \sigma) = \sum_{l_i=1}^{s_i} C_{l_i i}(v_0, w, \sigma)e^{\imath\hat{\omega}_{l_i i}t} \tag{10.62}$$

where $C_{l_i i} \in \mathbb{C}$, $\hat{\omega}_{1i}, \ldots, \hat{\omega}_{s_i i}$ are distinct, $\hat{\omega}_{l_i i} = -\hat{\omega}_{(1+s_i-l_i)i}$ and $C_{l_i i} = C^*_{(1+s_i-l_i)i}$.

We are now ready to state the following result.

Theorem 10.4 *Under Assumptions 10.1–10.6, suppose that the internal model (10.10) for each subsystem in (10.4) is of minimal order, and v_0, w, and σ are* *such that none of* $C_{l_i i}(v_0, w, \sigma)$ *is zero. Then, the distributed control law (10.55) is such that* $\lim_{t\to\infty}(\hat{\Gamma}_i(t) - \Gamma_i(\sigma)) = 0,\ i = 1, \ldots, N$.

Proof By Lemma 10.6, it suffices to show that $\tau_i(v(t), w, \sigma)$ is PE. From (10.62) and (10.7), it holds that

$$\lim_{T_0\to\infty} \frac{1}{T_0} \int_t^{t+T_0} \tau_i(v(t), w, \sigma) e^{-\imath\hat{\omega}_{l_i i} s} \mathrm{d}s = C_{l_i i}(v_0, w, \sigma) \bar{f}(\hat{\omega}_{l_i i}),\ l_i = 1, \ldots, s_i$$

uniformly in t where

$$\bar{f}(\hat{\omega}_{l_i i}) = \mathrm{col}(1, \imath\hat{\omega}_{l_i i}, (\imath\hat{\omega}_{l_i i})^2, \ldots, (\imath\hat{\omega}_{l_i i})^{s_i - 1}).$$

Thus, according to Definition A.4 in Sect. A.3, $\tau_i(v(t), w, \sigma) \in \mathbb{R}^{s_i}$ has spectral lines at s_i frequencies $\hat{\omega}_{l_i i}$, $l_i = 1, \ldots, s_i$.

Since, for each i, $\hat{\omega}_{l_i i}$ are distinct for $l_i = 1, \ldots, s_i$, and v_0, w, and σ are such that none of $C_{l_i i}(v_0, w, \sigma)$ is zero, the vectors $C_{l_i i}(v_0, w, \sigma)\bar{f}(\hat{\omega}_{l_i i})$, $l_i = 1, \ldots, s_i$, are linearly independent. By Lemma A.3 in Sect. A.3, $\tau_i(v(t), w, \sigma)$ is PE. The proof is thus completed. □

10.5 Numerical Examples

The Lorenz system takes the following form:

$$\dot{x}_1 = -x_1 + b_1 x_2 \tag{10.63a}$$

$$\dot{x}_2 = b_2 x_1 - x_2 - x_1 x_3 \tag{10.63b}$$

$$\dot{x}_3 = -b_3 x_3 + x_1 x_2 \tag{10.63c}$$

where b_i, $i = 1, 2, 3$, are three positive parameters. The Lorenz system was first studied by Edward Lorenz, and he showed that, with $b_1 = 10$, $b_2 = 28$, and $b_3 = 8/3$, the system (10.63) exhibits chaotic behavior. In fact, the well-known "butterfly effect" is a vivid explanation of the chaotic behavior in atmospheric dynamics.

Later, the so-called generalized Lorenz system as shown below

$$\dot{x}_1 = a_1 x_1 + a_2 x_2 \tag{10.64a}$$

$$\dot{x}_2 = a_3 x_1 + a_4 x_2 - x_1 x_3 \tag{10.64b}$$

$$\dot{x}_3 = a_5 x_3 + x_1 x_2 \tag{10.64c}$$

where a_i, $i = 1, \ldots, 5$, are constant parameters with $a_1 < 0$, $a_5 < 0$, was further introduced to study the chaotic behavior.

In this section, we will consider multiple generalized controlled Lorenz systems whose dynamics are given as follows:

$$\dot{x}_{1i} = a_{1i}x_{1i} + a_{2i}x_{2i} \tag{10.65a}$$
$$\dot{x}_{2i} = a_{3i}x_{1i} + a_{4i}x_{2i} - x_{1i}x_{3i} + b_i u_i \tag{10.65b}$$
$$\dot{x}_{3i} = a_{5i}x_{3i} + x_{1i}x_{2i} \tag{10.65c}$$
$$y_i = x_{2i},\ i = 1, \ldots, N \tag{10.65d}$$

where $a_i \triangleq \text{col}(a_{1i}, \ldots, a_{5i}, b_i)$ is a constant parameter vector that satisfies $a_{1i} < 0$, $a_{5i} < 0$, and $b_i > 0$. Express the uncertain parameter vector a_i as $a_i = \bar{a}_i + w_i$ where $\bar{a}_i \triangleq \text{col}(\bar{a}_{1i}, \ldots, \bar{a}_{5i}, \bar{b}_i)$ is the nominal value of a_i and $w_i \triangleq \text{col}(w_{1i}, \ldots, w_{5i}, w_{bi})$ is the uncertainty of a_i, and it is assumed that the uncertainty $w_i \in \mathbb{W} = \{w_i \in \mathbb{R}^6, \bar{a}_{1i} + w_{1i} < 0, \bar{a}_{5i} + w_{5i} < 0, \bar{b}_i + w_{bi} > 0, i = 1, \ldots, N\}$.

Performing the coordinate transformation $(z_{1i}, z_{2i}, y_i) = (x_{1i}, x_{3i}, x_{2i})$ on each subsystem of system (10.65) yields the following system:

$$\dot{z}_{1i} = a_{1i}z_{1i} + a_{2i}y_i \tag{10.66a}$$
$$\dot{z}_{2i} = a_{5i}z_{2i} + z_{1i}y_i \tag{10.66b}$$
$$\dot{y}_i = a_{3i}z_{1i} - z_{1i}z_{2i} + a_{4i}y_i + b_i u_i,\ i = 1, \ldots, N. \tag{10.66c}$$

Note that system (10.66) takes the form of (10.4) with $z_i = \text{col}(z_{1i}, z_{2i})$.

Also, let the exosystem take the form (10.5) with

$$S(\sigma) = \begin{bmatrix} 0 & \sigma \\ -\sigma & 0 \end{bmatrix}, \quad y_0 = v_1$$

where $v(0) \in \mathbb{V}_0 = \mathbb{R}^2$, and $\sigma \in \mathbb{S} \subseteq \{\sigma \in \mathbb{R}, \sigma > 0\}$, and let the regulated error be defined as

$$e_i = y_i - y_0 = x_{2i} - v_1.$$

It can be verified that the exosystem satisfies Assumption 10.1. Also, Assumption 10.2 holds with the function $\mathbf{z}_i(v, w, \sigma)$ being given by

$$\mathbf{z}_i(v, w, \sigma) = \begin{bmatrix} \mathbf{z}_{1i}(v, w, \sigma) \\ \mathbf{z}_{2i}(v, w, \sigma) \end{bmatrix} = \begin{bmatrix} r_{1i}v_1 + r_{2i}v_2 \\ r_{3i}v_1^2 + r_{4i}v_2^2 + r_{5i}v_1v_2 \end{bmatrix}$$

where

$$r_{1i} = -\frac{a_{1i}a_{2i}}{\sigma^2 + a_{1i}^2}, \quad r_{2i} = -\frac{a_{2i}\sigma}{\sigma^2 + a_{1i}^2}$$
$$r_{3i} = -\frac{(a_{5i}^2 + 2\sigma^2)r_{1i} - a_{5i}\sigma r_{2i}}{a_{5i}(a_{5i}^2 + 4\sigma^2)}$$
$$r_{4i} = \frac{r_{5i}\sigma}{a_{5i}}, \quad r_{5i} = -\frac{r_{2i}a_{5i} + 2\sigma r_{1i}}{a_{5i}^2 + 4\sigma^2}.$$

Thus, simple calculation gives, for $i = 1, \ldots, N$,

$$\mathbf{y}_i(v, w, \sigma) = v_1$$
$$\mathbf{u}_i(v, w, \sigma) = c_{1i}v_1 + c_{2i}v_2 + c_{3i}v_1^3 + c_{4i}v_2^3 + c_{5i}v_1^2v_2 + c_{6i}v_1v_2^2$$

where

$$c_{1i} = -b_i^{-1}(a_{4i} + a_{3i}r_{1i}),\ c_{2i} = b_i^{-1}(\sigma - a_{3i}r_{2i}),\ c_{3i} = b_i^{-1}r_{1i}r_{3i}$$
$$c_{4i} = b_i^{-1}r_{2i}r_{4i},\ c_{5i} = b_i^{-1}(r_{2i}r_{3i} + r_{1i}r_{5i}),\ c_{6i} = b_i^{-1}(r_{1i}r_{4i} + r_{2i}r_{5i}).$$

Then it can be verified that, for $i = 1, \ldots, N$, $\mathbf{u}_i(v, w, \sigma)$ satisfies

$$\frac{d^4\mathbf{u}_i(v, w, \sigma)}{dt^4} = -9\sigma^4\mathbf{u}_i(v, w, \sigma) - 10\sigma^2\frac{d^2\mathbf{u}_i(v, w, \sigma)}{dt^2}.$$

Thus, Assumption 10.3 holds.

From (10.8), $\Phi_i(\sigma)$ and Ψ_i are given by

$$\Phi_i(\sigma) = \begin{bmatrix} 0 & 1 & 0 & 0 \\ 0 & 0 & 1 & 0 \\ 0 & 0 & 0 & 1 \\ -9\sigma^4 & 0 & -10\sigma^2 & 0 \end{bmatrix},\ \Psi_i = \begin{bmatrix} 1 \\ 0 \\ 0 \\ 0 \end{bmatrix}^{\mathrm{T}}.$$

Let

$$M_i = \begin{bmatrix} 0 & 1 & 0 & 0 \\ 0 & 0 & 1 & 0 \\ 0 & 0 & 0 & 1 \\ -4 & -12 & -13 & -6 \end{bmatrix},\ Q_i = \begin{bmatrix} 0 \\ 0 \\ 0 \\ 1 \end{bmatrix}. \tag{10.67}$$

Then the equation (10.9) admits a unique solution $T_i(\sigma)$, such that

$$\Gamma_i(\sigma) = \Psi_i T_i^{-1}(\sigma) = \begin{bmatrix} 4 - 9\sigma^4 & 12 & 13 - 10\sigma^2 & 6 \end{bmatrix}. \tag{10.68}$$

Performing the transformations (10.12) and (10.16) gives the augmented error system as follows:

$$\dot{\bar{z}}_{1i} = a_{1i}\bar{z}_{1i} + a_{2i}e_i \tag{10.69a}$$
$$\dot{\bar{z}}_{2i} = a_{5i}\bar{z}_{2i} + \bar{z}_{1i}e_i + \bar{z}_{1i}v_1 + \mathbf{z}_{1i}e_i \tag{10.69b}$$
$$\dot{\bar{\eta}}_i = M_i\bar{\eta}_i + M_iQ_ib_i^{-1}e_i - Q_ib_i^{-1}\bar{g}_i(\bar{z}_{1i}, \bar{z}_{2i}, e_i, \mu) \tag{10.69c}$$
$$\dot{e}_i = \bar{g}_i(\bar{z}_{1i}, \bar{z}_{2i}, e_i, \mu) + \Gamma_i(\sigma)(b_i\bar{\eta}_i + Q_ie_i) + b_i(u_i - \Gamma_i(\sigma)\eta_i) \tag{10.69d}$$

where

$$\bar{g}_i(\bar{z}_{1i}, \bar{z}_{2i}, e_i, \mu) = a_{3i}\bar{z}_{1i} + a_{4i}e_i - \bar{z}_{1i}\bar{z}_{2i} - \bar{z}_{1i}\mathbf{z}_{2i} - \mathbf{z}_{1i}\bar{z}_{2i}. \tag{10.70}$$

We now verify that system (10.69) satisfies Assumption 10.4. Since system (10.69) contains only the polynomial nonlinearity, we propose a proper positive definite polynomial function as follows:

$$V_{0i}(\bar{z}_i) = \frac{\hbar_1}{2}\bar{z}_{1i}^2 + \frac{\hbar_1}{4}\bar{z}_{1i}^4 + \frac{\hbar_1}{8}\bar{z}_{1i}^8 + \frac{\hbar_2}{2}\bar{z}_{2i}^2 + \frac{\hbar_2}{4}\bar{z}_{2i}^4$$

for some $\hbar_1, \hbar_2 > 0$, whose derivative along the trajectory of (10.69a) and (10.69b) satisfies

$$\begin{aligned}\dot{V}_{0i}(\bar{z}_i) &= \hbar_1\bar{z}_{1i}\dot{\bar{z}}_{1i} + \hbar_1\bar{z}_{1i}^3\dot{\bar{z}}_{1i} + \hbar_1\bar{z}_{1i}^7\dot{\bar{z}}_{1i} + \hbar_2\bar{z}_{2i}\dot{\bar{z}}_{2i} + \hbar_2\bar{z}_{2i}^3\dot{\bar{z}}_{2i}\\ &= \hbar_1 a_{1i}\bar{z}_{1i}^2 + \hbar_1 a_{2i}\bar{z}_{1i}e_i + \hbar_1 a_{1i}\bar{z}_{1i}^4 + \hbar_1 a_{2i}\bar{z}_{1i}^3 e_i + \hbar_1 a_{1i}\bar{z}_{1i}^8 + \hbar_1 a_{2i}\bar{z}_{1i}^7 e_i\\ &\quad + \hbar_2 a_{5i}\bar{z}_{2i}^2 + \hbar_2\bar{z}_{2i}\bar{z}_{1i}e_i + \hbar_2\bar{z}_{2i}\bar{z}_{1i}v_1 + \hbar_2\bar{z}_{1i}\mathbf{z}_{1i}e_i + \hbar_2 a_{5i}\bar{z}_{2i}^4\\ &\quad + \hbar_2\bar{z}_{2i}^3\bar{z}_{1i}e_i + \hbar_2\bar{z}_{2i}^3\bar{z}_{1i}v_1 + \hbar_2\bar{z}_{2i}^3\mathbf{z}_{1i}e_i. \end{aligned} \tag{10.71}$$

Applying the mean square inequality and the Young's inequality[1] to each term of (10.71) gives

$$\hbar_1 a_{2i}\bar{z}_{1i}e_i \le \frac{\hbar_1^2 a_{2i}^2}{0.02}e_i^2 + \frac{0.01}{2}\bar{z}_{1i}^2 \tag{10.72a}$$

$$\hbar_1 a_{2i}\bar{z}_{1i}^3 e_i \le \frac{3}{4}\bar{z}_{1i}^4 + \frac{\hbar_1^4 a_{2i}^4}{4}e_i^4 \le \frac{3}{4}\bar{z}_{1i}^4 + \frac{\hbar_1^4 a_{2i}^4}{4}\left(\frac{1}{3}e_i^8 + \frac{2}{3}e_i^2\right) \tag{10.72b}$$

$$\hbar_1 a_{2i}\bar{z}_{1i}^7 e_i \le \frac{7}{8}\bar{z}_{1i}^8 + \frac{\hbar_1^8 a_{2i}^8}{8}e_i^8 \tag{10.72c}$$

$$\hbar_2\bar{z}_{2i}\bar{z}_{1i}e_i \le \frac{1}{2}\bar{z}_{2i}^2 + \frac{1}{4}\bar{z}_{1i}^4 + \frac{\hbar_2^4}{4}e_i^4 \le \frac{1}{2}\bar{z}_{2i}^2 + \frac{1}{4}\bar{z}_{1i}^4 + \frac{\hbar_2^4}{4}\left(\frac{1}{3}e_i^8 + \frac{2}{3}e_i^2\right) \tag{10.72d}$$

$$\hbar_2\bar{z}_{2i}\bar{z}_{1i}v_1 \le \frac{v_1^2\hbar_2^2}{2}\bar{z}_{1i}^2 + \frac{1}{2}\bar{z}_{2i}^2 \tag{10.72e}$$

$$\hbar_2\bar{z}_{1i}\mathbf{z}_{1i}e_i \le \frac{1}{2}\bar{z}_{1i}^2 + \frac{\hbar_2^2\mathbf{z}_{1i}^2}{2}e_i^2 \tag{10.72f}$$

$$\hbar_2\bar{z}_{2i}^3\bar{z}_{1i}e_i \le \frac{3}{4}\bar{z}_{2i}^4 + \frac{\hbar_2^8}{8}e_i^8 + \frac{1}{8}\bar{z}_{1i}^8 \tag{10.72g}$$

$$\hbar_2\bar{z}_{2i}^3\bar{z}_{1i}v_1 \le \frac{3}{4}\bar{z}_{2i}^4 + \frac{\hbar_2^4 v_1^4}{4}\bar{z}_{1i}^4 \tag{10.72h}$$

$$\hbar_2\bar{z}_{2i}^3\mathbf{z}_{1i}e_i \le \frac{3}{4}\bar{z}_{2i}^4 + \frac{\hbar_2^4\mathbf{z}_{1i}^4}{4}e_i^4 \le \frac{3}{4}\bar{z}_{2i}^4 + \frac{\hbar_2^4\mathbf{z}_{1i}^4}{4}\left(\frac{1}{3}e_i^8 + \frac{2}{3}e_i^2\right) \tag{10.72i}$$

Let $\bar{\mathbb{V}}$ and $\mathbb{W}_0$ be any fixed compact subsets of $\mathbb{R}^2$ and $\mathbb{W}$, respectively. Let $\mathbb{S}_0 = \{\sigma\}$ (i.e., $\mathbb{S}_0$ consists of a known singleton) for cases 1 and 2, and $\mathbb{S}_0$ be any fixed

[1] For any $a, b \in \mathbb{R}$ and any $\frac{1}{p} + \frac{1}{q} = 1$, $|ab| \le \frac{|a|^p}{p} + \frac{|b|^q}{q}$.

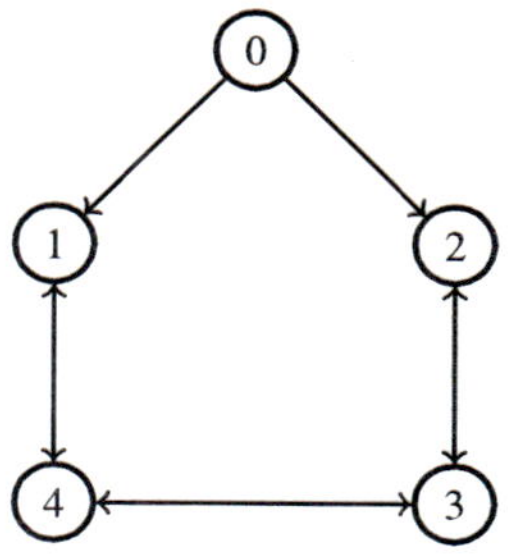

Fig. 10.1 The static communication graph $\bar{\mathscr{G}}$

compact subset of $\mathbb{S}$ for case 3. Denote $\Sigma = \bar{\mathbb{V}} \times \mathbb{W}_0 \times \mathbb{S}_0$, which is compact. Then inequalities (10.72a)–(10.72i) together with (10.71) yields, for all $(v, w, \sigma) \in \Sigma$,

$$\dot{V}_{0i}(\bar{z}) \leq -l_{1i}\bar{z}_{1i}^2 - l_{2i}\bar{z}_{1i}^4 - l_{3i}\bar{z}_{1i}^8 - l_{4i}\bar{z}_{2i}^2 - l_{5i}\bar{z}_{2i}^4 + l_{6i}e_i^2 + l_{7i}e_i^8, \tag{10.73}$$

where

$$\begin{aligned}
l_{1i} &\triangleq \hbar_1 \min_{w\in\mathbb{W}_0}\{-a_{1i}\} - \frac{1.01}{2} - \frac{\hbar_2^2 \max_{v\in\bar{\mathbb{V}}}\{v_1^2\}}{2} \\
l_{2i} &\triangleq \hbar_1 \min_{w\in\mathbb{W}_0}\{-a_{1i}\} - 1 - \frac{\hbar_2^4 \max_{v\in\bar{\mathbb{V}}}\{v_1^4\}}{4} \\
l_{3i} &\triangleq \hbar_1 \min_{w\in\mathbb{W}_0}\{-a_{1i}\} - 1 \\
l_{4i} &\triangleq \hbar_2 \min_{w\in\mathbb{W}_0}\{-a_{5i}\} - 1 \\
l_{5i} &\triangleq \hbar_2 \min_{w\in\mathbb{W}_0}\{-a_{5i}\} - \frac{9}{4} \\
l_{6i} &\triangleq \frac{\hbar_1^2 \max_{w\in\mathbb{W}_0}\{a_{2i}^2\}}{0.02} + \frac{\hbar_1^4 \max_{w\in\mathbb{W}_0}\{a_{2i}^4\}}{6} + \frac{\hbar_2^4}{6} \\
&\quad + \frac{\hbar_2^2 \max_{\mu\in\Sigma}\{\mathbf{z}_{1i}^2\}}{2} + \frac{\hbar_2^4 \max_{\mu\in\Sigma}\{\mathbf{z}_{1i}^4\}}{6} \\
l_{7i} &\triangleq \frac{\hbar_1^4 \max_{w\in\mathbb{W}_0}\{a_{2i}^4\}}{12} + \frac{\hbar_1^8 \max_{w\in\mathbb{W}_0}\{a_{2i}^8\}}{8} + \frac{\hbar_2^4}{12} + \frac{\hbar_2^8}{8} + \frac{\hbar_2^4 \max_{\mu\in\Sigma}\{\mathbf{z}_{1i}^4\}}{12}.
\end{aligned}$$

Since $a_{1i} < 0$, $a_{5i} < 0$, we can choose some proper $\hbar_1 > 0$ and $\hbar_2 > 0$ such that $l_{si} > 0, s = 1, \ldots, 7, i = 1, \ldots, N$. Here l_{si} are known for case 1, but unknown for cases 2 and 3. Thus, Assumption 10.4 holds.

The communication graph of this system is depicted in Fig. 10.1 which satisfies Assumptions 10.5 and 10.6. Thus, by Theorems 10.1, 10.2, and 10.3, all the three cases of the cooperative output regulation problem for the group of generalized Lorenz systems are solvable.

In the rest of this section, we will deal with these three cases by the approaches detailed in Sects. 10.2–10.4, respectively. For all three cases, we assume that $\bar{a}_i = \text{col}(-10, 10, 28, -1, -8/3, 1)$, and the nominal value of $\sigma = 0.8$.

We will synthesize control laws according to Remark 10.6. For this purpose, notice that system (10.69) is of the from (10.21) with $Z_i = \text{col}(\bar{z}_{1i}, \bar{z}_{2i}, \tilde{\eta}_i)$ and $Z = \text{col}(Z_1, \ldots, Z_N)$. From (10.70),

$$\|\bar{g}_i(\bar{z}_{1i}, \bar{z}_{2i}, e_i, \mu)\|^2 \leq c(c_{1i}(\bar{z}_{1i}^2 + 1)\bar{z}_{1i}^2 + c_{2i}(\bar{z}_{2i}^2 + 1)\bar{z}_{2i}^2 + c_{3i}e_i^2) \tag{10.74}$$

where $c_{si} > 0$, $s = 1, 2, 3$, are known constants, and, for case 1, $c = 1$ and, for cases 2 and 3, c is an unknown positive constant. Inequality (10.74) together with (10.24) and (10.69d) gives

$$\begin{aligned}\|G(Z, e, \mu)\|^2 &= \sum_{i=1}^{N} \|\bar{g}_i(\bar{z}_{1i}, \bar{z}_{2i}, e_i, \mu) + \Gamma_i(\sigma)(b_i\tilde{\eta}_i + Q_i e_i)\|^2 \\ &\leq c_0 \left(\sum_{i=1}^{N} \pi_{1i}(\bar{z}_{1i})\bar{z}_{1i}^2 + \sum_{i=1}^{N} \pi_{2i}(\bar{z}_{2i})\bar{z}_{2i}^2 + \sum_{i=1}^{N} \pi_{3i}\|\tilde{\eta}_i\|^2 + \sum_{i=1}^{N} \phi_i e_i^2 \right)\end{aligned} \tag{10.75}$$

where $\pi_{1i}(\bar{z}_{1i}) = \hat{c}_{1i}(\bar{z}_{1i}^2 + 1)$, $\pi_{2i}(\bar{z}_{2i}) = \hat{c}_{2i}(\bar{z}_{2i}^2 + 1)$, $\pi_{3i} = \hat{c}_{3i}$ for some known $\hat{c}_{si} > 0$, $s = 1, 2, 3$, $i = 1, \ldots, N$, $\phi_i > 0$ is known for cases 1 and 2 and unknown for case 3, and $c_0 = 1$ for case 1 and is an unknown positive constant for cases 2 and 3. Thus, (10.41) holds.

Let $V_2(e) = \frac{1}{2}e^{\mathsf{T}}He$. Then, similar to the derivation of (10.29), we have

$$\begin{aligned}\dot{V}_2(e) = &\sum_{i=1}^{N} \pi_{1i}(\bar{z}_{1i})\bar{z}_{1i}^2 + \sum_{i=1}^{N} \pi_{2i}(\bar{z}_{2i})\bar{z}_{2i}^2 + \sum_{i=1}^{N} \pi_{3i}\|\tilde{\eta}_i\|^2 \\ &+ \sum_{i=1}^{N} \phi_i(e_i)e_i^2 - \sum_{i=1}^{N} \left(b_i(w)k_i\rho_i(e_{vi}) - \frac{c_0}{4}\right) e_{vi}^2.\end{aligned}$$

Thus, if we can find a Lyapunov function $V_1(Z)$ for (10.69a)–(10.69c) such that

$$\dot{V}_1(Z) \leq -\sum_{i=1}^{N} (\Theta_{1i}(\bar{z}_{1i})\bar{z}_{2i}^2 + \Theta_{2i}(\bar{z}_{2i})\bar{z}_{2i}^2 + \Theta_{3i}\|\tilde{\eta}_i\|^2) + \hat{\delta}\sum_{i=1}^{N} \hat{\gamma}_i(e_i)e_i^2$$

with $\Theta_{li} \geq \pi_{li} + 1, l = 1, 2, 3$, then, from (10.42), it suffices to let k_i and $\rho_i(\cdot)$ satisfy (10.36).

For this purpose, let $P_i \in \mathbb{R}^{4\times 4}$ be symmetric positive definite matrix such that $P_i M_i + M_i^{\mathsf{T}} P_i = -I_4$, and $V_{1i}(\tilde{\eta}_i) = \tilde{\eta}_i^{\mathsf{T}} P_i \tilde{\eta}_i$. Then, along the trajectory of $\tilde{\eta}_i$ subsystem (10.69c)

$$\begin{aligned}
\dot{V}_{1i}(\tilde{\eta}_i) =& -\|\tilde{\eta}_i\|^2 + 2\tilde{\eta}_i^{\top} P_i (M_i Q_i b_i^{-1} e_i - Q_i b_i^{-1} \bar{g}_i(\bar{z}_{1i}, \bar{z}_{2i}, e_i, \mu)) \\
\leq& -\frac{1}{2}\|\tilde{\eta}_i\|^2 + 4\|P_i M_i Q_i b_i^{-1}\|^2 e_i^2 \\
& +4\|P_i Q_i b_i^{-1}\|^2 c\left(c_{1i}(\bar{z}_{1i}^2+1)\bar{z}_{1i}^2 + c_{2i}(\bar{z}_{2i}^2+1)\bar{z}_{2i}^2 + c_{3i}e_i^2\right) \\
=& -\frac{1}{2}\|\tilde{\eta}_i\|^2 + (\bar{l}_{1i}(\bar{z}_{1i}^2+1)\bar{z}_{1i}^2 + \bar{l}_{2i}(\bar{z}_{2i}^2+1)\bar{z}_{2i}^2 + \bar{l}_{3i}e_i^2
\end{aligned} \tag{10.76}$$

where

$$\begin{aligned}
\bar{l}_{1i} &\triangleq 4cc_{1i} \max_{w\in\mathbb{W}_0}\{\|P_i Q_i b_i^{-1}\|^2\} > 0 \\
\bar{l}_{2i} &\triangleq 4cc_{2i} \max_{w\in\mathbb{W}_0}\{\|P_i Q_i b_i^{-1}\|^2\} > 0 \\
\bar{l}_{3i} &\triangleq 4 \max_{w\in\mathbb{W}_0}\{\|P_i M_i Q_i b_i^{-1}\|^2 + cc_{3i}\|P_i Q_i b_i^{-1}\|^2\} > 0.
\end{aligned}$$

Let $U_{1i}(Z_i) = \varpi_i V_{0i}(\bar{z}_i) + V_{1i}(\tilde{\eta}_i)$ where ϖ_i are some positive constants to be determined. Then, from (10.73) and (10.76)

$$\begin{aligned}
\dot{U}_{1i}(Z_i) \leq& \varpi_i(-l_{1i}\bar{z}_{1i}^2 - l_{2i}\bar{z}_{1i}^4 - l_{3i}\bar{z}_{1i}^8 - l_{4i}\bar{z}_{2i}^2 - l_{5i}\bar{z}_{2i}^4 + l_{6i}e_i^2 + l_{7i}e_i^8) \\
& -\frac{1}{2}\|\tilde{\eta}_i\|^2 + \bar{l}_{1i}(\bar{z}_{1i}^2+1)\bar{z}_{1i}^2 + \bar{l}_{2i}(\bar{z}_{2i}^2+1)\bar{z}_{2i}^2 + \bar{l}_{3i}e_i^2 \\
\leq& -\left((\varpi_i l_{1i} - \bar{l}_{1i}) + (\varpi_i l_{2i} - \bar{l}_{1i})\bar{z}_{1i}^2 + \varpi_i l_{3i}\bar{z}_{1i}^6\right)\bar{z}_{1i}^2 \\
& -\left((\varpi_i l_{4i} - \bar{l}_{2i}) + (\varpi_i l_{5i} - \bar{l}_{2i})\bar{z}_{2i}^2\right)\bar{z}_{2i}^2 - \frac{1}{2}\|\tilde{\eta}_i\|^2 \\
& +\left((\varpi_i l_{6i} + \bar{l}_{3i}) + \varpi_i l_{7i}e_i^6\right)e_i^2.
\end{aligned} \tag{10.77}$$

Letting

$$\varpi_i \geq \left\{\frac{\bar{l}_{1i}}{l_{1i}}, \frac{\bar{l}_{1i}}{l_{2i}}, \frac{1}{l_{2i}}, \frac{\bar{l}_{2i}}{l_{4i}}, \frac{\bar{l}_{2i}}{l_{5i}}\right\} + 1$$

gives

$$\begin{aligned}
\dot{U}_{1i}(Z_i) \leq& -\left(1 + \bar{z}_{1i}^2 + \bar{z}_{1i}^6\right)\bar{z}_{1i}^2 - \left(1 + \bar{z}_{2i}^2\right)\bar{z}_{2i}^2 - \frac{1}{2}\|\tilde{\eta}_i\|^2 \\
& +\left((\varpi_i l_{6i} + 2\bar{l}_{3i}) + \varpi_i l_{7i}e_i^6\right)e_i^2.
\end{aligned} \tag{10.78}$$

Let

$$V_1(Z) = \sum_{i=1}^{N} \hat{l}_i U_{1i}(Z_i)$$

where $\hat{l}_i$, $i = 1, \ldots, N$, are some positive constants to be determined. Then, from (10.78),

$$\dot{V}_1(Z) \leq -\sum_{i=1}^{N} \left(\Theta_{1i}(\bar{z}_{1i})\bar{z}_{2i}^2 + \Theta_{2i}(\bar{z}_{2i})\bar{z}_{2i}^2 + \Theta_{3i}\|\bar{\eta}_i\|^2\right)$$
$$+\sum_{i=1}^{N} \hat{l}_i \left((\varpi_i l_{6i} + 2\bar{l}_{3i}) + \varpi_i l_{7i} e_i^6\right) e_i^2 \qquad (10.79)$$

where $\Theta_{1i}(\bar{z}_{1i}) = \hat{l}_i(1 + \bar{z}_{1i}^2 + \bar{z}_{1i}^6)$, $\Theta_{2i}(\bar{z}_{2i}) = \hat{l}_i(1 + \bar{z}_{2i}^2)$, and $\Theta_{3i} = \frac{1}{2}\hat{l}_i$. Then, for sufficiently large $\hat{l}_i$, $\Theta_{li} \geq \pi_{li} + 1$, $l = 1, 2, 3$.

Since $\hat{l}_i \left((\varpi_i l_{6i} + 2\bar{l}_{3i}) + \varpi_i l_{7i} e_i^6\right)$ is a degree 6 polynomial, for any $\hat{l}_i$, there exists $\hat{\delta}_i > 0$ and $\hat{\gamma}_i(e_i) = 1 + e_i^6$ such that $\hat{l}_i \left((\varpi_i l_{6i} + 2\bar{l}_{3i}) + \varpi_i l_{7i} e_i^6\right) \leq \hat{\delta}_i \hat{\gamma}_i(e_i)$.

Let

$$\tilde{\phi}_i(e_{vi}) = 1 + e_{vi}^6. \qquad (10.80)$$

Then, for some $\delta > 0$,

$$\sum_{i=1}^{N} \phi_i e_i^2 + \sum_{i=1}^{N} \hat{\delta}_i \hat{\gamma}_i(e_i) e_i^2 \leq \delta \sum_{i=1}^{N} \tilde{\phi}_i(e_{vi}) e_{vi}^2.$$

Thus, with $\rho_i(e_{vi}) = 1 + e_{vi}^6$, (10.36) is satisfied for a sufficiently large number k_i.

Example 10.1 This example considers the case where the exosystem is known and the compact set Σ_0 is also known. To be specific, let $\Sigma_0 = \bar{\mathbb{V}}_0 \times \mathbb{W}_0$ with $\bar{\mathbb{V}}_0 = [-8, 8] \times [-8, 8]$ and $\mathbb{W}_0 = [-1, 1] \times [-1, 1] \times [-1, 1] \times \{0\} \times [-1/3, 1/3] \times \{0\}$. Also, let $\mathbb{S} = \{0.8\}$. For this case, there is a known k_i such that (10.36) is satisfied with $\rho_i(e_{vi}) = 1 + e_{vi}^6$. We choose $k_i = 640$.

The simulation is conducted with the actual values of the uncertain parameters being $w_1 = \text{col}(1, -1, -1, 0, -1/6, 0)$, $w_2 = \text{col}(0, 0, -1, 0, 1/6, 0)$, $w_3 = \text{col}(-1, 1, 0, 0, -1/3, 0)$, $w_4 = \text{col}(-1, 1, -1, 0, 1/3, 0)$, and the initial conditions of the closed-loop system being $v(0) = \text{col}(0, 8)$, $z_1(0) = \text{col}(3, -1)$, $y_1(0) = -2$, $z_2(0) = \text{col}(1, 1)$, $y_2(0) = 0$, $z_3(0) = \text{col}(3, 2)$, $y_3(0) = 0$, $z_4(0) = (2, -1)$, $y_4(0) = -1$, and $\eta_1(0) = \eta_2(0) = \eta_3(0) = \eta_4(0) = \text{col}(0, 0, 0, 0)$. Figure 10.2 shows the response of $y_i(t)$. Figure 10.3 shows the response of $e_i(t)$. Satisfactory tracking performance of our control design is observed.

Example 10.2 This example considers the case where the exosystem is known and the compact subset Σ_0 can be any compact subset of $\mathbb{R}^2 \times \mathbb{W}$. In this case, our control law works for any $v(0) \in \mathbb{R}^2$ and any $w \in \mathbb{W}$. By Remark 10.8, we can choose the same $\rho_i(e_{vi})$ as in Example 10.1.

The simulation is conducted with the actual values of the uncertain parameters and the initial conditions of the closed-loop system being the same as those in Example 10.1, and $k_1(0) = k_2(0) = k_3(0) = k_4(0) = 1$. Figure 10.4 shows the response of $y_i(t)$. Figure 10.5 shows the response of $e_i(t)$. Figure 10.6 shows the response of the adaptive gain $k_i(t)$. Satisfactory tracking performance of the control law is observed.

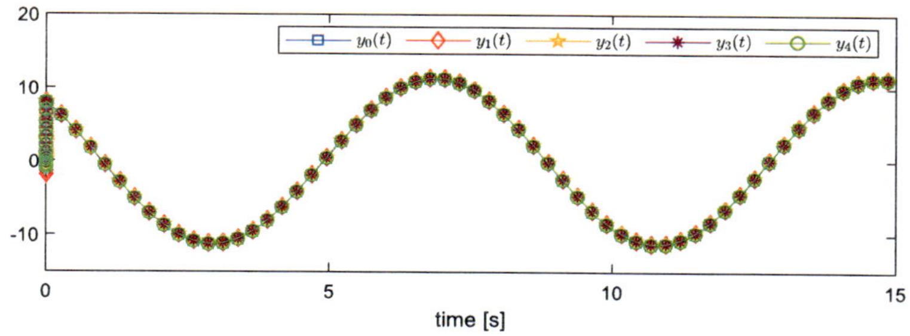

Fig. 10.2 The response of $y_0(t)$ and the response of $y_i(t)$ under the control law (10.39)

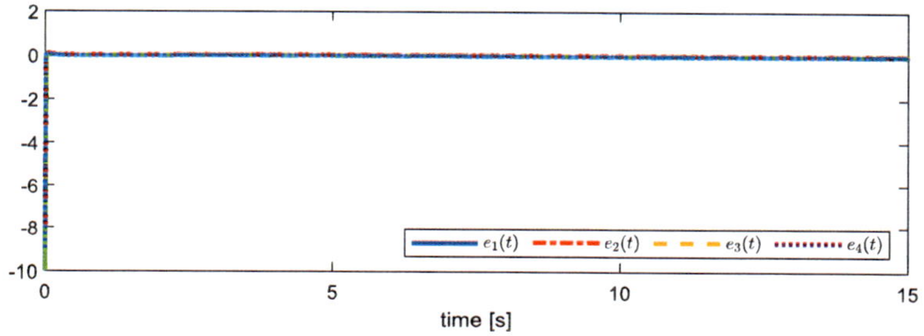

Fig. 10.3 The response of $e_i(t)$ under the control law (10.39)

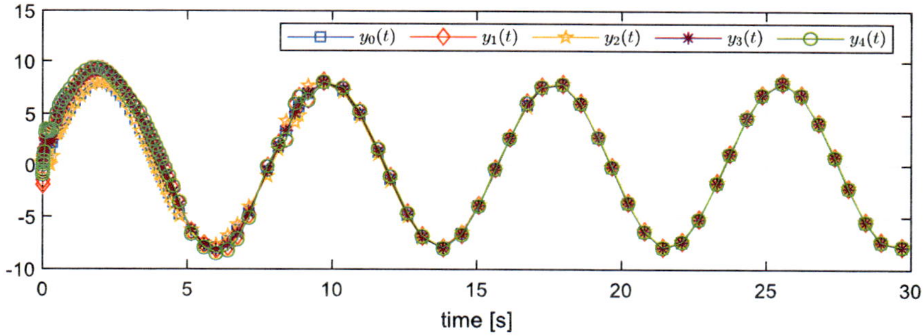

Fig. 10.4 The response of $y_0(t)$ and the response of $y_i(t)$ under the control law (10.53)

Example 10.3 This example further considers the case where the exosystem is uncertain with $\mathbb{S} = \{\sigma \in \mathbb{R}, \sigma > 0\}$ and Σ_0 can be any unknown compact subset of $\mathbb{V}_0 \times \mathbb{W} \times \mathbb{S}$. For this case, on top of the control law, for Example 10.2, we need to further employ the parameter updating law (10.55c) to estimate the unknown parameter σ.

By Remark 10.8, it is possible to design a distributed controller of the form (10.55) with $\rho_i(e_{vi}) = 5(e_{vi}^6 + 1)$. The simulation is conducted with the actual values of the uncertain parameters and the initial conditions of the closed-loop system

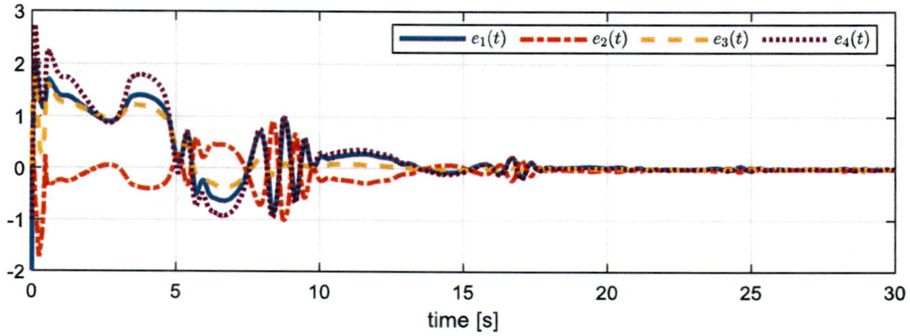

Fig. 10.5 The response of $e_i(t)$ under the control law (10.53)

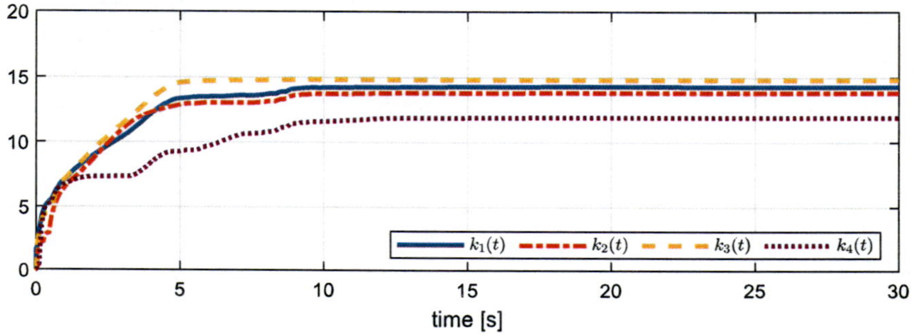

Fig. 10.6 The response of $k_i(t)$ under the control law (10.53)

being the same as those in Example 10.1, $k_1(0) = k_2(0) = k_3(0) = k_4(0) = 1$, and $\hat{\Gamma}_1(0) = \hat{\Gamma}_2(0) = \hat{\Gamma}_3(0) = \hat{\Gamma}_4(0) = \text{col}(1, 2, 3, 4)$. Figure 10.7 shows the response of $y_i(t)$. Figure 10.8 shows the response of $e_i(t)$. Figure 10.9 shows the response of the adaptive gain $k_i(t)$. Satisfactory tracking performance of our control design can be observed.

Next we address the convergence of the estimation $\hat{\Gamma}_i$. To this end, we will show that (10.67) is a minimal internal model. In fact, from

$$
\begin{aligned}
v_1(t) &= \left(\frac{v_{10}}{2} + \frac{v_{20}}{2\imath}\right) e^{\imath\sigma t} + \left(\frac{v_{10}}{2} - \frac{v_{20}}{2\imath}\right) e^{-\imath\sigma t} \\
v_2(t) &= \left(\frac{v_{10}}{2} + \frac{v_{20}}{2\imath}\right) \imath e^{\imath\sigma t} - \left(\frac{v_{10}}{2} - \frac{v_{20}}{2\imath}\right) \imath e^{-\imath\sigma t}
\end{aligned}
$$

we can put $\mathbf{u}_i(v, w, \sigma)$ in the form of (10.62) with $s_i = 4$, $\hat{\omega}_{1i} = \sigma$, $\hat{\omega}_{2i} = -\sigma$, $\hat{\omega}_{3i} = 3\sigma$, $\hat{\omega}_{4i} = -3\sigma$, $i = 1, \ldots, 4$, and none of the coefficients $C_{li}(v_0, w, \sigma)$, $i = 1, 2, 3, 4$, is zero. Thus, by Theorem 10.4, any four dimensional internal model is of minimal order. From (10.68), the estimated frequency $\hat{\sigma}_i$ is related to the first component $\hat{\Gamma}_{1i}$ of $\hat{\Gamma}_i$ as $\hat{\sigma}_i = \left((4 - \hat{\Gamma}_{1i})/9\right)^{1/4}$, $i = 1, 2, 3, 4$. Figure 10.10 shows that the estimated frequency $\hat{\sigma}_i$ approaches the actual frequency $\sigma = 0.8$ asymptotically.

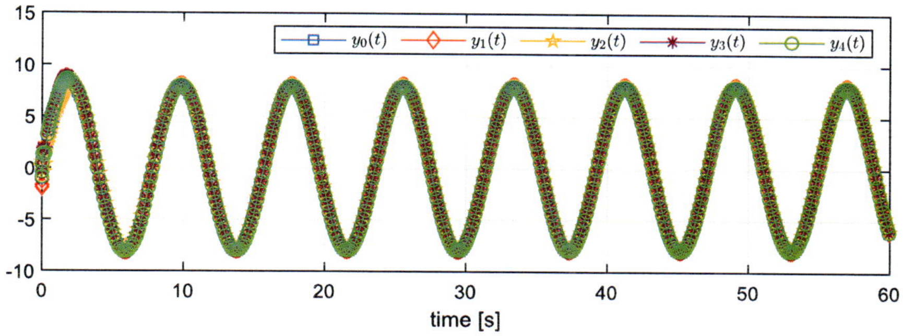

Fig. 10.7 The response of $y_0(t)$ and the response of $y_i(t)$ under the control law (10.55)

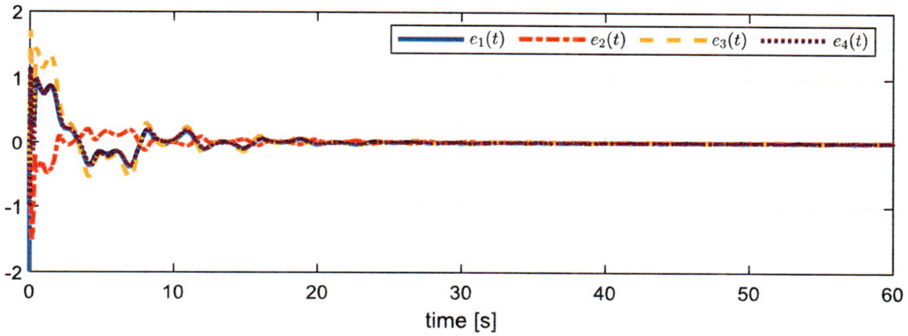

Fig. 10.8 The response of $e_i(t)$ under the control law (10.55)

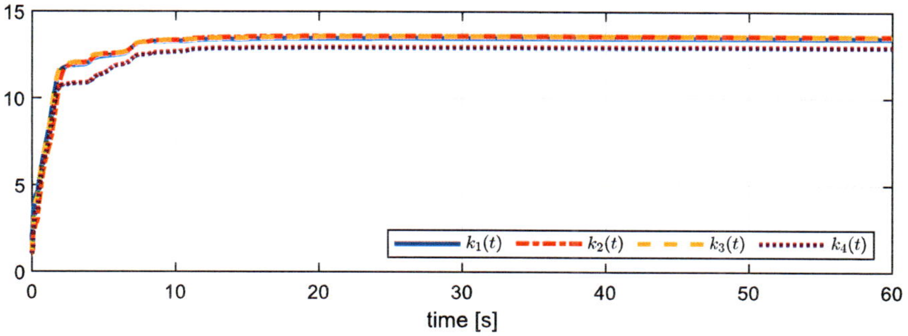

Fig. 10.9 The response of $k_i(t)$ under the control law (10.55)

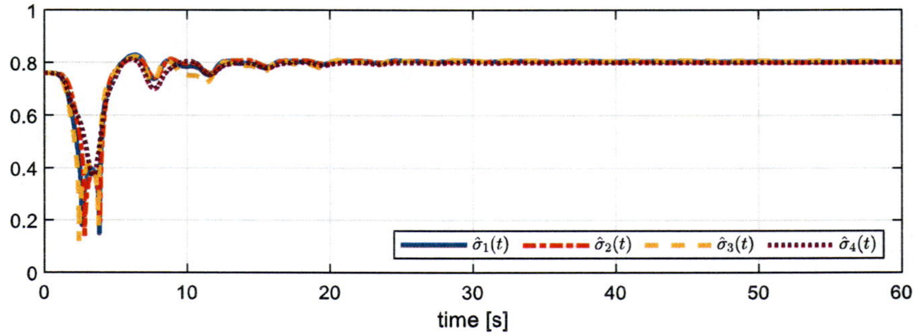

Fig. 10.10 The response of $\hat{\sigma}_i(t)$ under the control law (10.55)

10.6 Notes and References

Like the classic nonlinear output regulation problem for a single system, the studies on the cooperative global output regulation problem also focused on nonlinear systems with various special forms. The problem was first formulated for the case where the exosystem is exactly known and linear, and was studied for multiple nonlinear systems in output feedback form with unity relative degree over undirected static networks in [1], with higher relative degree over undirected static networks in [2], for second-order multiple nonlinear systems in [3], and in strict feedback form over directed static networks in [4]. The case where the exosystem is linear and contains uncertain parameters was first studied for multiple nonlinear systems in output feedback form with unity relative degree over undirected static networks in [5], and for second-order multiple nonlinear systems in [6]. The result in Sect. 10.2 was first studied in [1]. For this case, it is possible to remove Assumption 10.6 by making use of the approach in [4]. The result in Sect. 10.4 was first reported in [5] and it contains the result of Sect. 10.3 as a special case by assuming that the exosystem is exactly known.

Assumption 10.4 can be weakened to integral-input-to-state stability, which was treated in [7]. The case where the exosystem is nonlinear was studied in [8]. The case where the network is switching and every time connected was studied for multiple nonlinear systems in strict feedback form with unity relative degree in [9] and for multiple nonlinear systems in output feedback form with any relative degree in [10]. The implementation of the control law in Sect. 10.2 by distributed event-triggered control law was given in [11].

Some variants of the cooperative output regulation problem for nonlinear multi-agent systems can be found in [12, 13]. Also, the robust output synchronization problem of heterogeneous nonlinear networked systems was studied in [14, 15] using the internal model approach.

The Lorenz system was first given in [16]. The generalized Lorenz systems can be found in [17]. Internal model approach for a single generalized Lorenz systems was studied in [18].

References

1. Dong Y, Huang J (2014) Cooperative global robust output regulation for nonlinear multi-agent systems in output feedback form. J Dyn Syst Meas Control-Trans ASME 136(3):031001
2. Dong Y, Huang J (2014) Cooperative global output regulation for a class of nonlinear multi-agent systems. IEEE Trans Autom Control 59(5):1348–1354
3. Su Y, Huang J (2013) Cooperative global output regulation of heterogeneous second-order nonlinear uncertain multi-agent systems. Automatica 49(11):3345–3350
4. Su Y, Huang J (2015) Cooperative global output regulation for nonlinear uncertain multi-agent systems in lower triangular form. IEEE Trans Autom Control 60(9):2378–2389
5. Su Y, Huang J (2013) Cooperative global output regulation for a class of nonlinear uncertain multi-agent systems with unknown leader. Syst Control Lett 62(6):461–467
6. Dong Y, Chen J, Huang J (2018) Cooperative robust output regulation problem for second-order nonlinear multi-agent systems with an unknown exosystem. IEEE Trans Autom Control 63(10):3418–3425
7. Su Y, Huang J (2015) Cooperative global output regulation for nonlinear uncertain multi-agent systems with iISS inverse dynamics. Asian J Control 17(1):14–22
8. Lu M, Huang J (2016) Cooperative global robust output regulation for a class of nonlinear multi-agent systems with a nonlinear leader. IEEE Trans Autom Control 61(11):3557–3562
9. Liu W, Huang J (2015) Cooperative global robust output regulation for a class of nonlinear multi-agent systems with switching network. IEEE Trans Autom Control 60(7):1963–1968
10. Liu W, Huang J (2017) Cooperative global robust output regulation for nonlinear output feedback multi-agent systems under directed switching networks. IEEE Trans Autom Control 62(12):6339–6352
11. Liu W, Huang J (2018) Cooperative global robust output regulation for a class of nonlinear multi-agent systems by distributed event-triggered control. Automatica 93:138–148
12. Xu D, Hong Y, Wang X (2014) Distributed output regulation of nonlinear multi-agent systems via host internal model. IEEE Trans Autom Control 59(10):2784–2789
13. Ding Z (2013) Consensus output regulation of a class of heterogeneous nonlinear systems. IEEE Trans Autom Control 58(10):2648–2653
14. Isidori A, Marconi L, Casadei G (2014) Robust output synchronization of a network of heterogeneous nonlinear agents via nonlinear regulation theory. IEEE Trans Autom Control 59(10):2680–2691
15. Zhu L, Chen Z, Middleton R (2015) A general framework for robust output synchronization of heterogeneous nonlinear networked systems. IEEE Trans Autom Control 61(8):2092–2107
16. Lorenz EN (1963) Deterministic nonperiodic flow. J Atmos Sci 20(2):130–141
17. Liang X, Zhang J, Xia X (2008) Adaptive synchronization for generalized Lorenz systems. IEEE Trans Autom Control 53(7):1740–1746
18. Xu D, Huang J (2010) Robust adaptive control of a class of nonlinear systems and its application. IEEE Trans Circuits Syst I: Regul Pap 57(3):691–702

Chapter 11
Cooperative Robust Output Regulation of Nonlinear Multi-agent Systems over Switching Networks

So far, we have studied two approaches to handling the cooperative output regulation problem, namely, the distributed observer approach and the distributed internal model approach. The distributed observer approach can be viewed as an extension of the feedforward control approach for the conventional output regulation problem and is able to handle the cooperative output regulation problem for general heterogeneous linear multi-agent systems and some weak nonlinear multi-agent systems over jointly connected switching networks. However, as this approach relies on the solution of the regulator equations, it is less capable of tackling the nonlinearity and uncertainty of the system. On the other hand, the distributed internal model approach is an extension of the classical internal model approach and is effective in dealing with the nonlinearity and uncertainty of the system. Nevertheless, as this approach makes use of the virtual tracking error of the system which requires the communication network to be connected at all time, it cannot handle jointly connected switching networks. In this chapter, we will integrate the distributed observer approach and the distributed internal model approach, thus leading to an approach capable of handling the nonlinearity and uncertainty of the system over jointly connected switching networks.

Recall that three cases of the cooperative robust output regulation problem for systems (10.4) were studied in Chap. 10. However, for simplicity, in this chapter, we will only consider the first two cases, i.e., the case where the exosystem is known exactly and the compact subset Σ_0 is known, and the case where the exosystem is known exactly and the compact subset Σ_0 is unknown.

This chapter is organized as follows. Section 11.1 formulates the cooperative robust output regulation problem for the same class of systems (10.4) over jointly connected switching networks. Section 11.2 presents the preliminaries for solving the problem. Sections 11.3 and 11.4 solve the two cases of the problem via the integrated approach, respectively. Section 11.5 illustrates the integrated approach through numerical examples.

H. Cai et al., *Cooperative Control of Multi-agent Systems*, Advances in Industrial Control,
https://doi.org/10.1007/978-3-030-98377-2_11

11.1 Problem Formulation

In this section, we will handle the cooperative robust output regulation problem for (10.4) studied in Chap. 10 over jointly connected switching communication graphs $\bar{\mathcal{G}}_{\sigma(t)}$. For convenience, we repeat Assumption 4.1 as follows:

Assumption 11.1 There exists a subsequence $\{i_k : k = 0, 1, 2, \dots\}$ of $\{i : i = 0, 1, 2, \dots\}$ with $t_{i_{k+1}} - t_{i_k} < \nu$ for some $\nu > 0$ and all $k = 0, 1, 2, \dots$ such that the union graph $\bar{\mathcal{G}}([t_{i_k}, t_{i_{k+1}}))$ contains a spanning tree with the node 0 as the root.

Let us first point out that, in Sect. 10.2, we used the property that the matrix H is nonsingular in several places. In particular, to obtain (10.33), we made use of the relation $e = H^{-1}e_v$. Thus, our approach in Sect. 10.2 fails if H is singular. Since a jointly connected switching communication graph can be disconnected at every time instant, we cannot use the virtual tracking error to handle jointly connected switching communication graphs. To overcome this difficulty, instead of the virtual tracking error e_v, we will make use of the so-called estimated tracking error to synthesize our control law. The estimated tracking error is defined as follows:

$$\hat{e}_i = y_i - h_0(\hat{v}_i), \ i = 1, \dots, N$$

where $\hat{v}_i$ is an estimation of the state of the leader system by any type of the distributed observer. To be specific, let us adopt the adaptive distributed observer presented as Case 3 of Sect. 4.3 with $y_{m0} = v$ as follows:

$$\dot{S}_i = \mu_1 \sum_{j=0}^{N} a_{ij}(t)(S_j - S_i) \tag{11.1a}$$

$$\dot{\hat{v}}_i = S_i \hat{v}_i + \mu_2 \sum_{j=0}^{N} a_{ij}(t)(\hat{v}_j - \hat{v}_i), \ i = 1, \dots, N \tag{11.1b}$$

where $S_i \in \mathbb{R}^{q\times q}$, $\hat{v}_i \in \mathbb{R}^q$, $S_0 = S$, $\hat{v}_0 = v$, $\mu_1, \mu_2 > 0$.

To formulate our problem, like in Chap. 10, let $\mathbb{V}_0 \subseteq \mathbb{R}^q$ and $\mathbb{W} \subseteq \mathbb{R}^{n_w}$ be any subsets containing the origin of their respective Euclidian spaces. Consider the following class of distributed output feedback integrated control laws:

$$u_i = \kappa_i(\xi_i, \eta_i, \hat{e}_i) \tag{11.2a}$$
$$\dot{\xi}_i = g_i(\hat{e}_i) \tag{11.2b}$$
$$\dot{\eta}_i = M_i\eta_i + Q_i u_i \tag{11.2c}$$
$$\dot{S}_i = \mu_1 \sum_{j=0}^{N} a_{ij}(t)(S_j - S_i) \tag{11.2d}$$
$$\dot{\hat{v}}_i = S_i\hat{v}_i + \mu_2 \sum_{j=0}^{N} a_{ij}(t)(\hat{v}_j - \hat{v}_i),\ i = 1, \ldots, N \tag{11.2e}$$

where M_i and Q_i are the same as those in (10.10), $\kappa_i(\cdot, \cdot, \cdot)$, and $g_i(\cdot)$ are smooth functions to be designed. Like in Chap. 10, we allow the dimension of ξ_i to be zero to accommodate the case where $u_i = \kappa_i(\eta_i, \hat{e}_i)$.

We now describe our problem as follows.

Problem 11.1 (*Nonlinear Global Cooperative Robust Output Regulation*) Given the follower system (10.4), the leader system (10.5), the jointly connected switching communication graph $\bar{\mathcal{G}}_{\sigma(t)}$, and any known (or unknown) compact subset $\Sigma_0 \subseteq \mathbb{V}_0 \times \mathbb{W}$, find a distributed feedback control law of the form (11.2) such that, for any initial state and any $\text{col}(v(0), w) \in \Sigma_0$, the closed-loop system has the following properties:

- **Property 11.1** The solution of the closed-loop system exists for all $t \geq 0$, and is uniformly bounded over $[0, \infty)$ for uniformly bounded $v_0(t)$;
- **Property 11.2** For $i = 1, \ldots, N$, $\lim_{t\to\infty} e_i(t) = 0$.

As pointed out in Chap. 10, under Assumption 10.1, for any initial condition $v(0)$, the exogenous signal $v(t)$ is uniformly bounded over $[0, \infty)$. Thus, given $\mathbb{V}_0 \subseteq \mathbb{R}^q$, there exists a subset $\mathbb{V} \subseteq \mathbb{R}^q$ such that $v(0) \in \mathbb{V}_0$ implies $v(t) \in \mathbb{V}$ for all $t \geq 0$. As a result, given any compact subset $\Sigma_0 = \bar{\mathbb{V}}_0 \times \mathbb{W}_0 \subseteq \mathbb{V}_0 \times \mathbb{W}$, there exists a compact subset $\Sigma = \bar{\mathbb{V}} \times \mathbb{W}_0 \subseteq \mathbb{V} \times \mathbb{W}$ such that $(v(0), w) \in \Sigma_0$ implies $(v(t), w) \in \Sigma$ for all $t \geq 0$.

It can be seen that the first three equations of the control law (11.2) is obtained from a purely decentralized internal model control law by replacing e_i with its estimation $\hat{e}_i$, and the last two equations of the control law (11.2) is simply the adaptive distributed observer (11.1). Since the control law (11.2) utilizes both the distributed internal model and the adaptive distributed observer, we call it an integrated control law.

Under Assumptions 10.1–10.3, we can obtain the same internal model as (10.10) and the same augmented system (10.11). Performing the same transformation as (10.12) yields the same augmented error system as (10.13). Nevertheless, instead of the transformation (10.16), which leads to (10.17), performing on the augmented error system (10.13) the following coordinate transformation:

$$\tilde{\eta}_i = \bar{\eta}_i - b_i^{-1}(w)Q_i\hat{e}_i,\ i = 1, \ldots, N \tag{11.3}$$

letting $\tilde{v}_i = \hat{v}_i - v$, and noticing that

$$\hat{e}_i(t) + h_0(v + \tilde{v}_i) = e_i(t) + h_0(v)$$

converts the augmented error system (10.13) to the following form:

$$\dot{\bar{z}}_i = \bar{f}_i(\bar{z}_i, \hat{e}_i, \tilde{v}_i, v, w) \tag{11.4a}$$
$$\dot{\tilde{\eta}}_i = M_i\tilde{\eta}_i + b_i^{-1}(w)M_i Q_i\hat{e}_i - b_i^{-1}(w)Q_i\bar{g}_i(\bar{z}_i, \hat{e}_i, \tilde{v}_i, \dot{\tilde{v}}_i, v, w) \tag{11.4b}$$
$$\dot{\hat{e}}_i = \bar{g}_i(\bar{z}_i, \hat{e}_i, \tilde{v}_i, \dot{\tilde{v}}_i, v, w) + b_i(w)\Gamma_i\tilde{\eta}_i + \Gamma_i Q_i\hat{e}_i + b_i(w)\bar{u}_i, \ i = 1, \ldots, N \tag{11.4c}$$

where $\Gamma_i \triangleq \Gamma_i(\sigma)$ is known since the exosystem is known, and

$$\begin{aligned}\bar{f}_i(\bar{z}_i, \hat{e}_i, \tilde{v}_i, v, w) &= f_i(\bar{z}_i + \mathbf{z}_i(v, w), \hat{e}_i + h_0(v + \tilde{v}_i), v, w)\\ &\quad - f_i(\mathbf{z}_i(v, w), h_0(v), v, w)\\ \bar{g}_i(\bar{z}_i, \hat{e}_i, \tilde{v}_i, \dot{\tilde{v}}_i, v, w) &= g_i(\bar{z}_i + \mathbf{z}_i(v, w), \hat{e}_i + h_0(v + \tilde{v}_i), v, w)\\ &\quad - g_i(\mathbf{z}_i(v, w), h_0(v), v, w) + \frac{\partial h_0(v)}{\partial v}Sv\\ &\quad - \frac{\partial h_0(v + \tilde{v}_i)}{\partial v}Sv - \frac{\partial h_0(v + \tilde{v}_i)}{\partial \tilde{v}_i}\dot{\tilde{v}}_i.\end{aligned}$$

Let $\hat{Z}_i = \text{col}(\bar{z}_i, \tilde{\eta}_i)$, $\mathbf{v}_i = \text{col}(\tilde{v}_i, \dot{\tilde{v}}_i)$, and $\mu = \text{col}(v, w)$. Then the augmented error system (11.4) is put in the following compact form:

$$\dot{\hat{Z}}_i = \tilde{F}_i(\hat{Z}_i, \hat{e}_i, \mathbf{v}_i, \mu) \tag{11.5a}$$
$$\dot{\hat{e}}_i = \tilde{g}_i(\hat{Z}_i, \hat{e}_i, \mathbf{v}_i, \mu) + b_i(w)\bar{u}_i, \ i = 1, \ldots, N \tag{11.5b}$$

where

$$\tilde{F}_i(\hat{Z}_i, \hat{e}_i, \mathbf{v}_i, \mu) = \begin{bmatrix} \bar{f}_i(\bar{z}_i, \hat{e}_i, \tilde{v}_i, v, w) \\ M_i\hat{\eta}_i + b_i^{-1}(w)M_i Q_i\hat{e}_i - b_i^{-1}(w)Q_i\bar{g}_i(\bar{z}_i, \hat{e}_i, \tilde{v}_i, \dot{\tilde{v}}_i, v, w)\end{bmatrix}$$
$$\tilde{g}_i(\hat{Z}_i, \hat{e}_i, \mathbf{v}_i, \mu) = \bar{g}_i(\bar{z}_i, \hat{e}_i, \tilde{v}_i, \dot{\tilde{v}}_i, v, w) + b_i(w)\Gamma_i\tilde{\eta}_i + \Gamma_i Q_i\hat{e}_i.$$

It can be verified that the augmented error system (11.5) has the property that, for any $\mu \in \mathbb{R}^q \times \mathbb{R}^{n_w}$,

$$\tilde{f}_i(0, 0, 0, \mu) = 0 \text{ and } \tilde{g}_i(0, 0, 0, \mu) = 0.$$

Thus, when $\mathbf{v}_i = 0$, for any $\mu \in \mathbb{R}^q \times \mathbb{R}^{n_w}$, the origin is an equilibrium point of the augmented error system (11.5), and $\mathbf{v}_i$ can be viewed as an unknown time-varying disturbance.

Remark 11.1 The transformation (11.3) is different from the transformation (10.16) in Sect. 10.1 in that we replace e_i by its estimation $\hat{e}_i = y_i - h_0(\hat{v}_i)$, where $\hat{v}_i$ is provided by the adaptive distributed observer (11.1). As a result, the augmented error system (11.5) is also different from (10.17). Since $\bar{u}_i$ can directly access $\hat{e}_i$, there is no need to introduce the virtual tracking error e_{vi}. That is why it is possible for us to allow the communication graph to be switched and disconnected.

11.2 Preliminaries

In this section, we will establish two lemmas to lay the foundation for the solvability of the problem. Since the augmented error system (11.5) is different from (10.17), we need to modify Assumption 10.4 to the following form:

Assumption 11.2 Consider the subsystem (11.4a). For any compact subset $\Sigma \subseteq \mathbb{V} \times \mathbb{W}$, and, for $i = 1, \ldots, N$, there exist continuously differentiable functions $V_{0i}(\cdot)$ such that

$$\underline{\alpha}_{0i}(\|\bar{z}_i\|) \leq V_{0i}(\bar{z}_i) \leq \overline{\alpha}_{0i}(\|\bar{z}_i\|)$$

for some class $\mathcal{K}_\infty$ functions $\underline{\alpha}_{0i}(\cdot)$ and $\overline{\alpha}_{0i}(\cdot)$, and, for all $\mu \in \Sigma$, along the trajectory of the subsystem (11.4a),

$$\dot{V}_{0i}(\bar{z}_i) \leq -\alpha_{0i}(\|\bar{z}_i\|) + \delta_i \gamma_i(\|(\hat{e}_i, \mathbf{v}_i)\|) \tag{11.6}$$

where $\alpha_{0i}(\cdot)$ are some known class $\mathcal{K}_\infty$ functions satisfying $\limsup_{s\to 0^+} (s^2/\alpha_{0i}(s)) < \infty$, $\gamma_i(\cdot)$ are some known class $\mathcal{K}$ functions satisfying $\limsup_{s\to 0^+} (\gamma_i(s)/s^2) < \infty$, and δ_i is some positive number depending on Σ.

We are ready to state the following result regarding the subsystem (11.5a).

Lemma 11.1 *Under Assumptions 10.1–10.3, and 11.2, given any smooth functions $\Theta_i(\hat{Z}_i) > 0$, there exist continuously differentiable functions $V_{1i}(\hat{Z})$ such that*

$$\underline{\alpha}_{1i}(\|\hat{Z}_i\|) \leq V_{1i}(\hat{Z}_i) \leq \overline{\alpha}_{1i}(\|\hat{Z}_i\|) \tag{11.7}$$

for some class $\mathcal{K}_\infty$ functions $\underline{\alpha}_{1i}(\cdot)$ and $\overline{\alpha}_{1i}(\cdot)$, and, for all $\mu \in \Sigma$, along the trajectory of (11.5a),

$$\dot{V}_{1i}(\hat{Z}) \leq -\Theta_i(\hat{Z}_i)\|\hat{Z}_i\|^2 + \hat{\delta}_i \hat{\gamma}_{1i}(\hat{e}_i)\hat{e}_i^2 + \hat{\delta}_i \hat{\gamma}_{2i}(\mathbf{v}_i)\|\mathbf{v}_i\|^2 \tag{11.8}$$

where $\hat{\gamma}_{1i}(\cdot)$ and $\hat{\gamma}_{2i}(\cdot)$, $i = 1, \ldots, N$, are some known smooth positive functions, and $\hat{\delta}_i$ is some positive number depending on Σ.

Proof Since (11.5a) consists of (11.4a) and (11.4b) with $\hat{Z}_i = \text{col}(\bar{z}_i, \hat{\eta}_i)$, which is in the form of (A.57) with $z_1 = \bar{z}_i$ and $z_2 = \hat{\eta}_i$, under Assumption 11.2, for $i =$

$1, \ldots, N$, applying Lemma A.12 in Sect. A.7 to (11.5a) concludes the existence of continuously differentiable functions $U_{1i}(\hat{Z}_i)$ such that

$$\underline{\beta}_{1i}(\|\hat{Z}_i\|) \leq U_{1i}(\hat{Z}_i) \leq \overline{\beta}_{1i}(\|\hat{Z}_i\|)$$

for some class $\mathscr{K}_\infty$ functions $\underline{\beta}_{1i}(\cdot)$ and $\overline{\beta}_{1i}(\cdot)$, and, for all $\mu \in \Sigma$, along the trajectory of (11.5a),

$$\dot{U}_{1i}(\hat{Z}_i) \leq -\|\hat{Z}_i\|^2 + \bar{\delta}_i \bar{\gamma}_i(\hat{e}_i, \mathbf{v}_i) \left\|(\hat{e}_i, \mathbf{v}_i)\right\|^2 \tag{11.9}$$

for some smooth positive functions $\bar{\gamma}_i(\cdot, \cdot)$, and some positive numbers $\bar{\delta}_i$ depending on Σ.

Let

$$a_i(s) = \sup_{0 \leq \left\|(\hat{e}_i, \mathbf{v}_i)\right\| \leq s} \bar{\gamma}_i(\hat{e}_i, \mathbf{v}_i).$$

Then $a_i(s)$ is continuous, and non-decreasing, and $a_i(\left\|(\hat{e}_i, \mathbf{v}_i)\right\|) \geq \bar{\gamma}_i(\hat{e}_i, \mathbf{v}_i)$. Thus, (11.9) implies

$$\dot{U}_{1i}(\hat{Z}_i) \leq -\|\hat{Z}_i\|^2 + \bar{\delta}_i a_i(\left\|(\hat{e}_i, \mathbf{v}_i)\right\|) \left\|(\hat{e}_i, \mathbf{v}_i)\right\|^2. \tag{11.10}$$

Let $\alpha_i(s) = s^2$, which is a class $\mathscr{K}_\infty$ function, and $\sigma_i(s) = a_i(s)s^2$, which is a class $\mathscr{K}$ function. Then, $\limsup_{s \to 0^+}(s^2/\alpha_i(s)) = 1 < \infty$, and $\limsup_{s \to 0^+}(\sigma_i(s)/s^2) = \bar{\gamma}(0, 0) < \infty$.

Thus, from (11.10), by the changing supply functions technique (Corollary A.2 in Sect. A.7), given any smooth functions $\Theta_i(\hat{Z}_i) > 0$, $i = 1, \ldots, N$, there exist continuously differentiable functions $V_{1i}(\hat{Z}_i)$ satisfying (11.7) for some class $\mathscr{K}_\infty$ functions $\underline{\alpha}_{1i}(\cdot)$ and $\overline{\alpha}_{1i}(\cdot)$ such that, for all $\mu \in \Sigma$, along the trajectory of (11.5a),

$$\dot{V}_{1i}(\hat{Z}_i) \leq -\Theta_i(\hat{Z}_i)\|\hat{Z}_i\|^2 + \hat{\delta}\hat{\gamma}_i(\hat{e}_i, \mathbf{v}_i)\|(\hat{e}_i, \mathbf{v}_i)\|^2 \tag{11.11}$$

for some smooth positive functions $\hat{\gamma}_i(\cdot, \cdot)$, $i = 1, \ldots, N$, and some positive numbers $\hat{\delta}_i$ depending on Σ.

By Part (ii) of Lemma A.8 in Sect. A.7,

$$\hat{\gamma}_i(\hat{e}_i, \mathbf{v}_i)\|(\hat{e}_i, \mathbf{v}_i)\|^2 \leq \hat{\gamma}_{1i}(\hat{e}_i)\hat{e}_i^2 + \hat{\gamma}_{2i}(\mathbf{v}_i)\|\mathbf{v}_i\|^2 \tag{11.12}$$

for some smooth positive functions $\hat{\gamma}_{1i}(\cdot)$ and $\hat{\gamma}_{2i}(\cdot)$. Combining (11.11) and (11.12) gives (11.8). The proof is thus completed. □

Lemma 11.2 *Consider the augmented error system (11.5). Under Assumptions 10.1–10.3, and 11.2, there exist continuously differentiable functions $V_i(\hat{Z}_i, \hat{e}_i)$ such that*

$$\underline{\alpha}_{2i}(\|(\hat{Z}_i, \hat{e}_i)\|) \leq V_i(\hat{Z}_i, \hat{e}_i) \leq \overline{\alpha}_{2i}(\|(\hat{Z}_i, \hat{e}_i)\|) \tag{11.13}$$

for some class $\mathcal{K}_\infty$ functions $\underline{\alpha}_{2i}(\cdot)$ and $\overline{\alpha}_{2i}(\cdot)$, and, for all $\mu \in \Sigma$, along the trajectory of the ith subsystem of (11.5),

$$\dot{V}_i(\hat{Z}_i, \hat{e}_i) \leq -\|\hat{Z}_i\|^2 + \psi_i(\hat{e}_i)\hat{e}_i^2 + \phi_i(\mathbf{v}_i)\|\mathbf{v}_i\|^2 + b_i(w)\hat{e}_i\bar{u}_i \tag{11.14}$$

where $\psi_i(\cdot)$ and $\phi_i(\cdot)$, $i = 1, \ldots, N$, are some smooth positive functions.

Proof Since $\tilde{g}_i(0, 0, 0, \mu) = 0$, for all $\mu \in \Sigma$, by Part (iii) of Lemma A.8 in Sect. A.7,

$$\|\tilde{g}_i(\hat{Z}_i, \hat{e}_i, \mathbf{v}_i, \mu)\|^2 \leq c_i(\pi_i(\hat{Z}_i)\|\hat{Z}_i\|^2 + \phi_{1i}(\hat{e}_i)\hat{e}_i^2 + \phi_{2i}(\mathbf{v}_i)\|\mathbf{v}_i\|^2) \tag{11.15}$$

for some smooth functions $\pi_i(\cdot) \geq 1$, $\phi_{1i}(\cdot) \geq 1$, $\phi_{2i}(\cdot) \geq 1$, and some positive constants c_i depending on Σ. Let

$$V_{2i}(\hat{e}_i) = \frac{1}{2}\hat{e}_i^2.$$

Then, using (11.15), the derivative of V_{2i} along the trajectory of the subsystem (11.5b) satisfies

$$\begin{aligned}
\dot{V}_{2i}(\hat{e}_i) &= \hat{e}_i\tilde{g}_i(\hat{Z}_i, \hat{e}_i, \mathbf{v}_i, \mu) + b_i(w)\hat{e}_i\bar{u}_i \\
&\leq \frac{c_i}{4}\hat{e}_i^2 + \frac{1}{c_i}\|\tilde{g}_i(\hat{Z}_i, \hat{e}_i, \mathbf{v}_i, \mu)\|^2 + b_i(w)\hat{e}_i\bar{u}_i \\
&\leq \frac{c_i}{4}\hat{e}_i^2 + \pi_i(\hat{Z}_i)\|\hat{Z}_i\|^2 + \phi_{1i}(\hat{e}_i)\hat{e}_i^2 + \phi_{2i}(\mathbf{v}_i)\|\mathbf{v}_i\|^2 + b_i(w)\hat{e}_i\bar{u}_i.
\end{aligned} \tag{11.16}$$

Now, let

$$V_i(\hat{Z}_i, \hat{e}_i) = V_{1i}(\hat{Z}_i) + V_{2i}(\hat{e}_i).$$

Then, by Lemma A.9 in Sect. A.7, (11.13) holds for some class $\mathcal{K}_\infty$ functions $\underline{\alpha}_{2i}(\cdot)$ and $\overline{\alpha}_{2i}(\cdot)$, and, by (11.8) and (11.16), for all $\mu \in \Sigma$, along the trajectory of the ith subsystem of (11.5),

$$\begin{aligned}
\dot{V}_i(\hat{Z}_i, \hat{e}_i) &\leq -\Theta_i(\hat{Z}_i)\|\hat{Z}_i\|^2 + \hat{\delta}_i\hat{\gamma}_{1i}(\hat{e}_i)\hat{e}_i^2 + \hat{\delta}_i\hat{\gamma}_{2i}(\mathbf{v}_i)\|\mathbf{v}_i\|^2 + \frac{c_i}{4}\hat{e}_i^2 \\
&\quad + \pi_i(\hat{Z}_i)\|\hat{Z}_i\|^2 + \phi_{1i}(\hat{e}_i)\hat{e}_i^2 + \phi_{2i}(\mathbf{v}_i)\|\mathbf{v}_i\|^2 + b_i(w)\hat{e}_i\bar{u}_i \\
&\leq -(\Theta_i(\hat{Z}_i) - \pi_i(\hat{Z}_i))\|\hat{Z}_i\|^2 + \left(\hat{\delta}_i\hat{\gamma}_{1i}(\hat{e}_i) + \phi_{1i}(\hat{e}_i) + \frac{c_i}{4}\right)\hat{e}_i^2 \\
&\quad + (\hat{\delta}_i\hat{\gamma}_{2i}(\mathbf{v}_i) + \phi_{2i}(\mathbf{v}_i))\|\mathbf{v}_i\|^2 + b_i(w)\hat{e}_i\bar{u}_i. \qquad (11.17)
\end{aligned}$$

Letting

$$\Theta_i(\hat{Z}_i) = \pi_i(\hat{Z}_i) + 1 \qquad (11.18a)$$

$$\psi_i(\hat{e}_i) = \hat{\delta}_i\hat{\gamma}_{1i}(\hat{e}_i) + \phi_{1i}(\hat{e}_i) + \frac{c_i}{4} \qquad (11.18b)$$

$$\phi_i(\mathbf{v}_i) = \hat{\delta}_i\hat{\gamma}_{2i}(\mathbf{v}_i) + \phi_{2i}(\mathbf{v}_i). \qquad (11.18c)$$

gives (11.14). The proof is thus completed. □

11.3 Solvability of Case 1 of the Problem

In this section, we will consider the first case of the problem. Since, in this case, $\Sigma_0 = \bar{\mathbb{V}}_0 \times \mathbb{W}_0$ and hence $\Sigma = \bar{\mathbb{V}} \times \mathbb{W}_0$ are known, we can assume that the constants δ_i in (11.6), $\hat{\delta}_i$ in (11.8), and c_i in (11.15) are all known and will let them be unity. Also, there exists a known positive number $b_{\min}$ such that $b_{\min} \leq b_i(w)$ for all $w \in \mathbb{W}_0$ and all $i = 1, \ldots, N$.

The following lemma shows the input-to-state stabilizability of the augmented error system (11.5) with $(\hat{Z}_i, \hat{e}_i)$ as the state and $\mathbf{v}_i$ as the input by a distributed control law of the following form:

$$\bar{u}_i = -k_i\rho_i(\hat{e}_i)\hat{e}_i \qquad (11.19)$$

where $\rho_i(\cdot) \geq 1$ are some smooth positive functions and k_i are some positive numbers.

Lemma 11.3 *Under Assumptions 10.1–10.3, and 11.2, for some smooth positive functions $\rho_i(\cdot) \geq 1$, and positive numbers k_i, the closed-loop augmented error system composed of the system (11.5) and the distributed static output feedback control law (11.19) has the property that there exist continuously differentiable functions $V_i(\cdot, \cdot)$ such that*

$$\underline{\alpha}_{2i}(\|(\hat{Z}_i, \hat{e}_i)\|) \leq V_i(\hat{Z}_i, \hat{e}_i) \leq \overline{\alpha}_{2i}(\|(\hat{Z}_i, \hat{e}_i)\|) \qquad (11.20)$$

for some class $\mathcal{K}_\infty$ functions $\underline{\alpha}_{2i}(\cdot)$ and $\overline{\alpha}_{2i}(\cdot)$, and, for all $\mu \in \Sigma$,

$$\dot{V}_i(\hat{Z}_i, \hat{e}_i) \leq -\|\hat{Z}_i\|^2 - \hat{e}_i^2 + \sigma_i(\|\mathbf{v}_i\|) \tag{11.21}$$

for some class $\mathcal{K}$ functions σ_i. Thus, the closed-loop augmented error system composed of (11.5) and the control law (11.19) is input-to-state stable with $(\hat{Z}_i, \hat{e}_i)$ as the state and $\mathbf{v}_i$ as the input.

Proof Under Assumptions 10.1–10.3, and 11.2, by Lemma 11.2, (11.14) holds. Substituting the control law (11.19) to (11.14) shows that, along the trajectory of the ith subsystem of the closed-loop augmented error system composed of (11.5) and (11.19),

$$\begin{aligned}\dot{V}_i(\hat{Z}_i, \hat{e}_i) &\leq -\|\hat{Z}_i\|^2 + \psi_i(\hat{e}_i)\hat{e}_i^2 + \phi_i(\mathbf{v}_i)\|\mathbf{v}_i\|^2 - b_i(w)k_i\rho_i(\hat{e}_i)\hat{e}_i^2 \\ &= -\|\hat{Z}_i\|^2 - \left(b_i(w)k_i\rho_i(\hat{e}_i) - \psi_i(\hat{e}_i)\right)\hat{e}_i^2 + \phi_i(\mathbf{v}_i)\|\mathbf{v}_i\|^2. \end{aligned} \tag{11.22}$$

Let k_i be some positive numbers, and $\rho_i(\hat{e}_i)$ be some smooth positive functions satisfying

$$k_i\rho_i(\hat{e}_i) \geq \frac{1}{b_{\min}}\left(\psi_i(\hat{e}_i) + 1\right). \tag{11.23}$$

Then, we have, for all $w \in \mathbb{W}_0$,

$$b_i(w)k_i\rho_i(\hat{e}_i) \geq b_{\min}k_i\rho_i(\hat{e}_i) \geq \psi_i(\hat{e}_i) + 1,$$

which gives, for all $\mu \in \Sigma$,

$$\dot{V}_i(\hat{Z}_i, \hat{e}_i) \leq -\|\hat{Z}_i\|^2 - \hat{e}_i^2 + \phi_i(\mathbf{v}_i)\|\mathbf{v}_i\|^2. \tag{11.24}$$

Let

$$\sigma_i(s) = \sup_{0\leq\|\mathbf{v}_i\|\leq s} \phi_i(\mathbf{v}_i)\,\|\mathbf{v}_i\|^2.$$

Then $\sigma_i(s)$ are class $\mathcal{K}$ functions satisfying $\sigma_i(\|\mathbf{v}_i\|) \geq \phi_i(\mathbf{v}_i)\|\mathbf{v}_i\|^2$, which together with (11.24) implies (11.21). □

We now state our main result in this section as follows.

Theorem 11.1 *Under Assumptions 10.1–10.3, 11.1 and 11.2, let $\rho_i(\cdot)$ and k_i be the same as those in Lemma 11.3, M_i, Q_i, Γ_i the same as those in (10.9). Then, the cooperative global output regulation problem for the original plant (10.4) is solvable by the following distributed dynamic output feedback control law:*

$$u_i = -k_i \rho_i(\hat{e}_i)\hat{e}_i + \Gamma_i \eta_i \tag{11.25a}$$

$$\dot{\eta}_i = M_i \eta_i + Q_i u_i \tag{11.25b}$$

$$\dot{S}_i = \mu_1 \sum_{j=0}^{N} a_{ij}(t)(S_j - S_i) \tag{11.25c}$$

$$\dot{\hat{v}}_i = S_i \hat{v}_i + \mu_2 \sum_{j=0}^{N} a_{ij}(t)(\hat{v}_j - \hat{v}_i), \; i = 1, \ldots, N. \tag{11.25d}$$

Proof Under Assumptions 10.1–10.3, by Lemma 11.3, for all $\mu \in \Sigma$, the closed-loop system composed of the augmented error system (11.5) and the control law (11.19) is input-to-state stable with $(\hat{Z}_i, \hat{e}_i)$ as the state and $\mathbf{v}_i$ as the input. By Remark A.7 in Sect. A.7, for any uniformly bounded $\mathbf{v}_i(t)$, both $\hat{Z}_i(t)$ and $\hat{e}_i(t)$ are uniformly bounded over $[0, \infty)$, and $\lim_{t\to\infty} \hat{Z}_i(t) = 0$ and $\lim_{t\to\infty} \hat{e}_i(t) = 0$ if $\lim_{t\to\infty} \mathbf{v}_i(t) = 0$.

As shown in Sect. 4.3, let $\tilde{S}_i = S_i - S$, $\tilde{S} = \mathrm{col}\left(\tilde{S}_1, \ldots, \tilde{S}_N\right)$, $\tilde{S}_d = \mathscr{D}(\tilde{S}_1, \ldots, \tilde{S}_N)$, $\tilde{v}_i = \hat{v}_i - v$, $\tilde{v} = \mathrm{col}\,(\tilde{v}_1, \ldots, \tilde{v}_N)$ and $\hat{v} = \mathrm{col}\left(\hat{v}_1, \ldots, \hat{v}_1\right)$. Then $\tilde{S}$ and $\tilde{v}$ are governed by the following system:

$$\dot{\tilde{S}} = -\mu_1(H_{\sigma(t)} \otimes I_q)\tilde{S} \tag{11.26a}$$

$$\dot{\tilde{v}} = \left(I_N \otimes S - \mu_2(H_{\sigma(t)} \otimes I_q)\right) \tilde{v} + \tilde{S}_d \hat{v}. \tag{11.26b}$$

By Theorem 4.8, under Assumptions 10.1 and 11.1, for any initial conditions $S_i(0)$, $\hat{v}_i(0)$, $i = 1, \ldots, N$, the solution of system (11.25d) and (11.25c) is such that

$$\lim_{t\to\infty} (S_i(t) - S) = 0 \;\text{ and }\; \lim_{t\to\infty} (\hat{v}_i(t) - v(t)) = 0$$

both exponentially. From the proof of Theorem 4.8, $\lim_{t\to\infty} \tilde{S}_d(t)\hat{v}(t) = 0$ exponentially. Thus, from (11.26b), $\lim_{t\to\infty} \dot{\tilde{v}}_i(t) = 0$ exponentially. As a result, $\lim_{t\to\infty} \mathbf{v}_i(t) = 0$ exponentially. Thus, the solution of system (11.25d) and (11.25c) and $\mathbf{v}_i(t)$ are uniformly bounded over $[0, \infty)$. By Remark A.7, we have $\lim_{t\to\infty} \hat{e}_i(t) = 0$. Since $e_i(t) = \hat{e}_i(t) + h_0(v(t) + \tilde{v}_i(t)) - h_0(v(t))$ and $h_0(\cdot)$ is continuous, we also have $\lim_{t\to\infty} e_i(t) = 0$. The proof is thus completed. □

Remark 11.2 In the special case where the communication network is static as studied from Chap. 10, $H_{\sigma(t)}$ reduces to the constant matrix H, and Assumption 11.1 reduces to Assumption 10.5. As a result, the distributed control law (11.25) with $a_{ij}(t) \equiv a_{ij}$ is still valid for such a special case.

Remark 11.3 In the same spirit as Remark 10.6, we can adopt a more pragmatic way to synthesize the control law (11.19) which is defined by positive constants k_i and known functions $\rho_i(\cdot)$. Recall that all we need to do is to choose k_i and $\rho_i(\hat{e}_i)$ to satisfy (11.23). Like what was pointed out in Remark 10.6, for a polynomial nonlinear system of the form (10.4), if Assumption 11.2 is satisfied by a positive definite polynomial function $V_{0i}(\hat{Z}_i)$, $i = 1, \ldots, N$, then, for each positive polynomial function $\Theta_i(\hat{Z}_i)$ satisfying $\Theta_i(\hat{Z}_i) \geq \pi_i(\hat{Z}_i) + 1$, one can find a positive definite polynomial function $V_{1i}(\hat{Z}_i)$ such that $\dot{V}_{1i}(\hat{Z}_i)$ satisfies (11.8) with $\hat{\gamma}_{1i}(\hat{e}_i)$ a polynomial. Once a polynomial $\psi_i(\hat{e}_i)$ with degree m_0 is obtained, one can find a sufficiently large number k_i and a polynomial function $\rho_i(s) = (1 + s^{2m})$, with m satisfying $2m \geq m_0$, such that (11.23) is satisfied.

Again, $V_{1i}(\hat{Z}_i)$ does not have to be found by the changing supply functions technique. It can be found by any way as long as $\dot{V}_{1i}(\hat{Z}_i)$ satisfies (11.8) such that $\Theta_i(\hat{Z}_i) \geq \pi_i(\hat{Z}_i) + 1$ for all $\hat{Z}_i$.

Also, it is more efficient to work on the components of $\hat{Z}_i$. Denote the dimension of $\hat{Z}_i$ by n_{Z_i} and partition $\hat{Z}_i$ as $\hat{Z}_i = \text{col}(\zeta_{1i}, \ldots, \zeta_{s_i i})$ with $\zeta_{li} \in \mathbb{R}^{n_{z_{li}}}$, $l = 1, \ldots, s_i$, such that $\sum_{l=1}^{s_i} n_{z_{li}} = n_{Z_i}$. Then, instead of (11.8), we aim to achieve the following:

$$\dot{V}_{1i}(\hat{Z}_i) \leq -\sum_{l=1}^{s_i} \Theta_{li}(\zeta_{li})\|\zeta_{li}\|^2 + \hat{\delta}_i\hat{\gamma}_{1i}(\hat{e}_i)\hat{e}_i^2 + \hat{\delta}_i\hat{\gamma}_{2i}(\mathbf{v}_i)\|\mathbf{v}_i\|^2. \tag{11.27}$$

Correspondingly, instead of (11.15), by making use of the following inequality:

$$\|\tilde{g}_i(\hat{Z}_i, \hat{e}_i, \mathbf{v}_i, \mu)\|^2 \leq c_i\left(\sum_{l=1}^{s_i} \pi_{li}(\zeta_{li})\|\zeta_{li}\|^2 + \phi_{1i}(\hat{e}_i)\hat{e}_i^2 + \phi_{2i}(\mathbf{v}_i)\|\mathbf{v}_i\|^2\right) \tag{11.28}$$

we obtain, instead of (11.22), the following inequality:

$$\begin{aligned}\dot{V}_i(\hat{Z}_i, \hat{e}_i) \leq &-\sum_{l=1}^{s_i}(\Theta_{li}(\zeta_{li}) - \pi_{li}(\zeta_{li}))\|\zeta_{li}\|^2 \\ &- \left(b_i(w)k_i\rho_i(\hat{e}_i) - \psi_i(\hat{e}_i)\right)\hat{e}_i^2 + \phi_i(\mathbf{v}_i)\|\mathbf{v}_i\|^2.\end{aligned} \tag{11.29}$$

Thus, as long as $\Theta_{li}(\zeta_{li}) \geq \pi_{li}(\zeta_{li}) + 1$ for all ζ_i, we can still use (11.23) to obtain the number k_i and the function $\rho_i(\cdot)$.

11.4 Solvability of Case 2 of the Problem

In this section, like in Sect. 10.3, we study the second case of the problem where the exosystem is known and the compact subset $\Sigma_0 = \bar{\mathbb{V}}_0 \times \mathbb{W}_0$ and hence $\Sigma = \bar{\mathbb{V}} \times \mathbb{W}_0$ are unknown. Thus, we cannot assume that the constants δ_i in (11.6), $\hat{\delta}_i$ in (11.8), and c_i in (11.15) are known. Also, the positive number $b_{\min}$ such that $b_{\min} \leq b_i(w)$ for all $w \in \mathbb{W}_0$ and all $i = 1, \ldots, N$ is unknown either. As a result, our control law cannot rely on δ_i, $\hat{\delta}_i$, c_i, and $b_{\min}$. Like in Sect. 10.3, we will use the so-called dynamic gain to deal with this case.

Consider the following distributed dynamic output feedback control law:

$$\bar{u}_i = -k_i \rho_i(\hat{e}_i)\hat{e}_i, \tag{11.30a}$$

$$\dot{k}_i = \rho_i(\hat{e}_i)\hat{e}_i^2, \; i = 1, \ldots, N \tag{11.30b}$$

where $\rho_i(\cdot) \geq 1$ are some smooth positive functions.

Lemma 11.4 *Under Assumptions 10.1–10.3, and 11.2, for some smooth positive functions $\rho_i(\cdot) \geq 1$, $i = 1, \ldots, N$, the closed-loop augmented error system composed of (11.5) and (11.30) has the property that there exist continuously differentiable functions $U_i(\cdot, \cdot, \cdot)$ such that*

$$\underline{\alpha}_{3i}(\|(\hat{Z}_i, \hat{e}_i, \tilde{k})\|) \leq U_i(\hat{Z}_i, \hat{e}_i, \tilde{k}_i) \leq \overline{\alpha}_{3i}(\|(\hat{Z}_i, \hat{e}_i, \tilde{k}_i)\|) \tag{11.31}$$

for some class $\mathcal{K}_\infty$ functions $\underline{\alpha}_{3i}(\cdot)$ and $\overline{\alpha}_{3i}(\cdot)$, and, for all $\mu \in \Sigma$,

$$\dot{U}_i(\hat{Z}_i, \hat{e}_i, \tilde{k}_i) \leq -\|\hat{Z}_i\|^2 - \hat{e}_i^2 + \phi_i(\mathbf{v}_i)\|\mathbf{v}_i\|^2 \tag{11.32}$$

where ϕ_i are as defined in (11.18c), $\tilde{k} = \mathrm{col}(\tilde{k}_1, \ldots, \tilde{k}_N)$ with $\tilde{k}_i(t) = k_i(t) - \bar{k}_i$, $i = 1, \ldots, N$, and $\bar{k}_i$ being some positive constants.

Proof Let

$$U_i(\hat{Z}_i, \hat{e}_i, \tilde{k}_i) = V_i(\hat{Z}_i, \hat{e}_i) + \frac{1}{2} b_i(w)\tilde{k}_i^2.$$

Then, by Lemma A.9 in Sect. A.7, (11.31) holds for some class $\mathcal{K}_\infty$ functions $\underline{\alpha}_{3i}(\cdot)$ and $\overline{\alpha}_{3i}(\cdot)$.

Using (11.14), the derivative of $U_i(\hat{Z}_i, \hat{e}_i, \tilde{k}_i)$ along the trajectory of the ith subsystem of the closed-loop system composed of (11.5) and (11.30) satisfies

$$\begin{aligned}
\dot{U}_i(\hat{Z}_i, \hat{e}_i, \tilde{k}_i) &\leq -\|\hat{Z}_i\|^2 + \psi_i(\hat{e}_i)\hat{e}_i^2 + \phi_i(\mathbf{v}_i)\|\mathbf{v}_i\|^2 + b_i(w)\bar{u}_i\hat{e}_i + b_i(w)\tilde{k}_i\dot{\tilde{k}}_i \\
&= -\|\hat{Z}_i\|^2 + \psi_i(\hat{e}_i)\hat{e}_i^2 + \phi_i(\mathbf{v}_i)\|\mathbf{v}_i\|^2 - b_i(w)k_i\rho_i(\hat{e}_i)\hat{e}_i^2 + b_i(w)\tilde{k}_i\dot{\tilde{k}}_i \\
&= -\|\hat{Z}_i\|^2 + \psi_i(\hat{e}_i)\hat{e}_i^2 + \phi_i(\mathbf{v}_i)\|\mathbf{v}_i\|^2 - b_i(w)\bar{k}_i\rho_i(\hat{e}_i)\hat{e}_i^2 \\
&\quad + b_i(w)\tilde{k}_i(\dot{\tilde{k}}_i - \rho_i(\hat{e}_i)\hat{e}_i^2) \\
&= -\|\hat{Z}_i\|^2 - \left(\bar{k}_i b_i(w)\rho_i(\hat{e}_i) - \psi_i(\hat{e}_i)\right)\hat{e}_i^2 + \phi_i(\mathbf{v}_i)\|\mathbf{v}_i\|^2.
\end{aligned} \tag{11.33}$$

Noting from (11.18b) that

$$\psi_i(\hat{e}_i) + 1 = \left(\hat{\delta}_i\hat{\gamma}_{1i}(\hat{e}_i) + \phi_{1i}(\hat{e}_i) + \frac{c_i}{4} + 1\right) \leq (\hat{\delta}_i + \frac{c_i}{4} + 2)\max\{\hat{\gamma}_{1i}(\hat{e}_i), \phi_{1i}(\hat{e}_i), 1\}$$

and letting $\bar{k}_i \geq \frac{1}{b_{\min}}(\hat{\delta}_i + \frac{c_i}{4} + 2)$ and

$$\rho_i(\hat{e}_i) \geq \max\{\hat{\gamma}_{1i}(\hat{e}_i), \phi_{1i}(\hat{e}_i), 1\}$$

shows that, for all $w \in \mathbb{W}_0$,

$$\bar{k}_i b_i(w)\rho_i(\hat{e}_i) \geq \bar{k}_i b_{\min}\rho_i(\hat{e}_i) \geq (\hat{\delta}_i + \frac{c_i}{4} + 2)\max\{\hat{\gamma}_{1i}(\hat{e}_i), \phi_{1i}(\hat{e}_i), 1\} \geq \psi_i(\hat{e}_i) + 1. \tag{11.34}$$

Thus, (11.32) holds. □

We now state our main result in this section as follows.

Theorem 11.2 *Under Assumptions 10.1–10.3, 11.1, and 11.2, suppose the functions $V_{0i}(\cdot)$ in (11.6) are twice differentiable. Then the cooperative global output regulation problem for the original plant (10.4) is solvable by the following distributed dynamic output feedback control law:*

$$u_i = -k_i\rho_i(\hat{e}_i)\hat{e}_i + \Gamma_i\eta_i \tag{11.35a}$$

$$\dot{k}_i = \rho_i(\hat{e}_i)\hat{e}_i^2 \tag{11.35b}$$

$$\dot{\eta}_i = M_i\eta_i + Q_iu_i \tag{11.35c}$$

$$\dot{S}_i = \mu_1\sum_{j=0}^{N} a_{ij}(t)(S_j - S_i) \tag{11.35d}$$

$$\dot{\hat{v}}_i = S_i\hat{v}_i + \mu_2\sum_{j=0}^{N} a_{ij}(t)(\hat{v}_j - \hat{v}_i), \ i = 1, \ldots, N \tag{11.35e}$$

where $\rho_i(\cdot)$ are the same as those in Lemma 11.4, and M_i, Q_i, Γ_i the same as those in (10.9).

Proof Let us first show that, for any initial condition, the solution of the closed-loop system is uniformly bounded over $[0, \infty)$. In fact, as pointed out in the proof of Theorem 11.1, under Assumptions 10.1 and 11.1, for any initial conditions $S_i(0)$, $\hat{v}_i(0)$, $i = 1, \ldots, N$, the solution of system (11.1) is such that

$$\lim_{t\to\infty}(S_i(t) - S) = 0 \text{ and } \lim_{t\to\infty}\mathbf{v}_i(t) = 0$$

both exponentially. Thus, the solution of system (11.1) and $\mathbf{v}_i(t)$ are uniformly bounded over $[0, \infty)$.

Next, note the fact that the functions $V_{0i}(\cdot)$ are twice differentiable implies that the functions $U_i(\cdot, \cdot, \cdot)$ in Lemma 11.4 are twice differentiable. By Lemma 11.4, under Assumptions 10.1–10.3, and 11.2, $U_i(\cdot, \cdot, \cdot)$ satisfy (11.32). Let

$$W_i(t) = W_i(\hat{Z}_i(t), \hat{e}_i(t), \tilde{k}_i(t)) = U_i(\hat{Z}_i(t), \hat{e}_i(t), \tilde{k}_i(t)) - \int_0^t \phi_i(\mathbf{v}_i(\tau))\|\mathbf{v}_i(\tau)\|^2 d\tau.$$

Since $\lim_{t\to\infty} \mathbf{v}_i(t) = 0$ exponentially, $\int_0^\infty \phi_i(\mathbf{v}_i(\tau))\mathbf{v}_i(\tau)d\tau$ exist and are finite. Thus, $W_i(t)$ are lower bounded. Since $\dot{W}_i \le -\|\hat{Z}_i\|^2 - \hat{e}_i^2$, $\lim_{t\to\infty} W_i(t)$ exist and are finite. Thus, from the definition of $U_i(\hat{Z}_i, \hat{e}_i, \tilde{k}_i)$, $\hat{Z}_i(t)$, $\hat{e}_i(t)$, and $\tilde{k}_i(t)$ are uniformly bounded over $[0, \infty)$.

Next, we will show $\lim_{t\to\infty} e_i(t) = 0$. In fact, since all the state variables of the closed-loop system are uniformly bounded over $[0, \infty)$, $\ddot{U}_i(t)$ and hence $\ddot{W}_i(t)$ are also uniformly bounded over $[0, \infty)$. Thus, by Barbalat's Lemma, i.e., Lemma 2.9, $\lim_{t\to\infty} \dot{W}_i(t) = 0$, which imply $\lim_{t\to\infty} \hat{Z}_i(t) = 0$ and $\lim_{t\to\infty} \hat{e}_i(t) = 0$. Since $e_i(t) = \hat{e}_i(t) + h_0(v(t) + \tilde{v}_i(t)) - h_0(v(t))$ and $h_0(\cdot)$ is continuous, we have $\lim_{t\to\infty} e_i(t) = 0$. The proof is thus completed. □

Remark 11.4 Due to the use of the dynamic gain $k_i(t)$, the closed-loop augmented error system composed of (11.5) and (11.30) is not input-to-state stable. Thus, unlike the proof of Theorem 11.1, we cannot conclude the solvability of the problem by Remark A.7 in Sect. A.7. Instead, we have resorted to Barbalat's Lemma. For this purpose, we need to assume that the functions $V_{0i}(\cdot)$ are twice differentiable.

Remark 11.5 The approach described in Remark 11.3 still applies to this case. As in Sect. 10.3, since $\bar{k}_i$ are unknown, there is no need to calculate them. Thus, it suffices to determine the functions $\rho_i(\cdot)$ according to (11.34) assuming $\bar{k}_i$ can be arbitrarily large. In particular, if $\psi_i(\hat{e}_i)$ is a polynomial of degree $2m$ for some integer $m > 0$, one can always choose $\psi_i(\hat{e}_i) = k(1 + \hat{e}_i^{2m})$ for any $k > 0$. It is noted that the value of k will affect the transient response of the closed-loop system.

11.5 Numerical Examples

Let us again consider the same group of generalized Lorenz systems (10.65) studied in Sect. 10.5 over a jointly connected switching communication graph $\bar{\mathcal{G}}_{\sigma(t)}$ dictated by the following switching signal:

$$\sigma(t) = \begin{cases} 1, & \text{if } sT_0 \le t < (s + \frac{1}{4})T_0 \\ 2, & \text{if } (s + \frac{1}{4})T_0 \le t < (s + \frac{1}{2})T_0 \\ 3, & \text{if } (s + \frac{1}{2})T_0 \le t < (s + \frac{3}{4})T_0 \\ 4, & \text{if } (s + \frac{3}{4})T_0 \le t < (s + 1)T_0 \end{cases}$$

where $s = 0, 1, 2, \ldots$ and $T_0 = 1$s. The four graphs $\bar{\mathcal{G}}_i$, $i = 1, 2, 3, 4$, are as described in Fig. 11.1. Then, Assumption 11.1 can be easily verified. We assume that $\bar{a}_i = \text{col}(-10, 10, 28, -1, -8/3, 1)$, $\sigma = 0.8$, and the uncertainty w_i belongs to a set $\mathbb{W} = \{w_i \in \mathbb{R}^7, \bar{a}_{1i} + w_{1i} < 0, \bar{a}_{5i} + w_{5i} < 0, i = 1, 2, 3, 4\}$.

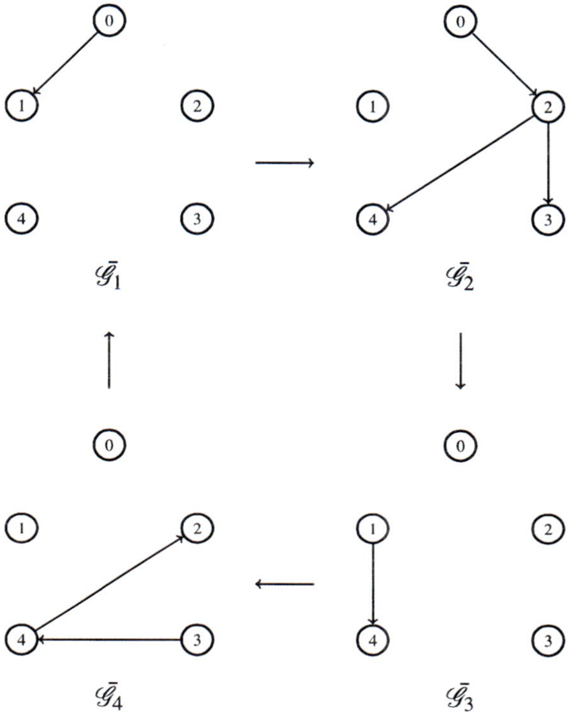

Fig. 11.1 The switching communication graph $\bar{\mathscr{G}}_{\sigma(t)}$

Performing the transformations (10.12) and (11.3) gives the augmented error system as follows:

$$\dot{\bar{z}}_{1i} = a_{1i}\bar{z}_{1i} + a_{2i}(\hat{e}_i + \tilde{v}_{1i}) \tag{11.36a}$$

$$\dot{\bar{z}}_{2i} = a_{5i}\bar{z}_{2i} + \bar{z}_{1i}(\hat{e}_i + \tilde{v}_{1i}) + \bar{z}_{1i}v_1 + \mathbf{z}_{1i}(\hat{e}_i + \tilde{v}_{1i}) \tag{11.36b}$$

$$\dot{\tilde{\eta}}_i = M_i\tilde{\eta}_i + M_iQ_ib_i^{-1}\hat{e}_i - Q_ib_i^{-1}\bar{g}_i(\bar{z}_{1i}, \bar{z}_{2i}, \hat{e}_i, \tilde{v}_{1i}, \dot{\tilde{v}}_{1i}, \mu) \tag{11.36c}$$

$$\dot{\hat{e}}_i = \bar{g}_i(\bar{z}_{1i}, \bar{z}_{2i}, \hat{e}_i, \tilde{v}_{1i}, \dot{\tilde{v}}_{1i}, \mu) + \Gamma_i(b_i\tilde{\eta}_i + Q_i\hat{e}_i) + b_i\bar{u}_i \tag{11.36d}$$

where

$$\bar{g}_i(\bar{z}_{1i}, \bar{z}_{2i}, \hat{e}_i, \tilde{v}_{1i}, \dot{\tilde{v}}_{1i}, \mu) = a_{3i}\bar{z}_{1i} + a_{4i}\hat{e}_i - \bar{z}_{1i}\bar{z}_{2i} - \bar{z}_{1i}\mathbf{z}_{2i} - \mathbf{z}_{1i}\bar{z}_{2i} + a_{4i}\tilde{v}_{1i} - \dot{\tilde{v}}_{1i}. \tag{11.37}$$

We now verify that system (11.36) satisfies Assumption 11.2. Since (11.36a) and (11.36b) are obtained from (10.69a) and (10.69b), respectively, by replacing e_i with $(\hat{e}_i + \tilde{v}_{1i})$, we can use the same $V_{0i}(\bar{z}_i)$ as we used in Sect. 10.5, that is,

$$V_{0i}(\bar{z}_i) = \frac{\hbar_1}{2}\bar{z}_{1i}^2 + \frac{\hbar_1}{4}\bar{z}_{1i}^4 + \frac{\hbar_1}{8}\bar{z}_{1i}^8 + \frac{\hbar_2}{2}\bar{z}_{2i}^2 + \frac{\hbar_2}{4}\bar{z}_{2i}^4$$

for some $\hbar_1, \hbar_2 > 0$, whose derivative along the trajectory of (11.36a) and (11.36b) satisfies

$$\begin{aligned}
\dot{V}_{0i}(\bar{z}_i) &= \hbar_1 \bar{z}_{1i}\dot{\bar{z}}_{1i} + \hbar_1 \bar{z}_{1i}^3 \dot{\bar{z}}_{1i} + \hbar_1 \bar{z}_{1i}^7 \dot{\bar{z}}_{1i} + \hbar_2 \bar{z}_{2i}\dot{\bar{z}}_{2i} + \hbar_2 \bar{z}_{2i}^3 \dot{\bar{z}}_{2i} \\
&= \hbar_1 a_{1i}\bar{z}_{1i}^2 + \hbar_1 a_{2i}\bar{z}_{1i}(\hat{e}_i + \tilde{v}_{1i}) \\
&\quad + \hbar_1 a_{1i}\bar{z}_{1i}^4 + \hbar_1 a_{2i}\bar{z}_{1i}^3(\hat{e}_i + \tilde{v}_{1i}) \\
&\quad + \hbar_1 a_{1i}\bar{z}_{1i}^8 + \hbar_1 a_{2i}\bar{z}_{1i}^7(\hat{e}_i + \tilde{v}_{1i}) \\
&\quad + \hbar_2 a_{5i}\bar{z}_{2i}^2 + \hbar_2 \bar{z}_{2i}\bar{z}_{1i}(\hat{e}_i + \tilde{v}_{1i}) + \hbar_2 \bar{z}_{2i}\bar{z}_{1i}v_1 + \hbar_2 \bar{z}_{1i}\mathbf{z}_{1i}(\hat{e}_i + \tilde{v}_{1i}) \\
&\quad + \hbar_2 a_{5i}\bar{z}_{2i}^4 + \hbar_2 \bar{z}_{2i}^3\bar{z}_{1i}(\hat{e}_i + \tilde{v}_{1i}) + \hbar_2 \bar{z}_{2i}^3\bar{z}_{1i}v_1 + \hbar_2 \bar{z}_{2i}^3\mathbf{z}_{1i}(\hat{e}_i + \tilde{v}_{1i}).
\end{aligned} \tag{11.38}$$

Applying the mean square inequality and the Young's inequality to each term of (11.38) gives

$$\hbar_1 a_{2i}\bar{z}_{1i}(\hat{e}_i + \tilde{v}_{1i}) \le \frac{\hbar_1^2 a_{2i}^2}{0.02}(\hat{e}_i + \tilde{v}_{1i})^2 + \frac{0.01}{2}\bar{z}_{1i}^2 \tag{11.39a}$$

$$\begin{aligned}
\hbar_1 a_{2i}\bar{z}_{1i}^3(\hat{e}_i + \tilde{v}_{1i}) &\le \frac{3}{4}\bar{z}_{1i}^4 + \frac{\hbar_1^4 a_{2i}^4}{4}(\hat{e}_i + \tilde{v}_{1i})^4 \\
&\le \frac{3}{4}\bar{z}_{1i}^4 + \frac{\hbar_1^4 a_{2i}^4}{4}\left(\frac{1}{3}(\hat{e}_i + \tilde{v}_{1i})^8 + \frac{2}{3}(\hat{e}_i + \tilde{v}_{1i})^2\right)
\end{aligned} \tag{11.39b}$$

$$\hbar_1 a_{2i}\bar{z}_{1i}^7(\hat{e}_i + \tilde{v}_{1i}) \le \frac{7}{8}\bar{z}_{1i}^8 + \frac{\hbar_1^8 a_{2i}^8}{8}(\hat{e}_i + \tilde{v}_{1i})^8 \tag{11.39c}$$

$$\begin{aligned}
\hbar_2 \bar{z}_{2i}\bar{z}_{1i}(\hat{e}_i + \tilde{v}_{1i}) &\le \frac{1}{2}\bar{z}_{2i}^2 + \frac{1}{4}\bar{z}_{1i}^4 + \frac{\hbar_2^4}{4}(\hat{e}_i + \tilde{v}_{1i})^4 \\
&\le \frac{1}{2}\bar{z}_{2i}^2 + \frac{1}{4}\bar{z}_{1i}^4 + \frac{\hbar_2^4}{4}\left(\frac{1}{3}(\hat{e}_i + \tilde{v}_{1i})^8 + \frac{2}{3}(\hat{e}_i + \tilde{v}_{1i})^2\right)
\end{aligned} \tag{11.39d}$$

$$\hbar_2 \bar{z}_{2i}\bar{z}_{1i}v_1 \le \frac{v_1^2 \hbar_2^2}{2}\bar{z}_{1i}^2 + \frac{1}{2}\bar{z}_{2i}^2 \tag{11.39e}$$

$$\hbar_2 \bar{z}_{1i}\mathbf{z}_{1i}(\hat{e}_i + \tilde{v}_{1i}) \le \frac{1}{2}\bar{z}_{1i}^2 + \frac{\hbar_2^2 \mathbf{z}_{1i}^2}{2}(\hat{e}_i + \tilde{v}_{1i})^2 \tag{11.39f}$$

$$\hbar_2 \bar{z}_{2i}^3\bar{z}_{1i}(\hat{e}_i + \tilde{v}_{1i}) \le \frac{3}{4}\bar{z}_{2i}^4 + \frac{\hbar_2^8}{8}(\hat{e}_i + \tilde{v}_{1i})^8 + \frac{1}{8}\bar{z}_{1i}^8 \tag{11.39g}$$

$$\hbar_2 \bar{z}_{2i}^3\bar{z}_{1i}v_1 \le \frac{3}{4}\bar{z}_{2i}^4 + \frac{\hbar_2^4 v_1^4}{4}\bar{z}_{1i}^4 \tag{11.39h}$$

$$\begin{aligned}
\hbar_2 \bar{z}_{2i}^3\mathbf{z}_{1i}(\hat{e}_i + \tilde{v}_{1i}) &\le \frac{3}{4}\bar{z}_{2i}^4 + \frac{\hbar_2^4 \mathbf{z}_{1i}^4}{4}(\hat{e}_i + \tilde{v}_{1i})^4 \\
&\le \frac{3}{4}\bar{z}_{2i}^4 + \frac{\hbar_2^4 \mathbf{z}_{1i}^4}{4}\left(\frac{1}{3}(\hat{e}_i + \tilde{v}_{1i})^8 + \frac{2}{3}(\hat{e}_i + \tilde{v}_{1i})^2\right)
\end{aligned} \tag{11.39i}$$

Let $\bar{\mathbb{V}}$ and $\mathbb{W}_0$ be any fixed compact subsets of $\mathbb{R}^2$ and $\mathbb{W}$, respectively. Denote $\Sigma = \bar{\mathbb{V}} \times \mathbb{W}_0$, which is compact. Then inequalities (11.39a) – (11.39i) together with (11.38) yields, for all $(v, w) \in \Sigma$,

$$\dot{V}_{0i}(\bar{z}) \leq -l_{1i}\bar{z}_{1i}^2 - l_{2i}\bar{z}_{1i}^4 - l_{3i}\bar{z}_{1i}^8 - l_{4i}\bar{z}_{2i}^2 - l_{5i}\bar{z}_{2i}^4 + l_{6i}(\hat{e}_i + \tilde{v}_{1i})^2 + l_{7i}(\hat{e}_i + \tilde{v}_{1i})^8 \tag{11.40}$$

where

$$\begin{aligned}
l_{1i} &\triangleq \hbar_1 \min_{w\in\mathbb{W}_0}\{-a_{1i}\} - \frac{1.01}{2} - \frac{\hbar_2^2 \max_{v\in\bar{\mathbb{V}}}\{v_1^2\}}{2}\\
l_{2i} &\triangleq \hbar_1 \min_{w\in\mathbb{W}_0}\{-a_{1i}\} - 1 - \frac{\hbar_2^4 \max_{v\in\bar{\mathbb{V}}}\{v_1^4\}}{4}\\
l_{3i} &\triangleq \hbar_1 \min_{w\in\mathbb{W}_0}\{-a_{1i}\} - 1\\
l_{4i} &\triangleq \hbar_2 \min_{w\in\mathbb{W}_0}\{-a_{5i}\} - 1\\
l_{5i} &\triangleq \hbar_2 \min_{w\in\mathbb{W}_0}\{-a_{5i}\} - \frac{9}{4}\\
l_{6i} &\triangleq \frac{\hbar_1^2 \max_{w\in\mathbb{W}_0}\{a_{2i}^2\}}{0.02} + \frac{\hbar_1^4 \max_{w\in\mathbb{W}_0}\{a_{2i}^4\}}{6} + \frac{\hbar_2^4}{6}\\
&\quad + \frac{\hbar_2^2 \max_{\mu\in\Sigma}\{\mathbf{z}_{1i}^2\}}{2} + \frac{\hbar_2^4 \max_{\mu\in\Sigma}\{\mathbf{z}_{1i}^4\}}{6}\\
l_{7i} &\triangleq \frac{\hbar_1^4 \max_{w\in\mathbb{W}_0}\{a_{2i}^4\}}{12} + \frac{\hbar_1^8 \max_{w\in\mathbb{W}_0}\{a_{2i}^8\}}{8} + \frac{\hbar_2^4}{12} + \frac{\hbar_2^8}{8} + \frac{\hbar_2^4 \max_{\mu\in\Sigma}\{\mathbf{z}_{1i}^4\}}{12}.
\end{aligned}$$

Since $a_{1i} < 0$, $a_{5i} < 0$, we can choose some proper $\hbar_1 > 0$ and $\hbar_2 > 0$ such that $l_{si} > 0$, $s = 1, \ldots, 7$, $i = 1, \ldots, N$. Here l_{si} are known for case 1, but unknown for case 2. Moreover, let $\hat{\gamma}_{1i}(\hat{e}_i) = \hat{e}_i^6 + 1$, $\hat{\gamma}_{2i}(\mathbf{v}_i) = \|\mathbf{v}_i\|^6 + 1$. Then it holds that

$$l_{6i}(\hat{e}_i + \tilde{v}_{1i})^2 + l_{7i}(\hat{e}_i + \tilde{v}_{1i})^8 \leq \bar{\delta}_i\hat{\gamma}_{1i}(\hat{e}_i)\hat{e}_i^2 + \bar{\delta}_i\hat{\gamma}_{2i}(\mathbf{v}_i)\|\mathbf{v}_i\|^2 \tag{11.41}$$

where $\bar{\delta}_i$ are some known positive constants for case 1, and some unknown positive constants for case 2. Then, from (11.40), for all $(v, w) \in \Sigma$,

$$\dot{V}_{0i}(\bar{z}) \leq -l_{1i}\bar{z}_{1i}^2 - l_{2i}\bar{z}_{1i}^4 - l_{3i}\bar{z}_{1i}^8 - l_{4i}\bar{z}_{2i}^2 - l_{5i}\bar{z}_{2i}^4 + \bar{\delta}_i\hat{\gamma}_{1i}(\hat{e}_i)\hat{e}_i^2 + \bar{\delta}_i\hat{\gamma}_{2i}(\mathbf{v}_i)\|\mathbf{v}_i\|^2. \tag{11.42}$$

Thus, Assumption 11.2 holds.

In the rest of this section, we will deal with the two cases by the approaches detailed in Sects. 11.3 and 11.4, respectively.

We will synthesize control laws according to Remark 11.3. For this purpose, notice that system (11.36) is of the from (11.5) with $\hat{Z}_i = \mathrm{col}(\bar{z}_{1i}, \bar{z}_{2i}, \hat{\eta}_i)$ and $\hat{Z} = \mathrm{col}(\hat{Z}_1, \ldots, \hat{Z}_N)$. From (11.37),

$$\|\bar{g}_i(\bar{z}_{1i}, \bar{z}_{2i}, \hat{e}_i, \tilde{v}_{1i}, \dot{\tilde{v}}_{1i}, \mu)\|^2 \leq c_{0i}(c_{1i}(\bar{z}_{1i}^2 + 1)\bar{z}_{1i}^2 + c_{2i}(\bar{z}_{2i}^2 + 1)\bar{z}_{2i}^2 + c_{3i}\hat{e}_i^2 + c_{4i}\|\mathbf{v}_i\|^2) \tag{11.43}$$

where $c_{si} > 0$, $s = 1, 2, 3, 4$, are known constants, and $c_{0i} = 1$ for case 1 and is an unknown positive constant for case 2. Inequality (11.43) together with (11.5b) and (11.36d) gives

$$\begin{aligned} \|\bar{g}_i(\hat{Z}_i, \hat{e}_i, \mathbf{v}_i, \mu)\|^2 &= \|\bar{g}_i(\bar{z}_{1i}, \bar{z}_{2i}, \hat{e}_i, \tilde{v}_{1i}, \dot{\tilde{v}}_{1i}, \mu) + \Gamma_i(b_i\bar{\eta}_i + Q_i\hat{e}_i)\|^2 \\ &\leq c_i\left(\pi_{1i}(\bar{z}_{1i})\bar{z}_{1i}^2 + \pi_{2i}(\bar{z}_{2i})\bar{z}_{2i}^2 + \pi_{3i}\|\bar{\eta}_i\|^2 + \phi_{1i}\hat{e}_i^2 + \phi_{2i}\|\mathbf{v}_i\|^2\right) \end{aligned} \tag{11.44}$$

where $\pi_{1i}(\bar{z}_{1i}) = \hat{c}_{1i}(\bar{z}_{1i}^2 + 1)$, $\pi_{2i}(\bar{z}_{2i}) = \hat{c}_{2i}(\bar{z}_{2i}^2 + 1)$, $\pi_{3i} = \hat{c}_{3i}$ for some known $\hat{c}_{si} > 0$, $s = 1, 2, 3$, $i = 1, \ldots, N$, $\phi_{1i}, \phi_{2i} > 0$ are known, and $c_i = 1$ for case 1 and is an unknown positive constant for case 2. Thus, (11.28) holds.

Let $V_{2i}(\hat{e}_i) = \frac{1}{2}\hat{e}_i^2$. Then, similar to the derivation of (11.16), we have

$$\dot{V}_{2i}(\hat{e}_i) = \frac{c_i}{4}\hat{e}_i^2 + \pi_{1i}(\bar{z}_{1i})\bar{z}_{1i}^2 + \pi_{2i}(\bar{z}_{2i})\bar{z}_{2i}^2 + \pi_{3i}\|\bar{\eta}_i\|^2 + \phi_{1i}\hat{e}_i^2 + \phi_{2i}\|\mathbf{v}_i\|^2 - b_ik_i\rho(\hat{e}_i)\hat{e}_i^2. \tag{11.45}$$

Thus, if we can find a Lyapunov function $V_{1i}(\hat{Z}_i)$ for (11.36a) – (11.36c) such that

$$\dot{V}_{1i}(\hat{Z}_i) \leq -\left(\Theta_{1i}(\bar{z}_{1i})\bar{z}_{1i}^2 + \Theta_{2i}(\bar{z}_{2i})\bar{z}_{2i}^2 + \Theta_{3i}\|\bar{\eta}_i\|^2\right) + \hat{\delta}_i\hat{\gamma}_{1i}(\hat{e}_i)\hat{e}_i^2 + \hat{\delta}_i\hat{\gamma}_{2i}(\mathbf{v}_i)\|\mathbf{v}_i\|^2 \tag{11.46}$$

with $\Theta_{li} \geq \pi_{li} + 1$, $l = 1, 2, 3$, then, $V_i(\hat{Z}_i, \hat{e}_i) = V_{1i}(\hat{Z}_i) + V_{2i}(\hat{e}_i)$ satisfies

$$\begin{aligned} \dot{V}_i(\hat{Z}_i, \hat{e}_i) \leq &-\left(\Theta_{1i}(\bar{z}_{1i}) - \pi_{1i}(\bar{z}_{1i})\right)\bar{z}_{1i}^2 - \left(\Theta_{2i}(\bar{z}_{2i}) - \pi_{2i}(\bar{z}_{2i})\right)\bar{z}_{2i}^2 \\ &- \left(\Theta_{3i} - \pi_{3i}\right)\|\bar{\eta}_i\|^2 - \left(b_i(w)k_i\rho_i(\hat{e}_i) - \psi_i(\hat{e}_i)\right)\hat{e}_i^2 \end{aligned}$$

where $\psi_i(\hat{e}_i) = \hat{\delta}_i\hat{\gamma}_{1i}(\hat{e}_i) + \phi_{1i}(\hat{e}_i) + \frac{c_i}{4}$. Thus, it suffices to let k_i and $\rho_i(\cdot)$ satisfy (11.23).

For this purpose, let $P_i \in \mathbb{R}^{4\times 4}$ be the symmetric positive definite matrix such that $P_iM_i + M_i^{\mathsf{T}}P_i = -I_4$, and $V_{1i}(\hat{\eta}_i) = \hat{\eta}_i^{\mathsf{T}}P_i\hat{\eta}_i$. Then, along the trajectory of $\hat{\eta}_i$ subsystem (11.36c)

$$\begin{aligned} \dot{V}_{1i}(\hat{\eta}_i) &= -\left\|\hat{\eta}_i\right\|^2 + 2\hat{\eta}_i^{\mathsf{T}}P_i(M_iQ_ib_i^{-1}\hat{e}_i - Q_ib_i^{-1}\bar{g}_i(\bar{z}_{1i}, \bar{z}_{2i}, \hat{e}_i, \tilde{v}_{1i}, \dot{\tilde{v}}_{1i}, \mu)) \\ &\leq -\frac{1}{2}\left\|\hat{\eta}_i\right\|^2 + 4\|P_iM_iQ_ib_i^{-1}\|^2\hat{e}_i^2 \\ &\quad +4\|P_iQ_ib_i^{-1}\|^2c_{0i}(c_{1i}(\bar{z}_{1i}^2 + 1)\bar{z}_{1i}^2 + c_{2i}(\bar{z}_{2i}^2 + 1)\bar{z}_{2i}^2 + c_{3i}\hat{e}_i^2 + c_{4i}\|\mathbf{v}_i\|^2) \\ &= -\frac{1}{2}\left\|\hat{\eta}_i\right\|^2 + (\bar{l}_{1i}(\bar{z}_{1i}^2 + 1)\bar{z}_{1i}^2 + \bar{l}_{2i}(\bar{z}_{2i}^2 + 1)\bar{z}_{2i}^2 + \bar{l}_{3i}\hat{e}_i^2 + \bar{l}_{4i}\|\mathbf{v}_i\|^2) \end{aligned} \tag{11.47}$$

where

$$\bar{l}_{1i} \triangleq 4c_{0i}c_{1i} \max_{w\in\mathbb{W}_0}\{\|P_iQ_ib_i^{-1}\|^2\} > 0$$
$$\bar{l}_{2i} \triangleq 4c_{0i}c_{2i} \max_{w\in\mathbb{W}_0}\{\|P_iQ_ib_i^{-1}\|^2\} > 0$$
$$\bar{l}_{3i} \triangleq 4 \max_{w\in\mathbb{W}_0}\{\|P_iM_iQ_ib_i^{-1}\|^2 + c_{0i}c_{3i}\|P_iQ_ib_i^{-1}\|^2\} > 0$$
$$\bar{l}_{4i} \triangleq 4c_{0i}c_{4i} \max_{w\in\mathbb{W}_0}\{\|P_iQ_ib_i^{-1}\|^2\} > 0.$$

Let $U_{1i}(\hat{Z}_i) = \varpi_i V_{0i}(\bar{z}_i) + V_{1i}(\hat{\eta}_i)$ where ϖ_i are some positive constants to be determined. Then, from (11.40) and (11.47),

$$\begin{aligned}
\dot{U}_{1i}(\hat{Z}_i) \le & \varpi_i(-l_{1i}\bar{z}_{1i}^2 - l_{2i}\bar{z}_{1i}^4 - l_{3i}\bar{z}_{1i}^8 - l_{4i}\bar{z}_{2i}^2 - l_{5i}\bar{z}_{2i}^4 + \bar{\delta}_i\hat{\gamma}_{1i}(\hat{e}_i)\hat{e}_i^2 + \bar{\delta}_i\hat{\gamma}_{2i}(\mathbf{v}_i)\|\mathbf{v}_i\|^2) \\
& - \frac{1}{2}\|\hat{\eta}_i\|^2 + \bar{l}_{1i}(\bar{z}_{1i}^2+1)\bar{z}_{1i}^2 + \bar{l}_{2i}(\bar{z}_{2i}^2+1)\bar{z}_{2i}^2 + \bar{l}_{3i}\hat{e}_i^2 + \bar{l}_{4i}\|\mathbf{v}_i\|^2 \\
\le & -\left((\varpi_i l_{1i} - \bar{l}_{1i}) + (\varpi_i l_{2i} - \bar{l}_{1i})\bar{z}_{1i}^2 + \varpi_i l_{3i}\bar{z}_{1i}^6\right)\bar{z}_{1i}^2 \\
& - \left((\varpi_i l_{4i} - \bar{l}_{2i}) + (\varpi_i l_{5i} - \bar{l}_{2i})\bar{z}_{2i}^2\right)\bar{z}_{2i}^2 - \frac{1}{2}\|\hat{\eta}_i\|^2 \\
& + \left((\varpi_i\bar{\delta}_i + \bar{l}_{3i}) + \varpi_i\bar{\delta}_i\hat{e}_i^6\right)\hat{e}_i^2 + \left((\varpi_i\bar{\delta}_i + \bar{l}_{4i}) + \varpi_i\bar{\delta}_i\|\mathbf{v}_i\|^6\right)\|\mathbf{v}_i\|^2.
\end{aligned} \tag{11.48}$$

Letting

$$\varpi_i \ge \left\{\frac{\bar{l}_{1i}}{l_{1i}}, \frac{\bar{l}_{1i}}{l_{2i}}, \frac{1}{l_{2i}}, \frac{\bar{l}_{2i}}{l_{4i}}, \frac{\bar{l}_{2i}}{l_{5i}}\right\} + 1$$

gives

$$\begin{aligned}
\dot{U}_{1i}(\hat{Z}_i) \le & -\left(1 + \bar{z}_{1i}^2 + \bar{z}_{1i}^6\right)\bar{z}_{1i}^2 - \left(1 + \bar{z}_{2i}^2\right)\bar{z}_{2i}^2 - \frac{1}{2}\|\bar{\eta}_i\|^2 \\
& + \left((\varpi_i\bar{\delta}_i + \bar{l}_{3i}) + \varpi_i\bar{\delta}_i\hat{e}_i^6\right)\hat{e}_i^2 + \left((\varpi_i\bar{\delta}_i + \bar{l}_{4i}) + \varpi_i\bar{\delta}_i\|\mathbf{v}_i\|^6\right)\|\mathbf{v}_i\|^2.
\end{aligned} \tag{11.49}$$

Let

$$V_{1i}(\hat{Z}_i) = \hat{l}_i U_{1i}(\hat{Z}_i)$$

where $\hat{l}_i$, $i = 1, \ldots, N$, are some positive constants to be determined. Then, from (11.49),

$$\begin{aligned}
\dot{V}_{1i}(\hat{Z}_i) \le & -\left(\Theta_{1i}(\bar{z}_{1i})\bar{z}_{2i}^2 + \Theta_{2i}(\bar{z}_{2i})\bar{z}_{2i}^2 + \Theta_{3i}\|\bar{\eta}_i\|^2\right) \\
& + \hat{l}_i\left((\varpi_i\bar{\delta}_i + \bar{l}_{3i}) + \varpi_i\bar{\delta}_i\hat{e}_i^6\right)\hat{e}_i^2 + \left((\varpi_i\bar{\delta}_i + \bar{l}_{4i}) + \varpi_i\bar{\delta}_i\|\mathbf{v}_i\|^6\right)\|\mathbf{v}_i\|^2
\end{aligned} \tag{11.50}$$

where $\Theta_{1i}(\bar{z}_{1i}) = \hat{l}_i(1 + \bar{z}_{1i}^2 + \bar{z}_{1i}^6)$, $\Theta_{2i}(\bar{z}_{2i}) = \hat{l}_i(1 + \bar{z}_{2i}^2)$, and $\Theta_{3i} = \frac{1}{2}\hat{l}_i$. Then, for sufficiently large $\hat{l}_i$, $\Theta_{li} \ge \pi_{li} + 1$, $l = 1, 2, 3$.

Since $\hat{l}_i\left((\varpi_i\bar{\delta}_i + \bar{l}_{3i}) + \varpi_i\bar{\delta}_i\hat{e}_i^6\right)$ is a degree 6 polynomial, for any $\hat{l}_i$, there exists $\hat{\delta}_i > 0$ and $\hat{\gamma}_{1i}(\hat{e}_i) = 1 + \hat{e}_i^6$ such that

$$\hat{l}_i\left((\varpi_i\bar{\delta}_i + \bar{l}_{3i}) + \varpi_i\bar{\delta}_i\hat{e}_i^6\right) \leq \hat{\delta}_i\hat{\gamma}_{1i}(\hat{e}_i). \tag{11.51}$$

Thus, with $\rho_i(\hat{e}_i) = 1 + \hat{e}_i^6$, (11.23) is satisfied for a sufficiently large number k_i.

Example 11.1 This example considers the case where the leader system is known with $S = \begin{bmatrix} 0 & 0.8 \\ -0.8 & 0 \end{bmatrix}$ and the compact set Σ_0 is also known, which was studied by the robust distributed design in Sect. 11.2. For this case, we set $\Sigma_0 = \bar{\mathbb{V}}_0 \times \mathbb{W}_0$ with $\bar{\mathbb{V}}_0 = [-8, 8] \times [-8, 8]$ and $\mathbb{W}_0 = [-1, 1] \times [-1, 1] \times [-1, 1] \times \{0\} \times [-1/3, 1/3] \times \{0\}$. Following Theorem 11.1, we can design the distributed controller (11.25) where $k_i(\hat{e}_i) = -k_i\rho_i(\hat{e}_i)\hat{e}_i$ with $\rho_i(\hat{e}_i) = (\hat{e}_i^6 + 1)$ and $k_i = 50$. The gains for the distributed observers are chosen to be $\mu_1 = 1$, $\mu_2 = 10$. The simulation is conducted with the actual values of the uncertain parameters being $w_1 = \text{col}(1, -1, -1, 0, -\frac{1}{3}, 0)$, $w_2 = \text{col}(0, 0, -1, 0, \frac{1}{3}, 0)$, $w_3 = \text{col}(-1, 1, 0, 0, 0, 0]^{\mathsf{T}}$, $w_4 = [-1, 1, -1, 0, 0, 0)$, and the initial conditions of the closed-loop system being $v(0) = \text{col}(0, 8)$, $z_1(0) = \text{col}(3, -1)$, $y_1(0) = -2$, $z_2(0) = \text{col}(1, 1)$, $y_2(0) = 0$, $z_3(0) = \text{col}(3, 2)$, $y_3(0) = 0$, $z_4(0) = \text{col}(2, -1)$, $y_4(0) = -1$, and zero for the others. Figure 11.2 shows the outputs of all subsystems. Figures 11.3 and 11.4 show the response of $e_i(t)$ and $\hat{e}_i(t)$, respectively. Satisfactory tracking performance of our control design can be observed.

Example 11.2 This example considers the case where the leader system is known with $S = \begin{bmatrix} 0 & 0.8 \\ -0.8 & 0 \end{bmatrix}$ and the compact subset Σ_0 can be any compact subset of $\mathbb{R}^2 \times \mathbb{W}$. In this case, our control law works for any $v(0) \in \mathbb{R}^2$ and any $w \in \mathbb{W}$. By Theorem 11.2, it is possible to design a distributed controller of the form (11.35) with $\rho_i(\hat{e}_i) = 5(\hat{e}_i^6 + 1)$ and the other design parameters are the same as those in Example 11.1. The simulation is conducted with the actual values of the uncertain

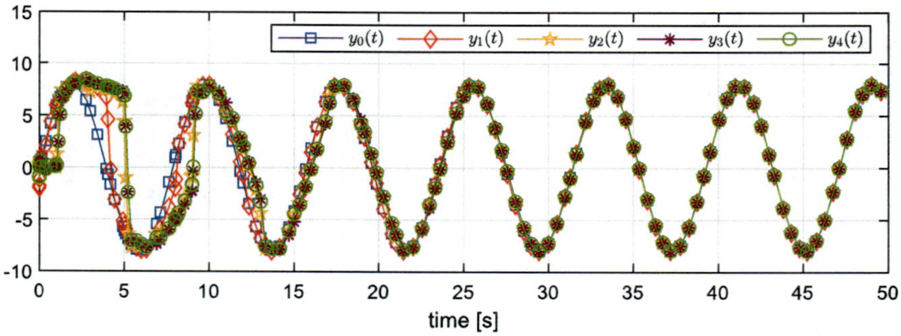

Fig. 11.2 The response of $y_0(t)$ and the response of $y_i(t)$ under the control law (11.25)

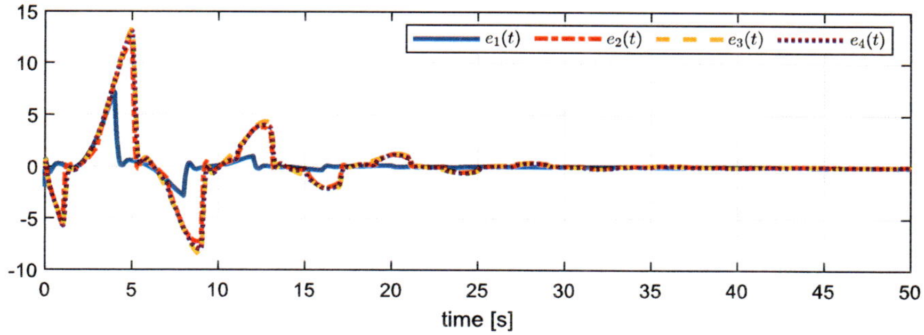

Fig. 11.3 The response of $e_i(t)$ under the control law (11.25)

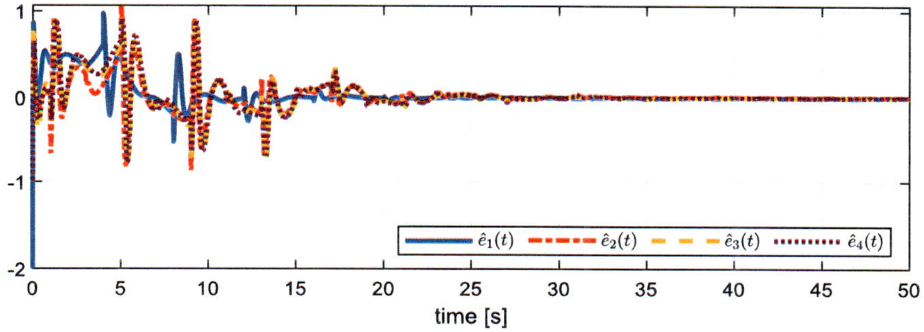

Fig. 11.4 The response of $\hat{e}_i(t)$ under the control law (11.25)

parameters, the gains for the distributed observers, and the initial conditions of the closed-loop system being the same as those in Example 11.1, and $k_1(0) = k_2(0) = k_3(0) = k_4(0) = 1$. Figure 11.5 shows the response of $y_i(t)$. Figures 11.6 and 11.7 show the response of $e_i(t)$ and $\hat{e}_i(t)$, respectively. Figure 11.8 shows the response of the adaptive gain $k_i(t)$. Satisfactory tracking performance of the control law is observed.

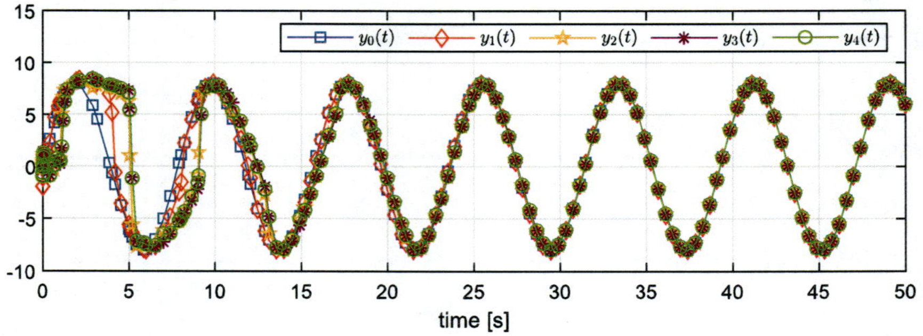

Fig. 11.5 The response of $y_0(t)$ and the response of $y_i(t)$ under the control law (11.35)

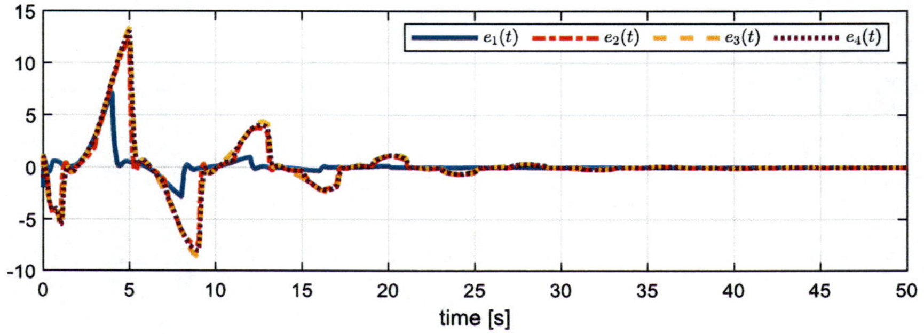

Fig. 11.6 The response of $e_i(t)$ under the control law (11.35)

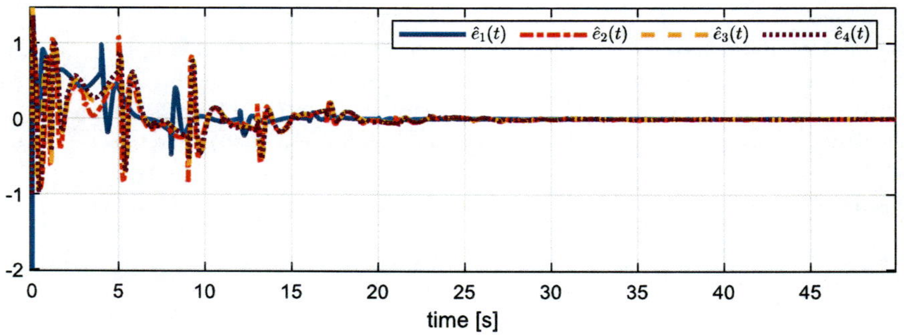

Fig. 11.7 The response of $\hat{e}_i(t)$ under the control law (11.35)

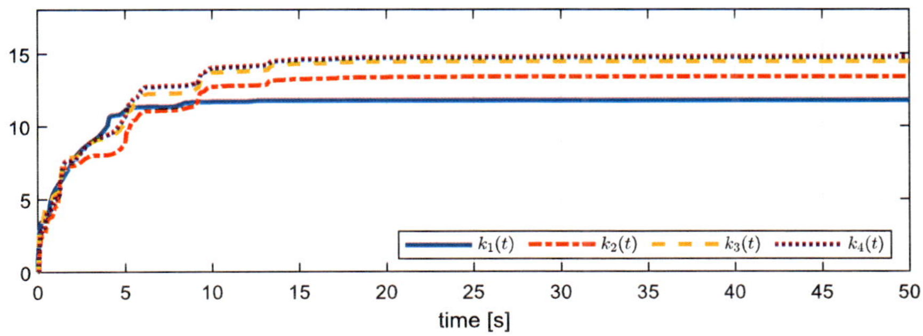

Fig. 11.8 The response of $k_i(t)$ under the control law (11.35)

11.6 Notes and References

The integrated approach has been applied to several classes of multi-agent systems such as linear minimum phase multi-agent systems in [1], nonlinear systems in normal form with unity relative degree in [2], a class of second-order nonlinear multi-agent systems in [3], and lower triangular nonlinear multi-agent systems in [4] which includes the systems studied in this chapter as a special case. Reference [5] further studied the cooperative robust output regulation for nonlinear systems in normal form with unity relative degree but with a nonlinear leader system.

References

1. Liu W, Huang J (2015) Cooperative robust output regulation of linear minimum-phase multi-agent systems under switching network. In: Proceedings of the 10th Asian control conference
2. Liu T, Huang J (2018) Cooperative output regulation for a class of nonlinear multi-agent systems with unknown control directions subject to switching networks. IEEE Trans Autom Control 63(3):783–790
3. Liu W, Huang J (2018) Cooperative adaptive output regulation for second-order nonlinear multi-agent systems with jointly connected switching networks. IEEE Trans Neural Netw Learn Syst 29(3):695–705
4. Liu W, Huang J (2020) Cooperative adaptive output regulation for lower triangular nonlinear multi-agent systems subject to jointly connected switching networks. IEEE Trans Neural Netw Learn Syst 31(5):1724–1734
5. Liu T, Huang J (2019) Cooperative robust output regulation for a class of nonlinear multi-agent systems subject to a nonlinear leader system. Automatica 108:108501

Appendix A

In this appendix, we summarize some results on Kronecker product, matrix theory, stability analysis of nonlinear systems, a design framework for nonlinear output regulation problem and stabilization techniques for nonlinear systems. These results are quite standard and the proofs of most results are thus omitted.

A.1 Kronecker Product and Sylvester Equation

Let $A = [a_{ij}] \in \mathbb{R}^{m\times q}$ and $B = [b_{ij}] \in \mathbb{R}^{p\times n}$. Then, the notation $A \otimes B$, called the Kronecker product of A and B, is defined as follows:

$$A \otimes B = \begin{bmatrix} a_{11}B & \cdots & a_{1q}B \\ \vdots & \ddots & \vdots \\ a_{m1}B & \cdots & a_{mq}B \end{bmatrix}.$$

It is straightforward to verify the following property.

Proposition A.1

1. *For any matrices A, B, C, D of conformable dimensions,*

$$\begin{aligned} (A \otimes B)(C \otimes D) &= (AC) \otimes (BD) \\ (A + B) \otimes (C + D) &= (A \otimes C) + (A \otimes D) + (B \otimes C) + (B \otimes D). \end{aligned}$$

2. *Let $A \in \mathbb{R}^{m\times q}$, $B \in \mathbb{R}^{p\times n}$, and $X \in \mathbb{R}^{n\times m}$. Then*

$$\text{vec}(BXA) = (A^\top \otimes B)\text{vec}(X).$$

Proposition A.2 *Let $A \in \mathbb{R}^{m\times m}$, $B \in \mathbb{R}^{n\times n}$. Then*

H. Cai et al., *Cooperative Control of Multi-agent Systems*, Advances in Industrial Control,
https://doi.org/10.1007/978-3-030-98377-2

$$\lambda(A \otimes B) = \{\lambda_i(A)\lambda_j(B), i = 1, \ldots, m, j = 1, \ldots, n\}$$
$$\lambda(A \otimes I_n + I_m \otimes B) = \{\lambda_i(A) + \lambda_j(B), i = 1, \ldots, m, j = 1, \ldots, n\}.$$

Proof For $i = 1, \ldots, m,\ j = 1, \ldots, n$, let $Ax_i = \lambda_i(A)x_i,\ By_j = \lambda_j(B)y_j$. By Proposition A.1, we have

$$(A \otimes B)(x_i \otimes y_j) = (Ax_i) \otimes (By_j) = (\lambda_i(A)x_i) \otimes (\lambda_j(B)y_j) = \lambda_i(A)\lambda_j(B)(x_i \otimes y_j)$$

which implies that $\lambda_i(A)\lambda_j(B)$ is an eigenvalue of $(A \otimes B)$. Moreover

$$\begin{aligned}(A \otimes I_n + I_m \otimes B)(x_i \otimes y_j) &= (Ax_i) \otimes y_j + x_i \otimes (By_j)\\ &= (\lambda_i(A)x_i) \otimes y_j + x_i \otimes (\lambda_j(B)y_j)\\ &= (\lambda_i(A) + \lambda_j(B))(x_i \otimes y_j)\end{aligned}$$

which implies that $\lambda_i(A) + \lambda_j(B)$ is an eigenvalue of $(A \otimes I_n + I_m \otimes B)$. □

Given $M, B \in \mathbb{R}^{p\times n}$, and $A, Q \in \mathbb{R}^{m\times q}$, and $C \in \mathbb{R}^{p\times q}$, using Proposition A.1, the following linear matrix equation:

$$MXA - BXQ = C \tag{A.1}$$

where $X \in \mathbb{R}^{n\times m}$ is an unknown matrix, can be converted into the following standard form:

$$(A^\top \otimes M - Q^\top \otimes B)\text{vec}(X) = \text{vec}(C). \tag{A.2}$$

When $q = m$, $p = n$, and $M = I_n$ and $Q = I_m$, (A.1) reduces to

$$XA - BX = C \tag{A.3}$$

and is called the Sylvester equation. Correspondingly, (A.2) reduces to the following form:

$$(A^\top \otimes I_n - I_m \otimes B)\text{vec}(X) = \text{vec}(C).$$

The Sylvester equation has the following properties.

Proposition A.3

1. *The Sylvester equation (A.3) has a unique solution if and only if A and B have no eigenvalues in common.*
2. *Given $A,\ B,\ C \in \mathbb{R}^{n\times n}$, suppose A and B have no common eigenvalues and there exist $\Phi \in \mathbb{R}^{n\times 1}$ and $\Psi \in \mathbb{R}^{1\times n}$ such that $C = \Phi\Psi$ with (B, Φ) controllable, and (Ψ, A) observable. Then the Sylvester equation (A.3) admits a unique solution $X \in \mathbb{R}^{n\times n}$ which is nonsingular.*

A.2 Some Results in Matrix Theory

Given any matrix $A \in \mathbb{C}^{m\times n}$, a matrix $X \in \mathbb{C}^{n\times m}$ is called the Moore–Penrose inverse [1] of A if it satisfies the following four equations:

$$AXA = A \tag{A.4a}$$
$$XAX = X \tag{A.4b}$$
$$(AX)^{\mathsf{H}} = AX \tag{A.4c}$$
$$(XA)^{\mathsf{H}} = XA. \tag{A.4d}$$

The Moore–Penrose inverse of any matrix A always exists and is denoted by $X = A^{\dagger}$.

Denote $\mathscr{R}(A)$ the range of a matrix A and $\mathscr{N}(A)$ the null space of A. Then it holds that $\mathscr{N}(A) = \mathscr{R}(A^{\mathsf{H}})^{\perp}$ and $\mathscr{N}(A^{\mathsf{H}}) = \mathscr{R}(A)^{\perp}$. We have the following result.

Lemma A.1 *For any $x \in \mathbb{C}^n$, $A^{\dagger}Ax$ is the orthogonal projection of x on $\mathscr{N}(A)^{\perp}$, and consequently, $x - A^{\dagger}Ax \in \mathscr{N}(A)$. Moreover, if $x \in \mathscr{N}(A)^{\perp}$, then $x = A^{\dagger}Ax$.*

Proof First of all, from (A.4b), it holds that $(A^{\dagger}A)^2 = (A^{\dagger}AA^{\dagger})A = A^{\dagger}A$. Hence, $A^{\dagger}A$ is idempotent. Then, on one hand, from (A.4d), $A^{\dagger}Ax = (A^{\dagger}A)^{\mathsf{H}}x = A^{\mathsf{H}}(A^{\dagger})^{\mathsf{H}}x \in \mathscr{R}(A^{\mathsf{H}}) = \mathscr{N}(A)^{\perp}$. On the other hand, $A^{\dagger}A(x - A^{\dagger}Ax) = A^{\dagger}Ax - (A^{\dagger}A)^2x = 0$, that is to say, $x - A^{\dagger}Ax \in \mathscr{N}(A)$. Thus, $A^{\dagger}Ax$ is the orthogonal projection of x on $\mathscr{N}(A)^{\perp}$.

If $x \in \mathscr{N}(A)^{\perp}$, then $x - A^{\dagger}Ax \in \mathscr{N}(A)^{\perp}$. Since it also holds that $x - A^{\dagger}Ax \in \mathscr{N}(A)$ and $\mathscr{N}(A) \bigcap \mathscr{N}(A)^{\perp} = \{0\}$, it must hold that $x - A^{\dagger}Ax = 0$. □

A square real matrix whose off-diagonal entries are nonnegative is called a Metzler matrix. A δ-graph associated with an $N \times N$ Metzler matrix A is the graph with the node set $\{1, \ldots, N\}$ and with an edge from l to k ($k \neq l$) if and only if the entry of A on the kth row and the lth column is strictly larger than δ.

Theorem A.1 (Lemma 2.4 of [2]) *An $N \times N$ Metzler matrix with zero row sum has at least one zero eigenvalue with 1_N as an eigenvector, and all the nonzero eigenvalues have negative real parts. Furthermore, a Metzler matrix with zero row sum has exactly one zero eigenvalue if and only if the associated graph has the property that there exists a node which can reach every other node.*

Theorem A.2 (Theorem 1 of [3]) *Consider the linear system*

$$\dot{x} = A(t)x \tag{A.5}$$

where for every time t, $A(t) \in \mathbb{R}^{n\times n}$ is Metzler with zero row sums, and every element of $A(t)$ is a uniformly bounded and piecewise continuous function of time over $[0, \infty)$. If there is an index $k \in \{1, \ldots, n\}$, a threshold value $\delta > 0$ and an interval length $T > 0$ such that for all $t \in \mathbb{R}$ the δ-graph associated with $\int_t^{t+T} A(s)ds$ has the property that all nodes may be reached from the node k, then the equilibrium set of *consensus states is uniformly exponentially stable. In particular, all components of any solution $x(t)$ of (A.5) converge to a common value as $t \to \infty$.*

A matrix A is called an $\mathscr{M}$-matrix if

$$A \in \left\{[a_{ij}] \in \mathbb{R}^{n\times n} : a_{ij} \leq 0 \text{ if } i \neq j\right\}$$

and all the eigenvalues of A have positive real parts. The following lemma provides a technique for evaluating the spectral set of $\mathscr{M}$-matrix.

Lemma A.2 *Let $A \in \mathbb{R}^{n\times n}$ be an $\mathscr{M}$-matrix and $D = \mathscr{D}(d_1, \ldots, d_n) \in \mathbb{R}^{n\times n}$ with $d_1, \ldots, d_n > 0$. Then $\underline{\lambda}_{DA} \geq \underline{\lambda}_A \min_i d_i$ and $\underline{\lambda}_{AD} \geq \underline{\lambda}_A \min_i d_i$.*

A.3 Some Stability Results

Definition A.1 A continuous function $\gamma : [0, a) \mapsto [0, \infty)$ is said to belong to class $\mathscr{K}$ if it is strictly increasing and satisfies $\gamma(0) = 0$, and is said to belong to class $\mathscr{K}_\infty$ if, additionally, $a = \infty$ and $\lim_{r\to\infty} \gamma(r) = \infty$.

Definition A.2 A continuous function $\beta : [0, a) \times [0, \infty) \mapsto [0, \infty)$ is said to belong to class $\mathscr{K}\mathscr{L}$ if, for each fixed s, the function $\beta(\cdot, s) : [0, a) \mapsto [0, \infty)$ is class $\mathscr{K}$ function defined on $[0, a)$, and, for each fixed r, the function $\beta(r, \cdot) : [0, \infty) \mapsto [0, \infty)$ is decreasing and $\lim_{s\to\infty} \beta(r, s) = 0$.

Theorem A.3 (Comparison Lemma) *Consider the scalar differential equation*

$$\dot{x} = f(x, t), \ \ x(t_0) = x_0$$

where $f(x, t)$ is continuous in t and locally Lipschitz in x, for all $t \geq t_0$ and all $x \in \mathbb{X} \subseteq \mathbb{R}$ where $\mathbb{X}$ is some interval. Let $[t_0, T)$ (T could be infinity) be the maximal interval of existence of the solution $x(t)$, and suppose $x(t) \in \mathbb{X}$ for all $t \in [t_0, T)$. Let $V(t)$ be a differentiable function whose derivative satisfies the differential inequality

$$\dot{V} \leq f(V, t), \ \ V(t_0) \leq x_0$$

with $V(t) \in \mathbb{X}$ for all $t \in (t_0, T]$. Then, $V(t) \leq x(t)$ for all $t \in [t_0, T)$.

Theorem A.4 (LaSalle–Yoshizawa Theorem) *Consider the system*

$$\dot{x} = f(x, d(t))$$

where $f(x, d(t))$ is locally Lipschitz in x uniformly in t. If there exists a continuously differentiable function $V : \mathbb{R}^n \times [t_0, \infty) \mapsto \mathbb{R}^+$ such that

$$W_1(x) \le V(x,t) \le W_2(x)$$
$$\dot{V}(x,t) \le -\alpha(x) \le 0,\ \forall x \in \mathbb{R}^n,\ \forall t \ge t_0$$

where $W_1(x)$ *and* $W_2(x)$ *are continuous positive definite and radially unbounded functions and* $\alpha(x)$ *is a continuous positive semidefinite function, then the state is uniformly bounded over* $[t_0, \infty)$ *and satisfies*

$$\lim_{t\to\infty} \alpha(x(t)) = 0.$$

Moreover, if $\alpha(x)$ *is positive definite, then the equilibrium point* $x = 0$ *is uniformly asymptotically stable.*

Remark A.1 Theorem A.4 holds with $W_1(x)$ and $W_2(x)$ being replaced by two class $\mathcal{K}_\infty$ functions $\underline{\alpha}(\|x\|)$ and $\bar{\alpha}(\|x\|)$, respectively.

Definition A.3 A uniformly bounded and piecewise continuous mapping $F : [t_0, \infty) \mapsto \mathbb{R}^{n\times m}$ is said to be persistently exciting (PE) if there exist positive constants ϵ, t_0, T_0 such that,

$$\frac{1}{T_0}\int_t^{t+T_0} F(s)F^{\mathsf{T}}(s)ds \ge \epsilon I_n, \quad \forall t \ge t_0.$$

Definition A.4 A piecewise continuous function $f : [t_0, \infty) \mapsto \mathbb{R}^n$ is said to have spectral lines at frequencies $\omega_1, \ldots, \omega_n$ if, for $k = 1, \ldots, n$,

$$\lim_{\delta\to\infty} \frac{1}{\delta}\int_t^{t+\delta} f(s)e^{-j\omega_k s}ds = \hat{f}(\omega_k) \ne 0$$

uniformly in t.

Lemma A.3 *If a function* $f : [t_0, \infty) \mapsto \mathbb{R}^n$ *has spectral lines at frequencies* $\omega_1, \ldots, \omega_n$, *that is, for* $k = 1, \ldots, n$,

$$\lim_{\delta\to\infty} \frac{1}{\delta}\int_t^{t+\delta} f(s)e^{-j\omega_k s}ds = \hat{f}(\omega_k) \ne 0$$

uniformly in t, *and* $\hat{f}(\omega_k)$, $k = 1, \ldots, n$ *are linearly independent in* $\mathbb{C}^n$, *then* $f(t)$ *is PE.*

Lemma A.4 *Consider a continuously differentiable function* $g : [t_0, \infty) \mapsto \mathbb{R}^n$ *and a uniformly bounded and piecewise continuous function* $f : [t_0, \infty) \mapsto \mathbb{R}^n$, *which satisfy*

$$\lim_{t\to\infty} g^{\mathsf{T}}(t)f(t) = 0.$$

Then,

$$\lim_{t\to\infty} g(t) = 0$$

holds under the following two conditions:

- $\lim_{t\to\infty} \dot{g}(t) = 0$;
- $f(t)$ *is PE.*

Lemma A.5 *Consider the linear time-varying system*

$$\dot{x} = Ax + \Omega^{\mathsf{T}}(t)z \tag{A.6a}$$

$$\dot{z} = -\Lambda\Omega(t)Px \tag{A.6b}$$

where $x \in \mathbb{R}^n$, $z \in \mathbb{R}^p$, $A \in \mathbb{R}^{n\times n}$ is Hurwitz, $P \in \mathbb{R}^{n\times n}$ is a symmetric and positive definite matrix satisfying $A^{\mathsf{T}}P + PA = -Q$ with Q being some symmetric and positive definite matrix, and $\Lambda \in \mathbb{R}^{p\times p}$ is a symmetric and positive definite matrix. If $\|\Omega(t)\|$, $\|\dot{\Omega}(t)\|$ are uniformly bounded and $\Omega(t)$ is PE, then the origin of system (A.6) is uniformly exponentially stable.

A.4 Proof of Lemma 3.4

The proof is based on the generalized Krasovskii–LaSalle theorem [4]. Consider the following linear time-varying system:

$$\dot{x} = \mathbf{A}(t)x + \mathbf{A}_1(t)x \tag{A.7a}$$

$$y = \mathbf{C}(t)x \tag{A.7b}$$

where $x \in \mathbb{R}^p$, $y \in \mathbb{R}^q$, $\mathbf{A}(t)$, $\mathbf{A}_1(t)$, and $\mathbf{C}(t)$ are all piecewise continuous in t and uniformly bounded for all $t \geq t_0 \geq 0$. We call x a solution of (A.7) if $x : [t_0, \infty) \mapsto \mathbb{R}^p$ is an absolutely continuous function satisfying the following integral equation:

$$x(t) = x(t_0) + \int_{t_0}^{t} (\mathbf{A}(\tau) + \mathbf{A}_1(\tau))x(\tau)\mathrm{d}\tau, \ \forall\, t_0 \leq t < \infty.$$

We call the system (A.7) in the output-injection form if, for some positive constant M, $\|\mathbf{A}_1(t)x\| \leq M\|\mathbf{C}(t)x\|$ for all $t \geq 0$ and all $x \in \mathbb{R}^p$.

A function $\bar{x} : \mathbb{R} \mapsto \mathbb{R}^p$ is called a limiting zeroing-output solution of (A.7) if there exist an unbounded sequence $\{t_l\}$ in $[0, \infty)$ with $t_l \geq 2l$, $l = 0, 1, 2, \ldots$, and a sequence $\{x_l\}$ of solutions of (A.7) such that the sequence $\{\hat{x}_l\}$ with $\hat{x}_l(t) \triangleq x_l(t + t_l)$ for all $t \geq -l$, converges uniformly to $\bar{x}$ on every compact subset of $\mathbb{R}$, and for almost all $t \in \mathbb{R}$, $\lim_{l\to\infty} \mathbf{C}(t + t_l)\bar{x}(t) = 0$. The following proposition is useful in illustrating the limiting zeroing-output solution.

Proposition A.4 *Consider the linear time-varying system (A.7) in the output-injection form. Then every uniformly bounded limiting zeroing-output solution $\bar{x} : \mathbb{R} \mapsto \mathbb{R}^p$ of (A.7) satisfies the following conditions:*

$$\bar{x}(t) = \bar{x}(0) + \lim_{l \to \infty} \int_0^t \mathbf{A}(\tau + t_l)\bar{x}(\tau)\mathrm{d}\tau, \text{ for all } t \in \mathbb{R} \tag{A.8}$$

and

$$\lim_{l \to \infty} \mathbf{C}(t + t_l)\bar{x}(t) = 0, \text{ for almost all } t \in \mathbb{R} \tag{A.9}$$

for some unbounded time sequence $\{t_l\}$ *with* $t_l \geq 2l$ *for all* $l = 1, 2, \ldots$.

Definition A.5 System (A.7) is said to be *weakly zero-state detectable* (WZSD) if every uniformly bounded limiting zeroing-output solution $\bar{x} : \mathbb{R} \mapsto \mathbb{R}^p$ of (A.7) satisfies $\inf_{t \in \mathbb{R}} \|\bar{x}(t)\| = 0$.

Specializing the generalized Krasovskii–LaSalle theorem from [4, Theorem 3] to linear time-varying system (A.7) gives the following result.

Theorem A.5 *Consider the linear time-varying system (A.7) in the output-injection form which is WZSD. Suppose that the origin is uniformly stable and the bounded-output-energy condition*

$$\int_s^\infty \|\mathbf{C}(\tau)x(\tau)\|^2 \mathrm{d}\tau \leq \alpha(x(s)), \ \forall s \geq 0 \tag{A.10}$$

holds with some continuous function $\alpha : \mathbb{R}^p \mapsto [0, \infty)$. *Then the origin is exponentially stable.*

The proof of Lemma 3.4 relies on the following lemma:

Lemma A.6 *Suppose that* $F_{\sigma(t)} \geq 0$ *for any* $t \geq 0$. *Then under Condition (ii), there exist* $T > 0$ *and* $\epsilon > 0$ *such that the inequality*

$$u^\mathsf{T} \left[\int_t^{t+T} \left(F_{\sigma(\tau)} \otimes I_n \right) \mathrm{d}\tau \right] u \geq \epsilon, \ \forall t \geq 0 \tag{A.11}$$

holds for any unit vector $u \in \mathbb{R}^{Nn}$.

Proof Let $T = 2\nu > 0$. Then for all $t \geq 0$, there exists a positive k such that $[t_{i_k}, t_{i_{k+1}}) \subseteq [t, t + T)$. For any unit vector $u \in \mathbb{R}^{Nn}$,

$$\begin{aligned} u^\mathsf{T} \left[\int_t^{t+T} \left(F_{\sigma(\tau)} \otimes I_n \right) \mathrm{d}\tau \right] u \geq & u^\mathsf{T} \left[\sum_{q=i_k}^{i_{k+1}-1} \left(F_{\sigma(t_q)} \otimes I_n \right) (t_{j+1} - t_j) \right] u \\ \geq & \tau u^\mathsf{T} \left[\sum_{q=i_k}^{i_{k+1}-1} \left(F_{\sigma(t_q)} \otimes I_n \right) \right] u \\ = & \tau \lambda_k \end{aligned}$$

where λ_k is the minimal eigenvalue of the matrix $\sum_{q=i_k}^{i_{k+1}-1} F_{\sigma(t_q)}$ which is positive since, for all $k = 1, 2, \ldots$, $\sum_{q=i_k}^{i_{k+1}-1} F_{\sigma(t_q)} > 0$. Since σ ranges over a finite set, and $t_{i_{k+1}} - t_{i_k} < \nu$ for some $\nu > 0$ and all $k = 0, 1, 2, \ldots$, there are only finitely many distinct matrices among $\sum_{q=i_k}^{i_{k+1}-1} F_{\sigma(t_q)}$ for all $k = 1, 2, \ldots$. Hence, there exists $\lambda > 0$ such that $\lambda_k \geq \lambda$. Therefore, (A.11) holds with $\varepsilon = \tau\lambda$. This completes the proof.□

Proof of Lemma 3.4

Proof Recall, with $\bar{K} = \mu \bar{B}^{\mathsf{T}}$ where $\mu > 0$, system (3.17) takes the following form:

$$\dot{\xi} = \left(I_N \otimes \bar{A} - \mu F_{\sigma(t)} \otimes (\bar{B}\bar{B}^{\mathsf{T}})\right) \xi \tag{A.12}$$

where $\bar{A} \in \mathbb{R}^{n\times n}$ is anti-symmetric, $\bar{B} \in \mathbb{R}^{n\times m}$ with $(\bar{A}, \bar{B})$ controllable, and $F_{\sigma(t)} \geq 0$.

Define a virtual output signal as follows:

$$y = (\mu F_{\sigma(t)})^{1/2} \otimes \bar{B}^{\mathsf{T}} \xi. \tag{A.13}$$

Then system (A.12) with output (A.13) is of the form (A.7) with $x(t) = \xi(t)$ and

$$\mathbf{A}(t) = I_N \otimes \bar{A} \tag{A.14a}$$
$$\mathbf{A}_1(t) = -\mu F_{\sigma(t)} \otimes (\bar{B}\bar{B}^{\mathsf{T}}) \tag{A.14b}$$
$$\mathbf{C}(t) = (\mu F_{\sigma(t)})^{1/2} \otimes \bar{B}^{\mathsf{T}}. \tag{A.14c}$$

Using the identity $F_{\sigma(t)} \otimes (\bar{B}\bar{B}^{\mathsf{T}}) = (F_{\sigma(t)}^{1/2} \otimes \bar{B}^{\mathsf{T}})^{\mathsf{T}} (F_{\sigma(t)}^{1/2} \otimes \bar{B}^{\mathsf{T}})$ gives

$$\|\mathbf{A}_1(t)x\| \leq \|(\mu F_{\sigma(t)})^{1/2} \otimes \bar{B}^{\mathsf{T}}\| \cdot \|\mathbf{C}(t)x\| \leq M\|\mathbf{C}(t)x\|$$

where $M \triangleq \max_{k=1,\ldots,\rho}\{\|(\mu F_k)^{1/2} \otimes \bar{B}^{\mathsf{T}}\|\}$. So the system (A.12) with output (A.13) is in the output-injection form. The conclusion of Lemma 3.4 can be obtained by Theorem A.5 with the following three steps:

Step-1: show the origin is uniformly stable. Indeed, let

$$V(\xi(t)) = \frac{1}{2}\xi^{\mathsf{T}}(t)\xi(t).$$

Then, the derivative of $V(\xi(t))$ along system (A.12) exists on every interval $[t_i, t_{i+1})$, $i = 0, 1, 2, \ldots$, and is given by, noting $\bar{A}^{\mathsf{T}} + \bar{A} = 0$,

$$\begin{aligned}
&\dot{V}(\xi(t))\,\big|_{(A.12)}\\
&= \frac{1}{2}\xi^{\mathsf{T}}(t)(I_N \otimes \bar{A}^{\mathsf{T}} - \mu F_{\sigma(t)} \otimes (\bar{B}\bar{B}^{\mathsf{T}}) + I_N \otimes \bar{A} - \mu F_{\sigma(t)} \otimes (\bar{B}\bar{B}^{\mathsf{T}}))\xi(t)\\
&= -\mu\xi^{\mathsf{T}}(t)(F_{\sigma(t)} \otimes (\bar{B}\bar{B}^{\mathsf{T}}))\xi(t) \leq 0.
\end{aligned} \tag{A.15}$$

That is to say, the origin is uniformly stable.

Step-2: verify the bounded-output-energy condition (A.10). By making use of (A.14c) and (A.15), we have

$$\begin{aligned}\int_s^t \|\mathbf{C}(\tau)\xi(\tau)\|^2 \mathrm{d}\tau &= \mu \int_s^t \xi^\mathsf{T}(\tau)(F_{\sigma(\tau)} \otimes (\bar{B}\bar{B}^\mathsf{T}))\xi(\tau)\mathrm{d}\tau \\ &= -\int_s^t \dot{V}(\xi(\tau))\mathrm{d}\tau \\ &= V(\xi(s)) - V(\xi(t)) \le V(\xi(s)),\ \forall s \ge 0\end{aligned}$$

for all $t \ge s$. Taking $t \to \infty$ and letting $\alpha = V$ gives the bounded-output-energy condition (A.10).

Step-3: show the weak zero-state detectablility of system (A.12). For this purpose, let $\bar{\xi} : \mathbb{R} \mapsto \mathbb{R}^{nN}$ be any uniformly bounded limiting zeroing-output solution of system (A.12). From Proposition A.4, by noticing that $\mathbf{A}(t)$ here is constant, $\bar{\xi}(t)$ satisfies

$$\dot{\bar{\xi}}(t) = (I_N \otimes \bar{A})\bar{\xi}(t), \text{ for all } t \in \mathbb{R} \tag{A.16}$$

and

$$\lim_{l\to\infty} [F^{1/2}_{\sigma(t+t_l)} \otimes (\bar{B}^\mathsf{T})]\bar{\xi}(t) = 0 \tag{A.17}$$

for almost all $t \in \mathbb{R}$, some unbounded time sequence $\{t_l\}$ with $t_l \ge 2l$ for all $l = 1, 2, \ldots$. Since $\bar{\xi}(t)$ is continuous, from (A.16), it holds that $\bar{\xi}(t) = (I_N \otimes e^{\bar{A}t})u_0$ for some constant $u_0 \in \mathbb{R}^{Nn}$. Let $u_l(t) = (F^{1/2}_{\sigma(t+t_l)} \otimes I_n)u_0$. Then (A.17) implies

$$(I_N \otimes \bar{B}^\mathsf{T})(I_N \otimes e^{\bar{A}t}) \lim_{l\to\infty} u_l(t) = 0 \tag{A.18}$$

for almost all $t \in \mathbb{R}$. Since $\left(\bar{B}^\mathsf{T}, \bar{A}\right)$ is observable and $e^{\bar{A}t}$ is nonsingular for all $t \in \mathbb{R}$, (A.18) further implies $\lim_{l\to\infty} u_l(t) = 0$, and hence

$$\lim_{l\to\infty} [F_{\sigma(t+t_l)} \otimes I_n]u_0 = 0$$

for almost all $t \in \mathbb{R}$. If $u_0 \neq 0$, according to Lemma A.6, under Condition (ii)

$$\begin{aligned}\epsilon &\le \lim_{l\to\infty} \frac{u_0^\mathsf{T}}{\|u_0\|} \left(\int_{t_l}^{t_l+T} \left(F_{\sigma(\tau)} \otimes I_n\right) \mathrm{d}\tau\right) \frac{u_0}{\|u_0\|} \\ &= \frac{1}{\|u_0\|^2} u_0^\mathsf{T} \left(\int_0^T \lim_{l\to\infty} \left(F_{\sigma(\tau+t_l)} \otimes I_n\right) u_0 \mathrm{d}\tau\right) = 0\end{aligned}$$

reaching a contradiction. Therefore, $u_0 = 0$, which results in that $\bar{\xi}(t) \equiv 0$ for all $t \in \mathbb{R}$. So system (A.12) with virtual output (A.13) is WZSD.

Finally, by applying Theorem A.5, the exponential stability of the origin of system (A.12) can be guaranteed. This completes the proof. □

A.5 Identities for Chap. 6

Identity A.1 *For any* $x \in \mathbb{R}^3$, $x^\times x = 0$, $x^\mathsf{T} x^\times = 0$.

Proof Let $x = \mathrm{col}(x_1, x_2, x_3)$. Then

$$x^\times x = \begin{bmatrix} 0 & -x_3 & x_2 \\ x_3 & 0 & -x_1 \\ -x_2 & x_1 & 0 \end{bmatrix} \cdot \begin{bmatrix} x_1 \\ x_2 \\ x_3 \end{bmatrix} = 0$$

and $x^\mathsf{T} x^\times = -(x^\times x)^\mathsf{T} = 0$. □

Identity A.2 *For any* $x, y \in \mathbb{R}^n$, $x^\mathsf{T} y x = x x^\mathsf{T} y$.

Proof Let $x = \mathrm{col}(x_1, \ldots, x_n)$ and $y = \mathrm{col}(y_1, \ldots, y_n)$. The mth entry of $x^\mathsf{T} y x$ is $\sum_{k=1}^n x_k y_k x_m$ and the mth entry of $x x^\mathsf{T} y$ is $\sum_{k=1}^n x_m x_k y_k$, which are the same. □

Identity A.3 *For any* $x, y \in \mathbb{R}^3$, $x^\times y^\times = y x^\mathsf{T} - x^\mathsf{T} y I_3$. *Therefore,* $x^\times x^\times = x x^\mathsf{T} - x^\mathsf{T} x I_3$.

Proof Let $x = \mathrm{col}(x_1, x_2, x_3)$ and $y = \mathrm{col}(y_1, y_2, y_3)$. Then

$$\begin{aligned} x^\times y^\times &= \begin{bmatrix} 0 & -x_3 & x_2 \\ x_3 & 0 & -x_1 \\ -x_2 & x_1 & 0 \end{bmatrix} \cdot \begin{bmatrix} 0 & -y_3 & y_2 \\ y_3 & 0 & -y_1 \\ -y_2 & y_1 & 0 \end{bmatrix} \\ &= \begin{bmatrix} -x_3 y_3 - x_2 y_2 & x_2 y_1 & x_3 y_1 \\ x_1 y_2 & -x_3 y_3 - x_1 y_1 & x_3 y_2 \\ x_1 y_3 & x_2 y_3 & -x_2 y_2 - x_1 y_1 \end{bmatrix} \\ &= y x^\mathsf{T} - x^\mathsf{T} y I_3. \end{aligned}$$

□

Identity A.4 *For any* $x, y \in \mathbb{R}^3$, $(x^\times y)^\times = y x^\mathsf{T} - x y^\mathsf{T}$.

Proof Let $x = \mathrm{col}(x_1, x_2, x_3)$ and $y = \mathrm{col}(y_1, y_2, y_3)$. Then

$$
\begin{aligned}
(x^\times y)^\times &= \begin{bmatrix} -x_3 y_2 + x_2 y_3 \\ x_3 y_1 - x_1 y_3 \\ -x_2 y_1 + x_1 y_2 \end{bmatrix}^\times \\
&= \begin{bmatrix} 0 & x_2 y_1 - x_1 y_2 & x_3 y_1 - x_1 y_3 \\ x_1 y_2 - x_2 y_1 & 0 & x_3 y_2 - x_2 y_3 \\ x_1 y_3 - x_3 y_1 & x_2 y_3 - x_3 y_2 & 0 \end{bmatrix} \\
&= y x^\mathsf{T} - x y^\mathsf{T}.
\end{aligned}
$$

□

Identity A.5 *For any* $x, y \in \mathbb{R}^3$, $-xy^\mathsf{T}x^\times + x^\mathsf{T}yx^\times = x^\mathsf{T}xy^\times - y^\times xx^\mathsf{T}$.

Proof Let $x = \mathrm{col}(x_1, x_2, x_3)$ and $y = \mathrm{col}(y_1, y_2, y_3)$. Then

$$
\begin{aligned}
& - xy^\mathsf{T}x^\times + x^\mathsf{T}yx^\times \\
&= \begin{bmatrix} x_2 y_2 + x_3 y_3 & -x_1 y_2 & -x_1 y_3 \\ -x_2 y_1 & x_1 y_1 + x_3 y_3 & -x_2 y_3 \\ -x_3 y_1 & -x_3 y_2 & x_1 y_1 + x_2 y_2 \end{bmatrix} \begin{bmatrix} 0 & -x_3 & x_2 \\ x_3 & 0 & -x_1 \\ -x_2 & x_1 & 0 \end{bmatrix} \\
&= \begin{bmatrix} -x_1 y_2 x_3 + x_1 y_3 x_2 & -x_2 y_2 x_3 - x_3^2 y_3 - x_1^2 y_3 & x_2^2 y_2 + x_2 x_3 y_3 + x_1^2 y_2 \\ x_1 y_1 x_3 + x_3^2 y_3 + x_2^2 y_3 & x_2 y_1 x_3 - x_1 x_2 y_3 & -x_2^2 y_1 - x_1^2 y_1 - x_1 x_3 y_3 \\ -x_3^2 y_2 - x_1 x_2 y_1 - x_2^2 y_2 & x_3^2 y_1 + x_1^2 y_1 + x_1 x_2 y_2 & -x_2 x_3 y_1 + x_1 x_3 y_2 \end{bmatrix}.
\end{aligned}
$$

$$
\begin{aligned}
x^\mathsf{T}xy^\times - y^\times xx^\mathsf{T} &= \begin{bmatrix} 0 & -(x_1^2 + x_2^2 + x_3^2) y_3 & (x_1^2 + x_2^2 + x_3^2) y_2 \\ (x_1^2 + x_2^2 + x_3^2) y_3 & 0 & -(x_1^2 + x_2^2 + x_3^2) y_1 \\ -(x_1^2 + x_2^2 + x_3^2) y_2 & (x_1^2 + x_2^2 + x_3^2) y_1 & 0 \end{bmatrix} \\
&\quad - \begin{bmatrix} 0 & -y_3 & y_2 \\ y_3 & 0 & -y_1 \\ -y_2 & y_1 & 0 \end{bmatrix} \cdot \begin{bmatrix} x_1^2 & x_1 x_2 & x_1 x_3 \\ x_2 x_1 & x_2^2 & x_2 x_3 \\ x_3 x_1 & x_3 x_2 & x_3^2 \end{bmatrix} \\
&= - xy^\mathsf{T}x^\times + x^\mathsf{T}yx^\times.
\end{aligned}
$$

□

Identity A.6 *For any* $q \in \mathbb{Q}_u$, $-\hat{q}^\times \hat{q}^\times + \bar{q}^2 I_3 + \hat{q}\hat{q}^\mathsf{T} = I_3$.

Proof By Identity A.3,

$$
-\hat{q}^\times \hat{q}^\times + \bar{q}^2 I_3 + \hat{q}\hat{q}^\mathsf{T} = -\hat{q}\hat{q}^\mathsf{T} + \hat{q}^\mathsf{T}\hat{q} I_3 + \bar{q}^2 I_3 + \hat{q}\hat{q}^\mathsf{T} = I_3.
$$

□

Identity A.7 *For any* $q_a, q_b \in \mathbb{Q}$, $\|q_a \odot q_b\| = \|q_a\| \cdot \|q_b\| = \|q_b \odot q_a\|$. *Hence, if* $q_a \in \mathbb{Q}_u$, *then* $\|q_a \odot q_b\| = \|q_b\|$.

Proof Let $q_a = \mathrm{col}(q_{a1}, q_{a2}, q_{a3}, q_{a4})$ and $q_b = \mathrm{col}(q_{b1}, q_{b2}, q_{b3}, q_{b4})$. Then

$$
q_a \odot q_b = \begin{bmatrix} q_{a4}q_{b1} + q_{b4}q_{a1} - q_{a3}q_{b2} + q_{a2}q_{b3} \\ q_{a4}q_{b2} + q_{b4}q_{a2} + q_{a3}q_{b1} - q_{a1}q_{b3} \\ q_{a4}q_{b3} + q_{b4}q_{a3} - q_{a2}q_{b1} + q_{a1}q_{b2} \\ q_{a4}q_{b4} - q_{a1}q_{b1} - q_{a2}q_{b2} - q_{a3}q_{b3} \end{bmatrix}
$$

and thus

$$
\begin{aligned}
\|q_a \odot q_b\|^2 &= (q_{a4}q_{b1} + q_{b4}q_{a1} - q_{a3}q_{b2} + q_{a2}q_{b3})^2 \\
&\quad + (q_{a4}q_{b2} + q_{b4}q_{a2} + q_{a3}q_{b1} - q_{a1}q_{b3})^2 \\
&\quad + (q_{a4}q_{b3} + q_{b4}q_{a3} - q_{a2}q_{b1} + q_{a1}q_{b2})^2 \\
&\quad + (q_{a4}q_{b4} - q_{a1}q_{b1} - q_{a2}q_{b2} - q_{a3}q_{b3})^2 \\
&= \|q_a\|^2 \cdot \|q_b\|^2
\end{aligned}
$$

which completes the proof. □

Identity A.8 *For any $q \in \mathbb{Q}$ and $x \in \mathbb{R}^3$, $q^* \odot \mathbf{Q}(x) \odot q = \mathbf{Q}(\mathbf{C}(q)x)$.*

Proof

$$
q^* \odot \mathbf{Q}(x) = \begin{bmatrix} \bar{q}x - \hat{q}^\times x \\ \hat{q}^\mathsf{T} x \end{bmatrix}.
$$

Then

$$
\begin{aligned}
q^* \odot \mathbf{Q}(x) \odot q &= \begin{bmatrix} \hat{q}^\mathsf{T} x \hat{q} + \bar{q}(\bar{q}x - \hat{q}^\times x) + (\bar{q}x - \hat{q}^\times x)^\times \hat{q} \\ \bar{q}\hat{q}^\mathsf{T} x - \hat{q}^\mathsf{T}(\bar{q}x - \hat{q}^\times x) \end{bmatrix} \\
&= \begin{bmatrix} \hat{q}^\mathsf{T} x \hat{q} + \bar{q}^2 x - \bar{q}\hat{q}^\times x - \hat{q}^\times(\bar{q}x - \hat{q}^\times x) \\ 0 \end{bmatrix} \\
&= \begin{bmatrix} \hat{q}^\mathsf{T} x \hat{q} + \bar{q}^2 x - 2\bar{q}\hat{q}^\times x + \hat{q}^\times \hat{q}^\times x \\ 0 \end{bmatrix} \\
&= \begin{bmatrix} \hat{q}^\mathsf{T} x \hat{q} + \bar{q}^2 x - 2\bar{q}\hat{q}^\times x + (\hat{q}\hat{q}^\mathsf{T} - \hat{q}^\mathsf{T}\hat{q} I_3)x \\ 0 \end{bmatrix} \\
&= \begin{bmatrix} ((\bar{q}^2 - \hat{q}^\mathsf{T}\hat{q})I_3 + 2\hat{q}\hat{q}^\mathsf{T} - 2\bar{q}\hat{q}^\times)x \\ 0 \end{bmatrix} \\
&= \begin{bmatrix} \mathbf{C}(q)x \\ 0 \end{bmatrix} \\
&= \mathbf{Q}(\mathbf{C}(q)x)
\end{aligned}
$$

where we have used Identities A.2 and A.3. □

Identity A.9 *For any $q \in \mathbb{Q}$ and $x \in \mathbb{R}^3$, $q^\mathsf{T}(q \odot \mathbf{Q}(x)) = 0$.*

Proof By Identity A.1

$$q^\top(q \odot \mathbf{Q}(x)) = \begin{bmatrix} \hat{q}^\top & \bar{q} \end{bmatrix} \begin{bmatrix} \bar{q}x + \hat{q}^\times x \\ -\hat{q}^\top x \end{bmatrix} = 0.$$

□

Identity A.10 *Consider the following systems:*

$$\begin{aligned} \dot{q}_a &= \frac{1}{2} q_a \odot \mathbf{Q}(\omega_a) + d \\ \dot{q}_b &= \frac{1}{2} q_b \odot \mathbf{Q}(\omega_b) \\ J\dot{\omega}_b &= -\omega_b^\times J\omega_b + u \end{aligned}$$

where $q_a, q_b, d \in \mathbb{Q}$, $\omega_a, \omega_b, u \in \mathbb{R}^3$, $J \in \mathbb{R}^{3\times 3}$ *is a constant matrix, and all* d, ω_a, u *are time-varying functions. Let*

$$\begin{aligned} q_e &= q_a^* \odot q_b \\ \omega_e &= \omega_b - \mathbf{C}(q_e)\omega_a. \end{aligned}$$

Then we have

$$\begin{aligned} \dot{q}_e &= \frac{1}{2} q_e \odot \mathbf{Q}(\omega_e) + d_q \\ J\dot{\omega}_e &= -\omega_b^\times J\omega_b + J\omega_e^\times \mathbf{C}(q_e)\omega_a - J\mathbf{C}(q_e)\dot{\omega}_a + u - Jd_C\omega_a \end{aligned}$$

where

$$\begin{aligned} d_q &= \frac{1}{2}(q_e^\top q_e - 1)\mathbf{Q}(\omega_a) \odot q_e + d^* \odot q_b \\ d_C &= 2\bar{q}_e\bar{d}_q I_3 - 2\hat{q}_e^\top \hat{d}_q I_3 + 2\hat{d}_q\hat{q}_e^\top + 2\hat{q}_e\hat{d}_q^\top - 2\bar{d}_q\hat{q}_e^\times - 2\bar{q}_e\hat{d}_q^\times. \end{aligned}$$

If $d = 0$ *and* $q_a, q_b \in \mathbb{Q}_u$, *then by Identity A.7,* $\|q_e\| = 1$. *Moreover*

$$\begin{aligned} \dot{q}_e &= \frac{1}{2} q_e \odot \mathbf{Q}(\omega_e) \\ J\dot{\omega}_e &= -\omega_b^\times J\omega_b + J\omega_e^\times \mathbf{C}(q_e)\omega_a - J\mathbf{C}(q_e)\dot{\omega}_a + u. \end{aligned}$$

Proof By Identities A.1 and A.8

$$
\begin{aligned}
\dot{q}_e &= \dot{q}_a^* \odot q_b + q_a^* \odot \dot{q}_b \\
&= (\frac{1}{2} q_a \odot \mathbf{Q}(\omega_a) + d)^* \odot q_b + q_a^* \odot (\frac{1}{2} q_b \odot \mathbf{Q}(\omega_b)) \\
&= \frac{1}{2}\mathbf{Q}(\omega_a)^* \odot q_a^* \odot q_b + d^* \odot q_b + \frac{1}{2} q_a^* \odot q_b \odot \mathbf{Q}(\omega_b) \\
&= -\frac{1}{2}\mathbf{Q}(\omega_a) \odot q_e + \frac{1}{2} q_e \odot \mathbf{Q}(\omega_b) + d^* \odot q_b \\
&= \frac{1}{2}(q_e \odot \mathbf{Q}(\omega_b) - q_e \odot q_e^* \odot \mathbf{Q}(\omega_a) \odot q_e) \\
&\quad + \frac{1}{2}(q_e \odot q_e^* \odot \mathbf{Q}(\omega_a) \odot q_e - \mathbf{Q}(\omega_a) \odot q_e) + d^* \odot q_b \\
&= \frac{1}{2} q_e \odot (\mathbf{Q}(\omega_b) - q_e^* \odot \mathbf{Q}(\omega_a) \odot q_e) + \frac{1}{2}(q_e \odot q_e^* - q_I) \odot \mathbf{Q}(\omega_a) \odot q_e + d^* \odot q_b \\
&= \frac{1}{2} q_e \odot q(\omega_b - \mathbf{C}(q_e)\omega_a) + \frac{1}{2}(q_e \odot q_e^* - q_I) \odot \mathbf{Q}(\omega_a) \odot q_e + d^* \odot q_b \\
&= \frac{1}{2} q_e \odot \mathbf{Q}(\omega_e) + \frac{1}{2}\begin{bmatrix} 0_{3\times 1} \\ q_e^\top q_e - 1 \end{bmatrix} \odot \mathbf{Q}(\omega_a) \odot q_e + d^* \odot q_b \\
&= \frac{1}{2} q_e \odot \mathbf{Q}(\omega_e) + \frac{1}{2}(q_e^\top q_e - 1)\mathbf{Q}(\omega_a) \odot q_e + d^* \odot q_b \\
&= \frac{1}{2} q_e \odot \mathbf{Q}(\omega_e) + d_q
\end{aligned}
$$

where

$$
d_q = \frac{1}{2}(q_e^\top q_e - 1)\mathbf{Q}(\omega_a) \odot q_e + d^* \odot q_b.
$$

Since $\mathbf{C}(q_e) = (\bar{q}_e^2 - \hat{q}_e^\top \hat{q}_e)I_3 + 2\hat{q}_e\hat{q}_e^\top - 2\bar{q}_e\hat{q}_e^\times$, by Identities A.1, A.3, A.4 and A.5, we have

$$
\begin{aligned}
\dot{\mathbf{C}}(q_e) &= 2\bar{q}_e\dot{\bar{q}}_e I_3 - 2\hat{q}_e^\top \dot{\hat{q}}_e I_3 + 2\dot{\hat{q}}_e\hat{q}_e^\top + 2\hat{q}_e\dot{\hat{q}}_e^\top - 2\dot{\bar{q}}_e\hat{q}_e^\times - 2\bar{q}_e\dot{\hat{q}}_e^\times \\
&= \bar{q}_e(-\hat{q}_e^\top\omega_e + 2\bar{d}_q)I_3 - \hat{q}_e^\top(\bar{q}_e\omega_e + \hat{q}_e^\times\omega_e + 2\hat{d}_q)I_3 + (\bar{q}_e\omega_e + \hat{q}_e^\times\omega_e + 2\hat{d}_q)\hat{q}_e^\top \\
&\quad + \hat{q}_e(\bar{q}_e\omega_e + \hat{q}_e^\times\omega_e + 2\hat{d}_q)^\top + (\hat{q}_e^\top\omega_e - 2\bar{d}_q)\hat{q}_e^\times - \bar{q}_e(\bar{q}_e\omega_e + \hat{q}_e^\times\omega_e + 2\hat{d}_q)^\times \\
&= -2\bar{q}_e\hat{q}_e^\top\omega_e I_3 + \bar{q}_e\omega_e\hat{q}_e^\top + \hat{q}_e^\times\omega_e\hat{q}_e^\top + \bar{q}_e\hat{q}_e\omega_e^\top \\
&\quad - \hat{q}_e\omega_e^\top\hat{q}_e^\times + \hat{q}_e^\top\omega_e\hat{q}_e^\times - \bar{q}_e^2\omega_e^\times - \bar{q}_e(\hat{q}_e^\times\omega_e)^\times + d_C \\
&= 2\bar{q}_e\left(\hat{q}_e\omega_e^\top - \hat{q}_e^\top\omega_e I_3 + \frac{1}{2}\omega_e\hat{q}_e^\top - \frac{1}{2}\hat{q}_e\omega_e^\top - \frac{1}{2}(\hat{q}_e^\times\omega_e)^\times\right) \\
&\quad - \omega_e^\times\hat{q}_e\hat{q}_e^\top - \hat{q}_e\omega_e^\top\hat{q}_e^\times + \hat{q}_e^\top\omega_e\hat{q}_e^\times - \bar{q}_e^2\omega_e^\times + d_C \\
&= 2\bar{q}_e\omega_e^\times\hat{q}_e^\times + \hat{q}_e^\top\hat{q}_e\omega_e^\times - 2\omega_e^\times\hat{q}_e\hat{q}_e^\top - \bar{q}_e^2\omega_e^\times + d_C \\
&= -\omega_e^\times\mathbf{C}(q_e) + d_C
\end{aligned}
$$

where

$$
d_C = 2\bar{q}_e\bar{d}_q I_3 - 2\hat{q}_e^\top\hat{d}_q I_3 + 2\hat{d}_q\hat{q}_e^\top + 2\hat{q}_e\hat{d}_q^\top - 2\bar{d}_q\hat{q}_e^\times - 2\bar{q}_e\hat{d}_q^\times.
$$

Then

$$J\dot{\omega}_e = -\omega_b^\times J\omega_b + J\omega_e^\times \mathbf{C}(q_e)\omega_a - J\mathbf{C}(q_e)\dot{\omega}_a + u - Jd_C\omega_a.$$

□

A.6 A General Framework for Robust Output Regulation Problem and the Canonical Internal Model

In this section, we give a brief review on the general framework for handling the robust output regulation problem.

Consider a plant described by

$$\dot{x} = f(x, u, v, w) \tag{A.24a}$$

$$e = h(x, u, v, w) \tag{A.24b}$$

where $x \in \mathbb{R}^n$, $u \in \mathbb{R}^m$, $e \in \mathbb{R}^p$ are the state, the control input, and the regulated output, respectively. $v \in \mathbb{R}^q$ is the exogenous signal representing the disturbance and/or the reference input, and $w \in \mathbb{R}^{n_w}$ is the plant uncertain parameter vector. The exogenous signal v is assumed to be generated by the following exosystem:

$$\dot{v} = \mathbf{a}(v, \sigma) \tag{A.25}$$

which may depend on a parameter vector $\sigma \in \mathbb{R}^{n_\sigma}$.

For technical convenience, all the functions involved here are assumed to be sufficiently smooth and defined globally on their appropriate Euclidean spaces, and it is assumed that $w = 0$ is the nominal value of the uncertain parameter w, $f(0, 0, 0, w) = 0$ and $h(0, 0, 0, w) = 0$ for all w, and $\mathbf{a}(0, \sigma) = 0$ for all σ.

For convenience, let us put together the plant (A.24) and the exosystem (A.25) to obtain the following so-called composite system:

$$\dot{x} = f(x, u, v, w) \tag{A.26a}$$

$$\dot{v} = \mathbf{a}(v, \sigma) \tag{A.26b}$$

$$e = h(x, u, v, w). \tag{A.26c}$$

We consider an error output feedback control law of the following form:

$$u = k(e, z) \tag{A.27a}$$

$$\dot{z} = g(e, z) \tag{A.27b}$$

where $z \in \mathbb{R}^{n_z}$ for some integer n_z, and k, g are globally defined smooth functions. For the control law (A.27) to be well defined, we assume either the function k is independent of e or the function h as defined in (A.24b) is independent of u.

Let $\mathbb{V}_0 \subseteq \mathbb{R}^q$, $\mathbb{W} \subseteq \mathbb{R}^{n_w}$, and $\mathbb{S} \subseteq \mathbb{R}^{n_\sigma}$, be some subsets with $\mathbb{V}_0$ and $\mathbb{W}$ containing the origin of their respective Euclidian spaces. We allow $\mathbb{S}$ to consist of a single known vector to accommodate the case where the exosystem (A.25) is known exactly.

We describe the (global) robust output regulation problem for system (A.26) as follows:

Problem A.1 (*Global Robust Output Regulation Problem*) Given the composite system (A.26), any compact subset $\Sigma_0 \subseteq \mathbb{V}_0 \times \mathbb{W} \times \mathbb{S}$, find a feedback control law of the form (A.27) such that, for any $\text{col}(v(0), w, \sigma) \in \Sigma_0$, and any initial state, the closed-loop system has the following properties:

- **Property A.1** The solution of the closed-loop system exists for all $t \geq 0$, and is uniformly bounded over $[0, \infty)$ for uniformly bounded $v(t)$;
- **Property A.2**

$$\lim_{t \to \infty} e(t) = 0. \tag{A.28}$$

For the above problem to be well posed, we make the following assumptions.

Assumption A.1 For any $\sigma \in \mathbb{S}$, and any initial condition $v(0)$, the solution $v(t)$ of the system $\dot{v}(t) = \mathbf{a}(v(t), \sigma)$ exists for all $t \geq 0$, and is uniformly bounded over $[0, \infty)$.

Assumption A.2 There exist smooth functions $\mathbf{x}(v, w, \sigma)$ and $\mathbf{u}(v, w, \sigma)$ with $\mathbf{x}(0, 0, 0) = 0$ and $\mathbf{u}(0, 0, 0) = 0$ that satisfy, for all $v(t)$ with $v(0) \in \mathbb{V}_0$, $w \in \mathbb{W}$, and $\sigma \in \mathbb{S}$, the following equations:

$$\frac{\partial \mathbf{x}(v, w, \sigma)}{\partial v} \mathbf{a}(v, \sigma) = f(\mathbf{x}(v, w, \sigma), \mathbf{u}(v, w, \sigma), v, w) \tag{A.29a}$$

$$0 = h(\mathbf{x}(v, w, \sigma), \mathbf{u}(v, w, \sigma), v, w). \tag{A.29b}$$

Remark A.2 Assumption A.1 guarantees that, for any initial condition $v(0)$, and any $\sigma \in \mathbb{S}$, the exogenous signal $v(t)$ is uniformly bounded over $[0, \infty)$. Thus, given $\mathbb{V}_0 \subseteq \mathbb{R}^q$, there exists a subset $\mathbb{V} \subseteq \mathbb{R}^q$ such that $v(0) \in \mathbb{V}_0$ implies $v(t) \in \mathbb{V}$ for all $t \geq 0$. As a result, given a compact subset $\Sigma_0 \subseteq \mathbb{V}_0 \times \mathbb{W} \times \mathbb{S}$, there exists a compact subset $\Sigma \subseteq \mathbb{V} \times \mathbb{W} \times \mathbb{S}$ such that $(v(0), w, \sigma) \in \Sigma_0$ implies $(v(t), w, \sigma) \in \Sigma$ for all $t \geq 0$. Equations (A.29) are called the regulator equations. The solvability of the regulator equations means the system (A.26) admits an output zeroing invariant manifold $\mathbb{M} = \{x = \mathbf{x}(v, w, \sigma)\}$ under the control $u = \mathbf{u}(v, w, \sigma)$.

In what follows, we will describe a general framework for handling the robust output regulation problem for the system (A.26) based on the so-called internal model approach. To introduce the concept of the internal model, consider a system of the following form:

$$\dot{\eta} = \alpha(\eta, u) \tag{A.30a}$$

$$\hat{\mathbf{u}} = \beta(\eta, \sigma) \tag{A.30b}$$

in which, for some positive integer d, $\alpha : \mathbb{R}^d \times \mathbb{R}^m \mapsto \mathbb{R}^d$ and $\beta : \mathbb{R}^d \times \mathbb{R}^{n_\sigma} \mapsto \mathbb{R}^m$ are some smooth functions vanishing at the origin.

Definition A.6 Under Assumptions A.1 and A.2, system (A.30) is called an *internal model* of system (A.26) with the estimated steady-state input $\hat{\mathbf{u}} = \beta(\eta, \sigma)$ if there exists a smooth mapping $\theta : \mathbb{V} \times \mathbb{W} \times \mathbb{S} \mapsto \mathbb{R}^d$ satisfying $\theta(0, 0, 0) = 0$ such that, for all $(v, w, \sigma) \in \mathbb{V} \times \mathbb{W} \times \mathbb{S}$,

$$\frac{\partial \theta(v, w, \sigma)}{\partial v}\mathbf{a}(v, \sigma) = \alpha(\theta(v, w, \sigma), \mathbf{u}(v, w, \sigma)), \tag{A.31a}$$

$$\mathbf{u}(v, w, \sigma) = \beta(\theta(v, w, \sigma), \sigma). \tag{A.31b}$$

To see the role of the internal model (A.30), define the following so-called augmented system:

$$\dot{x} = f(x, u, v, w) \tag{A.32a}$$

$$\dot{\eta} = \alpha(\eta, u) \tag{A.32b}$$

$$e = h(x, u, v, w). \tag{A.32c}$$

Then, (A.29) and (A.31) imply

$$\frac{\partial \mathbf{x}(v, w, \sigma)}{\partial v}\mathbf{a}(v, \sigma) = f(\mathbf{x}(v, w, \sigma), \mathbf{u}(v, w, \sigma), v, w) \tag{A.33a}$$

$$\frac{\partial \theta(v, w, \sigma)}{\partial v}\mathbf{a}(v, \sigma) = \alpha(\theta(v, w, \sigma), \mathbf{u}(v, w, \sigma)) \tag{A.33b}$$

$$0 = h(\mathbf{x}(v, w, \sigma), \mathbf{u}(v, w, \sigma), v, w). \tag{A.33c}$$

That is, the regulator equations associated with the augmented system (A.32) admit a solution pair $(\mathbf{x}(v, w, \sigma), \theta(v, w, \sigma))$ and $\mathbf{u}(v, w, \sigma)$. Alternatively, we say that the augmented system (A.32) contains an output zeroing invariant manifold

$$\mathbb{M} = \{(x, \eta, v) : x = \mathbf{x}(v, w, \sigma),\ \eta = \theta(v, w, \sigma), v \in \mathbb{V}, w \in \mathbb{W}, \sigma \in \mathbb{S}\}$$

under the control $u = \mathbf{u}(v, w, \sigma)$.

Now performing the following coordinate and input transformation:

$$\bar{x} = x - \mathbf{x}(v, w, \sigma) \tag{A.34a}$$
$$\bar{\eta} = \eta - \theta(v, w, \sigma) \tag{A.34b}$$
$$\bar{u} = u - \beta(\eta, \sigma) \tag{A.34c}$$

converts (A.32) to the following so-called augmented error system:

$$\dot{\bar{x}} = \bar{f}(\bar{x}, \bar{\eta}, \bar{u}, v, w, \sigma) \tag{A.35a}$$
$$\dot{\bar{\eta}} = \bar{\alpha}(\bar{x}, \bar{\eta}, \bar{u}, v, w, \sigma) \tag{A.35b}$$
$$e = \bar{h}(\bar{x}, \bar{\eta}, \bar{u}, v, w, \sigma) \tag{A.35c}$$

where

$$\begin{aligned}
\bar{f}(\bar{x}, \bar{\eta}, \bar{u}, v, w, \sigma) &= f(\bar{x} + \mathbf{x}(v, w, \sigma), \bar{u} + \beta(\bar{\eta} + \theta(v, w, \sigma), \sigma), v, w) \\
&\quad - f(\mathbf{x}(v, w, \sigma), \mathbf{u}(v, w, \sigma), v, w) \\
\bar{\alpha}(\bar{x}, \bar{\eta}, \bar{u}, v, w, \sigma) &= \alpha(\bar{\eta} + \theta(v, w, \sigma), \bar{u} + \beta(\bar{\eta} + \theta(v, w, \sigma), \sigma)) \\
&\quad - \alpha(\theta(v, w, \sigma), \mathbf{u}(v, w, \sigma)) \\
\bar{h}(\bar{x}, \bar{\eta}, \bar{u}, v, w, \sigma) &= h(\bar{x} + \mathbf{x}(v, w, \sigma), \bar{u} + \beta(\bar{\eta} + \theta(v, w, \sigma), \sigma), v, w) \\
&\quad - h(\mathbf{x}(v, w, \sigma), \mathbf{u}(v, w, \sigma), v, w).
\end{aligned}$$

It is ready to verify that system (A.35) has the property that, for all trajectories $v(t)$ of the exosystem, all $w \in \mathbb{R}^{n_w}$ and all $\sigma \in \mathbb{R}^{n_\sigma}$,

$$\begin{aligned}
\bar{f}(0, 0, 0, v, w, \sigma) &= 0 \\
\bar{\alpha}(0, 0, 0, v, w, \sigma) &= 0 \\
\bar{h}(0, 0, 0, v, w, \sigma) &= 0.
\end{aligned}$$

Consider an output feedback control law of the following form:

$$\bar{u} = \bar{k}(e, \xi) \tag{A.36a}$$
$$\dot{\xi} = \bar{g}_\xi(e, \xi) \tag{A.36b}$$

where $\xi \in \mathbb{R}^{n_\xi}$ for some integer n_ξ, and $\bar{k}$ and $\bar{g}_\xi$ are smooth functions vanishing at their respective origins. We say (A.36) globally stabilizes the zero solution of the augmented error system (A.35) on the compact subset $\Sigma_0 \subseteq \mathbb{V} \times \mathbb{W} \times \mathbb{S}$, if for any $(v(0), w, \sigma) \in \Sigma_0$, the zero solution of the closed-loop system composed of (A.35) and (A.36) is globally asymptotically stable.

Remark A.3 For the control law (A.36) to be well defined, we assume either the function $\bar{k}$ is independent of e or the function h as defined in (A.24b) is independent of u.

Due to the critical role played by the internal model, we make the following explicit assumption:

Assumption A.3 The composite system (A.26) admits an internal model of the form (A.30).

Proposition A.5 *Under Assumptions A.1–A.3, if a dynamic error output feedback control law of the form (A.36) globally stabilizes the zero solution of the augmented error system (A.35) on the compact subset $\Sigma_0 \subseteq \mathbb{V}_0 \times \mathbb{W} \times \mathbb{S}$, then the following control law:*

$$u = \bar{k}(e, \xi) + \beta(\eta, \sigma) \tag{A.37a}$$

$$\dot{\eta} = \alpha(\eta, u) \tag{A.37b}$$

$$\dot{\xi} = \bar{g}_{\xi}(e, \xi) \tag{A.37c}$$

solves the global robust output regulation problem for the composite system (A.26).

Proof Under Assumptions A.1–A.3, the augmented error system (A.35) is well defined, and, for any $v(0) \in \mathbb{V}_0$, the solution of the exosystem exists and is uniformly bounded over $[0, \infty)$. If (A.36) globally stabilizes the zero solution of the augmented error system (A.35) on the compact subset $\Sigma_0 \subseteq \mathbb{V}_0 \times \mathbb{W} \times \mathbb{S}$, then, for any $\bar{x}(0)$, $\bar{\eta}(0)$, and $\xi(0)$, and any $(v(0), w, \sigma) \in \Sigma_0$, $\lim_{t\to\infty} \bar{x}(t) = 0$, $\lim_{t\to\infty} \bar{\eta}(t) = 0$, and $\lim_{t\to\infty} \xi(t) = 0$. Note that the closed-loop system composed of (A.35) and (A.36) is equivalent to, up to the coordinate and input transformation (A.34), the closed-loop system composed of the original plant (A.24) and the control law (A.37). Thus, for any $x(0)$, $\eta(0)$ and $\xi(0)$, and any $(v(0), w, \sigma) \in \Sigma_0$, the solution of the closed-loop system composed of the original plant (A.24) and the control law (A.37) exists and is uniformly bounded over $[0, \infty)$. Moreover, by Remark A.3, either the function h is independent of u or the function $\bar{k}$ is independent of e. In the former case, $\lim_{t\to\infty} e(t) = \lim_{t\to\infty} \bar{h}(\bar{x}(t), \bar{\eta}(t), v, w, \sigma) = \bar{h}(0, 0, v, w, \sigma) = 0$. In the latter case, $\lim_{t\to\infty} e(t) = \lim_{t\to\infty} \bar{h}(\bar{x}(t), \bar{\eta}(t), \bar{k}(\xi(t)), v, w, \sigma) = \bar{h}(0, 0, 0, v, w, \sigma) = 0$. Thus, the control law (A.37) solves the global robust output regulation problem for the composite system (A.26). □

Remark A.4 The control law (A.37) is feasible only if the function β is independent of the unknown parameter vector σ. This is always the case when the exosystem is known exactly or, what is the same, $\mathbb{S}$ consists of a single known vector. In this case, we have $\beta(\eta, \sigma) = \beta(\eta)$. If the exosystem does contain some unknown parameter vector σ, it is difficult to find a function β satisfying (A.31) and is independent of σ. To overcome this difficulty, one can further employ the adaptive control technique to handle the unknown parameter vector σ. Doing so will lead to the following adaptive dynamic error output feedback control law:

$$u = \bar{k}(e, \xi) + \beta(\eta, \hat{\sigma}) \tag{A.38a}$$

$$\dot{\xi} = \bar{g}_{\xi}(e, \xi) \tag{A.38b}$$

$$\dot{\hat{\sigma}} = \bar{g}_{\hat{\sigma}}(e, \xi) \tag{A.38c}$$

where $\hat{\sigma}$ is considered as the estimation of σ which is updated via (A.38c) called the parameter update law. In Chap. 10, we will detail how to synthesize a specific parameter update law to solve the cooperative output regulation problem for a class of nonlinear multi-agent systems subject to a linear uncertain exosystem.

If (A.38) solves the global (adaptive) stabilization problem of the augmented error system (A.35) on the compact subset Σ, then we say that the following control law:

$$\begin{aligned} u &= \bar{k}(e, \xi) + \beta(\eta, \hat{\sigma}) \\ \dot{\eta} &= \alpha(\eta, u) \\ \dot{\xi} &= \bar{g}_{\xi}(e, \xi) \\ \dot{\hat{\sigma}} &= \bar{g}_{\hat{\sigma}}(e, \xi) \end{aligned}$$

solves the global (adaptive) robust output regulation problem for the original system (A.26).

From the previous argument, we see that finding an appropriate internal model for (A.26) is the key for solving the robust output regulation problem and the problem has been extensively studied in the literature (see [5] and the references therein). A good internal model should be made so that the zero solution of the augmented error system is stabilizable. Here we will show how to design a so-called canonical internal model for a scenario characterized by the following two assumptions.

Assumption A.4 $\mathbf{a}(v, \sigma) = S(\sigma)v$ for some matrix $S(\sigma) \in \mathbb{R}^{q \times q}$.

To introduce the next assumption, for $i = 1, \ldots, N$, denote the ith component of $\mathbf{u}(v, w, \sigma)$ by $\mathbf{u}_i(v, w, \sigma)$. Recall from Sect. 10.1 that $L_{\mathbf{a}}\mathbf{u}_i$ denotes the Lie derivative of $\mathbf{u}_i$ along the vector field $\mathbf{a}(v, \sigma) = S(\sigma)v$, i.e., $L_{\mathbf{a}}\mathbf{u}_i = \frac{\partial \mathbf{u}_i(v,w,\sigma)}{\partial v} S(\sigma)v$, and, for $l \geq 2$, $L_{\mathbf{a}}^l \mathbf{u}_i = L_{\mathbf{a}}(L_{\mathbf{a}}^{l-1}\mathbf{u}_i)$.

Assumption A.5 For $i = 1, \ldots, m$, there exist an integer s_i and real scalars $a_{i1}(\sigma)$, $a_{i2}(\sigma), \ldots, a_{is_i}(\sigma)$ such that $\mathbf{u}_i = \mathbf{u}_i(v, w, \sigma)$ satisfies, for all $w \in \mathbb{W}$ and all $\sigma \in \mathbb{S}$

$$L_{\mathbf{a}}^{s_i}\mathbf{u}_i = -a_{i1}(\sigma)\mathbf{u}_i - a_{i2}(\sigma)L_{\mathbf{a}}\mathbf{u}_i - \cdots - a_{is_i}(\sigma)L_{\mathbf{a}}^{s_i-1}\mathbf{u}_i. \tag{A.39}$$

We are now ready to construct the so-called canonical internal model as follows. Let

$$\tau_i(v, w, \sigma) = \mathrm{col}(\mathbf{u}_i, L_{\mathbf{a}}\mathbf{u}_i, \ldots, L_{\mathbf{a}}^{s_i-1}\mathbf{u}_i)$$

and

$$\Phi_i(\sigma) = \begin{bmatrix} 0 & 1 & \dots & 0 \\ \vdots & \vdots & \ddots & \vdots \\ 0 & 0 & \dots & 1 \\ -a_{i1}(\sigma) & -a_{i2}(\sigma) & \dots & -a_{is_i}(\sigma) \end{bmatrix}, \quad \Psi_i = \begin{bmatrix} 1 \\ 0 \\ \vdots \\ 0 \end{bmatrix}^{\mathrm{T}}_{s_i \times 1}.$$

Then it can be verified that the triplet $(\tau_i, \Phi_i(a), \Psi_i)$ satisfies

$$\begin{aligned} \frac{\partial \tau_i(v, w, \sigma)}{\partial v} S(\sigma)v &= \Phi_i(\sigma)\tau_i(v, w, \sigma) \\ \mathbf{u}_i(v, w, \sigma) &= \Psi_i \tau_i(v, w, \sigma). \end{aligned}$$

Let $M_i \in \mathbb{R}^{s_i \times s_i}$ and $Q_i \in \mathbb{R}^{s_i \times 1}$ be any controllable pair with M_i being Hurwitz. Then, if the spectra of the matrices $\Phi_i(\sigma)$ and M_i are disjoint, then, by Proposition A.3, there exists a unique nonsingular matrix $T_i(\sigma)$ that satisfies the following Sylvester equation:

$$T_i(\sigma)\Phi_i(\sigma) = M_i T_i(\sigma) + Q_i \Psi_i.$$

Let $M = \mathscr{D}(M_1, \dots, M_m)$, $Q = \mathscr{D}(Q_1, \dots, Q_m)$, $T(\sigma) = \mathscr{D}(T_1(\sigma), \dots, T_l(\sigma))$, $\Psi = \mathscr{D}(\Psi_1, \dots, \Psi_m)$. Let $\alpha(\eta, u) = M\eta + Qu$, $\beta(\eta, \sigma) = \Psi T^{-1}(\sigma)\eta$, and $\theta(v, w, \sigma) = T(\sigma)\tau(v, w, \sigma)$. Then it can be verified that

$$\begin{aligned} \frac{\partial \theta(v, w, \sigma)}{\partial v} S(\sigma)v &= \alpha(\theta(v, w, \sigma), \beta(\theta(v, w, \sigma)) \\ \mathbf{u}(v, w, \sigma) &= \Psi T^{-1}(\sigma)\theta(v, w, \sigma). \end{aligned}$$

That is, the following system:

$$\dot{\eta} = M\eta + Qu \tag{A.40a}$$

$$\hat{\mathbf{u}} = \Psi T^{-1}(\sigma)\eta \tag{A.40b}$$

is an internal model of (A.26) with the estimated steady-state input $\hat{\mathbf{u}} = \Psi T^{-1}(\sigma)\eta$. (A.40) is referred to as the canonical internal model. It is noted that if all the eigenvalues of $S(\sigma)$ have zero real parts for all $\sigma \in \mathbb{S}$, then, for any Hurwitz matrix M_i, the spectra of the matrices $\Phi_i(\sigma)$ and M_i are disjoint.

If the matrix $S(\sigma)$ is known exactly, that is, Σ consists of a known singleton σ, then the function $\beta(\eta, \sigma)$ is also known exactly. Thus, for this case, the internal model (A.40) can be directly employed to handle the robust output regulation problem. Moreover, even if the matrix $S(\sigma)$ depends on the unknown parameter vector σ, (A.40) is still independent of σ, and the function $\beta(\eta, \sigma)$ is linearly dependent on the unknown matrix $\Psi T^{-1}(\sigma)$. Thus, it is possible to further employ some adaptive control technique to handle the unknown matrix $\Psi T^{-1}(\sigma)$.

We now close this section by an interesting result. For simplicity, we assume that $m = 1$ without loss of generality, and omit the subscript 1. For any $v \in \mathbb{R}^q$, let

$$\begin{aligned} v^{[0]} &= 1 \\ v^{[1]} &= v = \mathrm{col}(v_1, \ldots, v_q) \\ v^{[k]} &= \mathrm{col}(v_1^k, v_1^{k-1}v_2, \ldots, v_1^{k-1}v_q, v_1^{k-2}v_2^2, \ldots, v_1^{k-2}v_2v_q, \ldots, v_q^k), \\ &\quad k = 2, 3, \ldots. \end{aligned}$$

It can be seen that the dimension of $v^{[k]}$ is given by the binomial coefficient

$$C_{q+k-1}^k = \frac{(q+k-1)!}{(q-1)!k!}.$$

Let Q_k be a row vector with dimension C_{q+k-1}^k whose entries are real numbers. Then we call $Q_k v^{[k]}$ a homogeneous polynomial of degree k. We call $\mathbf{u}(v, w, \sigma) \in \mathbb{R}$ a polynomial in v of degree $\ell > 0$ if there exist constant row vectors $Q_k(w, \sigma)$ which may depend on (w, σ), $k = 1, \ldots, \ell$, such that

$$\mathbf{u}(v, w, \sigma) = \sum_{k=1}^{\ell} Q_k(w, \sigma) v^{[k]} \tag{A.41}$$

where $Q_\ell \neq 0$.

Proposition A.6 *Under Assumptions A.2 and A.4, if* $\mathbf{u}(v, w, \sigma)$ *takes the form (A.41), then* $\mathbf{u}(v, w, \sigma)$ *satisfies Assumptions A.5.*

Proof Since, for $k \geq 1$, the components of $v^{[k]}$ consist of all products of the variables $v_1, \ldots, v_q$ taken k at a time. Therefore, any homogeneous polynomials in $v_1, \ldots, v_q$ of degree k is a linear combination of the components of $v^{[k]}$. Since every component of $\frac{\partial v^{[k]}}{\partial v}$ is a homogeneous polynomial in $v_1, \ldots, v_q$ of degree $k-1$, every component of $\frac{\partial v^{[k]}}{\partial v} S(\sigma) v$ is a homogeneous polynomial in $v_1, \ldots, v_q$ of degree k. Therefore, there exists a matrix denoted by $S^{[k]}(\sigma)$ such that

$$\frac{\partial v^{[k]}}{\partial v} S(\sigma) v = S^{[k]}(\sigma) v^{[k]}.$$

Thus, for any $j \geq 1$,

$$L_{\mathbf{a}}^j \mathbf{u} = \sum_{k=1}^{\ell} Q_k(w, \sigma) (S^{[k]})^j v^{[k]}.$$

Let the minimal polynomial of the matrix $\mathcal{D}(S^{[1]}, \ldots, S^{[\ell]})$ be as follows:

$$P(\lambda, \sigma) = \lambda^s + a_1(\sigma) + a_2(\sigma)\lambda + \cdots + a_s(\sigma)\lambda^{s-1}.$$

Then, since, for $k = 1, \ldots, \ell$, the minimal polynomials of $S^{[k]}$ divides $P(\lambda, \sigma)$, by the Cayley–Hamilton theorem, $P(S^{[k]}, \sigma) = 0$, $k = 1, \ldots, \ell$. Thus

$$\begin{aligned}
& a_1(\sigma)\mathbf{u} + a_2(\sigma)L_{\mathbf{a}}\mathbf{u} + \cdots + a_s(\sigma)L_{\mathbf{a}}^{s-1}\mathbf{u} + L_{\mathbf{a}}^{s}\mathbf{u} \\
&= \sum_{k=1}^{\ell} Q_k(w,\sigma)\left(a_1(\sigma) + a_2(\sigma)S^{[k]} + \cdots + a_s(\sigma)(S^{[k]})^{s-1} + (S^{[k]})^{s}\right) v^{[k]} \\
&= \sum_{k=1}^{\ell} Q_k(w,\sigma)P(S^{[k]},\sigma)v^{[k]} = 0.
\end{aligned}$$

□

Remark A.5 For any $k \geq 0$, the explicit expression for the matrix $S^{[k]}$ is given in [6, Chap. 4]. Moreover, denote the eigenvalues of $S(\sigma)$ by λ_i, $i = 1, \ldots, q$, and let

$$\Omega_k = \{\lambda : \lambda = k_1\lambda_1 + \cdots + k_q\lambda_q,\, k_1 + \cdots + k_q = k,\ k_1, \ldots, k_q = 0, 1, \ldots, k\ \}.$$

Then the eigenvalues of $S^{[k]}$ are given by the elements of Ω_k. Thus, if, for all $\sigma \in \mathbb{S}$, all the eigenvalues of $S(\sigma)$ have zero real parts, then, for all $\sigma \in \mathbb{S}$, all the eigenvalues of $S^{[k]}$ and hence all the roots of the minimal polynomial $P(\lambda, \sigma)$ have zero real parts. In particular, if, for all $\sigma \in \mathbb{S}$, all the eigenvalues of $S(\sigma)$ are distinct, then, for all $\sigma \in \mathbb{S}$, all the eigenvalues of $S^{[k]}$ and hence all the roots of the minimal polynomial $P(\lambda, \sigma)$ are distinct.

A.7 Stabilizing Techniques for Nonlinear Systems

This section reviews some stabilizing techniques for nonlinear control systems. Let us first recall the following lemma which is adapted from [7, Lemma 11.1].

Lemma A.7 *(i) For any continuous function $f : \mathbb{R}^n \times \mathbb{R}^l \mapsto \mathbb{R}$, there exist smooth functions $a(x), b(d) \geq 0$, such that*

$$|f(x,d)| \leq a(x)b(d).$$

Moreover, if f is continuously differentiable with $f(0,d) = 0$, then there exists smooth functions $m(x), b(d) \geq 0$, such that

$$|f(x,d)| \leq \|x\|m(x)b(d).$$

(ii) Partition x as $x = \mathrm{col}(x_1, \ldots, x_s)$ with $x_i \in \mathbb{R}^{n_i}$, $i = 1, \ldots, s$, for some $s > 0$, and $n_1 + \cdots + n_s = n$. Then, there exist smooth functions $a_i(x_i), b(d) \geq 0$, $i = 1, \ldots, s$, such that

$$|f(x_1, \ldots, x_s, d)| \leq b(d)\sum_{i=1}^{s} a_i(x_i).$$

Moreover, if f is continuously differentiable with $f(0, d) = 0$, then there exist smooth functions $b(d) \geq 0$ and $m_i(x_i) \geq 0$, $i = 1, \ldots, s$, such that

$$|f(x_1, \ldots, x_s, d)| \leq b(d) \sum_{i=1}^{s} \|x_i\| m_i(x_i).$$

Lemma A.8 *(i) Suppose $f : \mathbb{R}^n \times \mathbb{R}^l \mapsto \mathbb{R}$ is twice continuously differentiable with $f(0, d) = 0$, and $\frac{\partial f(0,d)}{\partial x} = 0$. Then there exist smooth functions $m(x), b(d) \geq 0$, such that*

$$|f(x, d)| \leq \|x\|^2 m(x) b(d). \tag{A.42}$$

(ii) Partition x as $x = col(x_1, \ldots, x_s)$ with $x_i \in \mathbb{R}^{n_i}$, $i = 1, \ldots, s$, for some $s > 0$, and $n_1 + \cdots + n_s = n$. Then, there exist smooth functions $b(d) \geq 0$ and $m_i(x_i) \geq 0$, $i = 1, \ldots, s$, such that

$$|f(x_1, \ldots, x_s, d)| \leq b(d) \sum_{i=1}^{s} \|x_i\|^2 m_i(x_i). \tag{A.43}$$

(iii) Suppose $f : \mathbb{R}^n \times \mathbb{R}^l \mapsto \mathbb{R}^m$ is continuously differentiable with $f(0, d) = 0$. Partition x as $x = col(x_1, \ldots, x_s)$ with $x_i \in \mathbb{R}^{n_i}$, $i = 1, \ldots, s$, for some $s > 0$, and $n_1 + \cdots + n_s = n$. Then, there exist smooth functions $b(d) \geq 0$ and $m_i(x_i) \geq 0$, $i = 1, \ldots, s$, such that

$$\|f(x_1, \ldots, x_s, d)\|^2 \leq b(d) \sum_{i=1}^{s} \|x_i\|^2 m_i(x_i). \tag{A.44}$$

Proof (i) Since $f(0, d) = 0$, and $\frac{\partial f(0,d)}{\partial x} = 0$, by Taylor's Theorem,

$$|f(x, d) - f(0, d)| \leq x^{\mathsf{T}} \left[\frac{\partial^2 f(cx, d)}{\partial x^2} \right] x$$

for some $0 \leq c \leq 1$. Let

$$f_1(x, d) = \max_{0 \leq c \leq 1} \left\| \frac{\partial^2 f(cx, d)}{\partial x^2} \right\|$$

which is continuous by construction, and hence bounded by $m(x)b(d)$ for some $m(x), b(d) \geq 0$ from Part (i) of Lemma A.7.

(ii) Let

$$f_i(x_i) = s \max_{\|x_j\| \leq \|x_i\|, \; \forall \; j \neq i} m(x_1, \ldots, x_s), \; i = 1, \ldots, s.$$

Then the functions $f_i(x_i)$ are continuous, and hence $f_i(x_i) \leq m_i(x_i)$ for some smooth functions $m_i(x_i)$. From (A.42), we conclude that, for $\|x_j\| \leq \|x_i\|$, $j \neq i$,

$$|f(x_1, \ldots, x_s, d)| \leq b(d)\|x\|^2 m(x) \leq b(d)\|x_i\|^2 m_i(x_i).$$

Thus, (A.43) holds for all $x \in \mathbb{R}^n$.

(iii) Since $f(x, d)$ is continuously differentiable with $f(0, d) = 0$, $\|f(x_1, \ldots, x_s, d)\|^2$ is twice continuously differentiable with $\|f(0, d)\|^2 = 0$, and $\frac{\partial \|f(0,d)\|^2}{\partial x} = 0$. Applying Part (ii) of this lemma to the function $\|f(x_1, \ldots, x_s, d)\|^2$ gives (A.44). This completes the proof. □

Remark A.6 In Lemmas A.7 and A.8, if the function $f(\cdot, \cdot)$ is known, then all functions on the right hand side of various inequalities of Lemmas A.7 and A.8 are also known. Also, if $d \in \mathbb{D}$ where $\mathbb{D}$ is compact, then $b(d) \leq b$ with $b = \max_{d \in \mathbb{D}} b(d)$. Moreover, if $\mathbb{D}$ is known, so is b. In this case, we can always assume that $b(d) = 1$ in all the inequalities of Lemmas A.7 and A.8.

Lemma A.9 *Suppose $V_1 : \mathbb{R}^n \mapsto [0, \infty)$ and $V_2 : \mathbb{R}^m \mapsto [0, \infty)$ are continuously differentiable such that*

$$\alpha_1(\|x\|) \leq V_1(x) \leq \beta_1(\|x\|), \ \forall x \in \mathbb{R}^n$$
$$\alpha_2(\|z\|) \leq V_2(z) \leq \beta_2(\|z\|), \ \forall z \in \mathbb{R}^m$$

where $\alpha_i(\cdot)$ and $\beta_i(\cdot)$, $i = 1, 2$, are some class $\mathcal{K}_\infty$ functions. Then, there exist class $\mathcal{K}_\infty$ functions $\alpha(\cdot)$ and $\beta(\cdot)$ such that

$$\alpha(\|col(x, z)\|) \leq V_1(x) + V_2(z) \leq \beta(\|col(x, z)\|), \ \forall col(x, z) \in \mathbb{R}^{n+m}. \quad \text{(A.45)}$$

Proof Let

$$\alpha(\|\text{col}(x, z)\|) = \inf_{\|\text{col}(\bar{x}, \bar{z})\| \geq \|\text{col}(x,z)\|} \{\alpha_1(\|\bar{x}\|) + \alpha_2(\|\bar{z}\|)\}$$

and

$$\beta(\|\text{col}(x, z)\|) = \sup_{\|\text{col}(\bar{x}, \bar{z})\| \leq \|\text{col}(x,z)\|} \{\beta_1(\|\bar{x}\|) + \beta_2(\|\bar{z}\|)\}.$$

Then it can be verified that $\alpha(\cdot)$ and $\beta(\cdot)$ are both class $\mathcal{K}_\infty$ functions, and $V_1(x) + V_2(z)$ satisfies (A.45). □

We now review the concept of input-to-state stability.

Definition A.7 (*Input-to-State Stability*) Consider the nonlinear system

$$\dot{x} = f(x, u, \mu) \quad \text{(A.46)}$$

where $x \in \mathbb{R}^n$, $u \in \mathbb{R}^m$, and $\mu : [0, \infty) \mapsto \mathbb{R}^{n_\mu}$ is uniformly bounded and piecewise continuous, and f is locally Lipschitz in $\mathrm{col}(x, u)$. System (A.46) is said to be input-to-state stable on a compact subset $\Sigma \subseteq \mathbb{R}^{n_\mu}$ with the state x and the input u if there exist a class $\mathcal{KL}$ function β and a class $\mathcal{K}$ function γ, independent of μ, such that for all $\mu \in \Sigma$, any initial state $x(t_0)$, and any uniformly bounded and piecewise continuous input function $u : [t_0, \infty) \mapsto \mathbb{R}^m$, the solution $x(t)$ exists and satisfies

$$\|x(t)\| \leq \max\left\{\beta(\|x(t_0)\|, t - t_0), \gamma\left(\sup_{t_0 \leq \tau \leq t} \|u(\tau)\|\right)\right\}.$$

Theorem A.6 *Consider the nonlinear system (A.46), where $x \in \mathbb{R}^n$, $u \in \mathbb{R}^m$, and $\mu : [0, \infty) \mapsto \mathbb{R}^{n_\mu}$ is a uniformly bounded and piecewise continuous function. System (A.46) is input-to-state stable on a compact subset $\Sigma \subseteq \mathbb{R}^{n_\mu}$ with the state x and the input u if there exists a continuously differentiable function $V : \mathbb{R}^n \mapsto [0, \infty)$ such that, for all $\mu \in \Sigma$, and any initial state $x(0)$,*

$$\underline{\alpha}(\|x\|) \leq V(x) \leq \bar{\alpha}(\|x\|) \tag{A.47a}$$
$$\dot{V}(x) \leq -\alpha(\|x\|) + \sigma(\|u\|) \tag{A.47b}$$

for some class $\mathcal{K}_\infty$ functions $\underline{\alpha}$, $\bar{\alpha}$, and α, and some class $\mathcal{K}$ function σ.

Remark A.7 It can be deduced from [8, Theorem 10.4.5] that, if system (A.46) is input-to-state stable on a compact subset $\Sigma \subseteq \mathbb{R}^{n_\mu}$ with the state x and the input u, then for all $\mu \in \Sigma$ and any initial state $x(0)$, the solution of system (A.46) satisfies

$$\limsup_{t\to\infty} \|x(t)\| \leq \gamma\left(\limsup_{t\to\infty} \|u(t)\|\right)$$

for some class $\mathcal{K}$ function γ. Thus, if $\lim_{t\to\infty} \|u(t)\| = 0$, then $\lim_{t\to\infty} \|x(t)\| = 0$.

Next we recall the so-called changing supply functions technique in [9].

Lemma A.10 (Changing Supply Functions Technique) *Suppose the nonlinear system (A.46) satisfies those conditions in Theorem A.6. Then, for any smooth class $\mathcal{K}_\infty$ function $\alpha' : [0, \infty) \mapsto [0, \infty)$ satisfying*

$$\limsup_{s\to 0^+} \frac{\alpha'(s)}{\alpha(s)} < \infty$$

there exists another continuously differentiable function $V' : \mathbb{R}^n \mapsto [0, \infty)$, class $\mathcal{K}_\infty$ functions $\underline{\alpha}'$ and $\bar{\alpha}'$, and a class $\mathcal{K}$ function σ' such that, for all $\mu \in \Sigma$ and any initial state $x(0)$,

$$\underline{\alpha}'(\|x\|) \leq V'(x) \leq \bar{\alpha}'(\|x\|) \tag{A.48a}$$
$$\dot{V}'(x) \leq -\alpha'(\|x\|) + \sigma'(\|u\|). \tag{A.48b}$$

The proof of Lemma A.10 relies on the following technical lemma:

Lemma A.11 *Let $\alpha(s)$ and $a(s)$ be class $\mathcal{K}_\infty$ functions. For any smooth function α' : $[0,\infty)\mapsto\mathbb{R}$ satisfying $\limsup_{s\to 0^+}[\alpha'(s)/\alpha(s)]<\infty$, there exists a smooth non-decreasing function $\rho:[0,\infty)\mapsto[0,\infty)$ satisfying $\rho(s)>0$ for $s>0$ such that $\rho(a(s))\alpha(s)\geq\alpha'(s)$ for all $s\geq 0$.*

Proof Since $\limsup_{s\to 0^+}[\alpha'(s)/\alpha(s)]<\infty$, there exists a function $\rho_1:[0,\infty)\mapsto[0,\infty)$ such that

$$\rho_1(s)=\sup_{t\in(0,s]}\frac{\alpha'(t)}{\alpha(t)},\quad s>0.$$

Also, we define $\rho_1(0)=\lim_{s\to 0^+}\rho_1(s)$. Clearly, $\rho_1(s)$ is a continuous non-decreasing function for $s\geq 0$. Since $\rho_1(s)\alpha(s)\geq\alpha'(s)$ for all $s\geq 0$, we only need to find a smooth non-decreasing function $\rho:[0,\infty)\mapsto[0,\infty)$ satisfying $\rho(s)>0$ for $s>0$ such that $\rho(a(s))\geq\rho_1(s)$, or equivalently, $\rho(s)\geq\rho_1(a^{-1}(s))$ since $a(s)$ is of class $\mathcal{K}_\infty$. Since $\rho_1(a^{-1}(s))$ is a continuous non-decreasing function, $\rho(s)$ always exists. □

Now let us give the proof of Lemma A.10.

Proof Let $\rho(\cdot)$ be a smooth non-decreasing function defined over $[0,\infty)$ with $\rho(s)>0$ for all $s>0$. Define

$$V'(x)=\int_0^{V(x)}\rho(s)\mathrm{d}s.$$

According to (A.47a), it holds that

$$\underline{\alpha}'(\|x\|)\leq\int_0^{\underline{\alpha}(\|x\|)}\rho(s)\mathrm{d}s\leq V'(x)\leq\int_0^{\bar{\alpha}(\|x\|)}\rho(s)\mathrm{d}s\leq\bar{\alpha}'(\|x\|)$$

for some class $\mathcal{K}_\infty$ functions $\underline{\alpha}'(\cdot)$ and $\bar{\alpha}'(\cdot)$.

Next, we show that, along the trajectory of (A.46)

$$\begin{aligned}\dot{V}'(x)&\leq\rho(V(x))[-\alpha(\|x\|)+\sigma(\|u\|)]\\&\leq-\frac{1}{2}\rho(\underline{\alpha}(\|x\|))\alpha(\|x\|)+\rho(\bar{\alpha}(\alpha^{-1}(2\sigma(\|u\|))))\sigma(\|u\|).\end{aligned}\tag{A.49}$$

In fact, (A.49) can be established by considering the following two cases:

- $\alpha(\|x\|)/2\geq\sigma(\|u\|)$: in this case, the claim follows from the fact that $\rho(V(x))$ $[-\alpha(\|x\|)+\sigma(\|u\|)]$ is bounded from above by $-\rho(V(x))\alpha(\|x\|)/2$ and hence bounded from above by $-\rho(\underline{\alpha}(\|x\|))\alpha(\|x\|)/2$.
- $\alpha(\|x\|)/2<\sigma(\|u\|)$: in this case, the following inequalities hold:

$$\rho(V(x))\leq\rho(\bar{\alpha}(\|x\|))\leq\rho(\bar{\alpha}(\alpha^{-1}(2\sigma(\|u\|)))).$$

Since $\limsup_{s\to 0^+}[\alpha'(s)/\alpha(s)] < \infty$, by Lemma A.11, there always exists a smooth non-decreasing function $\rho : [0,\infty) \mapsto [0,\infty)$ satisfying $\rho(s) > 0$ for $s > 0$ such that

$$\frac{1}{2}\rho(\underline{\alpha}(s))\alpha(s) \geq \alpha'(s).$$

Also, there exists a class $\mathscr{K}$ function σ' such that

$$\sigma'(s) \geq \rho(\bar{\alpha}(\alpha^{-1}(2\sigma(s))))\sigma(s).$$

The proof is thus completed. □

Lemma A.10 leads to the following two corollaries that have been used in Chaps. 10 and 11.

Corollary A.1 *Consider the nonlinear system*

$$\dot{x} = f(x, u, \mu) \tag{A.50}$$

where $x \in \mathbb{R}^n$, $u \in \mathbb{R}^m$, and $\mu : [0,\infty) \mapsto \mathbb{R}^{n_\mu}$ is piecewise continuous and uniformly bounded over $[0,\infty)$. Assume, for some given compact subset $\Sigma \subseteq \mathbb{R}^{n_\mu}$, there exists a continuously differentiable function $V : \mathbb{R}^n \mapsto [0,\infty)$ such that, for all $\mu \in \Sigma$,

$$\underline{\alpha}(\|x\|) \leq V(x) \leq \bar{\alpha}(\|x\|) \tag{A.51a}$$
$$\dot{V}(x) \leq -\alpha(\|x\|) + \delta\sigma(\|u\|) \tag{A.51b}$$

for some class $\mathscr{K}_\infty$ functions $\underline{\alpha}$, $\bar{\alpha}$, and α which is known, some constant δ depending on Σ, and some known class $\mathscr{K}$ function σ. Then given any smooth function $\Delta(x)$, there exists another continuously differentiable function $V' : \mathbb{R}^n \mapsto [0,\infty)$, class $\mathscr{K}_\infty$ functions $\underline{\alpha}'$ and $\bar{\alpha}'$, some positive constant $\bar{\delta}$ depending on Σ and some known smooth positive function $b : \mathbb{R}^m \mapsto (0,\infty)$ such that, for all $\mu \in \Sigma$

$$\underline{\alpha}'(\|x\|) \leq V'(x) \leq \bar{\alpha}'(\|x\|) \tag{A.52a}$$
$$\dot{V}'(x) \leq -\Delta(x)\alpha(\|x\|) + \bar{\delta}b(u)\sigma(\|u\|). \tag{A.52b}$$

Proof Applying Lemma A.10 to the system (A.50), noting the difference between (A.51b) and (A.47b) and using (A.49) gives the following inequality:

$$\dot{V}'(x) \leq -\frac{1}{2}\rho(\underline{\alpha}(\|x\|))\alpha(\|x\|) + \rho(\bar{\alpha}(\alpha^{-1}(2\delta\sigma(\|u\|))))\delta\sigma(\|u\|).$$

By Part (i) of Lemma A.7, there exist smooth functions $a(\delta) > 0$ and $b(u) > 0$ such that $\rho(\bar{\alpha}(\alpha^{-1}(2\delta\sigma(\|u\|)))) \leq a(\delta)b(u)$. Thus,

$$\dot{V}'(x) \le -\frac{1}{2}\rho(\underline{\alpha}(\|x\|))\alpha(\|x\|) + \bar{\delta}b(u)\sigma(\|u\|)$$

where $\bar{\delta} = a(\delta)\delta$ which depends on Σ since δ depends on Σ. Finally, like in the proof of Lemma A.10, choosing a smooth non-decreasing function $\rho : [0, \infty) \mapsto [0, \infty)$ satisfying $\rho(s) > 0$ for $s > 0$ such that

$$\frac{1}{2}\rho(\underline{\alpha}(s)) \ge \Delta(s)$$

completes the proof. □

If we further assume the class $\mathscr{K}_\infty$ function α and the class $\mathscr{K}$ function σ in (A.51b) satisfy the following condition:

$$\limsup_{s\to 0^+} \frac{s^2}{\alpha(s)} < \infty, \quad \limsup_{s\to 0^+} \frac{\sigma(s)}{s^2} < \infty \tag{A.53}$$

then we have the following result.

Corollary A.2 *Consider the same system as (A.50), which admits a continuously differentiable function $V : \mathbb{R}^n \mapsto [0, \infty)$ satisfying, for all $\mu \in \Sigma$, (A.51) for some class $\mathscr{K}_\infty$ functions $\underline{\alpha}$, $\bar{\alpha}$, and α which is known, some constant δ depending on Σ, and some known class $\mathscr{K}$ function σ. Moreover, the class $\mathscr{K}_\infty$ function α and the class $\mathscr{K}$ function σ satisfy the condition (A.53). Then given any smooth function $\Delta(x)$, there exists another continuously differentiable function $V' : \mathbb{R}^n \mapsto [0, \infty)$, class $\mathscr{K}_\infty$ functions $\underline{\alpha}'$ and $\bar{\alpha}'$, some positive constant $\bar{\delta}$ depending on Σ and some known smooth positive function ς such that, for all $\mu \in \Sigma$*

$$\underline{\alpha}'(\|x\|) \le V'(x) \le \bar{\alpha}'(\|x\|) \tag{A.54a}$$

$$\dot{V}'(x) \le -\Delta(x)\,\|x\|^2 + \bar{\delta}\varsigma(u)\,\|u\|^2 . \tag{A.54b}$$

Proof We first claim that there exist smooth functions $\alpha_0(x)$ and $\sigma_0(u)$ such that

$$\alpha_0(x)\alpha(\|x\|) \ge \|x\|^2, \ \sigma_0(u)\|u\|^2 \ge \sigma(\|u\|). \tag{A.55}$$

On one hand, since α satisfies the first inequality of (A.53), there exits a constant $l_1 \ge 1$ such that $\alpha(\|x\|) \ge \|x\|^2/l_1^2$ for $\|x\| \le 1$, and since α is of class $\mathscr{K}_\infty$, there exists a constant $l_2 > 0$ such that $\alpha(\|x\|) \ge l_2$ for $\|x\| \ge 1$. As a result, the first inequality of (A.55) holds for any smooth function α_0 satisfying

$$\alpha_0(x) \ge l_1^2 + \frac{1}{l_2}\|x\|^2.$$

On the other hand, since σ satisfies the second inequality of (A.53), we can define a function $l : [0, \infty) \mapsto [0, \infty)$ such that $l(s) = \sigma(s)/s^2$ for all $s > 0$ and $l(0) =$

$\limsup_{s\to 0^+} l(s) < \infty$. Let $\sigma_0(u)$ be a smooth function such that $\sigma_0(u) \geq l(\|u\|)$. Then

$$\sigma_0(u)\|u\|^2 \geq l(\|u\|)\|u\|^2 = \sigma(\|u\|).$$

Next, let $\bar{\Delta}(x)$ be any smooth function such that $\bar{\Delta}(x) \geq \Delta(x)\alpha_0(x)$.

With this function $\bar{\Delta}(x)$, by Corollary A.1, there exist another continuously differentiable function $V'(\cdot)$, class $\mathcal{K}_\infty$ functions $\underline{\alpha}'(\cdot)$ and $\bar{\alpha}'(\cdot)$, some positive constant $\bar{\delta}$ depending on Σ and some known smooth positive function $b(\cdot)$ such that, for all $\mu \in \Sigma$,

$$\underline{\alpha}'(\|x\|) \leq V'(x) \leq \bar{\alpha}'(\|x\|) \quad \text{(A.56a)}$$
$$\dot{V}'(x) \leq -\bar{\Delta}(x)\alpha(\|x\|) + \bar{\delta}b(u)\sigma(\|u\|) \quad \text{(A.56b)}$$

which yields (A.54b) upon noting

$$\bar{\Delta}(x)\alpha(\|x\|) \geq \Delta(x)\alpha_0(x)\alpha(\|x\|) \geq \Delta(x)\,\|x\|^2$$

and letting $\varsigma(\cdot)$ be a smooth function such that

$$\varsigma(u) \geq \sigma_0(u)b(u).$$

□

Remark A.8 If the compact set Σ is known, then the constant δ is considered known, so is the constant $\bar{\delta}$.

Corollary A.2 leads to the following stability property for a class of cascaded nonlinear systems.

Lemma A.12 *Consider the system*

$$\dot{z}_1 = \varphi_1(z_1, u, \mu) \quad \text{(A.57a)}$$
$$\dot{z}_2 = Az_2 + \varphi_2(z_1, u, \mu) \quad \text{(A.57b)}$$

where $z_1 \in \mathbb{R}^{n_1}$, $z_2 \in \mathbb{R}^{n_2}$, $u \in \mathbb{R}^{n_u}$, $\mu : [0, \infty) \mapsto \mathbb{R}^{n_\mu}$ is piecewise continuous and uniformly bounded over $[0, \infty)$. $A \in \mathbb{R}^{n_2 \times n_2}$ is a Hurwitz matrix, and $\varphi_1(\cdot)$ and $\varphi_2(\cdot)$ are sufficiently smooth with $\varphi_1(0, 0, \mu) = 0$ and $\varphi_2(0, 0, \mu) = 0$ for all $\mu \in \mathbb{R}^{n_\mu}$.

Assume, for some compact subset $\Sigma \subset \mathbb{R}^{n_\mu}$, there exists a continuously differential function $V_{z_1} : \mathbb{R}^{n_1} \mapsto [0, \infty)$ such that

$$\underline{\alpha}_{z_1}(\|z_1\|) \leq V_{z_1}(z_1) \leq \bar{\alpha}_{z_1}(\|z_1\|)$$

for some class $\mathcal{K}_\infty$ functions $\underline{\alpha}_{z_1}$ and $\bar{\alpha}_{z_1}$, and, for any $\mu \in \Sigma$, along the trajectory of system (A.57a),

$$\dot{V}_{z_1}(z_1) \leq -\alpha_{z_1}(\|z_1\|) + \delta_u\sigma_u(\|u\|) \quad \text{(A.58)}$$

for some class $\mathcal{K}_\infty$ function α_{z_1} satisfying $\limsup_{s\to 0^+} \frac{s^2}{\alpha_{z_1}(s)} < \infty$, *some positive number δ_u depending on Σ, and some known class $\mathcal{K}$ satisfying* $\limsup_{s\to 0^+} \frac{\sigma_u(s)}{s^2} < \infty$. *Let $z = col(z_1, z_2)$. Then there exists a continuously differential function $V_z : \mathbb{R}^{n_1+n_2} \mapsto [0, \infty)$ satisfying*

$$\underline{\alpha}_z(\|z\|) \le V_z(z) \le \bar{\alpha}_z(\|z\|)$$

for some class $\mathcal{K}_\infty$ functions $\underline{\alpha}_z$ and $\bar{\alpha}_z$ such that, for any $\mu \in \Sigma$, along the trajectory of system (A.57),

$$\dot{V}_z(z) \le -\|z_1\|^2 - \|z_2\|^2 + \tilde{\delta}_u \varsigma_u(u)\|u\|^2 \tag{A.59}$$

where $\tilde{\delta}_u$ is some positive number depending on Σ and $\varsigma_u(\cdot)$ is some known smooth positive function.

Proof From (A.58), by Corollary A.2, given any smooth function $\Delta(z_1) \ge 0$, there exists a continuously differentiable function $\bar{V}_{z_1}(z_1)$ satisfying $\underline{\alpha}_{1z_1}(\|z_1\|) \le \bar{V}_{z_1}(z_1) \le \bar{\alpha}_{1z_1}(\|z_1\|)$ for some class $\mathcal{K}_\infty$ functions $\underline{\alpha}_{1z_1}(\cdot)$ and $\bar{\alpha}_{1z_1}(\cdot)$ such that, for any $\mu \in \Sigma$, along the trajectory of system (A.57a),

$$\dot{\bar{V}}_{z_1}(z_1) \le -\Delta(z_1)\|z_1\|^2 + \bar{\delta}_u \bar{\gamma}_u(u)\|u\|^2 \tag{A.60}$$

for some positive number $\bar{\delta}_u$ and some known smooth positive function $\bar{\gamma}_u(\cdot)$.

Next, since $\varphi_2(z_1, u, \mu)$ is smooth and satisfies $\varphi_2(0, 0, \mu) = 0$, by Part (ii) of Lemma A.7, there exist some constant $c > 0$ depending on Σ, and smooth positive functions $\psi_1(z_1)$ and $\psi_2(u)$ such that, for all $z_1 \in \mathbb{R}^{n_1}$, $u \in \mathbb{R}^{n_u}$, and $\mu \in \Sigma$,

$$\|\varphi_2(z_1, u, \mu)\| \le c\Big(\psi_1(z_1)\|z_1\| + \psi_2(u)\|u\|\Big). \tag{A.61}$$

Let $V_z(z_1, z_2) = l\bar{V}_{z_1}(z_1) + 2z_2^\mathsf{T} P z_2$ where l is some positive number to be specified later, and P is the positive definite solution of the Lyapunov equation

$$PA + A^\mathsf{T} P = -I_{n_2}. \tag{A.62}$$

By Lemma A.9, $V_z(z_1, z_2)$ is continuously differentiable satisfying $\underline{\alpha}_z(\|(z_1, z_2)\|) \le V_z(z_1, z_2) \le \bar{\alpha}_z(\|(z_1, z_2)\|)$ for some class $\mathcal{K}_\infty$ functions $\underline{\alpha}_z(\cdot)$ and $\bar{\alpha}_z(\cdot)$. Using (A.60) and (A.62), the time derivative of $V_z(z_1, z_2)$ satisfies, for any $\mu \in \Sigma$, along the trajectory of system (A.57),

$$\begin{aligned}
\dot{V}_z(z) \le\ & -l\Delta(z_1)\|z_1\|^2 + l\bar{\delta}_u\bar{\gamma}_u(u)\|u\|^2 - 2\|z_2\|^2 \\
& +4Pz_2\varphi_2(z_1, u, \mu) \\
\le\ & -l\Delta(z_1)\|z_1\|^2 + l\bar{\delta}_u\bar{\gamma}_u(u)\|u\|^2 - 2\|z_2\|^2 \\
& +(z_2^2 + 8c^2\|P\|^2\psi_1^2(z_1)z_1^2 + 8c^2\|P\|^2\psi_2^2(u)\|u\|^2) \\
=\ & -\big(l\Delta(z_1) - 8c^2\psi_1^2(z_1)\|P\|^2\big)\|z_1\|^2 - \|z_2\|^2 \\
& +\Big(l\bar{\delta}_u\bar{\gamma}_u(u) + 8c^2\psi_2^2(u)\|P\|^2\Big)\|u\|^2.
\end{aligned}$$

Letting $\Delta(z_1) \geq 1 + \psi_1^2(z_1)$, $\varsigma_u(u) \geq \bar{\gamma}_u(u) + \psi_2^2(u)$ and

$$l \geq \max\left\{1, 8c^2 \|P\|^2\right\},\ \tilde{\delta}_u \geq \max\left\{l\bar{\delta}_u, 8c^2 \|P\|^2\right\}$$

yields (A.59). □

Remark A.9 Again, if the compact set Σ is known, the number δ_u is also known. From the proof of Lemma A.12, the constants c and $\bar{\delta}_u$ are also known, and hence, $\tilde{\delta}_u$ is known. In this case, both δ_u and $\tilde{\delta}_u$ can be chosen as one by viewing $\delta_u\sigma_u(\|u\|)$ and $\tilde{\delta}_u\varsigma_u(u)$ as a whole, respectively.

A.8 Notes and References

A good reference for detailed discussions on Kronecker product, Sylvester equation, Metzler matrix, and $\mathscr{M}$-matrix is [10]. Ben-Israel and Greville [1] is a convenient reference on Moore–Penrose inverse. Property 2 of Proposition A.3 is a special case of [11, Theorem 7–10]. Theorems A.1 and A.2 are adapted from [2, Lemma 2.4] and [3, Theorem 1], respectively. Lemma A.2 is extracted from [10, Page 131]. Theorem A.3 is a special case of [12, Lemma 3.4]. Theorem A.4, i.e., LaSalle–Yoshizawa Theorem, is rephrased from [13, Theorem 2.1] due to LaSalle [14] and Yoshizawa [15]. Lemma A.3 is adapted from [16, Lemma 3.4], and Lemma A.4 can be found in [17]. Lemma A.5 is adapted from [18, Lemma B.2.3]. The nonlinear robust output regulation problem is extensively studied in [6, 19, 20]. A recent overview on the internal model approach can be found in [5]. The textbooks [7, 8, 12] gave in-depth studies on stabilizing techniques for nonlinear systems in Sect. A.7. The technique of changing supply functions was developed in [9]. Some variants of this technique introduced in Sect. A.7 are adapted from [7, Sect. 2.5].

References

1. Ben-Israel A, Greville NE (2003) Generalized inverse: theory and applications. Springer, NewYork
2. Ren W, Beard RW (2008) Distributed consensus in multi-vehicle cooperative control: theory and application. Springer, London
3. Moreau L (2004) Stability of continuous-time distributed consensus algorithms. In: Proceedings of the 41st IEEE conference on decision and control, pp 3998–4003
4. Lee TC, Jiang ZP (2005) A generalization of Krasovskii-LaSalle theorem for nonlinear time-varying systems: converse results and applications. IEEE Trans Autom Control 50(8):1147–1163
5. Huang J, Isidori A, Marconi L, Mischiati M, Sontag E, Wonham WM (2018) Internal models in control, biology and neuroscience. In: Proceedings of 2018 IEEE conference on decision and control, pp 5370–5390

6. Huang J (2004) Nonlinear output regulation problem: theory and applications. SIAM, Philadelphia
7. Chen Z, Huang J (2015) Stabilization and regulation of nonlinear systems: a robust and adaptive approach. Springer International Publishing, Switzerland
8. Isidori A (1999) Nonlinear control systems II. Springer, London
9. Sontag E, Teel A (1995) Changing supply functions in input/state stable systems. IEEE Trans Autom Control 40(8):1476–1478
10. Horn RA, Johnson CR (1991) Topics in matrix analysis. Cambridge University Press, New York
11. Chen CT (1984) Linear system theory and design. Holt, Rinehart and Winston, New York
12. Khalil HK (2002) Nonlinear systems. Prentice Hall, NJ
13. Krstic M, Kanellakopoulos I, Kokotović P (1955) Nonlinear and adaptive control design. Wiley, New York
14. LaSalle JP (1968) Stability theory for ordinary differential equations. J Diff Equ 4(1):57–65
15. Yoshizawa T (1966) Stability theory by Lyapunov's second method. The Mathematical Society of Japan, Tokyo
16. Boyd S, Sastry S (1983) On parameter convergence in adaptive control. Syst Control Lett 3(6):311–319
17. Liu L, Chen Z, Huang J (2009) Parameter convergence and mimimal internal model with an adaptive output regulation problem. Automatica 45(5):1306–1311
18. Marino R, Tomei P (1995) Nonlinear control design: geometric. Adaptive and robust. Prentice Hall, Englewood Cliffs
19. Huang J, Chen Z (2004) A general framework for tackling the output regulation problem. IEEE Trans Autom Control 49(12):2203–2218
20. Isidori A, Marconi L, Serrani A (2003) Robust autonomous guidance: an internal model-based approach. Springer, London

Index

H. Cai et al., *Cooperative Control of Multi-agent Systems*, Advances in Industrial Control,
https://doi.org/10.1007/978-3-030-98377-2